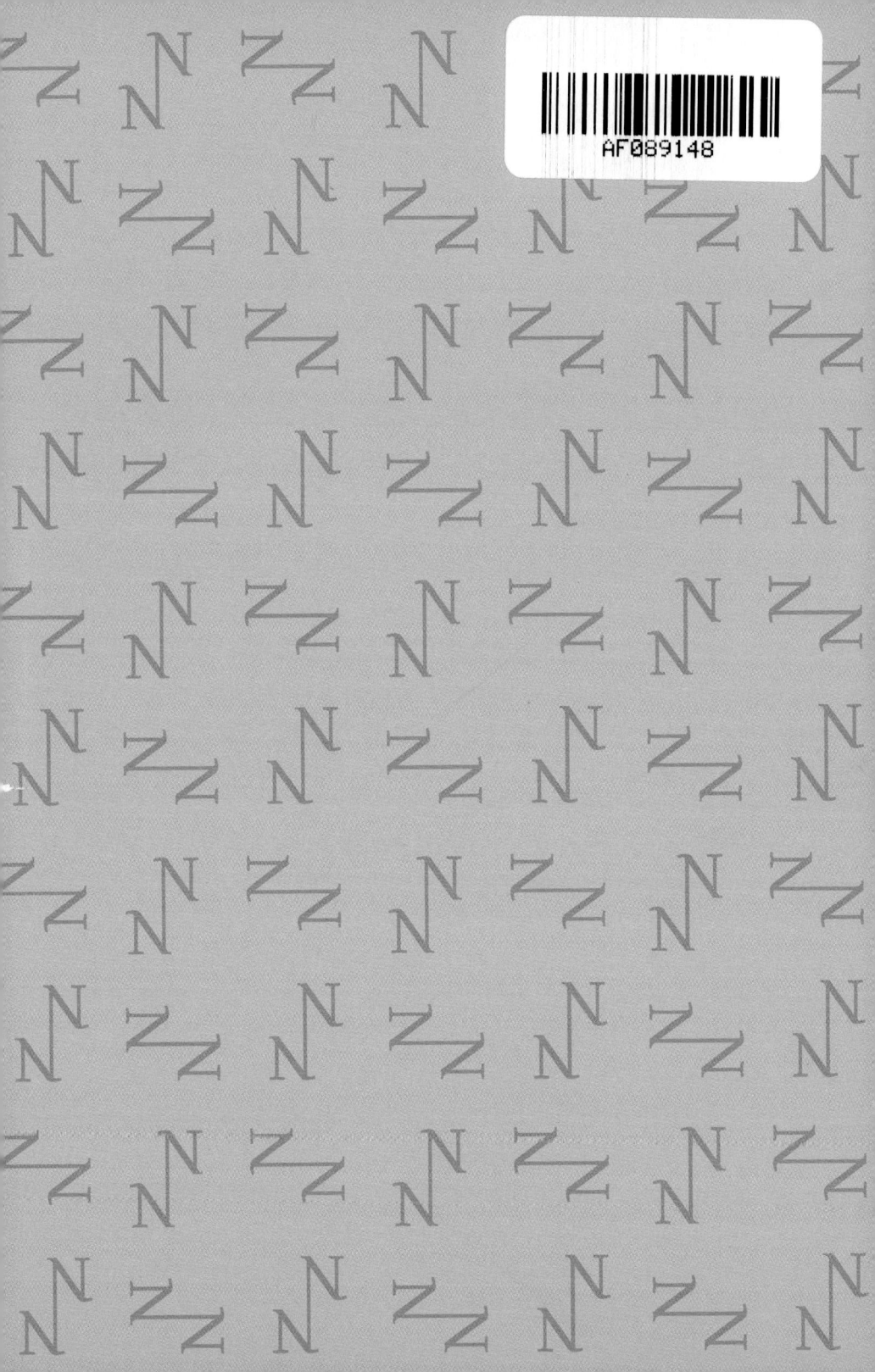

THE NEW NATURALIST LIBRARY

A SURVEY OF BRITISH NATURAL HISTORY

SOLITARY BEES

EDITORS
SARAH A. CORBET, ScD
DAVID STREETER, MBE, FRSB
JIM FLEGG, OBE, FIHort
Prof. JONATHAN SILVERTOWN
Prof. BRIAN SHORT

*

The aim of this series is to interest the general reader in the wildlife of Britain by recapturing the enquiring spirit of the old naturalists. The editors believe that the natural pride of the British public in the native flora and fauna, to which must be added concern for their conservation, is best fostered by maintaining a high standard of accuracy combined with clarity of exposition in presenting the results of modern scientific research.

THE NEW NATURALIST LIBRARY

SOLITARY BEES

TED BENTON AND NICK OWENS

WILLIAM
COLLINS

This edition published in 2023 by William Collins,
an imprint of HarperCollins*Publishers*

HarperCollins*Publishers*
1 London Bridge Street
London SE1 9GF

WilliamCollinsBooks.com

HarperCollins*Publishers*
Macken House, 39/40 Mayor Street Upper
Dublin 1, D01 C9W8, Ireland

First published 2023

© Ted Benton and Nick Owens, 2023
Photographs © Ted Benton and Nick Owens unless otherwise credited

Illustrations by Nick Owens except those on pp. 95, 329, 357, 365, 378, 407 (top), 413, 414, 417, 422, 423, 426, 444 and 454 by Martin Brown

All rights reserved. No part of this publication may be reproduced, stored in a retrieval system or transmitted in any form or by any means, electronic, mechanical, photocopying, recording or otherwise, without the prior written permission of the copyright owner.

A CIP catalogue record for this book is available from the British Library.

Set in Nexus Serif Pro and Nexus Mix Pro

Edited and designed by
D & N Publishing
Baydon, Wiltshire

Printed in Bosnia-Herzegovina by GPS Group

Hardback
ISBN 978-0-00-830455-3

Paperback
ISBN 978-0-00-830457-7

All reasonable efforts have been made by the authors to trace the copyright owners of the material quoted in this book and of any images reproduced in this book. In the event that the authors or publisher are notified of any mistakes or omissions by copyright owners after publication of this book, the authors and the publisher will endeavour to rectify the position accordingly for any subsequent printing.

Contents

Editors' Preface   vi
Authors' Acknowledgements   viii

Introduction   1
1   The Diversity of Solitary Bees   27
2   Sex and the Solitary Bee   59
3   The Life Cycle: Nesting Behaviour and Development   101
4   From Solitary to Social and Back   159
5   Bees and Flowers, Part I   177
6   Bees and Flowers, Part II   235
7   Parasites and Predators   303
8   Cuckoo Bees   361
9   Time, Space and Temperature   405
10   Ecology and Conservation   455

List of British Solitary Bees   531
Glossary   538
References and Bibliography   540
Species Index   568
General Index   582

Editors' Preface

The social bees – honeybees and bumblebees – have attracted much more attention than their solitary relatives, although the latter are more numerous and more diverse and sometimes no less important as pollinators of crops and wild flowers. The solitary bees exhibit a wonderful diversity of structure and behaviour. Some are as large as bumblebees, and some are only a few millimetres long. Some are marked with red, white or yellow, and some are iridescent green. Some species collect oil from flowers; some build nests in abandoned snail shells; some use a secretion to waterproof their brood cells; some cut leaf discs to line their cells; some build turrets in the soil around the entrance to their burrows; some lay their eggs in the nests of other species; some sleep in regular roosting places; some regularly patronise bee hotels; and some that are generally included in the category of solitary bees show a degree of sociality, with queens and workers.

An abiding dilemma for ecologists studying the lives of insects of such a species-rich group is the problem of naming the insect. Identification often requires capture and examination under a microscope, which is often incompatible with behavioural observations on living individuals. Bumblebees and honeybees can mostly be identified without capture, but until recently, solitary bees have been tantalisingly inaccessible because their identification has proved such a challenge. The recent publication of up-to-date keys means that we can at last name the bees that we find and link our behavioural observations to a body of knowledge about the species. Many species can be named in the field or from photographs, and the illustrations in this book, together with contextual information such as season of appearance, flowers visited and nest type, will make it easier for readers to name the species they encounter. This book will show how rich the behavioural inventory of solitary bees is, how much remains to be discovered about them, and how patient observations of named species can contribute to our knowledge of their behavioural profiles. As the authors say, 'we can find out more simply by watching and waiting'. Your garden may be the best place to start.

This book is timely because the recent publication of user-friendly keys has made the study of British species of solitary bees much more accessible. This book will complement these works on identification by providing the ecological and behavioural background of our British species and by drawing the attention of naturalists to aspects of these bees' natural history that are poorly known and await investigation.

Authors' Acknowledgements

The authors wish to express their appreciation for the work of our distinguished editor, Sarah Corbet. As always, she combined warm encouragement with rigorous and scholarly critical attention to our drafts of each chapter of the book. Of course, she is not to be blamed for any remaining shortcomings. Myles Archibald has been generous in his tolerance of our delays in delivery of the book and in his prompt replies to our queries. We greatly appreciate the careful and expert attention to the work by the editorial team, notably Jennifer Dixon and David Price-Goodfellow.

Both authors benefited greatly from our early involvement in the Bees, Wasps and Ants Recording Society (BWARS) and are very appreciative of the advice and encouragement we have received from George Else and Mike Edwards, in particular, but also from Paul Brock, Jeremy Early, Rosie Earwaker, George Pilkington, Steven Falk and the late, much missed, David Baldock. Both authors wish to thank George and Anna Else for their hospitality, and TB would also like to thank Martin and Maria Jenner and Sue and Mike Edwards for their hospitality.

Ian Beavis, Ben Hargreaves, Louise Hislop, Paddy Saunders, Liam Olds and Karen McCartney provided invaluable information, and images, concerning habitats and localities in various parts of Britain that were outside the experience of the authors. We feel we have not been able to do full justice to their contributions within the limits of one book. TB expresses his appreciation of the help kindly given by Bryan Danforth in response to an enquiry about sociality in furrow bees. Claire Carvell, Graham Stone, Rob Parker, Martin Jenner, Adrian Knowles, Peter Harvey and Taina Conrad, among others, also generously shared their expert knowledge.

TB offers both thanks and apologies to his indispensable life partner, Shelley Pennington, for her unyielding critical support and generous tolerance of his prolonged immersion in the writing of this book, and his inability to keep going in a straight line or avoid distracting forays into adjacent vegetation when supposedly out 'for a walk'. Apologies are also due to her for the state of her 'garden' (or 'bee habitat', as it has become). Apologies, too, to the rest of his family, to Jay and Rowan especially, for the somewhat eccentric pastimes and commitments of father and grandfather.

Colchester Natural History Society has provided an indispensable and convivial setting for many years of enjoying, learning about and fighting for wild nature, without which TB would not have been able to make his contribution to this book. Carla Davis, Sonya Lindsell, David Nichols, Bob Seago, Russell Leavett, Liz Cutting, Pete and Carole Hewitt, Peter Beard, Rowena MacAulay and Jane Hindley have all been important in various ways, as have the late Joe Firmin, doyen of the society, and my companion on so many blissful days 'in the field', the late Roy Cornhill. Maria Fremlin has been an expert and generous contributor to our studies of bee life-histories. Sean Nixon gets special mention as former colleague and for the shared pleasure of walks by the local riverside and some notable field trips further abroad. As always, John Kramer and Simon Randolph.

TB is especially indebted to his fellow author, Nick Owens, for many companionable and educative excursions in search of insects, for his learning, and for his generosity in sharing his knowledge and expertise in our work on this volume. I have been saved from numerous errors as well as rescued from occasional despair by his persistent enthusiasm for our work. Thanks, too, to Frankie for her generosity and hospitality.

NWO expresses his gratitude to Frankie, for her patience and generous support throughout the book's rather long gestation. Family members Katie, Gina, Ramón, Miguel and Silvia have also been extremely supportive and have offered bee observations of their own, while Robert Owens and Susan Morrison have been inspiring role models. Thanks are due to a wide range of naturalists and friends who have provided ideas and support, many being fellow members of the Norfolk and Norwich Naturalists' Society. I especially thank Mark Welch for his enthusiastic encouragement and for sharing his knowledge of diptera during many productive field excursions. Tony Irwin also assisted with identification of parasitic flies. Vanna and Jeremy Bartlett have been of particular help in providing first-hand observations and photographs of bee behaviour and have produced several very helpful publications. Rowan Edwards kindly provided maps of bee distribution from the BWARS data. Steven Falk has cheerfully answered a multitude of questions on identification and new discoveries. Tim Strudwick was responsible for triggering NWO's interest in solitary bees in 2009 by asking for records and tirelessly providing advice and assistance. I am also most grateful to the other friends and correspondents who have shared their knowledge and experience, including (in addition to those already mentioned by TB) Mike Ball, Clare Boyes, Ian Cheeseborough, Jean Claessens, Pip Collyer, Ian Cross, Michael Engel, Perry Fairman, Francis Farrow, Mike Fogden, Adrian Gardiner, Rob Hawkes, Grant Hazlehurst, Alice Hughes, Tom Ings, Nigel Jones, Jeyaraney Kathirithamby, Chris Kirby-Lambert, William Kirk, Hauke Koch, Alain Livory, Tracy Money, Ash Murray, Bill Neill, Clive Nuttman, Carlo Polidori, Chris Preston, Manuela Sann, Simon Saxton, Ljubisa Stanisavljević, Roger Tidman, Stephen Tomkins, John Walters, Rob Walton, Paul Westrich, Keith Wilson, Tom Wood and Rob Yaxley. The late David Baldock provided great encouragement. There are others who deserve my thanks, and I apologise to anyone I have not individually mentioned.

TB bravely and generously took NWO on board as a co-author in 2018 and has been remarkably tolerant and supportive throughout. His deep knowledge of insect natural history, coupled with his eminent status as an academic sociologist, have provided an exciting balance and perspective to our studies. Fieldwork has at times been frustrated by events but always associated with good camaraderie. Shelley has been a kind host in Colchester.

Introduction

Historically, solitary bees have hardly been noticed except by those beleaguered souls who devoted their lives to the study of them. For most of us, 'bees' means honeybees, and in many popular representations these, too, are confused with bumblebees. Cuddly, furry, black and yellow bumblebees are figured entering or leaving beehives. So, the existence of these 'other bees' has passed under the radar. Until very recently there has been no popular citizen science project to raise the alarm about their possible decline, as there has been for butterflies, farmland and garden birds and once-familiar mammals such as water voles and hedgehogs.

Keen gardeners will have noticed symmetrical holes cut into the leaves of roses and other garden plants, and this may be the most familiar sign of the existence of these small insects: in this case, the work of a leafcutter bee selecting materials to line her nest. Alongside paths, and on bare banks, bees entering and leaving holes in the ground, often in large numbers, may attract attention. Other bees can be seen entering or leaving holes in masonry or even in window frames and garden furniture. Most of us have seen these activities but probably thought little more about what the bees were up to. In this book we hope to explain current knowledge of British solitary bees as far as we can, drawing on a large international literature as well as reports from amateur bee-watchers and our own observations to build an account of the diversity, source of fascination and ecological importance of these too-often ignored insects.

Perhaps the first thing we need to explain is that (nearly!) all the species of bees we'll be discussing are 'solitary' in the sense that they do not form social groups with a division of labour between a queen and non-reproductive workers. The majority of species to be discussed here are truly solitary: males and females mate, and after that the female goes on to make a nest, with a variable number of brood cells, each of which she stocks with a store of food for her offspring.

She does this by herself, without the help of a worker caste. Unfortunately for her (but of great interest to us), there are other bees which have evolved a rather different lifestyle. These nest-parasites (or cleptoparasites, or inquilines) manage to sneak into the nests of their more industrious relatives and lay eggs in their brood cells. When they hatch, the resulting larvae consume the store of food provided by the host female. Informally, these bees are often called cuckoo bees. As they do not collect food for their offspring, they do not have the pollen-collecting brushes (known as a scopa, plural scopae) of the nest-making species, but they are frequently related to the species whose nests they invade (hosts). We devote Chapter 8 to a detailed discussion of them.

There are currently some 244 species of solitary bees in Britain (excluding the Channel Islands). Of those, 67 species (27 per cent), belonging to six genera, are cleptoparasites (cuckoos) on other solitary bees. Just to make things more confusing, several species in the family Halictidae are usually included as solitary bees, even though they show complex patterns of social life (eusocial behaviour). They are of special interest and will be discussed in Chapter 4. Next, we need to introduce some 'basics' of the study of bees in general and of solitary bees in particular.

WHAT IS A BEE?

Bees are members of the large insect order Hymenoptera, which also includes wasps, ants, parasitic wasps, sawflies and related groups. There are around 6,500 species of Hymenoptera in the British Isles, of which about 272 are bees. Hymenoptera have two pairs of membranous wings, coupled together by minute hooks. The wings are supported by a network of veins which, in Hymenoptera, produce a relatively simple wing venation (the pattern of compartments, or 'cells') compared with many other insect groups. Females of bees, wasps and ants have a sting, and so these groups are termed Aculeata (aculeates). No other Hymenoptera have a sting, though the long ovipositor (egg-laying tube) of some parasitic wasps can resemble one. Bees, technically referred to as Anthophila, belong to a subgroup of the Aculeata known as Apoidea. This includes both bees and apoid (bee-like) hunting wasps (Fig. 1). Bees are believed to have evolved from one particular type of apoid wasp which switched from hunting animal prey to collecting pollen. Bees can be regarded as a specialised type of apoid wasp and are more similar to this group of wasps than some wasps are to each other. Evidence from fossil bees indicates that bees originated in the Cretaceous period over 100 million years ago, in tandem with the appearance of flowering plants.

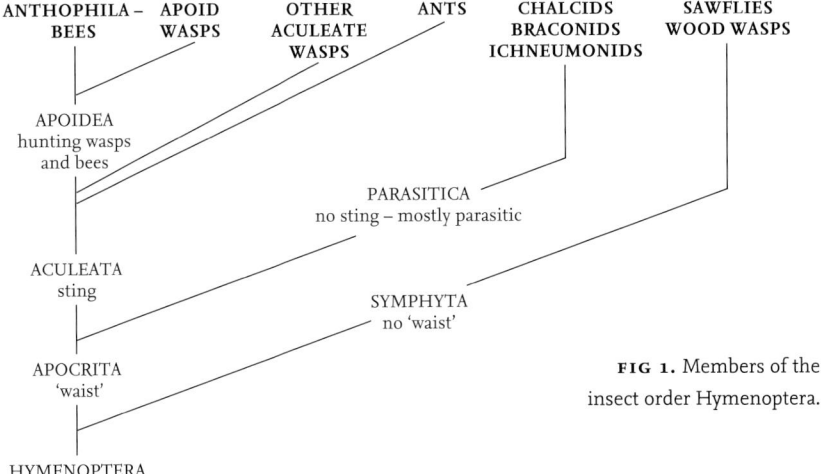

FIG 1. Members of the insect order Hymenoptera.

Bees can be distinguished from wasps by their covering of branched (plumose) hairs, which are unique to bees. Wasps have hairs but they are not generally branched. Bees also possess a more heart-shaped face and a longer tongue than most wasps. Many hunting wasps have dense silvery or golden hairs covering the lower face, whereas bees have softer, longer hairs. Areas of specialised long hairs on a female bee's body, the scopae, are used for collecting clumps of pollen which are transported back to the nest as a source of protein-rich food for her larvae. In both sexes the long tongue assists in imbibing nectar from flowers. Male bees do not collect pollen but, as in females, their body hairs play a part in temperature control. The hairs can also develop an electrostatic charge during flight, which assists in pollen adhesion. Male bees lack scopae, but the attachment of pollen to their general body hair means that males as well as females can play an important role in pollen transfer between flowers. In some male bees, long hairs can also be involved in courtship and mating. Wasps take nectar from plants but feed their larvae on paralysed insects or other invertebrates and only in rare cases on pollen. Thus, nearly all wasps are omnivores, obtaining nutrition from both animal prey and nectar, whereas almost all bees are herbivores, consuming only plant material in the form of pollen and nectar and sometimes plant oils. The food-store may, however, be transformed by microorganisms before it is consumed by a bee larva. The larva will then be ingesting microbial as well as plant protein.

Worker honeybees and bumblebees (but not cuckoo bumblebees) are recognisable by their possession of a pollen basket, technically known as a corbicula (plural corbiculae), in which pollen is carried back to the hive or nest. The pollen basket is formed by the shiny bare outer surface of one part of the hind leg, the tibia, which has a fringe of stiff hairs along its front and hind edges. Pollen is squeezed into the pollen basket using a pollen press at the junction between the tibia and the broad basitarsus below it. As it is collected, the pollen is mixed with nectar and the mixture forms a sticky, roundish, glistening mass. Together, honeybees and bumblebees are termed corbiculate bees. All other British bees lack corbiculae, but instead, as mentioned, the nest-making species use areas of long, branched hairs, the scopae, for carrying pollen. The scopae can be on the hind legs and/or parts of the main body, such as the sides of the mid-body (on a structure termed the propodeum) or the underside of the abdomen (Fig. 15). In the solitary bees, pollen is not generally mixed with nectar, though there are exceptions. Instead, nectar is generally ingested and carried back to the nest in the solitary bee's crop, where it is released onto the pollen already deposited inside the nest.

Some solitary bees are quite similar in size to honeybees and easily confused at first glance. Honeybees have a slow, hovering flight, accompanied by a high-pitched continuous buzz. With notable exceptions, solitary bees are fairly quiet to the human ear. The abdomen of a honeybee is usually orange-yellow and somewhat stripy at the base, though some feral forms are entirely dark. The body of solitary bees is usually black, but parts of the abdomen are red in some species. On close inspection or with the aid of a photograph, a honeybee's eyes can be

TABLE 1. Solitary bees and corbiculate bees compared.

Solitary bee female	Solitary bees	Scopa (pollen-collecting hairs) on legs and/or underside	Mostly non-social, some eusocial
Solitary bee male		No pollen-collecting structures	Do not collect pollen
Cuckoo solitary bee		No pollen-collecting structures	Nest parasite of other solitary bees
Honeybee worker	Not solitary bees	Corbicula (pollen basket)	Eusocial
Bumblebee queen or worker		Corbicula (pollen basket)	Eusocial
Male honeybee or bumblebee		No pollen-collecting structures	Do not collect pollen
Cuckoo bumblebee		No pollen-collecting structures	Nest parasite of other bumblebees

FIG 2. Solitary mining bee Impunctate Mini-miner *Andrena subopaca* female with pollen carried on scopae on several parts of her body.

seen to be hairy, and there is a long, narrow, marginal cell formed by the veins on the outer front edge of the forewing. Very few solitary bees have hairy eyes, and the marginal cell is generally quite broad. The basitarsus of a honeybee is as wide as the tibia above it, whereas in a solitary bee it is narrower (Fig. 3).

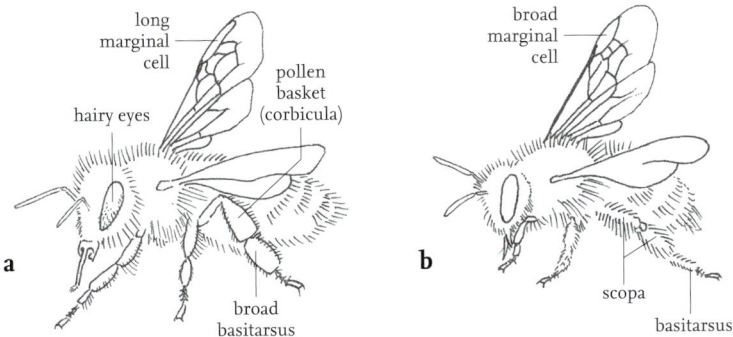

FIG 3. Comparison of (a) a honeybee worker and (b) an *Andrena* solitary bee female.

FIG 4. (a) An unusual honeybee nest in a hedgerow. The dark central areas of the nest consist of brood cells, occupied by growing larvae, and the outer pale cells contain honey as a food-store. The brood cells are hexagonal and made of wax, secreted by worker bees. (© E. Dack) (b) Western Honey Bee *Apis mellifera* on a Gorse *Ulex europaea* flower with pollen in her pollen basket, on the hind tibia.

FIG 5. Buff-tailed Mining Bee *Andrena humilis* female, a solitary bee, (a) at her nest burrow in the ground and (b) returning with a full pollen load.

A BEE'S BODY PLAN

Adult bees show the typical features shared by most insects. The body is encased in an exoskeleton (external skeleton) composed of chitin, a tough, plastic-like material. The body surface, known as the cuticle, is patterned and sculptured by ridges and shallow indentations known as punctures and is modified in a great variety of ways to form the mouthparts, hairs, pointed spurs, cleaning structures, external genitalia and a variety of minute sense organs, as well as the sting in females.

The body of most insects is formed of three clearly distinct sections: head, thorax and abdomen. However, in bees, the first segment of the abdomen is fused with the thorax, to form a structure called the propodeum. To the rear of this is a narrow waist followed by the remaining segments of the abdomen. Strictly speaking, the mid- and rear parts of the bee body are termed the mesosoma and metasoma, respectively. We will use these terms when necessary but will generally speak more loosely of 'thorax' and 'abdomen' where it will not cause confusion.

The head of a bee bears a pair of antennae and complex mouthparts which are used for cutting, excavating and manipulating materials, as well as for feeding. There is a pair of large compound eyes and between them are three simple light-sensitive eyes, or ocelli. The antennae of bees are highly sensitive. The first segment, known as the scape, attaches the antenna to the head between the eyes. The second segment, or pedicel, serves as a versatile joint, while the long outer section of the antenna, the flagellum, carries a large number of minute sensory receptors. Vibrations are used extensively in bee communication and are detected by a cluster of receptors known as a Johnston's organ within the pedicel, which registers any deflections of the antenna by air waves or other impacts. Receptors in the flagellum also provide touch and the all-important chemical senses. The bee's brain is situated in the head and is served by incoming and outgoing nerves. The main nerve cord of a bee runs close to the ventral (lower) side of the body, and each body segment has its own nerve interchange, known as a ganglion.

Each of the three (true) thorax segments bears a pair of legs and the second and third segments additionally support a pair of wings. The thorax contains the wing muscles, which operate by distorting the shape of the mesosoma (mid-body), indirectly tilting the wings up and down rather than pulling directly on the wing bases themselves. The buzz of bees is created partly by the beating of the wings and perhaps also by the distortion of the thorax. The narrow waist of bees and wasps allows the body to curve around when stinging. It may also reduce excessive heat caused by muscle action in the thorax being transferred to the main part of the abdomen.

FIG 6. (a) Anatomy of *Andrena* species female. Mauve = head, blue = thorax, green = abdomen. Yellow areas represent groups of hairs. (b) Structure of the face. (c) Structure of a solitary bee's leg. Pale green parts together comprise the tarsus.

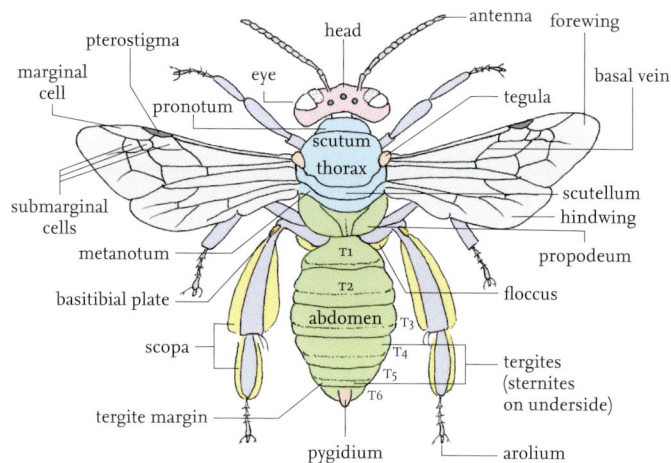

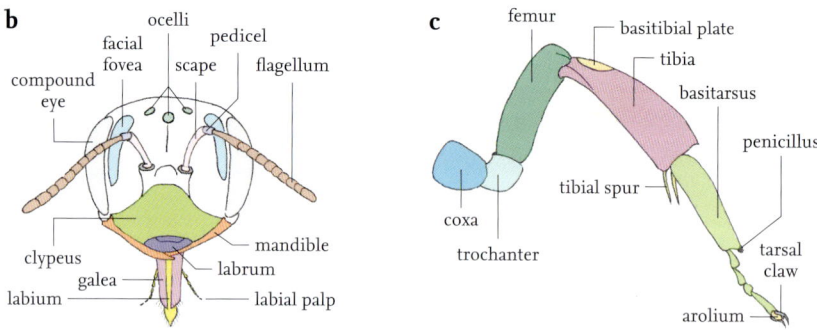

The metasoma (abdomen) has six visible segments in female bees and seven in males. Each metasoma segment is covered by a pair of plates: a tergite above and a sternite below, usually numbered T1, T2 or S1, S2, etc. The terminal abdomen segment (T6) of females covers the sting and egg-laying structures. Its upper surface often includes a raised, triangular, toughened plate known as a pygidium, used for excavating and smoothing the nest burrow. The final abdomen segment in males protects the complex genitalia, which are extruded during mating.

APPEARANCES CAN BE DECEPTIVE

Some solitary bees look remarkably like wasps. This is particularly the case in nomad bees (*Nomada* species), which are cuckoo bees, with about 35 British species. Nomad bees have very little hair, and many species have black and yellow (or red) warning colour patterns resembling those seen in some wasps. However, behaviour can sometimes betray their true identity; if prey is being carried, the insect in question is almost certainly a wasp, whereas a similar insect inspecting the nest of a bee is more likely to be a cuckoo bee. In the absence of such clues, wing venation and other details may be needed to distinguish wasps from bees.

FIG 7. (a) Sand Tailed Digger Wasp *Cerceris arenaria* with weevil prey. (b) Female cuckoo bee, Gooden's Nomad Bee *Nomada goodeniana*, inspecting an *Andrena* bee nest burrow.

FIG 8. (a) Furry Dronefly *Eristalis intricarius* male (a hoverfly with no sting).
(b) Tawny Mining Bee *Andrena fulva* female (a solitary bee with a sting).

One small difference is that the basitarsus is usually wider in relation to the tibia (the leg joint above it) in bees than in wasps.

Flies (Diptera) can also resemble bees. Hoverflies (Syrphidae) in particular can look very similar to bees or wasps and perhaps gain protection by doing so. Their colours, size and behaviour can closely match those of bees, and some have a bee-like buzz. However, flies are readily distinguished from bees (and wasps), since flies have just one pair of wings (*Diptera* means 'two-winged'), their antennae are generally much shorter, and the legs are not so broad and hairy as those of a bee (they do not harvest pollen). The differences in head shape can also be a useful clue: the compound eyes are large, almost covering the head in most flower-visiting flies such as hoverflies, but the eyes are relatively smaller in bees.

Bumblebees are relatively easy to distinguish from solitary bees by their large size and loud buzz, especially in early spring when only queen bumblebees are present. There is, however, one very common solitary bee in parts of Britain in which the male is easily mistaken for the similarly coloured Common Carder Bumblebee and which also has a loud buzz. This is the charmingly named Hairy-footed Flower Bee *Anthophora plumipes*. The rich-brown males are seen on the first warm spring days, darting between flowers and imbibing nectar with their long tongue as they seek the black-coated females. A bumblebee worker has a pollen basket, whereas the male flower bee has narrow legs adorned with long hairs. Bee-flies (Bombyliidae) are also confusing spring flower visitors, with mid-brown hair. Their larvae parasitise the nests of some solitary bees, and the adults are often seen in association with solitary bee nest aggregations. The most common bee-fly, the Dark-edged Bee-fly *Bombylius major*, can be recognised by the dark

FIG 9. Hairy-footed Flower Bee *Anthophora plumipes* (a) male and (b) female. (c) Common Carder Bumblebee *Bombus pascuorum* worker. (d) Dark-edged Bee-fly *Bombylius major* female.

front edge to its wings. When flying between flowers, the long tongue often projects forwards, rather than being withdrawn beneath the head, as is usual in bees. Bee-fly behaviour is described further in Chapter 7.

FINDING AND OBSERVING SOLITARY BEES

For many naturalists, discovering solitary bees can be a revelation, with a whole range of species living unnoticed in one's own neighbourhood. Up to 50 or more solitary bee species can be found in a flower-rich garden and 80 or so in some well-established parks and botanic gardens. But they cannot all be seen at once. Solitary bees are much more seasonal than bumblebees and honeybees, with a spectrum of different species appearing then disappearing throughout the spring, summer and autumn. It is very interesting and exciting to watch and wait for each species to appear. Adult solitary bees may live for just 4–6 weeks, and during other phases of their life cycle they exist only as eggs, immatures or hibernating adults, hidden from view. Solitary bees are also more particular about the weather than honeybees and bumblebees, many appearing only when the air temperature approaches 15°C. Sunny days, when bees can bask to warm up, are much more rewarding for the solitary bee-watcher, though a few of the larger bee species can generate sufficient heat within their bodies to fly in cooler conditions. Many of the diagnostic features of adult bees can be observed as they visit flowers or attend their nests. Some solitary bees are just a few millimetres long, while our largest species, the Violet Carpenter Bee, which is an intermittent coloniser, can exceed the size of a queen bumblebee. The largest bee in the world is about 4 cm in length and is a solitary leafcutter bee (*Megachile pluto*), discovered by Alfred Russel Wallace.

Digital photography has revolutionised insect studies in recent years, and with mobile phones now likely to include an excellent camera, it is even easier to take an opportunistic shot of a bee. It is possible to identify many, though by no means all, of our 240-plus solitary bee species from photographs. Differences between species are often quite subtle, and for confirmation many solitary bees require examination with the aid of a hand lens or stereo microscope using a

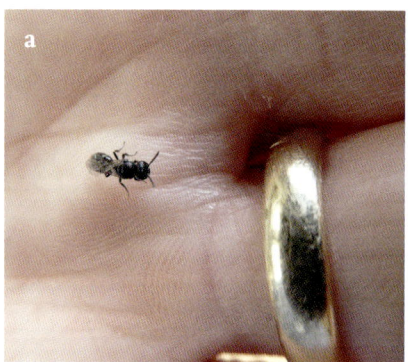

FIG 10. (a) Little Yellow-face Bee *Hylaeus pictipes*, one of our smallest bees. (© V. Bartlett) (b) Violet Carpenter Bee *Xylocopa violacea*, our largest bee. (© R. Few)

magnification of up to 20×. Temporarily placing a bee in a plastic magnifying pot can be a useful means of general inspection, without causing harm. Alternatively, a bee can be restrained inside a glass tube using soft paper and examined with a hand lens in the field, perhaps with reference to a field guide. An alternative is to place a captured bee in a refrigerator for a few minutes before examining it with a microscope. As the bee will recover quickly, it is useful to be clear before you start about which particular features you would like to check. The bee can be released unharmed, ideally at the location where it was found.

Since our main interest in this book is in the behaviour and ecology of solitary bees, we will be focusing on the activities of living insects. However, for these observations to have scientific value, it is important to know which species is being observed. There are some groups of bees that pose major challenges to the methods of identification just mentioned. Many mining bees (*Andrena*), and especially the males, are very similar to each other and some can be definitively identified only by microscopic examination of a dead specimen. Many species of furrow bees (*Lasioglossum*) and blood bees (*Sphecodes*) are similarly very difficult to separate and require microscopic examination, even dissection.

Many readers will not wish to kill specimens, and there are clearly ethical questions to be addressed. Our view is that killing bees should be avoided as far as possible, but there are important justifications. For example, providing evidence for conservation of their habitat or making a contribution to greater knowledge could outweigh the loss of a small number of individual insects. This is particularly clear where a site is threatened with a development which would eliminate the habitat of the whole invertebrate community.

There are now two excellent books that deal with the detailed classification of bees and provide keys to the identification of British species (Falk and Lewington, *Field Guide to the Bees of Great Britain and Ireland*, 2015; Else and Edwards, *Handbook of the Bees of the British Isles*, 2018). A further recent publication is *The Solitary Bees*, by Danforth, Minckley and Neff (2019), which describes bee behaviour, evolution and conservation from a North American perspective. In Chapter 1 we will offer some guidance on how to begin distinguishing the different bee groups, and also outline the diagnostic features of the bee genera that can be found in Britain.

LIFE CYCLE

All insects, including bees, possess an external skeleton, a feature shared with members of the very diverse invertebrate group known as Arthropods (Arthropoda) which, in addition to insects, includes animals such as crabs, spiders and millipedes and their many relatives. Arthropods grow by periodically shedding their external covering, but, once the adult skeleton has hardened, an arthropod cannot increase in size. This means that a small bee should not be assumed to be a young bee; it is a small bee species or a poorly nourished larger one.

The immature stages of some insect groups, including dragonflies (Odonata) and grasshoppers and crickets (Orthoptera), look quite similar to the adults. The division of the body into head, thorax and abdomen is clearly visible, as are the six legs. In species that have wings in the adult stage, small 'stubs' are present in the immature stages. Development occurs by repeated shedding of the exoskeleton to allow the insect to grow towards the adult stage. This pattern of development is sometimes referred to as 'incomplete' metamorphosis.

By contrast, bees share with other groups of insects, such as flies (Diptera), butterflies and moths (Lepidoptera) and beetles (Coleoptera), a very different pattern of development: 'complete' metamorphosis. There are four radically different phases in their development: egg, larva, pupa and adult. The bee's egg is relatively large and sausage-shaped. It is generally laid on or close to the food-store provided. The larva is a maggot-like creature without legs or eyes, which can move only by wriggling. The larva is a feeding machine with simple mouthparts and a large gut but lacking reproductive organs. When it is fully grown, the bee larva evacuates its faeces and moults to enter into a new phase as a prepupa. It does not feed during this phase, and, in many species, this is the dormant stage (diapause) in which it passes the winter, though other solitary bees pass the winter as hibernating adults. The prepupa still has the appearance of the earlier larval stages but has a more thickened and rugged cuticle. Eventually a final moult leads to the emergence of a pupa, in which the adult organs, including the reproductive organs and external features, are all present, but the cuticle remains white and soft and lacks hair. Metamorphosis is completed within one or two weeks by the hardening of the cuticle, the growth of hair and the expansion of the wings, leading to the fully formed adult (Fig. 11).

OPPOSITE: **FIG 11.** Life cycle of the Hairy-footed Flower Bee *Anthophora plumipes*.

YEAR 1
a Female excavating a nest burrow (February–May)
b+c Cell being stocked by female (February–May)
d Sealed completed cell, with egg (February–May)
e First instar larva feeding on pollen/nectar provision (March–April)
f Second/third instar larva (April–May)
g Fourth instar larva (May–June)
h Prepupa (June)
i Pupa (July)
j Adult (female) in diapause (July/August and overwinter)

YEAR 2
k Empty cell after adult emergence
l Adult (male) emerging from a brood cell (February–March)
m Male pursuing female for mating (February–May)

(Then back to a)

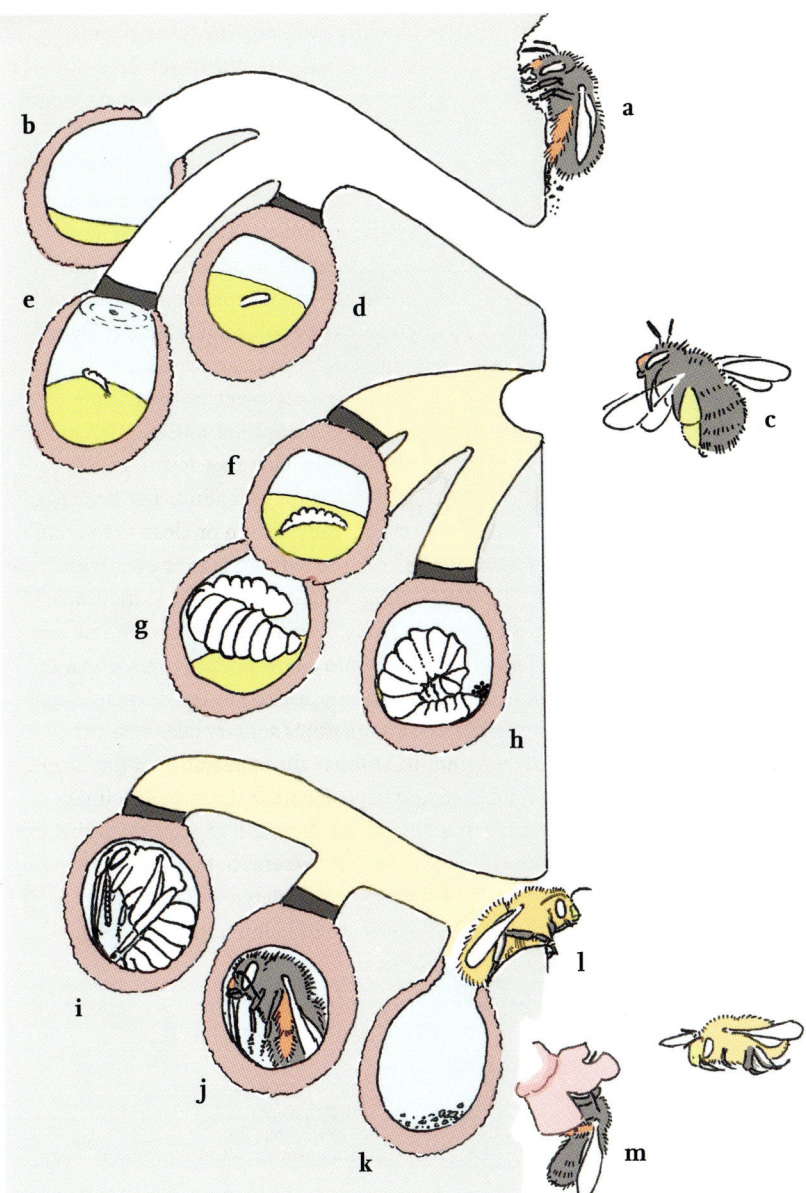

These processes of development all take place within a nest made by the female bee. This consists of a variable number of brood cells, each of which is stocked with a supply of food (usually a mixture of nectar and pollen) to power the development of the offspring. In the case of the cuckoo bees, of course, the food supply is delivered not by the mother but by the host female whose nest has been invaded.

The astonishing effort, skill and diversity displayed by nest-making bees has evoked wonder and admiration in those who have taken the trouble to study it. A century ago, the entomological writer Edward Step gave pride of place to the nest-making achievements of solitary bees in his *Insect Artizans and their Work*. He celebrated the achievements of many kinds of insects in paralleling the craft skills of humans – as spinners and weavers, masons, carpenters, upholsterers, miners and many more. Of the miners he has this to say: 'One of the most astonishing things to an observant entomologist is the sight of the little bees of the genus *Andrena* busy sinking their vertical mineshafts into a path that has been beaten down to make it uniformly firm and level, and trodden by feet innumerable. Try with your fingers or your pocket-knife to excavate such a hole yourself, with all a man's strength, and you will acknowledge that your best efforts only make a very sorry job of it.' (Step, no date, p. 21.)

Step even recognises the nest-parasites, or cuckoo bees, as artisans, comparing them to human burglars or housebreakers. He is quick to reply to the objection against elevating burglars to the status of artisans: 'a popular dictionary…defines "artisan" as one skilled in any art, mystery, or trade, and surely there is both art and mystery about the proceedings of the human burglar, and the Insect Burglar does not fall short of her human prototype in these respects.' (Step, p. 279.)

As suggested by the terms 'miners', 'masons' and 'carpenters', Step recognised the great diversity of places chosen and materials used by bees in their nesting activities. The majority, including the Andrenas, dig their nests in underground burrows, often using looser and more friable soils or sand, as well as the edges of paths. A common nest architecture is a vertical burrow with side branches which terminate in hollow cells. The cells usually have waterproofed walls, and each of them is stocked by the female bee with a mixture of pollen and nectar, onto which she lays a single egg. The resulting larva goes through its developmental stages, and the adult eventually digs its way up to the surface. Some bees sink their mineshafts in great clusters, termed aggregations, numbering hundreds or even thousands of individual nests. Others nest singly or in small groupings in sparsely vegetated ground. In either case, each female has her own nest, and there is no social division of labour: the strictly solitary bees are often gregarious, but not social. There are two main variations on this theme. In a few species, several individuals may share a common nest entrance, but within the burrow each female has her own distinct brood cells, and there is no division of labour. These are termed communal species. The second variation was briefly mentioned above: in the family Halictidae, there are several species in which the female produces one or more broods of sterile female offspring which act as workers to provide food for her own subsequent progeny. A small number of these species is capable of switching between social and solitary ways of life, depending on environmental conditions. We take the discussion of nesting behaviour and the life cycle further in Chapters 3 and 4.

An alternative to the underground nest is to make use of available aboveground holes and crevices. The bees which adopt this mode of life are termed 'aerial' nesters, but, again, there are many variations. Some bees make their nests in the exit holes of beetles in dead wood, others in hollow stems of plants, and a few in empty snail shells or even vacated galls of other insects. Usually, the aerial nesters are opportunist users of already-existing niches, hollows or crevices, but a few will use their mandibles to remove pith or modify the internal space they are using. Another possibility is to use sharp and powerful mandibles to dig out the necessary cavity, in dead wood or a similar substrate. The exemplars of this are the carpenter bees of the genus *Xylocopa*, of which Britain has just one species that appears to be establishing itself. However, there are flower bees (*Anthophora*) and mason bees (*Osmia*) that also might qualify as carpenters.

MALES AND FEMALES

Sex determination in Hymenoptera (bees, wasps and ants) is very different from the system we are familiar with in mammals, such as ourselves. Mammals produce two types of sperm cells in about equal numbers. Eggs fertilised by a sperm bearing a Y chromosome result in males, and eggs fertilised by a sperm bearing an X chromosome result in females. Male mammals have XY sex chromosomes and females XX. Mammal eggs which are not fertilised cannot develop.

In Hymenoptera such as bees, sperm cells are all of one type in regard to the determination of sex. Eggs which are fertilised by any sperm cell produce female offspring, while eggs laid without being fertilised develop into males. Females are able to 'decide' whether or not to fertilise each egg, which allows them to control the ratio of males to females and the position of male and female brood cells within the nest. In most bee species, females are larger than males. Brood cells in which fertilised eggs are placed tend to be more spacious and stocked with more food.

Male bees (and other male Hymenoptera) have just one set of chromosomes rather than the usual two. As a consequence, the millions of sperm that a male bee produces are all genetically identical to himself and to each other (barring copying mistakes). This type of breeding system is known as haplodiploidy because males are haploid (having one set of chromosomes) and females are diploid (having two sets of chromosomes). If a female bee is mated by only one male, her female offspring will share 75 per cent of their genes with each other but 50 per cent with their mother. The consequences of haplodiploidy are complex and far-reaching, influencing sex ratios and the tendency towards cooperation and social behaviour.

Male bees usually emerge one or two weeks before females, allowing males to compete for mates at the earliest opportunity, as soon as the females appear. This phased timing is a result of differing male and female responses to temperature. Males may live for several weeks and can potentially mate many times. They adopt a range of different strategies in their search for virgin females, and we discuss these differences and possible explanations for them in Chapter 2. Some males possess specialised structures used in restraining a potential mate, such as thickened muscular legs and long, pointed mandibles. In some species, males possess extra-large heads, perhaps giving them a competitive advantage in some circumstances. Long hairs or expanded white areas on the legs, which play a role during courtship and mating, occur in some male leafcutter bees. In addition, males (and some females) of a number of species have white or yellow markings on the face which in some cases are thought to be involved in territorial competition.

Males have no scopa or sting. They visit flowers and take nectar for their own nutrition but do not take part in any aspect of nesting behaviour. In some species, the colour of the male's hair differs from that of females, as we have seen in

FIG 12. Grey-patched Mining Bee *Andrena nitida* (a) female and (b) male.

FIG 13. Coast Leafcutter Bee *Megachile maritima* (a) female and (b) male.

Anthophora plumipes. The females of two mason bee species have horns on the face, used in nesting activities and absent in males. The scopa of females, which males lack, is often brightly coloured or has contrasting black and white hairs. Males often have a tuft of dense erect hair on the face. They also have a slimmer build and longer antennae than females (with 13 rather than 12 antenna segments). In some cases, the mandibles of males are longer than those of females and have a triangular projection at the base. Careful observation of all such features will assist in distinguishing males from females and can also help with identification of species. Solitary bees are best observed soon after emergence, as the colour of their hair can be bleached by the sun in just a few days.

On very rare occasions a bee turns up with a mixture of male and female body parts. These individuals are gynandromorphs and can be a different sex on each side of the body or a patchwork of the two sexes. Figure 14 shows a gynandromorph of *Anthophora plumipes* discovered in Norfolk in 2021. The head has male features on the right side (left side of the image) and female features on the other side. The male side of the face shows the typical patterned yellowish colour and the antenna has 13 segments, whereas on the female side the face is entirely black and there are 12 antenna segments. The tongue, too (not shown), is divided down the middle into male and female, only the female side bearing the stiff spines used in

FIG 14. Gynandromorph Hairy-footed Flower Bee *Anthophora plumipes*. (© M. D. Welch)

extracting pollen. The genitalia of this individual are male, but most of the body is a patchwork of the two sexes (M. D. Welch pers. comm. based on dissection).

This very striking example illustrates the extent of sexual dimorphism in the external appearance of bees. But sex differences extend beyond physical appearance: male and female bees behave in almost entirely different ways, as explained in later chapters. Sex-specific genes are operating throughout a bee's development, leading to contrasting behavioural roles and chemical signalling.

COLLECTING AND TRANSPORTING FOOD

Flowering plants and bees share a long history of coevolution, during which plants developed forms of display that attracted bees and other insects to the 'rewards' they had on offer for visitors to their blooms. If these insect flower visitors were to carry out the function of pollination for the plant, then their behaviour had to be manipulated. Some plants have evolved specialised floral structures that attract specific groups of insects (or other pollinators) while, at the same time, encouraging them to visit plants of the same species on subsequent visits. Meanwhile, bees, having become dependent on plants for their nutritional needs, evolved both anatomical and behavioural adaptations enabling them to access and reap the rewards offered by the plants. The visual abilities of bees are important in recognising flowers at a distance and, at close range, they have acute chemical senses in their feet, mouthparts and, especially, antennae. They also have high cognitive and learning abilities, enabling them to find and re-find patches of the appropriate flowers for their needs, negotiate often complicated floral structures, and navigate the return journeys between the flowers and their nests.

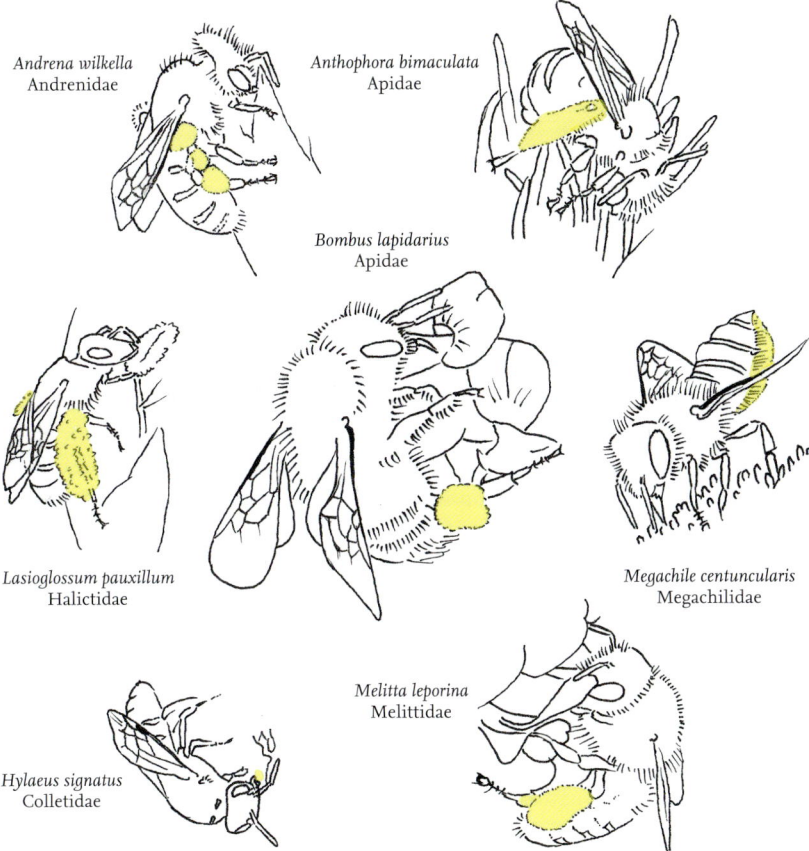

FIG 15. The position of the pollen-collecting structures in representatives of each of the six bee families, drawn from photographs. The Red-tailed Bumblebee *Bombus lapidarius* in the centre has a corbicula (true pollen basket), with pollen gathered into a single roundish lump on the hind tibia. The other six are non-corbiculate and are therefore solitary bees. The position of the scopa differs between families and genera. The Large Yellow-face Bee *Hylaeus signatus* is a solitary bee, but members of this genus have no scopa and instead swallow pollen, transporting it in the crop. Other members of the Colletidae (not shown) do however possess a scopa on the hind legs.

Apart from their sensory organs, bees have anatomical specialisations both for harvesting pollen and nectar from flowers and for transporting these foodstuffs back to the nest to stock the brood cells of their future offspring. As we mentioned above, bees have very complex mouthparts, but these are basically of two sorts: a 'tongue' and a pair of jaws (mandibles). The tongue is itself a very complex organ, articulated so that it can be folded away when not in use. Its main function in most bees is to lap or suck up nectar from flowers, but in some bees (in the family Colletidae) it is also used to deliver secretions in nest construction. There is considerable variation among the bee families in the form taken by the tongue, and this relates to the different floral structures they can access. The mandibles are used for digging, cutting and carrying nesting materials and sometimes for grappling with rivals or potential mates. However, they are also used for holding on to flowers while the bee is foraging or resting and can also play a part in obtaining pollen from anthers. Sometimes, too, pollen is collected on minute hairs on parts of the tongue, or on the face of the bee.

Mostly, however, pollen is collected among the branched hairs on the body surface of the bee as it forages. In this way, it is incidentally transported from flower to flower and may contribute to the pollination of the plants that are visited. In female bees of the nest-making species, brushes on the bees' foot segments are used to scrape the dispersed pollen into the scopae, and the pollen becomes unavailable for pollination. In most species, the scopae are located on one or more of the segments of the hind legs, or on the sides of the propodeum, but in members of the family Megachilidae they cover the underside of the abdomen.

These and other aspects of the diverse and complex relations between flowers and bees are discussed in more depth in Chapters 5 and 6.

WINGS AND FLIGHT

Flight is essential to almost every aspect of a bee's life, and no bee species is naturally wingless, though there are some solitary wasp species in which the female lacks wings, and both sexes of most ant species possess only temporary wings, used in mating flights. The wingbeats of bees are quite slow compared with those of Diptera (flies) but in some cases have a high enough frequency to make a buzz during flight. Some solitary bees, notably the flower bees, *Anthophora* species and the Wool Carder Bee *Anthidium manicatum*, are able to hold their station while hovering in front of a flower or potential mate. However, most other solitary bees are less agile and maintain a forward motion. Some bees are able to generate heat by shivering their wing muscles. The pattern of veins on the wings of bees (venation) is highly distinctive and of considerable use in classification and identification. More details follow in Chapter 1.

STINGS

Bees, along with wasps and ants, are famous for their stings, and honeybees can be dangerous in large numbers. Solitary bees, even social species, do not attack and sting people, and there is little danger from the increasingly popular bee hotels which sometimes house hundreds of solitary bees. Solitary bees do not attempt to sting even when their nests are disturbed. When physically restrained or trodden on, female solitary bees will sting, but in the authors' experience the pain is slight and lasts only a minute or two. However, some people do react strongly to stings, and very rarely deaths have been reported from solitary bees. Honeybees have a barbed sting which causes the sting apparatus to be left in the victim if the bee is pushed aside. No other British bees have this feature and their stings simply enter and exit, depositing a cocktail of proteins which cause pain but leave the sting's owner intact. The sting of a bee evolved from a modified egg-laying tube, the ovipositor, which is now bypassed during oviposition. Male bees, naturally, do not have an ovipositor or a sting.

BEE STUDIES

Students of solitary bees today are very lucky. They have two excellent guides for identification which also include illustrations and information about the species that occur in Britain (Falk & Lewington 2015, Else & Edwards 2018). There are also Internet sources, most notably Steven Falk's Flickr feature (https://www.flickr.com/photos/63075200@N07/collections/72157631518508520) and the website of the Bees, Wasps and Ants Recording Society (BWARS) (http://www.bwars.com).

Until the beginning of the 19th century in Britain, the aspiring student of bees would have had very little literature to go on. There are passing references in the works of the celebrated naturalists Gilbert White, John Ray and Francis Willughby, but the first major work on the bees ('Anthophyla') was by William Kirby (1759–1850), who was rector of Barham in Suffolk. *Monographia Apum Angliae* (Monograph on British Bees) was published in 1802. Kirby described 221 bee species, placing them in just two genera, *Melitta* and *Apis*, representing short-tongued and long-tongued bees. This was the first significant account of bees of any area in the world (Michener 2007). Kirby's intact bee collection survives in the Natural History Museum in London. It contains the type specimens of many British bee species and continues to be consulted (Notton & Dathe 2008, Praz et al. 2022).

Frederick Smith published a *Catalogue of the Hymenoptera*, including bees, in the collections of the British Museum in 1853 and 1854 (Smith 1858), and W. E. Shuckard's *British Bees* was eventually published in 1866, just a couple of years before his death (Shuckard 1866, Baker 1998). The Surrey entomologist Edward Saunders published numerous papers on British Hymenoptera from 1880 onwards, culminating in his major work on the bees, wasps and ants in 1896 (Saunders 1896). Though nomenclature has changed a great deal since then, and many new species have been added to the British list, Saunders' book remains an important reference for entomologists today. R.C.L. Perkins published valuable papers in leading entomological journals in the first two decades of the 20th century, describing many of the species of British bees, and providing keys to the

FIG 16. William Kirby (Wikimedia Commons).

identification of several difficult genera. Edward Step's work on insect artisans, mentioned above, seems to have been written just after the end of World War I, but it was not until rather later, in 1932, that his work on the British bees, wasps and ants was published (posthumously) in the popular Wayside and Woodland series of introductory books on natural history published by Frederick Warne (Step 1932).

It was not until the early 1970s that the resurgence of scientific and popular interest in solitary bees (and their relatives) began. In late 1969, George Else was appointed bee curator at the London Natural History Museum. Largely through his enthusiasm for the group, and his association with Mike Edwards, moves were made through the 1970s towards establishing a national recording scheme for Aculeate Hymenoptera. This was set up as the Bees, Wasps and Ants Recording Scheme in 1977, and in 1995 what we now know as the Bees, Wasps and Ants Recording Society (BWARS) was formally established. This gave access to a library of key works for members and also to specimens and expert help in identification. The society promoted recording activity on selected groups of species on an annual basis and in 1997 began publishing a series of atlases. Not only were these atlases of great interest in themselves, but their production was aided by Else's draft identification keys, which he generously distributed to local recorders. This greatly increased the numbers of skilled recorders and enthusiasts, and it also allowed for the keys to go through a series of corrections and revisions in the light of feedback from users.

During the prolonged anticipation of Else's *magnum opus*, Steven Falk published his excellent and user-friendly field guide (Falk & Lewington 2015), and this, together with Falk's online photographic collection, added further stimulation and accessibility to the growing awareness of and enthusiasm for solitary bees. The classic work on British bees, by George Else, now joined by Mike Edwards as co-author, followed in 2018.

THE NAMES OF BEES

Over very many years, people who studied bees learned to use their scientific names. Each species has two parts to its name, and this applies to bees as to other groups of living things. The first part allocates a species of bee to a group of related bees (its genus), and the second part is its specific name. For example, *Osmia bicornis* belongs to the genus *Osmia*, and *bicornis* is its specific name (which refers to the two small projections from its face). This system of naming has the advantage that it is a kind of common language, understood by researchers internationally. It means that students of bees who come from different language communities can exchange knowledge while being reasonably sure they are talking about the same species. Scientific names often include a description of the species named, as in the case of *Osmia bicornis* (meaning 'two-horned'). They also indicate evolutionary relationships between species: thus, all *Osmia* species (*Osmia bicornis*, *Osmia bicolor*, etc.) are considered to be related to each other.

Unfortunately, these benefits of scientific names are not obvious to people (most of us) who do not have a classical education! For many of us, being able to name a species using our own language is highly desirable. Historically, a great many wild flowers, birds, mammals, butterflies, grasshoppers and other living beings had familiar names in our own language, but sadly these names are slowly being lost in the wider public. They do, however, continue to be understood and used by groups of people with special interests in, for example, birds or butterflies.

The difficulty with bees, especially solitary bees, is that the great majority of them never had common names that were shared, even by special interest groups. In part this is because there are so many species and only a small proportion of the British bee fauna can be reliably identified on sight. It seems likely that this gap, and the consequent need to rely on scientific names, has discouraged many people from getting into the study of these fascinating insects. Steven Falk has made a Herculean effort to change this situation by giving all the British species names in the English language in his excellent field guide (Falk & Lewington 2015). Many of these names have been readily taken up in the community of bee students and seem genuinely to have brought in many new enthusiasts. Therefore, with the exception of some lists and tables, we use common names at first mention of a species followed by the scientific name. Subsequently, we generally use just the scientific name. We hope that this compromise will suit most readers.

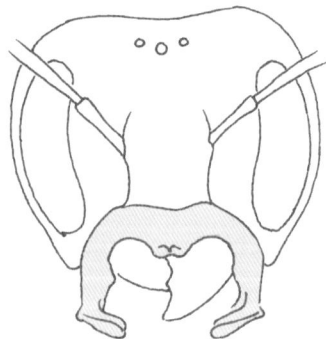

FIG 17. The face of a female Red Mason Bee *Osmia bicornis*. The two horns are thought to be used to fashion mud inside the nest burrow.

CHAPTER 1

The Diversity of Solitary Bees

Some solitary bees can be definitively identified on sight 'in the field' (or in the garden!). With time and experience, it becomes possible to expand this number, but even with the methods of close inspection we mentioned in the previous chapter, many of our 240-plus species demand microscopic examination and, in a few cases, dissection if we are to be certain.

However, there is no need to be discouraged! The first step for beginners will be to learn to recognise the differences between the families and genera of bees. Once this skill has been achieved, a great deal of appreciation and understanding opens up to the observer: miners, masons, leafcutters, wool carders and the rest can be recognised and watched with fascination. For those who want to take their studies further, being able to recognise the genus of a specimen takes you a long way in the direction of following up one of the keys in Falk and Lewington or Else and Edwards to secure an identification at the level of species.

The sheer numbers of species (sometimes with their variations, the differences between males and females, and so on) can be bewildering. However, the bewilderment can be eased by remembering that many species are quite local in their distribution, so not all of them are likely to be seen in your area. In addition, most fly only for relatively limited periods of time during the year; each has its own flight period (phenology). Finally, some species of solitary bees are particularly associated with one or a small number of flowering plant species. Taking all these clues into account gives you a rough guide as to what you are likely to see in your local spot and at a given time of year. Here, we will begin by introducing the sorts of pollen-collecting bees likely to be seen as we go through the year. Later in the chapter, we will give a more systematic account of the different families and genera, including the cuckoo bee species, and how to recognise them.

A good time to begin a study of solitary bees is the early spring. Often the first bee to make itself known is the Hairy-footed Flower Bee, which we already met in the Introduction. The males appear first and resemble small brown bumblebees as they dart about in search of females. These look like a different species, with entirely black coats except for the orange pollen-collecting brushes (scopae) on the hind legs. Spring is also a good time to learn to recognise some of the *Andrena* species (mining bees). There may be 10 or more spring Andrenas in a typical garden, visiting tree blossom, dandelions, celandines and other spring flowers, as well as garden plants. Most bees do not show any discrimination between wild and cultivated flowers. Andrenas vary in size from the mini-miners (subgenus *Micrandrena*, small dark bees with typical wing length of up to 5.5 mm), to bees about 1 cm long. Andrenas have distinctive hair coloration or bands on the abdomen but can be quite confusing on first acquaintance. As with all solitary bees, they are much easier to identify when they are newly emerged, as the colours often fade quite quickly. Recognising that you are looking at an *Andrena* is a good first achievement. *Andrena* bees belong to the family Andrenidae, which carry pollen on their hind legs. Identification is easier when no pollen is being carried so that the colour of the scopa hairs can be seen.

Mason bees (genus *Osmia*) emerge from April to May. The most well known is the Red Mason Bee *Osmia bicornis*, formerly *Osmia rufa*, one of the most frequent occupants of artificial bee nests (bee hotels) and familiar to many garden wildlife enthusiasts. The Red Mason Bee is a representative of the Megachilidae, a family

FIG 18. Orange-tailed Mining Bee *Andrena haemorrhoa* female on Coltsfoot *Tussilago farfara*.

FIG 19. Red Mason Bee *Osmia bicornis* male and female coupled.

FIG 20. Common Furrow Bee *Lasioglossum calceatum* female. A furrow is visible on the tip of her abdomen. Some *Lasioglossum* species have pale hair bands or patches at the upper edge of each abdomen segment (tergite).

whose members carry pollen under the abdomen rather than on the legs. Mason bee nests are made in various hollow structures using mud or plant material.

More similar to *Andrena* are members of the family Halictidae, which become numerous from late spring and nest in the ground. In the early weeks, only females are present. A common example is the Common Furrow Bee *Lasioglossum calceatum*. Nearly all female halictids have a narrow hairless area on the fifth segment of the abdomen, giving the group name furrow bees. Halictid bees are mostly small and dark, sometimes with narrow white abdominal bands, and collect pollen on their hind legs. Some, including the Common Furrow Bee, are eusocial. *Lasioglossum* means 'hairy tongue'. Though furrow bees are first seen in late spring, most of them have long flight periods through spring and summer.

Members of the genus *Colletes*, known as plasterer bees, include one early spring species but it is uncommon. Several other species appear from late spring onwards. Most are medium-sized and have rich-brown hair on the thorax and pale marginal bands of flattened hairs on the abdomen. They are attracted to ragworts and other yellow flowers in the Asteraceae family. They usually nest in the ground.

From the middle of May into June, many more solitary bee species are emerging, so identification, even of the main groups, becomes more difficult. More mason bee species are taking over from the Red Mason Bee, and these are now joined by the leafcutters (genus *Megachile*). Males appear first, and then the females, with their distinctive 'tail in the air' posture when they land on a flower. The scopa is beneath the abdomen. From May onwards is a good time to look for the yellow-face bees (genus *Hylaeus*). These are small and black, often with a few yellow areas on the legs, and most of them with yellow patterns on the face: these are usually paired patches in the females but form a large yellow (or cream-coloured) 'shield' over the face in males. In gardens with patches of Lamb's Ear *Stachys byzantina*, you are likely to see males of the Wool Carder Bee *Anthidium manicatum* patrolling and darting at females. If watched closely, the females can be observed stripping leaves or stems of their fine hairs. Two solitary bee species that once had very localised distributions in southeastern England are becoming more widespread. The Large-headed Resin Bee *Heriades truncorum* is very small and is best recognised by the way the female 'vibrates' its abdomen as it collects pollen from a yellow-flowered plant in the Asteraceae family. Through June and July, the Bryony Mining Bee *Andrena florea* can be seen at patches of White Bryony

FIG 21. Bellflower Blunthorn Bee *Melitta haemorrhoidalis* female on Harebell *Campanula rotundifolia*. Purple pollen is being collected, and the gold tip of the tail can just be seen. The hairs on the face meet the eyes, indicating that it is not an *Andrena*. It is one of the six British members of the family Melittidae.

Bryonia dioica in gardens and hedgerows (it could be mistaken for a honeybee or one of the other mining bees which also visit this plant – look for the narrow red bands across the abdomen). June is also a good time to look out for the specialist bees that visit *Campanula* flowers (including both wild and garden varieties). The Small Scissor Bee *Chelostoma campanularum* and the much larger member of the Melittidae, the Bellflower Blunthorn Bee *Melitta haemorrhoidalis*, can often be seen visiting the same flower. More species of flower bees (*Anthophora*) are most frequently seen in June and July. The Four-banded Flower Bee *Anthophora quadrimaculata* and the Green-eyed Flower Bee *Anthophora bimaculata* are quite distinctive species but are restricted to southern Britain.

From early August onwards, the summer-flying bees become less evident and are replaced by a small number of late-flying species. These are often pollen specialists, so can be easily found within their geographical range. Three of these late bees are plasterer bees (*Colletes*). They are all very similar in appearance, with reddish brown hair on the thorax and bands of flattened pale hair on the abdominal segments. The Sea Aster Bee *Colletes halophilus* has the most restricted distribution, mostly in coastal districts of southern and eastern Britain, but it is often abundant where it occurs. The Heather Colletes *Colletes succinctus* is primarily a species of

FIG 22. Sea Aster Bee *Colletes halophilus* male. Males of this bee usually take nectar from the same plant used by females to collect pollen.

heather heathland, and it flies during August, timed with the flowering of Ling *Calluna vulgaris*. The third species in this group is the Ivy Bee *Colletes hederae*, a relative newcomer which has spread very rapidly across Britain. Within its geographical range it can be seen along with numerous wasps and hoverflies at Ivy blossom from late August into the autumn. Another late species is also a pollen specialist. This is the Red Bartsia Blunthorn Bee *Melitta tricincta* (Fig. 24), which, as the name suggests, collects pollen only from Red Bartsia *Odontites vernus*.

We now turn to a more systematic introduction to the different families and genera that are represented among the British bees.

TABLE 2. The families and genera of British bees. The species number excludes those found only in the Channel Islands and includes some species which may be extinct or never well-established in the British Isles.

Family	*Genus*	*Common name*	*British species*
Melittidae	*Dasypoda*	Pantaloon bees	1
	Macropis	Oil-collecting bees	1
	Melitta	Blunthorn bees	4
Apidae	*Anthophora*	Flower bees	5
	Apis	Honeybees	1
	Bombus	Bumblebees	27
	Ceratina	Small carpenter bees	1
	Epeolus*	Variegated cuckoo bees	2
	Eucera	Long-horned bees	2
	Melecta*	Mourning bees	2
	Nomada*	Nomad bees	35
	Xylocopa	Carpenter bees	1
Megachilidae	*Anthidium*	Wool carder bees	1
	Chelostoma	Scissor bees	2
	Coelioxys*	Sharp-tail bees	6

BEE FAMILIES AND GENERA

The majority of modern authors classify British bees into six families: the Melittidae, Apidae, Megachilidae, Andrenidae, Halictidae and Colletidae, which are further divided into 29 genera (Table 2). Some authors rank these six families as subfamilies within just one overarching family, the Apidae, which includes all bees. Here we use the conventional and more familiar classification used by Michener (2007), the Bees, Wasps and Ants Recording Society (BWARS) and Falk and Lewington (2015), though there are good arguments for the alternative arrangement (Else & Edwards 2018). Of the approximately 272 British bee species (excluding those found only in the Channel Islands), about 244 species of solitary bees are currently recognised (all except *Apis* and *Bombus* species).

Family	Genus	Common name	British species
	Heriades	Resin bees	2
	Hoplitis	Lesser mason bees	3
	Megachile	Leafcutter bees	9
	Osmia	Mason bees	13
	Stelis*	Dark bees	5
Andrenidae	Andrena	Mining bees	68
	Panurgus	Shaggy bees	2
Halictidae	Dufourea	Short-faced bees	2
	Halictus	End-banded furrow bees	6
	Lasioglossum	Base-banded furrow bees	32
	Rophites	Bristle-headed bees	1
	Sphecodes*	Blood bees	17
Colletidae	Colletes	Plasterer bees	9
	Hylaeus	Yellow-face bees	12
Total species			272

*Cleptoparasitic (cuckoo bee/inquiline) solitary bee genera.

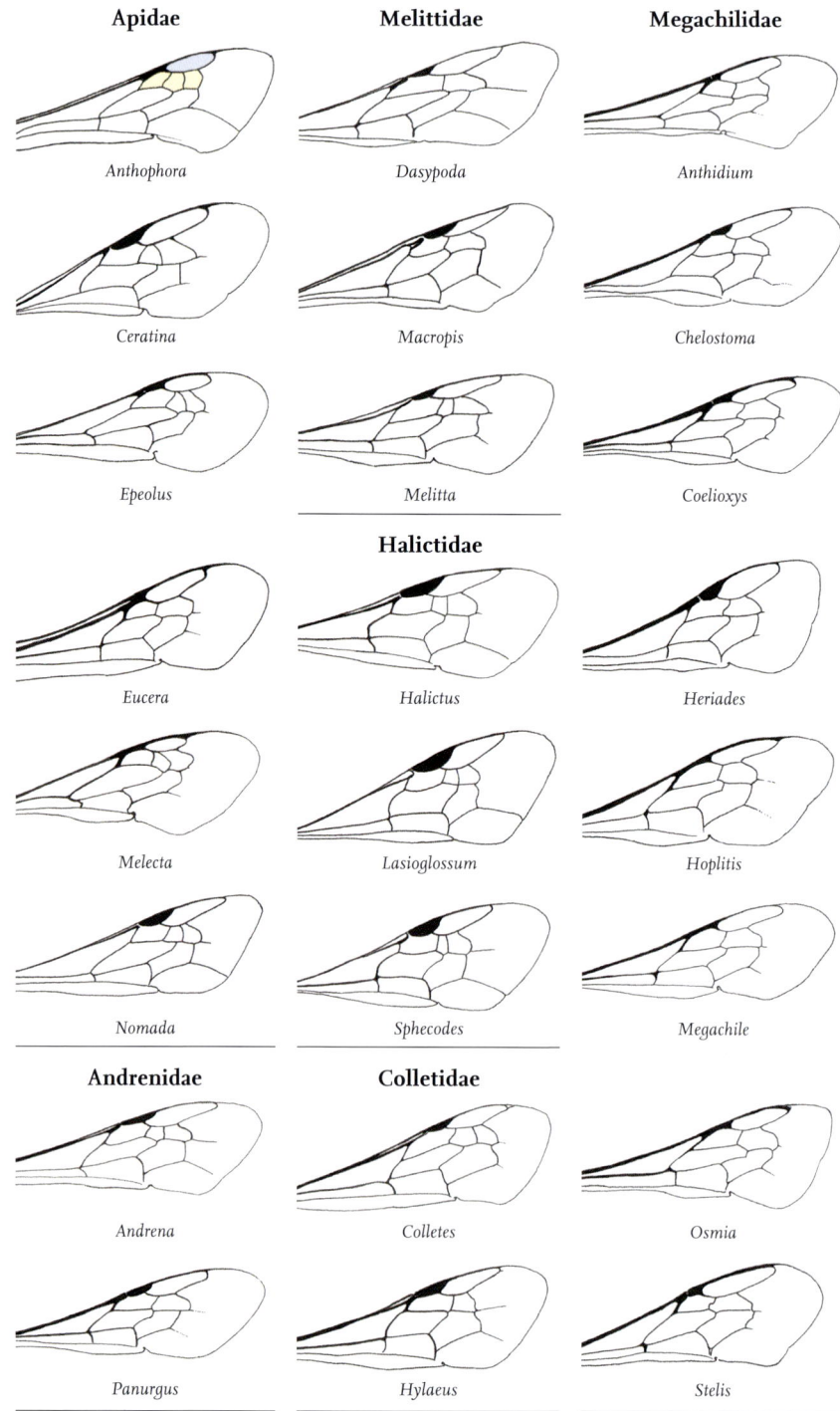

OPPOSITE: **FIG 23.** Forewing venation of the main British solitary bee genera (not to scale). The black area on the leading edge of each wing is the pterostigma. The large cell to its right is the marginal cell, and the cells (either two or three) immediately below are the submarginal cells (see *Anthophora*, top left).

Family Melittidae
This is a small family of ground-nesting bees with several features which make them a distinct group. Pollen is collected on the hind legs, usually on just the tibia and tarsus and not on the femur. The tips of the antennae have an oblique shiny end, as if sliced off with a knife. The tongue is short and the mandibles have two points. All six British members of the family are of particular interest in being pollen specialists, with one species also taking plant oils. Most are rather uncommon or local in their distribution and are most easily found by searching around their forage plants. The family does not include any parasitic species. Recent interpretations place Melittidae close to the ancestral form of all bees.

Dasypoda pantaloon bees
There is just one British species, the Pantaloon Bee *Dasypoda hirtipes* whose name refers to the spectacular bushy scopae, which are bright orange in colour. Nests with long, shallow tunnels are made in sandy ground. *Dasypoda* can be distinguished from *Andrena* and *Melitta* species by the presence of two rather than three submarginal cells on the forewing (Fig. 23). The tongue is quite short and both sexes are very hairy. They are usually seen at yellow Asteraceae flowers. *Dasypoda hirtipes* is one of the few hosts of a very small nest parasite, the flesh fly *Miltogramma germari*.

Melitta blunthorn bees
There are four British species in this genus, each specialised in their choice of pollen and most easily found by watching their particular food plants. They are medium-sized bees with a short tongue and can be separated from the rather similar *Andrena* species by their use of only the scopae on their tibia and tarsus to carry pollen (not significantly by the femur, trochanter or propodeum), a lack of facial foveae (see Figure 6 b in the Introduction) and the swollen final segment of the tarsi. Unlike most other solitary bees, the females add nectar to the pollen load, making it sticky. The Clover Blunthorn Bee *Melitta leporina* can be tempted into gardens by White Clover *Trifolium repens* and the Bellflower Blunthorn Bee *Melitta haemorrhoidalis* by bellflowers (Campanulaceae). The only known British site for the Sainfoin Bee *Melitta dimidiata* is Salisbury Plain, where it uses Sainfoin *Onobrychis viciifolia*, while the Red Bartsia Blunthorn Bee specialises on Red Bartsia, an annual

FIG 24. Red Bartsia Blunthorn Bee *Melitta tricincta* female.

plant often found in compacted earth at field gateways and tracks. Several *Melitta* species are parasitised by the Blunthorn Nomad Bee *Nomada flavopicta*.

Macropis oil-collecting bees

The Yellow Loosestrife Bee *Macropis europaea* is the only representative of this genus in the British Isles. It is a very unusual bee in that it uses floral oils as well as collecting pollen from Yellow Loosestrife *Lysimachia vulgaris*. The bee and the plant appear to be mutually dependent. Yellow Loosestrife is a tall wetland plant which grows in scattered clumps in reedbeds. Pollen is collected on the hind tibiae, which have white hairs, contrasting with the black hairs on the basitarsi. There are two submarginal cells on the forewings, and the tongue is short. Males have very stout legs and a yellow face and can be found tirelessly pursuing females around the food plant or quietly taking nectar from other plant species.

Pantaloon Bee
Dasypoda hirtipes f

Sainfoin Bee
Melitta dimidiata f

Bellflower Blunthorn Bee *Melitta haemorrhoidalis* f

Yellow Loosestrife Bee
Macropis europaea f

FIG 25. Examples of the genera of the family Melittidae (f = female).

Family Apidae
This is a very diverse family which has been the subject of several taxonomic revisions. It includes Honeybees *Apis mellifera* and Bumblebees *Bombus* species, which are not described here. Social behaviour is confined to these two genera in the British Apidae, though solitary Apidae do often nest in aggregations of individual nests. The Apidae is one of the two bee families whose members have a long tongue (relative to the body), the other being the Megachilidae. Three out of the nine genera of the Apidae, *Epeolus*, *Melecta* and *Nomada*, are entirely cleptoparasitic on other bees.

Anthophora flower bees
There are five British *Anthophora* species, one of them (the Potter Flower Bee *Anthophora retusa*) now being very rare. The forewings have three submarginal cells with each cell being of similar size (in other genera, the first is often larger than the other two). (See Figure 23.) *Anthophora* species have a round body shape and a long tongue. The body hair is long, giving a resemblance to a small bumblebee, and they may make a loud buzz in flight. Flowers with a deep corolla, especially Lamiaceae, are often used as a food source, and pollen is collected in scopae on the hind tibia and tarsus. Flight is quick and darting, interspersed with hovering. The males of some species have distinctive hairs on their middle legs, and there can be white or yellow markings on the face in one or both sexes. Some species have bright green eyes. Large nesting aggregations can occur in soft cliffs or mortar, though some species nest in stems or wood. The Hairy-footed Flower Bee *Anthophora plumipes* can be tempted to nest in home-made mud bricks, and

FIG 26. Hairy-footed Flower Bee *Anthophora plumipes* female entering her nest.

the somewhat smaller Fork-tailed Flower Bee *Anthophora furcata* occasionally uses hollow stems in bee boxes, where it can be recognised by the reddish hairs on the tip of the abdomen. Bees in the genera *Melecta* and *Coelioxys* are cleptoparasites of some members of the genus.

Ceratina small carpenter bees
The Little Carpenter Bee *Ceratina cyanea* is the only British representative of this diverse genus of small, shiny, metallic-blue bees. The tongue is long and the abdomen somewhat globular, giving the bee a distinctive shape. Males have a white face marking. The body has very little hair. Pollen is collected on the hind tibiae and tarsi, but the scopae are poorly developed. Nesting occurs mostly in hollow plant material, such as the broken tips of bramble stems. The bee is largely confined to southeast England and the Bristol area.

Epeolus variegated cuckoo bees
This genus of smallish dumpy bees is represented by two quite similar British species. They are cleptoparasites of *Colletes* species and can often be seen at their nesting aggregations, crouched on the ground or entering nest burrows. The eyes are reddish brown or greenish, and the black body bears various paired white bars and spots composed of flattened short white hairs. The legs are red and black, and the tongue is moderately long.

Eucera long-horned bees
These are large and conspicuous bees with some resemblance to *Anthophora* or *Bombus* species. They collect pollen in scopae on their hind legs and sometimes nest in aggregations. The forewings have two submarginal cells (three in *Anthophora* and *Bombus*). There are two British species, with one thought now to be extinct. Females are large and broad with pale hair on most body areas and reddish brown hair on the thorax. The pollen hairs on the hind legs are long and pale, and there are also long, pale hairs on the sides of the tergites. Nectar is usually added to the pollen loads on the scopae as the female forages. Males have a similar colour pattern and are instantly recognisable by the extreme length of their antennae as well as by their white face. The tongue is long, and both sexes are usually seen on Fabaceae flowers.

Melecta mourning bees
There are two British species, one of which is probably extinct. They are cleptoparasites, attacking the nests of *Anthophora* species, and are correspondingly quite large. The body has long brown, black and white hairs, giving the bee a

FIG 27. Common Mourning Bee *Melecta albifrons* male on Green Alkanet *Pentaglossis sempervirens*.

spotted appearance. There are three submarginal cells, and the tongue is long. *Melecta* species are most easily found around the nest sites of their host but also visit flowers, sometimes in gardens. Like many cleptoparasites, they are relatively slow moving.

Nomada nomad bees

Nomad bees collect no pollen and most have very little body hair. All are cleptoparasites of other bees, laying eggs in their nests and exploiting the pollen reserves collected by the host. The host is often an *Andrena* but can be of other genera such as *Colletes*, *Lasioglossum* or *Melitta*. Nomad bees are typically seen in low wandering flight seeking the nests of their hosts and often visiting the same flowers. The cuticle is brightly coloured, usually with wasp-like black and yellow or black and red markings. They can, however, be distinguished from wasps by the very different pattern of veins on the wings. Some can be identified from good images or even in the field from their characteristic pattern of spots, bands and stripes, though others require the inspection of microscopic details. Attendance at the nests of a particular host can give a useful clue. There are 35 species in the British Isles, some having arrived only recently. Spring is the best time to find a range of species, but some fly later in the season, in synchrony with their host, which can be a second generation of a spring-nesting species.

Xylocopa carpenter bees

These spectacular iridescent blue-black bees are the size of a queen bumblebee. There are several species in mainland Europe, with one, the Violet Carpenter Bee *Xylocopa violacea*, occurring sporadically in Britain, with some recent evidence of successful breeding. Nests are made in hollowed-out dead wood.

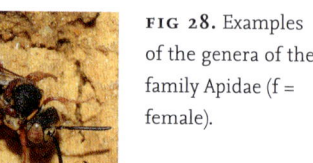

FIG 28. Examples of the genera of the family Apidae (f = female).

Green-eyed Flower Bee *Anthophora bimaculata* f

Little Carpenter Bee *Ceratina cyanea* f

Red-thighed Epeolus *Epeolus cruciger* f

Long-horned Bee *Eucera longicornis* f

Common Mourning Bee *Melecta albifrons* f

Painted Nomad Bee *Nomada fucata* f

Violet Carpenter Bee *Xylocopa violacea* f

Western Honey Bee *Apis mellifera* f

Red-tailed Bumblebee *Bombus lapidarius* f

Family Megachilidae

Members of this family use plant materials or mud rather than glandular secretions in the construction of their nests. The name refers to the large mandibles of leafcutter bees (genus *Megachile*) which are used to snip pieces out of plant leaves. All family members, other than cuckoo species, carry pollen under the abdomen and all have two submarginal cells on the forewing. Nests are usually in cavities of various kinds above ground, including empty snail shells, though some species excavate burrows. The mature larva spins a silken cocoon before pupating. The tongue is long, allowing access to flowers with a deep corolla or long spur.

Anthidium wool carder bees
The one British species in this genus, the Wool Carder Bee *Anthidium manicatum*, is a large, lively, broadly built bee with a long tongue, big eyes and prominent yellow markings. Pollen is collected beneath the abdomen, often from Lamiaceae species. Males are larger than females and have spines on the tip of the abdomen. Larger males chase away smaller ones from flower patches that females might visit. Nests are made in pre-existing cavities.

FIG 29. Wool Carder Bee *Anthidium manicatum* female on Lamb's Ear *Stachys byzantina*.

Chelostoma scissor bees

These are narrow-bodied black bees with long mandibles and a scopa beneath the abdomen. Males have a pair of projections on the tip of the abdomen, appearing somewhat like scissors. Two species occur in the British Isles: the very small Small Scissor Bee *Chelostoma campanularum* and the medium-sized Large Scissor Bee *Chelostoma florisomne*. Nests are made in cavities in wood. Both species are pollen specialists, the former using members of the family Campanulaceae (bellflowers) and the latter Ranunculaceae (buttercup) species.

Coelioxys sharp-tail bees

Bees in this genus are cleptoparasites of various species of *Anthophora* and *Megachile*. The name sharp-tail bee refers to the female's abdomen, which tapers to a sharp point and is used to insert eggs into the nest of its host. The male abdomen has an array of projecting spikes. Details of these features are helpful in identification. The face has a mixture of pale hairs and dark bristles, and the

FIG 30. Shiny-vented Sharp-tail Bee *Coelioxys inermis* inspecting a partly finished nest of a Wood-carving Leafcutter Bee *Megachile ligniseca* in a bee hotel, with completed nests nearby.

eyes are hairy. The thorax has pale brown hairs fading to white with age. The abdomen has prominent wedge-shaped white hair bands. The bees are difficult to spot because they spend a lot of time waiting near nest sites for an opportunity to enter. The area around bee boxes is a good place to look for them. Six species occur in the British Isles.

Heriades resin bees
Two members of this genus are known in the British Isles, the Large-headed Resin Bee *Heriades truncorum* and the Small-headed Resin Bee *Heriades rubicola*, the latter with just two records, though possibly established but overlooked (Cross & Notton 2017). These bees are 5–6 mm long and can be recognised by their habit of rapidly dipping their abdomen up and down as they brush pollen onto the scopa, usually from a yellow Asteraceae flower such as Ragwort *Jacobaea vulgaris*. Nests are made in old beetle holes in wood, using resin for partitions between the brood cells and resin impregnated with pebbles to seal the entrance. The genus is characterised by a minute transverse ridge near the front edge of the first tergite.

Hoplitis lesser mason bees
This genus is very similar to *Osmia* and can be distinguished mainly by fine details of sculpturing on the thorax, not easily visible in a photograph. Nests are made vertically in rotting wood or hollow stems. Three species occur in the British Isles, one a recent arrival.

Megachile leafcutter bees
The robust toothed mandibles of leafcutter bees work like scissors to cut pieces out of leaves or petals to form their nests, which are made in crevices or holes in timber or in sandy ground. The tongue is quite long, allowing access to flowers with a deep corolla such as those of the family Fabaceae. The long tarsal claws are unusual in lacking an arolium (small pad) between them. A leafcutter's characteristic body posture, with tail in the air on alighting, makes these bees easy to spot. There are seven established species in the British Isles which can often be identified from good photographs using the colour of the scopa hairs, but this needs to be seen without pollen. The males of three species have distinctive white swollen extensions fringed with hairs on the front tarsi. The eyes of some species are bright green. Bees in the genus *Coelioxys* are cleptoparasites of some *Megachile* species.

FIG 31. Willughby's Leafcutter Bee *Megachile willughbiella* male.

- no arolium between claws
- expanded front tarsus

BELOW: FIG 32. Wood-carving Leafcutter Bee *Megachile ligniseca* female cutting a leaf.

FIG 33. Orange-vented Mason Bee *Osmia leaiana* female, with orange scopa visible.

Osmia mason bees

The Red Mason Bee *Osmia bicornis* is one of the easiest bees to recognise, since it so readily nests in bee boxes or holes in mortar. Thirteen species of *Osmia* occur in the British Isles, of which three nest in vacated snail shells (Walters 2022). Nest cells are sealed with either mud or leaf mastic. Those species using mastic have a large head containing the necessary muscles for chewing leaves. As with *Megachile*, the colour of the scopa hairs beneath the abdomen provides a clue as to identification. The tongue is quite long, enabling flowers with a deep corolla to be used, especially members of the Fabaceae. Males tend to return frequently to the same basking spots. As with many bee species, the bright colours of newly emerged individuals quickly fade to grey, making identification more difficult. *Osmia* species can be distinguised from *Megachile* by the presence of an arolium between the claws in *Osmia* and also by a round smooth patch (rather than a line) among the punctures on each side of the scutum.

Stelis dark bees

Members of this genus are cleptoparasites of other Megachilidae and they have no scopa. Like many cleptoparasites, they have a very strong cuticle, punctured with minute indentations. Some species have abdominal bands or spots, but the overall appearance is dark grey, and the bees are quite difficult to find. Most are seemingly rare. Five species occur in the British Isles, one appearing recently.

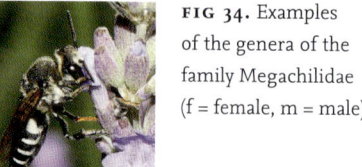

Wool Carder Bee
Anthidium manicatum m
(© L. Olds)

Small Scissor Bee
Chelostoma campanularum f

Large Sharp-tail Bee
Coelioxys conoidea f

Large-headed Resin Bee
Heriades truncorum f

Welted Lesser Mason Bee
Hoplitis claviventris f

Silvery Leafcutter Bee
Megachile leachella f

Willughby's Leafcutter Bee
Megachile willughbiella f

Blue Mason Bee
Osmia caerulescens f

Banded Dark Bee
Stelis punctulatissima f
(© M. Fogden)

FIG 34. Examples of the genera of the family Megachilidae (f = female, m = male).

Family Andrenidae

The family comprises two genera: *Andrena* and *Panurgus*. The tongue is short and pointed, and there are depressions (facial foveae) beside the inner margins of the eyes, more obvious in females than in males. The basal vein is almost straight (curved in Halictidae), and there are three submarginal cells in *Andrena* species, whereas *Panurgus* species have two. Pollen is carried on the hind legs and the propodeum. The legs have conspicuous basitibial plates that help to brace the bee against the walls of the nest burrow. Nest burrows are often in close proximity with others, sometimes revealed by a tumulus of soil around each entrance. Periods of dormancy are usually passed as a prepupa, and the larva does not spin a cocoon.

ABOVE: **FIG 35.** Grey-patched Mining Bee *Andrena nitida* female, showing the three submarginal cells in the forewing (red shading added).
RIGHT: **FIG 36.** Large Scabious Mining Bee *Andrena hattorfiana* female.

Andrena mining bees

All British Andrenas have three submarginal cells on the forewing, as in the somewhat similar genus *Melitta*. *Andrena* females have a facial fovea alongside the inner margin of each eye, a depression with a short hair pile, particularly noticeable in female mini-miners (subgenus *Micrandrena*). Another characteristic is the narrow final tarsus segment of *Andrena*, this being rather bulbous in *Melitta*. The last joint in the antennae of *Andrena* is rounded, while, as mentioned, this joint has a slanted end in *Melitta*. Sixty-eight *Andrena* species have been recorded as established species in the British Isles, of which 61 have been recorded post-2000, including the Small Gorse Mining Bee *Andrena ovatula* which has recently been recognised as comprising two very similar (cryptic) species: *Andrena ovatula* and *Andrena afzeliella* (Praz et al. 2022). The term 'mining bee' describes their habit of making nest burrows in the ground (they are fossorial), but this behaviour is not exclusive to the genus. Sites selected differ between species and include grassland, root plates of trees and very loose sand. Nest burrows can often be detected by mini-volcanoes of spoil around the entrances, in lawns or footpaths, sometimes in very large aggregations. Pollen is carried by hairs on the hind legs and also among curved hairs at the sides of the propodeum. The short, pointed tongue generally limits Andrenas to flowers with a short corolla, though some species can take pollen from clovers and about half show strong preferences for particular plant species. A few species are typically bivoltine (having two generations a year) and others are partially so. Bivoltine species can show differences in the colour of the cuticle or hair between the two generations.

Andrena species are grouped into subgenera, the most distinctive being the mini-miner subgenus *Micrandrena*. Species within this group are tricky to identify, and some European populations have yet to be fully described. Many *Andrena* species are targeted by cuckoo bees in the genus *Nomada*, with varying degrees of specialism between cleptoparasite and host. The time of year an *Andrena* bee is seen, the flowers it visits and *Nomada* species associated can all assist with identification. Many species visit spring-flowering shrubs, and others emerge in late spring or summer.

Panurgus shaggy bees

Two members of this genus occur in the British Isles. They are medium to small bees with sparse but long black hair, making them look rather unkempt. There are contrasting orange hairs on the legs, forming a bushy scopa on the female's tibia. They have two submarginal cells and moderately long tongues and specialise on Asteraceae flowers. Nesting can occur in large aggregations, usually on bare, fairly level ground. No cuckoo bees of *Panurgus* have been recorded in mainland Britain.

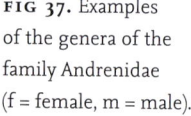

FIG 37. Examples of the genera of the family Andrenidae (f = female, m = male).

Small Scabious Mining Bee *Andrena marginata* f (© L. Olds)

Grey-patched Mining Bee *Andrena nitida* f

Oak Mining Bee *Andrena ferox* f

Buff-banded Mining Bee *Andrena simillima* f (© P. Saunders)

Large Shaggy Bee *Panurgus banksianus* f

Large Shaggy Bee *Panurgus banksianus* m

Family Halictidae

Halictids are often referred to as sweat bees, through their habit of being attracted to perspiring skin, especially in warmer climates. The alternative name, furrow bees, derives from the 'rima', present in most halictids. The rima looks rather like a hair parting at the tip of tergite 5 of females, overlapping the pygidium (terminal triangular projection). The basal vein of the forewing is strongly curved (straight or slightly curved in other families). Parts of the legs are orange in the females of some species. Pollen is collected on the hind legs and to a varying extent on the sternites. The antennae of the males of some species are very long, and some males have red markings on the abdomen. Most species are generalist in their choice of pollen sources and some are eusocial. Nests are usually made in the ground, often in large aggregations. The tongue is short.

Dufourea short-faced bees

There are two British species, last recorded in 1920 and 1956 and thought to be extinct. However, they are quite possibly overlooked owing to their small size and similarity to *Lasioglossum*. The bees are small (wing length 4 mm) and dark, with two submarginal cells in the forewings, as in *Rophites* but differing from all other British halictids, which have three. There is no rima present on the abdomen of the female, in contrast to *Halictus* and *Lasioglossum*. The face is very short from top to bottom, as in *Rophites*.

FIG 38. White-zoned Furrow Bee *Lasioglossum leucozonium* female, showing rima.

Halictus end-banded furrow bees
Six *Halictus* species have been recorded in mainland Britain, four of them being well established. All *Halictus* species have a rima and most have white hair bands on the hind margins of the tergites. The females of some *Halictus* species have orange legs. Males often have pale leg markings and a pale patch on the clypeus, while two of the established species have a slightly greenish cuticle. The tongue is short, and pollen is collected on scopae on the hind legs. Several members of the genus are eusocial, having egg-laying females (queens) and sterile females (workers) which help at the nest. Nesting occurs in the ground, sometimes in aggregations, with adult mated females surviving through the winter. The genus is parasitised by *Sphecodes* (blood bees).

Lasioglossum base-banded furrow bees
Thirty-two *Lasioglossum* species have been recorded in the British Isles. The majority of species are small and difficult to identify without using microscopic features. The name means 'hairy tongue', a feature shared with *Halictus*. The genus consists mostly of small dark bees which have a rima at the end of the

FIG 39. Furrow bee *Lasioglossum* sp. female. (© Liam Olds)

FIG 40. *Halictus* sp. female on Field Scabious *Knautia arvensis*.

hairs on hind margin of tergites

female abdomen (see Halictidae). They differ from *Halictus* in that any hair bands present occur only at the base of the tergites and not on the hind margin of the segments. The hair bands (pressed closely down on the cuticle) become more exposed as the abdomen is stretched. As in *Halictus*, males often have white markings on the legs and a pale spot on the clypeus and/or the labrum. Four species have a metallic green or blue cuticle. The tongue is short. Pollen is collected on the hind legs, and nesting occurs in the ground or on root plates, often in aggregations, with some species being eusocial. Mated females survive the winter, and males usually first appear in early summer. Bees in the genus *Sphecodes* are often cleptoparasites of *Lasioglossum* species, as is Sheppard's Nomad Bee *Nomada sheppardana*. The Ornate Bee Fox *Cerceris rybyensis* preys particularly on *Halictus* and *Lasioglossum* species (see Chapter 7).

Rophites bristle-headed bees

There is just one British representative of this genus, the Five-spined Rophites *Rophites quinquespinosus*, the sole evidence for its British status being two female bees captured in the 19th century in Sussex. *Rophites* species share some features with *Dufourea* but are much larger. Pollen is collected from Lamiaceae with the assistance of bristles situated between the eyes of the rather flattened head.

Sphecodes blood bees

All members of this genus are cleptoparasites. The hosts are usually other halictids but include some *Andrena* species. The hosts are not all yet completely

ABOVE: **FIG 41.** Five-spined Rophites *Rophites quinquespinosus* female. (© P. Westrich)

FIG 42. Square-headed Blood Bee *Sphecodes monilicornis* male.

known. The abdomen of females is usually red with a dark tip. Males, too, usually have red markings, sometimes interrupted by black bars, and one or two species can be entirely black. Black *Sphecodes* males can be distinguished from *Lasioglossum* species by the knobbly appearance of male *Sphecodes'* antennae. As with other cleptoparasitic bees, the cuticle has deep punctures and is very robust. Seventeen species have been recorded in the British Isles but they are difficult to identify without recourse to a microscope. Males can be identified by details of the distribution of hairs on the antenna segments and the anatomy of their genitalia.

Family Colletidae

This family is represented by two genera in the British Isles. All species are characterised by the possession of a short, bilobed tongue which is used to spread secretions from Dufour's and salivary glands. The secretions solidify to form a thin transparent lining to the brood cells. The food-store in the brood cells is unusual in that it is often semi-liquid. *Colletes* are ground-nesting bees, often nesting in aggregations, while *Hylaeus* construct nests in hollow stems or similar cavities. There are no cleptoparasitic members of the family in Britain. *Colletes* species are parasitised by cuckoo bees in the genus *Epeolus*, by the flesh fly *Miltogramma punctata* and by the anthomyiid fly *Leucophora grisella*.

Colletes plasterer bees

Nine species occur in the British Isles, mostly of medium size. *Colletes* look superficially like *Andrena*, but differ in having a characteristic curved outer cross-vein on the forewing and a lobed tongue, rather than a pointed one. There are prominent white marginal bands of flattened hairs on the tergites of all but one species, and the features that distinguish species are rather subtle. Pollen is

FIG 43. Heather Colletes *Colletes succinctus* brood cells washed out of a cliff.

collected on the hind legs and the propodeum. Nests are made in the ground, sometimes in aggregations numbering in the thousands. The tongue is used to spread a cellophane-like layer, produced by a very large Dufour's gland, round the nest chamber, which contains a semi-liquid pollen food-store. Several species are pollen specialists.

Hylaeus (formerly *Prosopis*) yellow-face bees
Hylaeus species are very small and black with pale yellow or white markings and very little body hair. Males of most species have large white or cream facial markings, while females usually have two white facial spots or triangles. Most also have pale body and leg markings. The males of some species have an enlarged scape (the first part of the antenna) which stores and releases pheromones. Females ingest pollen rather than collecting it on a scopa and they carry both pollen and nectar in the crop to the nest. Nests are often in plant stems or holes in dead wood but can be in the ground. The nest is lined with a waterproof secretion which is similar but not identical to the cell linings of *Colletes* (Almeida 2008). The pattern of facial and body markings can often allow identification in the field or from photographs. Adults of most species first appear in May. Twelve species occur in the British Isles. Most of them collect pollen from a wide range of flowers, but one is a pollen specialist.

FIG 44. Examples of the genera of the family Colletidae (f = female, m = male).

Early Colletes
Colletes cunicularius f

Hairy-saddled Colletes
Colletes fodiens f

Margined Colletes
Colletes marginatus f

Ivy Bee
Colletes hederae m

Large Yellow-face Bee
Hylaeus signatus f

Little Yellow-face Bee
Hylaeus pictipes f
(© V. Bartlett)

THE DIVERSITY OF SOLITARY BEES · 55

TABLE 3. Characteristics of bee genera.

	Female features											Male features					Both sexes								
	Scopa beneath abdomen	Scopa on hind legs	No scopa	Corbicula (true pollen basket)	Very broad metatarsus	Pointed tip to abdomen	Furrow (rima) at tip of abdomen	Facial foveae (reduced in males)	Red abdomen markings	Very long scopa hairs	White face marking	Spines on tip of abdomen	Knobbly antennae	Red abdomen markings	White expanded structure on forelegs	White face markings	Long tongue	Two submarginal cells	Three submarginal cells	Metallic cuticle	Hairy eyes	Construct nest in ground	Construct nest above ground	Nest in snail shell	Black and white pattern
Family Melittidae																									
Dasypoda		✓								✓						✓		✓				✓			
Macropis		✓																✓				✓			
Melitta		✓																				✓			
Family Apidae																									
Anthophora		✓									✓					✓			✓			✓	✓		
Apis				✓	✓												✓		✓		✓		✓		
Bombus				✓	✓												✓		✓			✓	✓		

continued overleaf

TABLE 3. *continued*

Feature	Ceratina	Epeolus	Eucera	Melecta	Nomada	Xylocopa	Family Megachilidae	Anthidium
Female features								
Scopa beneath abdomen								✓
Scopa on hind legs	✓		✓			✓		
No scopa		✓		✓	✓			
Corbicula (true pollen basket)								
Very broad metatarsus								
Pointed tip to abdomen				✓				
Furrow (rima) at tip of abdomen								
Facial foveae (reduced in males)								
Red abdomen markings				✓				
Very long scopa hairs								
White face marking								
Male features								
Spines on tip of abdomen								✓
Knobbly antennae								
Red abdomen markings					✓			
White expanded structure on forelegs								
White face markings	✓		✓					✓
Both sexes								
Long tongue	✓	✓	✓	✓	✓	✓		✓
Two submarginal cells				✓				✓
Three submarginal cells	✓	✓			✓	✓	✓	
Metallic cuticle	✓					✓		
Hairy eyes								
Construct nest in ground			✓					
Construct nest above ground	✓					✓		✓
Nest in snail shell								
Black and white pattern		✓			✓			

THE DIVERSITY OF SOLITARY BEES · 57

continued overleaf

Chelostoma
Coelioxys
Heriades
Hoplitis
Megachile
Osmia
Stelis
Family Andrenidae
Andrena
Panurgus
Family Halictidae
Dufourea
Halictus
Lasioglossum
Rophites
Sphecodes

TABLE 3. *continued*

	Female features											Male features					Both sexes								
Family Colletidae	Scopa beneath abdomen	Scopa on hind legs	No scopa	Corbicula (true pollen basket)	Very broad metatarsus	Pointed tip to abdomen	Furrow (rima) at tip of abdomen	Facial foveae (reduced in males)	Red abdomen markings	Very long scopa hairs	White face marking	Spines on tip of abdomen	Knobbly antennae	Red abdomen markings	White expanded structure on forelegs	White face markings	Long tongue	Two submarginal cells	Three submarginal cells	Metallic cuticle	Hairy eyes	Construct nest in ground	Construct nest above ground	Nest in snail shell	Black and white pattern
Colletes		✓						✓											✓			✓			
Hylaeus			✓					✓			✓					✓		✓				✓	✓		

✓ = present in at least some members of the genus.

CHAPTER 2

Sex and the Solitary Bee

'... (T)he advantages which favoured males derive from conquering other males in battle or courtship, and thus leaving a numerous progeny, are in the long run greater than those derived from rather more perfect adaptation to the conditions of life. We shall further see . . . that the power to charm the female has sometimes been more important than the power to conquer other males in battle.'

—Charles Darwin (1874)

INTRODUCTION

Charles Darwin noticed a widespread difference in the behaviour of males and females of many species of mammals and birds, with males eagerly seeking and competing with each other for opportunities to mate, while females were much more 'coy' and resistant to their advances. This suggested to him that the most competitive males, or the ones more attractive to the females, would be the ones most likely to be successful in contributing to the next generation. Darwin regarded what he called sexual selection as a 'second agency' of evolutionary change, distinct from natural selection. Darwin used this idea to explain the evolution of many physical differences between males and females that occur in numerous species: horns, antlers, spurs in males of many mammals, and magnificent plumage and displays in males of some birds, such as peacocks and pheasants. However, he had relatively little to say on the question of why it was males, rather than females, that competed for possession of the opposite sex. Though Darwin's concept of sexual selection has been a topic of prolonged controversy, the generalisation that it is males that compete with each other for mating opportunities, while females are relatively resistant, is not in dispute.

Though males and females have a common interest in mating and reproducing, this mutualism is consistent with significant disparities between the sexes in reproductive strategy. Females are said to have a greater investment in each potential offspring, as eggs require more energy and resources than do sperm, and females usually invest more, too, in the further development of the embryo and infant offspring. These differences generally result in the potential for males to fertilise large numbers of eggs, while females will usually be more limited in the numbers of eggs, and therefore of offspring, that they can produce. This is a plausible explanation of the behavioural differences between males and females in relation to sex: females will be more concerned with the viability of their limited numbers of offspring and choosy about the 'quality' of the males they mate with. Males, by contrast, will have a reproductive interest in securing as many mating opportunities as possible.

Of course, there are many exceptions to this generalisation. Males of the Three-spined Stickleback fish *Gasterosteus aculeatus* invest in nest-making and entice females to enter and lay eggs which they fertilise. Some bush-cricket males offer highly attractive 'nuptial gifts' for which females compete (Benton 2012). In many bird species, males and females cooperate in nest-making and caring for their offspring and this is associated with (approximate!) monogamy. But these counter-instances are still compatible with the investment idea: sex differences in reproductive behaviour seem to correlate with shifts of parental investment across the sex divide.

As we have seen, the majority of solitary bees make nests within which brood cells are stocked with enough nutrients to sustain a larva through its whole development. This work of foraging, nest construction and, sometimes, defence is carried out exclusively by the females. Also, physiological limits to the numbers of eggs that may be laid are combined with limits imposed by the availability in the local environment of necessary resources and the bee's own capacity to collect and make use of them. By contrast, males play no part in nest-making or provisioning but are centrally preoccupied with maximising their mating opportunities. The result, for most bee species, is intense competition among males for mating opportunities. The reproductive priorities of the females are quite different. As they lay relatively few eggs, they may acquire enough sperm in a single mating, and subsequent sexual harassment from males could hinder their nest-making and foraging activity. It is believed (though the evidence is limited) that in most species of solitary bees, females mate only once (they are monandrous), so the priority for males is to focus their mating attempts early in the emergence period of the females and to locate and pounce on receptive females as soon as they emerge (in some cases, even before). At any moment, the ratio of mate-searching males (sexually active through their adult lives) to receptive females (available to only one male, for a very brief period early in adulthood) will be very high. So, competitive, selective pressures on the males are likely to be especially intense in the nest-making solitary bees.

This disparity between male and female reproductive interests is very widespread among solitary bees but not universal. In a few species, females are known to mate more than once (they are polyandrous), and this shifts the focus of male competition. The priority for them is to be the last male with which a female mates before she lays her eggs, as his sperm are likely to be used to fertilise the next eggs to be laid. Also, as we have seen, the females of many species do not construct their own nests but invade the nests of others, utilising the food-store there for their own offspring. These cleptoparasitic, or cuckoo, species do not have quite the same disparities of reproductive interest, and so we might expect their behaviour patterns to be different from those of the nest-makers.

For now, we will focus on the reproductive behaviour of the nest-builders. The general pattern of divergent reproductive interests is shared by most animal species, but there is great diversity in the way behaviour (as well as anatomy and physiology) of the sexes has evolved in each species in response to this broad constraint – and the topic is full of fascination for the observer. The particular pattern of interaction between males and females developed within each animal species is often referred to as its 'mating system'. Several phases are involved, including 'rapprochement' (males and females coming into close proximity), courtship (male activities to stimulate female receptivity), copulation, insemination, fertilisation and, in some species, care and nurture of the offspring. The timing and sequencing of these phases differ greatly between the main groups of sexually reproducing animals. In the case of most solitary bees, courtship does occur but it is very cryptic. Unlike the singing and dancing of some species of grasshoppers, what passes for courtship among the bees appears to be little more than a series of thrusts and buzzes after the male has already grasped and mounted the female. This phase, during which the male has mounted the female but not yet made genital contact, is sometimes prolonged, and females can continue to resist unwanted mating attempts and often break free.

Copulation and insemination take place if the male is successful, but fertilisation of the eggs occurs later (if it occurs at all), when the eggs are laid. As we mentioned in the Introduction, in Hymenoptera, females result from fertilised eggs and males from unfertilised ones (a system known as haplodiploidy). Females store sperm internally in a small sac (the spermatheca) and so they can determine which eggs have the potential to become male or female as they lay them. Finally, the mating systems of solitary bees do not generally include continuing care of the offspring (larvae) as in, for example, bumblebees and honeybees. However, the life's work of the female solitary bee is devoted to providing food and protection for her future offspring in advance (mass provisioning). This is very different from the situation with most insects, where eggs are laid singly or in clusters without specific provision other than the nutrients present within the egg.

Human observers are at a great disadvantage in our attempts to understand the behaviour of bees, and this applies especially to their mating systems. Bees, like us, use visual and tactile information in the organisation of their behavioural responses. Auditory cues are also important for us, and while many species of bees certainly produce buzzing noises, they appear to lack specialised hearing organs, although, as mentioned in the Introduction, a cluster of receptors in the second segment of the antennae (Johnston's organ) are sensitive to air waves. Recent research has discovered the important role of vibratory communication in the mating systems of at least some bee species and it seems likely that this modality will turn out to be more significant than so far recognised.

The role of chemical communication has long been recognised, but detailed research on this important aspect of insect mating systems had to wait for the development of sophisticated methods of chemical analysis. Since 1975, the *Journal of Chemical Ecology* has been a crucially important source of information on this developing field, and researchers such as M. Ayasse, R. J. Paxton and J. Tengö and their associates have greatly enhanced our understanding of chemical communication in insects – and, in particular, in the solitary bees.

Scents and tastes play a significant part in human lives, sometimes subliminally, but in the case of bees (and especially in their mating systems), they appear to play a decisive role. Sometimes the scents produced by bees can be detected by human observers (for example, the 'lemony' scent associated with nest aggregations of the plasterer bee *Colletes cunicularius*). However, as techniques of chemical analysis have progressed in recent years, it has become clear that bees secrete a huge range of volatile compounds, some of these (pheromones) having an identifiable role in modifying behaviour. These are secreted from glands located in the head, labrum, mandibles, on the abdominal (metasomal) tergites and elsewhere, depending on sex and species. The antennae are equipped with numerous sensory cells and are the main receptors for chemical communication.

Experimental work has shown, in the case of the small number of species so far studied, that bee chemical senses are able to make extremely fine distinctions among chemical signals. Even these studies often fall short of explaining the exact functional role of each compound in the mating system of the bee. Nevertheless, important progress has been made, and pheromones have been isolated that act as sexual attractants to each sex, acting at close quarters or at a distance, rendering females receptive to mating and stimulating copulation in males. In some species, there are also pheromones delivered by males to female mates that reduce their receptivity to subsequent male advances (antiaphrodisiacs). Complex scent 'bouquets' enable recognition of individuals, kin and species and are likely to play an important part in female choice of mates. Scents can also be utilised as deceptive signals. Some species of cleptoparasitic bees, including some species of *Nomada* (nomad bees), mimic the scent of their hosts to gain access to nests. In the extraordinary pollination mechanisms of some orchids, the flower emits scents that mimic the female sexual attractant of a bee species, inducing 'pseudocopulation' on the part of a duped visiting male bee (see Chapter 5). Even more remarkable are the neotropical orchid bees (euglossines), whose males collect and store scents from various sources, including orchids. They use these to attract females by adopting a perch from which they fan their specific blend of scents into the air, upwind of the females.

MATING SYSTEMS

Male mate-finding strategies

Now, to a more detailed view of the elements of the solitary bee mating systems and how they differ from species to species. The first phase, rapprochement, is mainly a matter for the males. They need to locate receptive females, and we take this phase to conclude with a successful attempt to mount the female. Several alternative search strategies are possible for them. Paxton (2005) offers a persuasive framework for understanding what shapes the adoption of different male mating strategies. Key variables are the reproductive life history of the females of the species (e.g. how many times they mate), the distribution of the receptive females in space and time, and the relative density of the males. Locations at which males encounter females and attempt to mate with them are often called 'rendezvous' sites. Paxton's framework suggests that where females mate only once, soon after emergence (most species of solitary bee), and where they nest in dense aggregations, males will be selected to emerge slightly before females and seek opportunities to mate with emerging females at their natal nest site. A second strategy may be more effective where nests are more dispersed and less easily located. In such cases, females may be easier to find where they collect essential resources. Patches of flowers where female bees collect nectar or pollen (or plant oils) are particularly suitable as rendezvous sites for species whose females collect pollen from one or a small number of plant species (oligolectic species – see Chapter 6), but males of pollen-generalist species may also use this strategy. Where large numbers of females of generalist species are attracted to particularly rich nectar or pollen sources, males of these species, too, will take advantage. Males typically gather around suitable patches, patrolling persistently, and attempt to mate with foraging females. As males may succeed in mating with more than one female, this strategy is known as 'resource-based polygyny'.

But there is a third strategy, particularly adopted by many species in the large genus of mining bees *Andrena*. These males scent-mark a trail with pheromones, which are attractive to both males and females of the species, resulting in numerous males patrolling a route along a hedge, bush or some other landscape feature, in expectation of the arrival of receptive females. This is known as 'non-resource-based polygyny'. There appear to be very few observations of female

FIG 45. Large-headed male of the Oak Mining Bee *Andrena ferox*. (© P. Brock)

responses to these male performances in the case of solitary bees, and the topic is worthy of further research. Males using any of these mate-seeking rendezvous sites may rely on speed, sharp perception or stamina in open competition with other males. This is known as 'scramble' competition, and sometimes results in what can easily be recognised as a scramble: where numbers of rival males grapple with one another to attach themselves to a single female. Males of some species are equipped with anatomical modifications such as thick muscular legs or pointed mandibles to aid their competitive efforts. In a few species, some males have exceptionally large heads, which may offer competitive advantages.

Alternatively, males may attempt control of a territory, monopolising mating opportunities by driving away other males. The males of only a few British species show overt territorial behaviour with frequent threat or direct physical conflict between rival bees. Paxton's framework predicts that territorial behaviour is most likely where the density of males is sufficiently low for one male to secure his dominance at a site, while scramble competition is to be expected where there is a high density of patrolling males. Among British solitary bees, territorial behaviour seems to be confined to males of a very few species as part of a resource-based strategy, and in the best-studied species it is also associated with multiple mating among the females (polyandry). Finally, there is a good deal of variability in the use of these strategies, not just between different species but also between different males (e.g. depending on size) and for individual males in different situations. Often, males may shift from one strategy to another in response to competitive pressures, at different times of day or during different phases of the flight season.

1. Patrolling the female emergence sites

The most commonly observed male strategy is to patrol nest aggregations, ready to pounce on females as they emerge. This is characteristic of those species of bees which nest in dense aggregations and is true of many ground-nesting bees, including species in genera such as *Andrena*, *Anthophora*, *Lasioglossum* and *Colletes*. Males and females respond differently to temperature changes, and males usually emerge before females (they are protandrous). They take up patrolling routes, flying low and rapidly over the nest site. Sometimes, many hundreds of males may be seen, criss-crossing one another's flight paths, occasionally stopping to investigate an entrance hole (or even departure holes made by other emerging males). The patrolling continues relentlessly, with males occasionally departing to replenish their energy from a nearby nectar source.

Four species of the genus *Colletes* (plasterer bees) illustrate this pattern very vividly. The Early Colletes *Colletes cunicularius* nests in large dense aggregations, usually on banks or dunes with loose, sandy soils. The emergence of the bees

FIG 46. Green-eyed Flower Bee *Anthophora bimaculata* male at a female emergence site.

coincides with the flowering of willows (*Salix* species), which provides most of the pollen used by the females to stock their nests. Males emerge a few days earlier than the females and patrol rapidly, low over the nest site, in large numbers. Although at first sight the movements of the males seem quite random, there do appear to be cross-cutting directional 'flows' across the site. Occasionally, individuals settle to investigate emergence holes. As females begin to emerge, they are immediately pounced upon by patrolling males, and 'mating balls' are formed as several males cling to one another and the hapless female until one succeeds in making genital contact, and the others eventually give up and continue patrolling. In this species, emerging females emit an attractant pheromone (linalool) which can be detected by males even before the female reaches the surface of the ground (Cane & Tengö 1980). Males can be observed digging down to intercept females underground.

Three other *Colletes* species (*Colletes hederae*, *Colletes halophilus* and *Colletes succinctus*) emerge in late summer and autumn, in time for the flowering of their main pollen sources: Ivy *Hedera helix*, Sea Aster *Tripolium pannonicum* and Ling *Calluna vulgaris*, respectively. Male search strategies appear very similar in all three species, with hordes of males patrolling low over nest aggregations in advance of female emergence. At a large coastal aggregation of the Sea Aster

FIG 47. Early Colletes *Colletes cunicularius*: (a) males digging to reach emerging females, (b) mating ball and (c) male mounting female.

Bee *Colletes halophilus* in Essex, large numbers of patrolling males are regularly observed in early August, before the emergence of the females. Individual males frequently settle on the ground but are immediately dived at by other males, presumably mistaking them for females. Here and there, huge mating balls of wriggling males are formed with loud buzzing sounds, males gripping one another and communicating with a distinctive 'wiggle' of the abdomen. Having reached a crescendo of intensity, these balls gradually disperse. As fresh females emerge, the males form smaller mating balls of approximately 6–10 individuals around a 'captive' female. The males gradually depart as one of them gains genital

FIG 48. Mating ball of Sea Aster Bee *Colletes halophilus*.

contact, or the mating pair fly out of the ball to settle on nearby vegetation. Even then, the successful male has no peace, and the pair continue to be 'hit on' by other males. Males continue to patrol and frequently make mating forays at females returning with pollen loads but almost immediately retreat and fly off. Later in the nesting period, fewer males can be seen patrolling the nest aggregation, and mating balls are less frequently seen. As the end of the female emergence period approaches, male competition for receptive females appears to be less intense, with as few as one or two males attached to a female.

Another group of ground-nesting bees, the furrow bees, belonging to the genus *Lasioglossum*, often nest in large, dense aggregations. The Sharp-collared Furrow Bee *Lasioglossum malachurum* is one of these, in which only fertilised females pass the winter and establish nests in the spring. As this is a social species, the adults of the first brood are non-reproductive worker females, and the sexually receptive females and males do not emerge until summer. As with the *Colletes* species, males patrol nest aggregations in large numbers, investigating cracks or holes for emerging females. These are immediately pounced upon by males, sometimes four or five together, producing a cluster not unlike the

FIG 49. Sharp-collared Furrow Bee *Lasioglossum malachurum* male patrolling for emerging females.

mating balls observed in the *Colletes* species but generally with fewer males. In this species, the sexually receptive female secretes sexual attractants from her Dufour's gland. These pheromones are subsequently located on the cuticle, where they are detected by males as they attempt to attach themselves to the body of a female. Compounds that have been analysed include isopentenyl esters, lactones and hydrocarbons (Ayasse *et al.* 1999, Smith & Ayasse 1980).

A genetic study of worker bees belonging to aggregations of *Lasioglossum malachurum* in Germany suggests that the overwintered fertile females of this social species generally make their nest burrows close to where they emerged (Friedel *et al.* 2017). It is suggested that this may avoid risks and costs involved in dispersal to other possible sites and offer benefits in terms of reduced antagonism between more closely related neighbours, or better protection from parasites or predators. However, the same study showed genetic uniformity at wider spatial scales within and between nesting aggregations. Some drifting of workers between nests could be a partial explanation for this, but there is behavioural evidence that males of *Lasioglossum malachurum* are able to discriminate between closely and more distantly related females and show a preference for unrelated females (Smith & Ayasse 1987). This points to the likelihood that, in this species, it is the males that are the dispersive sex: the males observed patrolling nesting aggregations may not all have emerged here but may include significant numbers from other local aggregations.

Males of many species of the genus *Andrena* also collectively patrol nest sites, often in large numbers. Males of *Andrena nigroaenea*, *Andrena flavipes*, *Andrena fulva*, *Andrena barbilabris*, *Andrena argentula*, *Andrena cineraria* and other species persistently patrol nest aggregations, investigating emergence holes and attempting to mount emerging females. Sometimes males attempt to mount

FIG 50. Grey-backed Mining Bee *Andrena vaga*: (a) female fighting off two males and (b) male mounted on female. Clarke's Mining Bee *Andrena clarkella*: (c) male at emergence site and (d) approaching female. (e) Mating 'scramble' of the Yellow-legged Mining Bee *Andrena flavipes*.

females indiscriminately whether or not they have full pollen loads – e.g. *Andrena vaga*, *Andrena clarkella* and *Andrena argentata* – and females may need to struggle for some time to shake off an unwanted male.

Some species of flower bees (*Anthophora*) are also ground-nesting. The well-known Hairy-footed Flower Bee often nests in large aggregations in earthen or muddy banks or slopes. Males patrol these sites and pounce on females as they enter or leave their nests. Some territorial behaviour is exhibited, with rival males 'chasing' one another. At a large aggregation on the Norfolk coast, males attempt to land on females the instant they alight at the nest entrance, often carrying pollen, but the female quickly enters the nest, and the male does not follow. Males also pursue virgin females as they leave the nest for the first time, usually heading out high and fast with several males in pursuit. At peak times there is a spinning ring of bees circling the nest mound at great speed, usually in a clockwise direction, with males chasing females, and sometimes coupled pairs are seen on the ground (NWO pers. obs.).

In Anthophora plumipes *the males' middle legs are elongated and bear long brush-like hairs but no odour glands. Before mounting the female, the male brushes secretions from glands in the abdomen onto his hind legs, then transfers the odour to the middle legs and finally brushes the secretions during mating onto the female's antennae. (Wittmann* et al. *2004)*

A display, using the long hairs on the mid-legs, is illustrated in a YouTube video. The male grips the female's wings with his forelegs and uses his mid-legs to tap the female's antennae alternately.

FIG 51. Hairy-footed Flower Bee *Anthophora plumipes* (a) male, with middle legs exposed, and (b) mating display (drawn from https//www.youtube.com/watch?v=SZ93YaHIsMg).

FIG 52. (a) Orange-vented Mason Bee *Osmia leaiana* male at emergence site of females. (b) Spined Mason Bee *Osmia spinulosa* male waiting at snail-shell nest. (c) Reed Yellow-face Bee *Hylaeus pectoralis* female at reed gall nest.

A close relative, the Green-eyed Flower Bee *Anthophora bimaculata*, also nests in large aggregations, where males congregate, flying fast and low and emitting a high-pitched buzz.

Although most British leafcutter bees (genus *Megachile*) are aerial nesters, some species, including the Silvery Leafcutter Bee *Megachile leachella*, nest in the ground. This species inhabits areas with sandy soil, often close to the coast, where the females make their nests in burrows. Their nest aggregations are dense and often extensive. As with other bees that nest in such aggregations, large numbers of males patrol, flying fast and low, on independent but overlapping routes. They frequently settle to investigate holes and crevices and, like males of the Green-eyed Flower Bee, emit a loud, high-pitched buzz. An alternative search method is adopted by some males, which take up perches, such as rabbit excavations, from which they dart out to intercept passing females or chase rival males for 2–3 m before returning to the perch.

Males of some aerial, or cavity, nesters also search for potential mates at the emergence sites of the females. Males of several species of mason bees (genus *Osmia*) do so. In an artificial bee hotel with nests of Blue, Orange-vented and Red Mason Bees (*Osmia caerulescens*, *Osmia leaiana* and *Osmia bicornis*), males repeatedly visited the nest boxes, patrolling, occasionally settling briefly, or basking for longer periods and investigating emergence holes from nests. Males of all three species spend nights as well as dull cool periods during the day in unused nest holes, facing out from their shelters. The Spined Mason Bee *Osmia spinulosa* is one of the three British species which nest in snail shells. In this species, males fly patrol routes around likely snail shells, often investigating a shell or resting for substantial periods next to one (TB pers. obs.). Males of the recent British colonist the Viper's Bugloss Mason Bee *Hoplitis adunca* show similar behaviour at their artificial nest site, as well as patrolling for females at patches of Viper's Bugloss *Echium vulgare* (TB). Males of the Large-headed Resin Bee *Heriades truncorum* also patrol nest sites, settling frequently and entering holes, even ones already occupied by nesting females. Females are repeatedly pounced on, especially when they are more exposed as they cap their final nest cells, and one episode was observed of a male reaching into a nest to drag out and mount the incumbent female.

Although resource-based searching seems to be a common male strategy in the cavity-nesting species of leafcutters (*Megachile*), one puzzling observation (TB) was of a male Wood-carving Leafcutter Bee *Megachile ligniseca* which was first observed investigating a bee hotel on 7 July, including it in its patrol route on 8 July, and persisting until 18 July, when a female arrived and began to nest there. This raises the intriguing speculation that the male may have located a potential

FIG 53. Wood-carving Leafcutter Bee *Megachile ligniseca* male visiting a potential nest site.

nest site and 'staked out' a patrol route to attract a female. As this species had not previously nested here, mere chance encounters seem relatively unlikely. Honey (2020) also reports a male of this species roosting in a bee hotel for 'a week or so' before a female turned up.

Finally, males of communally nesting species have a rather different set of mating opportunities. In these species, several individual females (sometimes hundreds) may share a single nest entrance, while constructing their own distinct nests within. Although females of these species nest communally, there is believed to be no social cooperation between them. Danforth *et al.* (2019) report that in some American communally nesting species, as many as 71–97 per cent of females are already mated by the time they emerge from the nest. As they put it: 'Large communal nests represent a reproductive bonanza for any male that can monopolise some or all of the resident female population.' Little is known about behaviour within the nests, but British bees that always or sometimes nest communally (*Andrena ferox*, *Andrena bucephala*, *Andrena scotica* and *Panurgus calcaratus*) have markedly variable males. Some males are much larger than others and have proportionately larger heads (Fig. 45). This trait could be explained in terms of sexual selection by females, but also in terms of an advantage on the part of the larger males in their competition with other males inside the nest. The Chocolate Mining Bee *Andrena scotica* sometimes nests communally. Paxton and colleagues studied the mating system of this species, showing very high levels of mating within the communal nest. However, communally nesting females showed low levels of relatedness, presumably limiting further evolution towards sociality (Paxton *et al.* 1996, Paxton & Tengö 1996).

2. Resource-based male mate-finding strategies

Resource-based strategies are commonly, but not exclusively, strategies used by the males of (oligolectic) species whose females, as we have seen, collect pollen from one or a few closely related species of flowering plants. Males actively patrol stands or patches of the appropriate plant, often hovering briefly before open flowers and moving swiftly from flower to flower. As with the first male strategy, males are likely to aggregate around the plant, resulting in intense competition among them. Usually this does not take the form of aggressive conflict but visual 'stand-offs' or swift darts at one or another. These apparent attacks appear to be identical to male approaches to females that they locate as they forage – a sudden pounce, usually resulting in the female flying off or dropping quickly to the ground. It is unclear whether males pounce on males and females indiscriminately as a way of optimising their chances of getting it right, or whether there are subtle differences in the mode of attack. Sometimes a pouncing male will manage to gain a foothold on a female and, if he manages to hold on, succeed in mating.

Even for bees with broad pollen diets (polylectic species), resource-based patrolling may still be a viable male strategy. Females often congregate around particular flower species. This may have to do with limitations in the range of suitable plants in flower at a particular time of year (e.g. sallows or dead-nettles for spring-flying species), or an abundance of resource-rich flowers available in a particular locality (e.g. bramble patches or, later in the year, blue-purple Asteraceae such as knapweeds, thistles and burdocks). Patches of such flowers may therefore be rewarding targets for patrolling males. In some cases, males patrol a particular flower patch, while in others they may follow a more extensive route that includes one or more flower patches. In some cases, males may follow individual patrol routes, while in others, there can be numerous males at a single patch, with cross-cutting routes round and through the plants. Where numerous males are active at a single patch, there may be direct interference competition among the males, or males may establish territories and aggressively attempt to exclude other males and/or other foragers. There appears to be considerable within-species variation in which each of these strategies is adopted, and individual males may shift from one to another strategy according to the density of competition or availability of alternatives.

Among oligolectic species, males converge on individual plants or patches of the favoured host plant, often patrolling in large numbers. Males of the Bryony Mining Bee *Andrena florea* patrol plants of the common climber White Bryony *Bryonia dioica*, diving at females as they collect pollen. From early to midday, these mating attempts are usually rejected. However, at one site, from mid- to late afternoon, males were observed patrolling and occasionally settling on leaves, possibly scent-marking, while females were much less in evidence. When a female did arrive, she was immediately pounced upon by a male who managed to secure his hold, and the pair disappeared into foliage (TB pers. obs.).

A distinct diurnal pattern was also observed in the Red Bartsia Blunthorn Bee *Melitta tricincta*. During the middle part of the day, males ceaselessly patrol patches of Red Bartsia in independent but overlapping routes, occasionally stopping briefly to take nectar. Females work the one-sided flowering spikes of Red Bartsia in a systematic way, seemingly rejecting the repeated male 'hits' on them. Later in the afternoon (approximately from 4:30 pm), females disappeared, and males exchanged patrolling for taking nectar (TB pers. obs.). In the Heather Mining Bee *Andrena fuscipes*, males actively patrol patches of Ling, often in large numbers, quickly diving at any female that arrives to forage. In cooler, cloudy periods, the pattern changes, as more females arrive to forage, while males shift their activity to nectaring. It may be that the females take advantage of lower male activity in cooler periods to forage with reduced disruption from amorous males. Males of the Long-horned Bee *Eucera longicornis* follow individual patrol routes through grassland rich in Fabaceae flowers, where the females forage. Where there are

FIG 54. (a) Bryony Mining Bee *Andrena florea* male at a patch of White Bryony *Bryonia dioica*. (b) Red Bartsia Blunthorn Bee *Melitta tricincta* male at a patch of Red Bartsia *Odontites vernus*.

FIG 55. Yellow Loosestrife Bee *Macropis europaea*: (a) males in a 'stand-off', (b) female signalling rejection to male and (c) mating attempt.

mixed arrays of flowers (for example, Red and White Clovers and Common Bird's-foot-trefoil), the females show a definite preference for Meadow Vetchling, while patrolling males also give most attention to patches of this plant (TB pers. obs.).

The Yellow Loosestrife Bee *Macropis europaea* is narrowly oligolectic on pollen of Yellow Loosestrife, from which it also collects plant oils. Nests of this species are solitary or in small aggregations and usually well concealed among vegetation. Accordingly, males congregate around stands of the host plant, flying circuits around and through clusters of the plant, frequently darting at females as they forage. Frequently females escape the male grasp, as their hind legs project back, obstructing attempts to mount. Sometimes they are able to continue foraging, but occasionally they drop to the ground after struggling to get free. There is frequent interaction among the males, too, with face-to-face 'stand-offs', the contrasting pale yellowish faces seeming to operate as a gender signal. Males occasionally settle on leaves, but if they visit flowers, they are darted at by other males, either as aggressive interactions or as a result of gender confusion.

Males of polylectic species also frequently patrol flowers in search of potential mates. Individual males of two common *Osmia* species (the Orange-vented Mason Bee *Osmia leaiana* and Blue Mason Bee *Osmia caerulescens*) which visit nest sites in search of females also patrol flower patches. In TB's garden, males of both species are observed to follow extended patrol routes, covering both non-flowering shrubs and patches of flowering herbaceous plants, with regular stopping points on the route, and occasionally taking nectar from *Geranium* cultivars. One male Orange-vented Mason Bee was observed to locate a female on the open corolla of a *Geranium* flower and succeed in mounting her. Such pairings on flower heads seem to be quite frequent in this species, but, as we shall see, they do not always lead to successful mating.

The Grey-patched Mining Bee *Andrena nitida* is another polylectic bee, often collecting pollen from spring-flowering shrubs. The nests of this species are dispersed, and not in aggregations, so according to Paxton's framework, it could be a more effective male strategy to visit likely flowers rather than to search for female emergence sites. Numerous males of this species were observed patrolling a flowering hawthorn bush in mid-May at Cambridge University Botanic Garden, for example (NWO). The Impunctate Mini-miner *Andrena subopaca* is another species which does not nest in dense aggregations and in which males patrol flowers in search of females. For example, approximately six males of this species were watched persistently patrolling a single inflorescence of Cow Parsley *Anthriscus sylvestris* until a female arrived. She was immediately pounced on by one of the males, followed by another, both attempts being unsuccessful (TB pers. obs.).

In many instances of resource-based searching, numerous males cluster around a single flower patch or shrub, following their individual routes without any obvious signs of overt conflict. However, in some species of flower bees (*Anthophora*), and in the Wool Carder Bee *Anthidium manicatum*, there is overt conflict between males, in a few cases amounting to territorial behaviour.

The common and widespread Hairy-footed Flower Bee *Anthophora plumipes* is a familiar spring-flying species in gardens. Although the males frequently congregate at nest sites, they are probably most frequently observed patrolling flowering shrubs or early herbaceous flowers, such as Primrose *Primula vulgaris* or Red Dead-nettle *Lamium purpureum*. Some males fly persistently around a large flowering shrub (Red Currant is a popular one) or adopt more extended patrol routes with occasional stopping points, including flower patches where they hover attentively before moving on. Occasionally, directly following a foraging trip by a female, two or more males arrive rapidly, and follow her, remaining a few centimetres behind until one makes a lunge at her.

Stone et al. (1995) report a 'queue' of as many as three or four males following a female, the lead male (usually the largest) being the most likely to be successful in mounting her. The pattern of events strongly suggests that scent communication along the route is involved. Females are dived at while in flight, while foraging and also, commonly, when they settle to brush pollen onto the scopae on their hind legs. The females appear to choose a concealed spot to carry out this operation. In one such instance, a male broke off patrolling and switched to a localised zigzag flight before lunging at the hidden female, who flew off directly (TB pers. obs.). It has been shown that the foraging of females can be seriously disrupted by male harassment. Stone (1995) recorded males making as many as 11 mating attempts per minute at a patch of comfrey, reducing female foraging efficiency by some 50 per cent. The females responded by foraging on flowers hidden under the leaves when males were present but reverted to outer flowers when males were experimentally removed (Stone et al. 1995).

As well as diving at females of their own species, patrolling males of *Anthophora plumipes* also strike at muscid flies, bumblebees, their cuckoo (*Melecta albifrons*) and other insects encountered on their patrols. It may be that at least some of these are misdirected mating attempts, but similar interactions with other males can be interpreted as territorial behaviour. Stone et al. (1995) observed territoriality in this species when the density of males at a foraging patch was low. As male density increased, the costs of chasing off other males outweighed any gains in monopolising potential mates – especially, they point out, as mating in this species takes some 40 minutes. At high male density, territoriality gave way to more routine resource-based searching.

Some other *Anthophora* species whose males search for females at flowers also show signs of territoriality. In the Green-eyed Flower Bee *Anthophora bimaculata*, males patrol flower patches, such as stands of Wood Sage *Teucrium scorodonia*, Spear Thistle *Cirsium vulgare* or Black Horehound *Ballota nigra*, that are frequented by females. Males attempt to mount foraging females, but also pounce upon intruders of other species. In one instance, a male of this species patrolling a patch of Black Horehound was observed to chase off an intruding male of the notoriously aggressive Wool Carder Bee *Anthidium manicatum* (TB pers. obs.). The male Four-banded Flower Bee *Anthophora quadrimaculata* behaves in a similar way, patrolling with a high-pitched buzz, hovering and switching direction with a zigzag motion as it hovers, facing flowers. Foraging females located in this way are dived at, as are insects of other species.

FIG 56. Contrasting facial pattern of the Four-banded Flower Bee *Anthophora quadrimaculata* male.

FIG 57. Coast Leafcutter Bee *Megachile maritima* male holding territory on a bramble stem. Note the enlarged hind tibia.

Males of Willughby's Leafcutter Bee *Megachile willughbiella* patrol extended routes across flower patches and around hedges or bushes. When they encounter foraging bees of other species, including much larger bumblebees, they hover and appear to 'fix' them visually before diving at them and driving them away. NWO observed a male of another leafcutter, the Coast Leafcutter Bee *Megachile maritima*, holding a territory at a bramble clump at Winterton, on the Norfolk coast. The bee 'dive-bombed' all-comers, including bumblebees, which were grabbed and wrestled to the ground.

The most widely reported example of territoriality among bees that occur in Britain is the Wool Carder Bee *Anthidium manicatum*. Females of this species collect plant hairs to line their brood cells, in addition to foraging for nectar and pollen. In parks and gardens, Lamb's Ear *Stachys byzantina* is commonly used as a source of 'wool' (as well as nectar), and elsewhere other lamiates are used, notably Black Horehound and woundworts. Males aggressively patrol patches of such plants, hovering briefly before flowers, as if searching visually for foraging females. Other insects, especially bumblebees, are dived at and driven away. Some of these even retreat at the approach of a territorial male. It is reported that males may grapple with intruders and spike them with the sharp points at the rear of the abdomen, even killing victims such as honeybees (Severinghaus *et al.* 1981). A female that flies in to forage is usually quickly located, and after a brief visual 'fix', the incumbent male mounts her. Commonly, but certainly not always, male mating attempts are accepted, and mating takes place on the plant where the female was intercepted.

It seems likely that the main advantage to males of holding territories is increased mating opportunities as females enter their domain to forage. However, to maintain the attractiveness of his patch, the male must resist depletion of its floral resources by other flower visitors. This may explain why honeybees and bumblebees are prime targets for aggressive removal, while other insects are often ignored. Notwithstanding the description of territorial rivalry between males given by Severinghaus *et al.*, it is not uncommon to see flower patches patrolled independently by two males with little evident aggressive interaction.

Male Wool Carder Bees *Anthidium manicatum* commonly patrol flower patches that harbour not just floral rewards (pollen and/or nectar) of interest to females but also dense plant hairs with potential for use in nest construction. On superficial observation, females seem rarely to visit these territories to collect 'wool'. On closer inspection, however, females can be observed flying close to the ground, swiftly and directly to the basal leaves of the plant (Lamb's Ear, especially). These females have already mated and are now engaged in nest-making. They proceed to clip

FIG 58. Patrolling Wool Carder Bee *Anthidium manicatum* male showing abdominal spines.

FIG 59. A patch of flowering Lamb's Ear *Stachys byzantina*. Wool Carder Bee *Anthidium manicatum* males patrol among the flowering stems, mating with females arriving to forage, while nest-making females fly in low to collect plant hairs from basal leaves (see Chapter 3, Fig. 110).

hairs from the underside of these leaves, and depart, staying 'below the radar' of the male who is focused on females attempting to forage on flowering stems at higher levels. In this way, they avoid the costs to their nest-making that would otherwise be imposed by the intense sexual attentions of males.

Unusually among bees, males of *Anthidium manicatum* are generally larger than the females, presumably as a result of sexual selection for the most effective defenders of territory. However, territorial defence is not the only available male strategy. Some males, often the smaller individuals, resort to non-territorial patrolling, relying on occasional encounters with females at flower patches not defended by a territorial male, or when the incumbent male at a patch is otherwise engaged. It seems that males holding a territory mate more frequently than others. O'Toole (2013) reported a male in his garden mating 16 times in a day, but this is presumably because of a higher probability of encountering females, as non-territorial males were not rejected more frequently by the (fewer) females they encountered.

So far, we have assumed that resource-based patrolling would be a viable male strategy for intercepting females as they forage at flowers. But females of some species require other, non-floral, resources, such as mud, resin or leaf cuttings for nest construction. As females collecting these resources will already be mated, it might seem that patrolling the sites where females collect these resources would not be a good strategy for the males. Despite this, males of the Silvery Leafcutter Bee *Megachile leachella* frequently harass females while they collect leaf cuttings, sometimes succeeding in mounting and mating with them (TB pers. obs.). This suggests that females in this species may mate more than once.

3. Non-resource-based male strategies

These seem to be confined to quite a limited range of solitary bees but are much more common among the bumblebees. The male bees patrol the contours of a bush, hedge or some other prominent feature, or sometimes may follow one another in a more extensive route such as, for example, a circuit of a garden. At intervals, especially early in the day, individual males settle on a leaf or twig and deposit a scent mark. The resulting pheromones are said to attract both males and females of the appropriate species, and numerous males either follow one another or patrol the route in clusters. Patrolling can seem relentless and often persists over several days at the same site. Reports of females actually arriving at patrol routes and mating occurring are rare to non-existent, but it is in any case quite rare to see mating bumblebees.

Charles Darwin famously observed patrolling behaviour by male bumblebees in his own garden, and even gave an account of his use of child labour in researching it:

> *The flight paths remain the same for a considerable time and the buzzing places are fixed within an inch. I was able to prove this by stationing five or six children on a number of separate occasions each close to a buzzing place, and telling the one farthest away to shout out 'here is a bee' as soon as one was buzzing around. The others followed this up, so that the same cry of 'here is a bee' was passed on from child to child without interruption until the bees reached the buzzing place where I myself was standing.* (C. Darwin, tr. R. B. Freeman 1965, cited Benton 2006)

Darwin, of course, could not have known of the role of pheromones in this behaviour, but subsequent research has shown it to be widespread among British bumblebee species, including *Bombus terrestris*, *Bombus hortorum*, *Bombus pratorum*, *Bombus lapidarius*, *Bombus pascuorum*, *Bombus jonellus*, *Bombus monticola* and the cuckoo bumblebee *Bombus vestalis*. The Swedish researcher Bo G. Svensson showed that males of *Bombus lapponicus* sometimes use shared paths for a part of their patrolling routes but diverge to follow individual flight paths for other segments. He also showed that different species occupying the same area could separate their patrolling routes by flying at different heights (Svensson 1979).

Among solitary bees, non-resource-based male patrolling is widespread in the mining bees belonging to the genus *Andrena*. Males fly close to, and appear to contour, the outline of a hedge or shrub or some other landscape feature. As with bumblebees, the males settle briefly at intervals, and this may be when they deposit a scent mark. Often dozens of males may be seen persistently patrolling such features, the same stopping points being used repeatedly. However, the males usually seem to be following independent, though intersecting, patrol routes, so that they may be spread out over much of the outline of the feature at any one time.

Males of many *Andrena* (mining bee) species (including *Andrena nigroaenea*, *Andrena fuscipes*, *Andrena nitida*, *Andrena dorsata*, *Andrena synadelpha* and others) commonly form collective patrols in this way. Up to six species of *Andrena* have been reported as patrolling the same shrubbery. Tengö (1979) argues that this indicates a high degree of specificity in the complex scent bouquets secreted by each species. NWO netted males of eight *Andrena* species (*Andrena nigroaenea*, *Andrena cineraria*, *Andrena haemorrhoa*, *Andrena nitida*, *Andrena flavipes*, *Andrena dorsata*, *Andrena tibialis* and *Andrena praecox*) and a female of another (*Andrena bimaculata*) in one hour at some willows and blackthorn in April 2016 near Welney, Norfolk. The bees were present in the hundreds. A holly bush in TB's garden also attracted several species of patrolling male Andrenas during the 'lockdown spring' of 2020. Males occasionally settled on leaves and performed a

FIG 60. (a) Gwynne's Mining Bee *Andrena bicolor* male. (b) Buffish Mining Bee *Andrena nigroaenea* male, apparently scent-marking.

FIG 61. (a) Grey-patched Mining Bee *Andrena nitida* female, apparently scent-marking. (b and c) Chocolate Mining Bee *Andrena scotica* female, apparently scent-marking at a rendezvous site. (d) Orange-tailed Mining Bee *Andrena haemorrhoa* female at a rendezvous site.

routine that was interpreted as scent-marking: the bee rubs its hind legs against its abdomen, often raising itself over the leaf surface, before bringing its feet down onto the leaf again – the whole operation taking only a few seconds. Fresh females with no pollen loads also arrived to settle briefly, often performing a sequence similar to that of the males.

On another occasion, TB observed at least four species of *Andrena* males patrolling around a single hedgerow of Field Maple near Harlow, Essex, with females of all four species making occasional brief visits. This spot appeared to have no distinguishing features, raising the question why several species should have simultaneously selected it for their patrols. It seems at least possible that the chemical blends include more general attractants that call in the other species, providing a more powerful signal than would be emitted by the scents of a single species.

In the case of Wilke's Mining Bee *Andrena wilkella*, males mark odour spots with a cephalic gland secretion and have been shown to have high sensitivity to specific compounds in it (Tengö *et al.* 1990). Some compounds (spiroacetals) are widespread in insect communication systems, and different isomers (specifically enantiomers, the molecules of which are mirror images of one another) have been shown to be crucial for species-specific scent discrimination. Tengö *et al.* carried out field tests over three years, observing the behavioural responses of patrolling males of *Andrena wilkella* to twigs marked with compounds containing enantiomers of spiroacetals *(E,E)*- and *(E,Z)*-DSU and mixtures of them. The bees' responses showed ability to discriminate between different enantiomers, and electrophysiological tests of antennal sensitivity in the laboratory produced similar results. The authors argue that this sensitivity indicates the importance of this compound in patrolling behaviour, and that it

FIG 62. Willughby's Leafcutter Bee *Megachile willughbiella* male patrolling. Note the wide, flattened front tarsi.

may enable both species-specific communication and also individual identification, signalling kinship status, for example.

Although non-resource-based patrolling has been most associated with *Andrena* species among solitary bees, one of us (TB) observed a female *Megachile willughbiella* foraging on dense patches of a *Campanula* species, to be followed several seconds after she had left by a male rapidly following the same route. This behaviour seemed strongly suggestive of chemical communication, possibly involving the male picking up scents left by the female, rather than this being an example of non-resource-based patrolling. Alternatively, the female may have been following a scent trail marked earlier by the male.

The distinction between resource-based and non-resource-based patrolling may not be clear-cut, as illustrated by the behaviour of males of several species of flower bees. *Anthophora plumipes*, *Anthophora bimaculata*, *Anthophora furcata* and *Anthophora quadrimaculata* have patrol routes that may be quite extensive, covering several adjacent features, as in the case of bumblebees. However, these usually include, at one or more locations on the route, patches of flowers that are visited by females. These cases rather cut across the distinction between resource-based and non-resource-based systems. Such routes include the possibility of locating stationary females sweeping pollen from their body hair onto their scopae, for example, as well as foraging from flowers.

TABLE 4. Male mate-searching strategies.

Nest distribution	Rendezvous	Descriptive term	Explanation
Aggregated	Nesting/emergence sites	Patrolling emergence sites	Males search around nest aggregation, often where they emerged. Scramble competition.
Dispersed	Female foraging sites	Resource-based polygyny	Males search around flowers or other resources used by females. Scramble competition.
Dispersed	Sunny glades, hedges, etc.	Non-resource-based polygyny	Scent-marking by males to which females and other males are attracted. Scramble competition.
Dispersed	Territory defended by male	Territorial behaviour	Territory may include food resources used by females. Male density low. Male pounces on females which visit.

MATING AND AFTER

Once a male has succeeded in clinging on to the back of a female (i.e. mounting), it may seem that he has succeeded in mating, but this is misleading. Having managed to locate a female and mount her, perhaps having already struggled with rivals in a mating ball, the male has further trials to face. The sequence of events from the moment the male mounts the female is generally divided into three phases: 'courtship', copulation and what is sometimes rather romantically described as 'post-copulatory embrace'. The courtship phase is very variable in duration and may involve the male emitting pheromones, vibrating and struggling to physically restrain the female. It ends when (and if) he succeeds in making genital contact. This is the beginning of the copulatory phase, when sperm are passed from male to female. Sometimes there is a further phase after the genitalia are separated, when the male remains mounted on the female. In some species, chemical signals are exchanged during this period.

The first phase is rather misleadingly termed 'courtship', as it is a variably mixed process of persuasion, stimulation and coercion. For some species, the initial pounce on the part of the male is directly followed by attempts at physically restraining the female, sometimes involving the use of specialised anatomical adaptations, but in others, physical restraint is much less in evidence. Although the initial attempt to mount the female has the appearance of an uninvited attack on the part of the male, there is evidence in some species at least that unmated females secrete 'calling' pheromones and also that during courtship, chemical communication flows both ways. Above, we noted this in the case of the Early Colletes *Colletes cunicularius*, and females of the genus *Andrena* that arrive among male patrol routes frequently appear to scent-mark their perches (TB pers. obs.). Males of at least some species do not attempt copulation unless this behaviour is triggered by a female sex pheromone.

Females may also use behavioural signals to dissuade males from mating attempts. Females of *Megachile* species (notably the Patchwork Leafcutter Bee *Megachile centuncularis*) have been observed, as they forage for pollen, or simply when they land on a flower, to raise their abdomen when approached by males (G. Pilkington https//:nurturing-nature.com). This would be an 'honest' signal, indicating that the female is already mated and engaged in nest-making. One of us (TB) observed a female *Megachile centuncularis* perform this behaviour when approached by a male Blue Mason Bee *Osmia caerulescens*.

Another possible example is provided by the behaviour of female Yellow Loosestrife Bees *Macropis europaea*. As noted above, they are subject to intense harassment by males as they forage. Their habit of displaying their full scopae

FIG 63. Patchwork Leafcutter Bee *Megachile centuncularis* female signalling rejection to an amorous Blue Mason Bee *Osmia caerulescens* male.

above and behind their bodies while entering flowers of the host plant may be another rejection signal to amorous males, indicating that they are stocking a brood cell and so already mated (see Figure 55 b, above). Females of *Andrena hattorfiana* also do this (NWO). Systematic tests of these suggestions would be worth trying.

In the Early Colletes *Colletes cunicularius*, it has been shown that although (as noted above) linalool emitted by the newly emerged female is attractive to males, at short range, other compounds (hydrocarbons) on the female cuticle are usually required to stimulate mating behaviour on the part of the male (Mant et al. 2005). Sex pheromones secreted by females have been identified in other species, including the Pantaloon Bee, the Large and Small Shaggy Bees and the Red Mason Bee (*Dasypoda hirtipes*, *Panurgus banksianus*, *Panurgus calcaratus* and *Osmia bicornis*) and several species of *Andrena* and *Lasioglossum*. In the Sharp-collared Furrow Bee *Lasioglossum malachurum*, for example, hydrocarbons on the female cuticle have been shown to be attractive to males, but actual attempts at copulation are stimulated by other compounds (isopentyl esters of unsaturated fatty acids and unsaturated macrocyclic lactones). However, the mix of compounds on the cuticle of females changes after mating, rendering them less attractive to males. The proportions of hydrocarbons in relation to lactones were found to be higher in nesting queens, rendering them less attractive to males, which were able to discriminate between mated and unmated females within three hours after mating (Ayasse *et al.* 1990, 1999). The authors consider that this change and the males' ability to detect it is likely to benefit both sexes, in enabling males to avoid inappropriate mating attempts and females to avoid sexual harassment.

A comparable shift in female chemical signals during the life course has been studied in the common Buffish Mining Bee *Andrena nigroaenea*. Cuticular extracts from newly emerged females have been shown to be high in alkenes and alkanes, which are known to elicit attempts at copulation on the part of males of this species. However, once mated and involved in nest-making, females secrete, from their Dufour's gland, increased amounts of compounds used in lining their brood cells. These compounds, farnesol and farnesyl hexanoate, have been shown, both separately and together, to inhibit male mating behaviour, especially greatly reducing the frequency of actual copulation attempts. It is not established whether these compounds are spread on the cuticle of the females incidentally during their nest construction, or whether they are smeared on by the female, or even secreted from distinct cuticular glands. In any case, it seems that chemical secretions originally evolved to play a part in nest construction have acquired a secondary role as signals enabling males to discriminate between virgin and mated females. It remains unclear whether this is behaviour learned by the males or is an evolved trait (Scheistl & Ayasse 2000).

Since, as we saw above, male sexual harassment of post-mating and unreceptive females can be costly to both sexes, it would not be surprising to find that strategies for avoiding unwanted male attentions have evolved. As these compounds (farnesol and farnesyl hexanoate) have been found to be produced by females of other species of *Andrena* for brood-cell lining, it is possible that their role as inhibitors of male sexual harassment is widespread in the genus. However, either it is not universal or it is of limited effectiveness for particularly amorous males. A female Sandpit Mining Bee *Andrena barbilabris* engaged in digging a nest burrow was filmed as she was set upon by a male, who struggled to maintain a hold on her thorax until she succeeded in throwing him off some five seconds later (TB). Females of the Grey-backed Mining Bee *Andrena vaga* were observed to adopt a side-to-side rolling movement to dislodge males (see Figure 50 a, above; TB & NWO, Dungeness).

Even more persistent were two males of the Small Sandpit Mining Bee *Andrena argentata*, both struggling to get a grip on the dorsum of a female with full pollen loads. After 10 seconds one flew off, but the second persisted, trapping her wings with his middle pair of legs and attaching his tarsi to her pollen-loaded hind tibiae, before finally abandoning the attempt 14 seconds later. In another encounter, a female *Andrena argentata* with full pollen loads struggled persistently to remove an unwanted male. He gripped her thorax with his forelegs, while trapping her wings with the 'embrace' of his middle pair. Meanwhile she raised and waved a middle leg as if to force him off, while running forward and then rolling over twice. He eventually moved backwards and reached down with the tip of his abdomen in what was presumably a copulation attempt. Her response was to curve her abdomen further down to make this impossible. He finally gave up and flew off after 1 minute 20 seconds (TB). One of us (NWO) observed instances of this in which females appeared to be physically harmed by their struggles with unwanted males. Still more gruesome is NWO's account of a female Cliff Mining Bee *Andrena thoracica* being mated in sequence by two or three males and dying in the process.

Another species whose mating behaviour has been thoroughly studied is the familiar Red Mason Bee *Osmia bicornis*. According to the description given by Seidelmann (1995), once a male of this species has mounted a female, he embraces her mesothorax with the first two pairs of his legs. His antennae point forwards, while hers are held out laterally. He uses his antennae to stroke hers, while covering her eyes with his forelegs, and makes a humming sound with each stroke of the antennae. He then moves backwards and lowers his abdomen in an attempt to make genital contact, while the humming becomes a buzz, and he drums on the female's face with his antennae. She may allow copulation or bend her abdomen further down so as to avoid it and attempt to shake him off. If rejected, the male may fly off or repeat the whole sequence. Ayasse and Dutzler (1998)

FIG 64. Mating attempt by a Small Sandpit Mining Bee *Andrena argentata*. The female has raised her middle legs in attempting to repulse the male. Some damaged females were found nearby.

analysed volatile compounds (unsaturated fatty acids, ethyl esters, hydrocarbons and aldehydes) taken from the cuticle of unmated females of this species and showed experimentally that when mixed and applied to 'dummy' females, they elicited copulatory attempts in males. It seems that courtship in this species involves both male physical activity to stimulate female receptivity and chemical signals on the female cuticle that elicit mating attempts on the part of the male.

Conrad *et al.* (2010) designed experiments to determine what criteria are used by female *Osmia bicornis* in accepting or rejecting males. They found that females were more likely to accept males that were slightly bigger than average, were able to produce longer bursts of vibratory communication, and were more closely related. The relationship between various components of male odour and female preference was more complex, with the suggestion that composition of the 'bouquet' might be used in kin discrimination as well as a source of evidence of male vigour. The authors conclude that more research would be needed to understand how these different criteria are integrated in female choice.

A female of another common mason bee, the Orange-vented Mason Bee *Osmia leaiana*, was observed apparently taking nectar from a flower of a *Geranium* cultivar in TB's garden. A patrolling male located her and swiftly mounted her, grasping her with his forelegs against her thorax, close to the tegulae, and using his mid-legs to trap her wings and hind femora. His hind legs were pressed round her lower abdomen. At the point of mounting, the male's genitalia were

FIG 65. Pair of Orange-vented Mason Bees *Osmia leaiana* showing (a) the male attempting to mount and (b) subsequent 'courtship', a process taking over two hours.

extruded, but the pair remained apparently motionless for one hour, with no sign of the male attempting copulation. After a further hour, they had become physically separated, the male stretched out across the corolla, forming a cage-like structure with her facing him, within the shallow bowl of the flower. After a further 25 minutes (i.e. 2 hours 25 minutes after the initial mounting), the male had finally managed genital contact, and the mating pair were arranged around the central receptacle of the flower (TB pers. obs.). On another occasion, a male was observed mounted on a female, using a similar grip to the one just described, but in this case the male was active, making jerking movements with his abdomen and flicking the female's head with his antennae. It could be that the female in the second example was more resistant to the male's attentions, requiring more active 'persuasion'. Both examples show the considerable gap either in time, or in persuasive activity on the part of the male, between mounting and mating.

Although males of the above species grasp the females in such a way as to reduce their ability to move, and certainly to prevent flight, there are no obvious anatomical modifications to enhance the male's capacity for coercion. However, there are three species of the related genus *Megachile* (leafcutters) which do have such anatomical features. These are the Coast Leafcutter Bee *Megachile maritima*, the Black-headed Leafcutter Bee *Megachile circumcincta* and Willughby's Leafcutter Bee *Megachile willughbiella*. The courting and mating behaviour of the last-named species has been well researched. We noted above the possible involvement of chemical cues in the males' mate-searching behaviour. When a male finally makes contact with a potential mate and mounts her, there are several features that come into play. The most obvious are the white flattened tarsi of the front legs, with their associated 'fan' of long hairs hanging ventrally from the curved plate formed by the tarsi. Wittmann and Blochtein (1994) have identified several other, less obvious, adaptations, each of which seems to play a part in physically restraining the female. A male positions himself on the back of the female, with his head above and slightly forward of hers, presses her wings down with his middle legs, and pulls up her abdomen with his rear legs. Meanwhile, he grasps the scapes of her antennae in a pair of complex structures formed by a hollow on each side of the head and a projection from the base of the mandibles, while pressing the antennal flagella into grooves on the inner edge of each of the fore tarsi. In doing this, he covers the compound eyes of the female with the fan of long hairs. Though not observable when the male is in position, there are pointed projections from the coxae of the front legs which serve to hold the female's head in place. Also, the tibiae of the middle and hind legs are slightly inflated and curved ventrally.

FIG 66. Coast Leafcutter Bee *Megachile maritima* male cleaning antenna with foreleg and showing flattened fore tarsus.

Wittmann and Blochtein also describe pores on the inner surface of the modified fore tarsi, especially in the grooves that hold the female antennal flagella. These pores are linked to glands in the cuticle which produce compounds including carbohydrates and esters. The dominant compounds across species of *Megachile* were 7-pentacosen and pentacosan. In addition, glands were found on the mid-tibiae and hind tarsi. Although the function of these compounds is unclear, it seems likely that they are chemical signals, which either identify species or stimulate sexual receptivity in the female. In *Megachile willughbiella*, courtship is a combination of coercion and chemical 'persuasion'. The latter is aided by anatomical features of the male used to constrain female movement, ensuring the passage of chemical signals to receptor cells in the female antennae. Despite the male's equipment for coercion, female rejection can still occur. The male faces a difficult task of coordination to place the female's flagella correctly, and she is able to throw the male off by jerking movements of her abdomen.

Interestingly, Wittmann and Blochtein examined males of other species of Megachile, in some species finding some of the anatomical adaptations of *Megachile willughbiella*, even without the remarkable features of the fore tarsi of that species. They do not mention other British *Megachile* species, but males of the Silvery Leafcutter Bee *Megachile leachella*, the Wood-carving Leafcutter Bee *Megachile ligniseca*, and the Patchwork Leafcutter Bee *Megachile centuncularis*, which do not have the 'boxing glove' front tarsi, do have 'tongue-and-groove' formations ventrally at the junction of the mandibles with the head cuticle, used

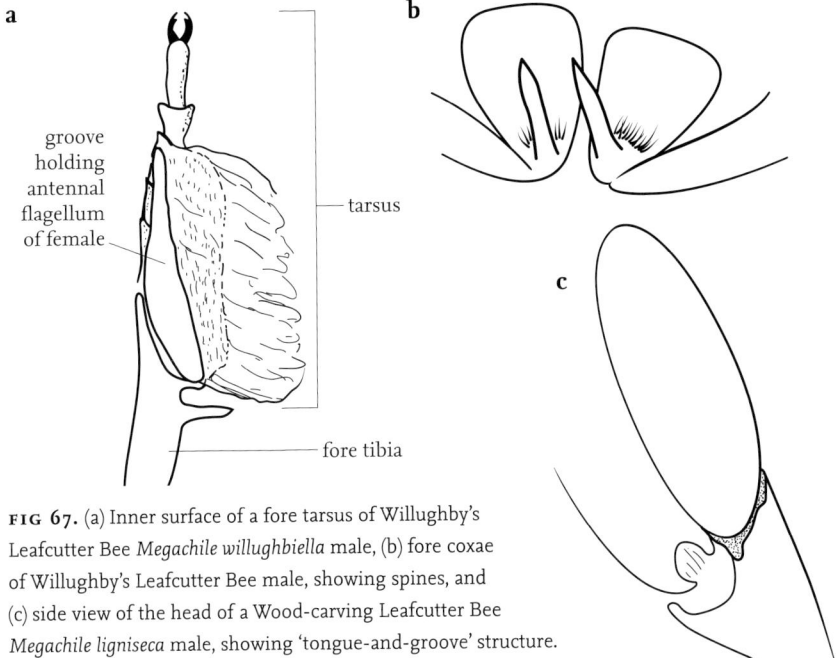

FIG 67. (a) Inner surface of a fore tarsus of Willughby's Leafcutter Bee *Megachile willughbiella* male, (b) fore coxae of Willughby's Leafcutter Bee male, showing spines, and (c) side view of the head of a Wood-carving Leafcutter Bee *Megachile ligniseca* male, showing 'tongue-and-groove' structure.

to confine the female antennae. *Megachile leachella* also shares with *Megachile willughbiella* forward-pointing projections from the front coxae that are supposed to constrain female movements, though they are less fully developed in the former species. This feature is lacking in *Megachile ligniseca* and *Megachile centuncularis*.

Males of the Spined Mason Bee *Osmia spinulosa* have a long spine projecting from the ventral surface of their first abdominal segment, a row of short points along the rear rim of the sixth tergite, and a small central spine on the seventh. The curvature of the abdomen in this species results in the spine on the seventh tergite pointing forwards, suggesting a possible role in combination with the spine on the first sternite in constraining a female. Close observation of mating in this species would test this possibility. It may be that the raised extrusion from the second sternite of the Welted Lesser Mason Bee *Hoplitis claviventris* and its terminal point on sternite seven also play a part in constraining a female once she has been mounted.

Males of the Large-headed Resin Bee *Heriades truncorum* investigate holes at nest sites, in search of females, sometimes dragging them out and attempting to mate with them. One male was filmed as he entered an occupied opening and attempted to drag out a female with his mandibles and forelegs. He initially

FIG 68. Large-headed Resin Bee *Heriades truncorum* female using open mandibles to fight off an unwanted male.

failed and was pulled into the hole by the female, still attached to her. Seconds later the pair re-emerged, the male now grappling the abdominal segments of the female with all three pairs of legs and set back so as to reach under the female's abdomen with the tip of his abdomen. The female actively resisted, pushing back with her rear legs and attempting to grasp the substrate with her mandibles. Eventually she succeeded in pulling the male towards an open hole, which she entered, the male still struggling to hold on, and occasionally flapping his wings, presumably to maintain balance. So far as could be seen, the male failed to achieve genital contact, and the female re-emerged from the hole, followed by the now dislodged male (video sequence lasting 1 minute 20 seconds). Apart from powerful mandibles, the male has no obvious anatomical adaptations to aid his control over a female. However, it seems the mating system of this bee includes a high degree of aggressive behaviour on the part of the male but strong resistance on the part of the female, too.

Males of the Wool Carder Bee *Anthidium manicatum* have no obvious anatomical adaptations for coercion of females apart from their greater size, although there is one account (Severinghaus *et al.* 1981) of the male lifting the female's abdomen with the spines on the tip of his. Typically, a male pounces on a female as she forages, trapping her forewings with his middle pair of legs, holding on to the plant or her abdomen with his hind legs, and placing his forelegs over the female's face – effectively 'caging' her. There appears to be little or no interval between the male mounting the female and copulation, though females are sometimes able to throw the males off. Mating is brief, lasting for as little as 10–30 seconds, after which the female resumes foraging, and the male continues his patrols or turns to foraging. NWO reports a male resting and ventilating his abdomen for at least a minute after dismounting. The female remained on the ground.

In the Early Colletes *Colletes cunicularius*, mating, which generally takes place at the nest aggregation, is much more prolonged, lasting at least several minutes. The male holds on to the female abdomen with all three pairs of legs, the front pair pressing down her wings. At intervals of 2.5–3 seconds, he brings down his antennae to tap the thoracic dorsum of the female with their tips. Sometimes this is a single tap, sometimes a 'vibratory' one, with four or five 'hits' of small amplitude each time the antennae are brought down. Between these taps, the male makes very rapid small-amplitude thrusting movements. In one instance, a

FIG 69. Mating pair of Wool Carder Bees *Anthidium manicatum*, the female captured as she foraged.

FIG 70. Mating pair of Early Colletes *Colletes cunicularius*. The male repeatedly taps the dorsal surface of the female with his antennae.

BELOW: **FIG 71.** Mating pair of Common Yellow-face Bee *Hylaeus communis*. The male 'vibrates' at intervals but does not physically restrain the female.

mating pair were intercepted by two males in succession attempting (but failing) to mount the female by placing themselves on the female's dorsum forward of the mating male. Mating in the Common Yellow-face Bee *Hylaeus communis* is also prolonged, the male holding on to the rear segments of the female, or sometimes just attached by the genitalia, motionless except for recurrent bouts of vibration.

As we have seen, in the Sharp-collared Furrow Bee *Lasioglossum malachurum* and the Buffish Mining Bee *Andrena nigroaenea* (and probably many other species in these genera), the cuticular surface of females carries chemical compounds whose composition changes during the life course. Early in their adult lives, females secrete sexual attractant pheromones, but after mating and beginning nest construction, the chemical mix changes. As we saw above, males are able to detect the new chemical aura and are accordingly less likely to approach or attempt to mate with those females. In species whose females mate only once, it can be argued that this avoids a waste of time and energy on the part of the male and also avoids costs to the foraging or nesting efficiency of the female. As we have seen, however, males are not always so easily discouraged!

In some species, there is an alternative which also results in a loss of sexual attractiveness on the part of the female after mating: there is a prolonged 'post-copulatory embrace', during which the male applies chemical compounds to the female. These inhibit mating attempts on the part of subsequent males she might encounter. These substances are termed 'antiaphrodisiacs'. In the Red Mason Bee *Osmia bicornis*, the post-copulatory phase may last for as long as 13 minutes, during which the male continuously strokes his abdomen over the female, applying an antiaphrodisiac pheromone as he does so (Siedelmann 1995). Ayasse *et al.* (1998) observed that the males secrete the antiaphrodisiac from their sternal glands and rub it onto the wings of the females. Ayasse and colleagues also demonstrated that females carrying extracts from male sternal glands were significantly less attractive to males. They were able to determine the nature of the compound (Z)-7-ethyl hexadecenoate (Ayasse *et al.* 2001).

Ayasse (1994) found a comparable male strategy in the Sharp-collared Furrow Bee *Lasioglossum malachurum*. During mating, males release an antiaphrodisiac (hexadecyl octadecenoate) which inhibits mating behaviour in other males. This presumably enhances the inhibitory effect, already mentioned, of post-mating changes in the compounds on the female cuticle. In this species, fertile females (gynes) have been shown to be capable of mating repeatedly under laboratory conditions and may mate up to three times in nature (Paxton *et al.* 2002). Reduced attractiveness of already mated females presumably favours the reproductive interest of the male by increasing the chances that the subsequent (fertilised) eggs laid by the female will be fertilised by his sperm.

CONCLUSION

Male mate-searching strategies fit well into the three broad categories suggested by Paxton (2005): patrolling a nest aggregation for newly emerged females, searching at flowers or sites for other resources used by females (resource-based strategies), or patrolling scent-marked routes (non-resource-based strategies). Adoption of these alternative strategies conforms to expectations about the likely availability of newly emerged females, though for males of many species there is great flexibility in the use of alternative strategies. In species that nest in large aggregations, for example, many males will also search for females at flowers. There is also evidence that individual males may vary their strategies according to immediate circumstances. For example, males of the Wool Carder Bee *Anthidium manicatum* may shift between territorial patrolling and opportunistic searching depending on the presence or absence of other males at a favoured patch. Territorial behaviour in males might be a candidate for being considered a distinct strategy in its own right, as it seems to be more widespread than often assumed. Aggressive behaviour amounting to territorialism is quite widespread among males of several flower bees (*Anthophora* species), for example. The distinction between non-resource-based and resource-based strategies is not always clear-cut, as in some species males patrol extended routes that include both stretches with and without flowers or other female attractants. Finally, the observations of male Wood-carving Leafcutter Bees *Megachile ligniseca* apparently 'staking out' a suitable nest site and waiting for females to arrive suggests the possibility of another variant of the resource-based strategy. The resource could be a suitable nest site rather than a likely flower.

The processes involved in mating after a male's initial grasp of the female are quite cryptic. In many species, males quickly adopt postures that restrict the movement of their captive female, especially holding down her wings. However, she still has behavioural opportunities for resistance and escape, and he frequently has considerable behavioural and chemical work to do to persuade her to allow actual copulation. Research on the chemical signals exchanged during these processes is producing fascinating insights but much remains unknown.

The mating systems of cuckoo bees are much less well known than those of the nest-makers. The sexual division of labour is rather different in the two groups. Female cuckoos have less parental investment in the offspring, not needing to construct a nest or stock brood cells, and lay more but smaller eggs than their nest-making hosts. This might lead us to expect that there would be fewer conflicts of reproductive interest and strategy between the two sexes in these species. Is the competitive intensity among males reduced? Do females resist mating attempts less vigorously? Are there differences in mating frequency? Systematic research addressing these questions might shed some light on the implications of the 'parental investment' view of mating systems. (For more on the cuckoo bees, see Chapter 8.)

CHAPTER 3

The Life Cycle: Nesting Behaviour and Development

'That reason did not extend itself with the bulk of the body: on the contrary, we observed in our country that the tallest persons were usually least provided with it. That among other animals, bees and ants had the reputation of more industry, art and sagacity than many of the larger kinds.'
—Jonathan Swift, Gulliver's Travels (1726)

INTRODUCTION

The life cycle

Once a female has mated, the next steps in the reproductive cycle are for her to begin constructing a nest, stock it with provisions and commence egg-laying. Most insects are very exacting in the placing of their eggs – many species with herbivorous larvae, such as butterflies and moths, lay their eggs on or close to the appropriate food plant. Sometimes the eggs are laid singly, attached to the plant, or inserted into its tissues, or scattered apparently randomly across sites dominated by the food plant. In other cases, such as the familiar Small Tortoiseshell Butterfly *Aglais urticae* and the Peacock Butterfly *Aglais io*, eggs may be laid in large clusters, and the resulting larvae spend at least part of their development in communal webs. Where the egg is the life stage that passes the winter, then it may be laid in a protective environment, such as a niche in the bark of a tree, or enclosed in a 'pod', as in many grasshoppers. Other species, including many flies, beetles and dragonflies, lay their eggs in decaying plant material, mud, animal dung, randomly in water, or attached to aquatic plants. Parasitic and parasitoid insects, including many of those discussed in Chapter 7, lay their eggs in or on the bodies of their hosts or, sometimes, in places where the favoured host is likely to encounter them.

In all these cases, the insects are dependent on locating already-existing situations that favour the future survival of their offspring. By contrast, the (non-parasitic) bees, and most species of aculeate wasps, are among the rather few insect groups that actively prepare a 'home' for their future offspring and stock it with provisions. In the British bees, these provisions are all derived from plants, usually a mixture of pollen and nectar, but in one British species, plant oils, too. Many social bees and wasps (and some solitary wasps, such as the Heath Sand Wasp *Ammophila pubescens*) go still further and continue to nurture their larvae through all or part of their development. Almost all solitary bees are referred to as 'mass provisioners', meaning that they provide a stock of resources for their offspring but then move on to other activities, usually constructing and stocking the next brood cell or beginning a new nest. However, this apparent lack of interest in 'hands-on' parenting takes nothing away from the astonishing variety of materials and methods they use, the ingenuity shown in their choice of nest sites, or their construction techniques and persistence in carrying out the labour of provisioning each cell.

Most solitary bees in Britain complete a single life cycle each year (they are univoltine), but a few of them manage two generations each year (bivoltine), while others which are usually univoltine have been shown to be able to stretch their life cycle over two years, or possibly even three (Else & Edwards 1996, Boyes 2020). All bees have a development pattern with four principal stages: egg, larva, pupa and adult. The eggs of solitary bees are usually laid singly in each brood cell, sometimes attached to the food mass, sometimes to the brood cell wall. Bee eggs are long and curved but few in number – perhaps just 8–30 are laid during a female solitary bee's lifetime – but the security of the nest with its food-store results in a relatively high survival rate. The larvae, which hatch in a few days, are grub-like and relatively lacking in external features. Their main activity is to consume food and grow, casting off their external 'skin' (cuticle) three or four times as they do so. When fully grown, they enter into a final larval stage, in which they have rather more external structure. This is known as the prepupa stage, and, as it moults into this stage, the larva vents its gut of waste products for the first time. This is also the stage at which larvae belonging to some groups (especially megachilids) spin a protective cocoon around themselves. In the pupal stage, most of the features of the adult bee are visible externally, while the internal structures are radically transformed, leading in due course to the splitting open of the pupal cuticle and the emergence of the adult bee (see also the Introduction).

Nesting habits

For the majority of bees, each of these developmental stages takes place within a brood cell that has previously been constructed by the female parent. The cell is itself usually one component in a nest with several other cells, separated by dividing walls and closed with an outer plug, or cap. Within each cell is a store of provisions with enough nutrients to enable the full development of the larva and the construction of the cocoon in those species which spin one. Resources may also be needed to take the insect through a protracted period of diapause, for it to make its way out of its nest, and to power its initial foraging trip. For a minority of species, however, this development takes place in a cell constructed by a host female of a different species. These are the cleptoparasites, or cuckoos, about which we say more in Chapter 8. For the rest of this chapter, we will be focusing on how the nest-making species provide for their offspring and on the subsequent development of the latter.

The brood cells, and the nests of which they are a part, have to serve many functions for the protection and development of the next generation. The cell walls need to be sufficiently robust to protect the larva from mechanical impacts and pressures, sufficiently impermeable to retain the store of food provided for it, and to protect it from fungal, viral or bacterial attack, but also sufficiently permeable for the exchange of gases necessary for the respiration of the incumbent. The siting and construction of the nest also has to offer security from attacks by a great range of parasites and other enemies for whom the rich store of food in the cells is a great prize. Some of these enemies are, as we noted above, other bees that have given up making their own nests and found ways of penetrating the stocked brood cells of host species. In a few species, the host may be a female of the same species, whose nest has been usurped by a neighbour.

In seasonal climates like our own, the nests also have to offer protection against inclement weather. In the case of ground-nesting species, this may include defence against long periods of waterlogging from flooding. The seasons impose other demands, too. Some species have evolved very close relationships to one or a small number of flowering plants as sources of pollen, so it is essential that the flight period of the bees coincides with the flowering of their host plants. Even species not so tightly linked to specific flowering plants frequently have limited flight periods. Therefore, ensuring that they are on the wing when floral resources are available and also that males and females can encounter one another impose important time constraints on the developmental rhythms of their offspring. Many species that fly in early spring spend the winter in their brood cells as fully developed adults, whilst, by contrast, a late-flying bee (the Ivy Bee *Colletes hederae*) overwinters as a part-grown larva, only completing its development the following spring and pupating in late July (Bischoff *et al.* 2005, Benton & Fremlin 2018, Müller & Weibel 2020; see also below).

In the great majority of the bees we are discussing in this book, each female makes her own nest, stocks each cell with a nectar-pollen mix and lays a single egg, before sealing the cell and, when the complement of cells has been completed, usually finishing off the nest with a protective cap, or plug. In some species the activity is genuinely solitary, but in others many females nest in clusters, sometimes called 'aggregations', and these may involve large numbers of bees – up to several thousand. However, they remain technically solitary, as each female makes and stocks her own nest independently of her neighbours. In a few species, nest entrances may be shared, though inside each female has her own compartment. These are referred to as 'communally' nesting species and include the Oak Mining Bee *Andrena ferox*, the Big-headed Mining Bee *Andrena bucephala* and the Chocolate Mining Bee *Andrena scotica*. Finally, a few species (notably in the family Halictidae) have evolved a form of eusocial behaviour: the fertilised females make a nest and then rear one or more subsequent generations of workers before finally producing a new generation of fertile females and males. The environmental and evolutionary conditions that make for one or other of these various ways of behaving are of great interest, and we will return to discuss the issues involved in Chapter 4.

Foraging and provisioning the nest

Once a nest site has been selected, the female bee devotes almost all her active time to nest construction, collecting and delivering nutrients to stock each cell, and egg-laying. Since the flight period of most solitary bees is quite short, and there is no guarantee of suitable weather for foraging, there is a strong time constraint on her completion of these tasks. Also, since many nest-parasites gain entry to their hosts' nests while the incumbent female is away foraging and the cell remains unsealed, there is pressure to complete each foraging trip and to provision each cell as rapidly as possible.

For female bees that have already mated, time and energy may be lost in avoiding unwanted male attentions (Stone 1993) or in avoidance of predators or parasites. Especially for bees that nest early in the spring, or live in cool climates, ability to forage at low temperatures is important. Stone and Willmer (1989) assessed the thermal regime to which 55 species of bees were adapted by investigating the relationships between body mass, the ability to use muscular 'shivering' to raise thoracic temperature, and the minimum temperature at which they were able to fly. The relationships were complex, but controlling the other factors, they found strong positive correlations between size and warm-up rate, and that bees adapted to cooler climates had faster warm-up rates and were able to fly at lower ambient temperatures (see Chapter 9).

These contingencies aside, the main limitations on a female bee's reproductive activity will be the availability of the relevant floral resources, the bee's load-carrying capacity, the availability of materials used in nest construction, the rate of maturation of her eggs, and the quantity of nutrients required to provide for the development of each offspring.

For most bees, the nutrients brought back to the nest consist of pollen and nectar. Nectar is a solution of sugars, including sucrose and also fructose and glucose, with small quantities of proteins and amino acids. These latter components may be significant for taste, as well as for nutrition. Pollen is generally recognised as providing necessary protein in the diet of the future larva, but it also contains starches, lipids and sterols. Nectar may vary considerably in concentration, especially in plants with superficial nectaries that allow evaporation in warm weather. The protein content of pollen from different plant taxa also varies greatly – from only 12 per cent up to 61 per cent in some plant groups (Roulston *et al.* 2000). It seems that bees are unable to detect the nutritional value of the pollen they collect, but it may be that it has a role in the evolution of generalisation or specialisation in bee pollen collection.

One British species, *Macropis europaea*, collects plant oils, a more frequent characteristic of solitary bees worldwide (Rasmussen & Olesen 2000). These oils are included in the nutritional mix, but they are also used in nest-lining. Other species provide nutritional secretions, so not all the provision mass comes directly from plant sources. Flower bees (*Anthophora*) add secretions from their Dufour's gland in lining their brood cells, and these secretions are also consumed by the larvae (Norden *et al.* 1980, Batra & Norden 1996).

Finally, though much attention has been given to the protection of the future larva from attack by harmful microorganisms, there is growing recognition of the importance of symbiotic bacteria in larval nutrition, development, and disease resistance. In the solitary bees, there is no obvious way for beneficial microbes to be transferred between generations. Inoculation of eggs might be one route, but recent research has explored the possible role of pollen as a transmission route for bacteria. The communities of bacteria in the nests of several species of solitary bees were linked to the pollen sources they used in a study by Voulgari-Kokota *et al.* (2019). Proteins synthesised by bacteria in pollen play a part in the nutrition of larvae (Steffan *et al.* 2019). This is a relatively new field of research and it promises to offer significant insights into foraging behaviour.

An important contribution to our understanding of the factors affecting the provisioning behaviour of solitary bees, and the constraints involved in determining the rate at which brood cells can be completed, was provided by John L. Neff (2008). Assuming that bees forage under strong time constraints, their provisioning success can be understood in terms of the rate at which they are able to stock a cell with sufficient nutrients to provide for the requirements of the future offspring. The ratio of dry weight of provisions to the dry weight of an adult bee is termed the 'conversion ratio' (that is, effectively, the quantity of provisions required for the foraging female to replace herself). Where possible, Neff's calculations take into account the different provisioning requirements of male and female offspring in most species. The rate at which a cell can be completed is a function of the amount of nutrients that need to be collected, the transport capacity of the bee (i.e. the amount it can deliver in one trip), the number of trips required to stock the cell, and the length of time it takes to complete each foraging trip (including the time spent in the nest, unloading). These calculations omit the time taken to construct the cell and also to lay an egg. These activities can be very time-consuming. Beyond the completion of the cell, the overall reproductive performance of the bee will be determined by how many cells can be completed, the rate at which her eggs mature, and the number of eggs she is able to lay in her lifetime.

The conversion ratio varies greatly between species and has been estimated as from 2.75:1–8.33:1 (dry weights), averaging 4.6:1 (Neff 2008). This means that on average a female bee has to collect between four and five times her own dry body weight of nutrients to replace herself or, what amounts to the same thing, to fully stock a brood cell. The ratio is rather less when measured in terms of fresh weight, but a female still has to deliver on average from 2–3 times her own fresh weight in provisions for each cell. The conversion ratio is still higher in those species whose larvae spin cocoons. For two species of *Osmia*, for example, Neff gives almost 50 per cent extra dry weight for cocoons in males and approximately 40 per cent in females. In most species, males are smaller than females, so the conversion ratio is less for cells destined for male larvae, though it seems that in terms of unit mass, males are more expensive to produce.

Studies suggest that the load-carrying capacity, as a percentage of body weight, is more or less constant across species, though larger bees carry proportionally smaller loads. The rate at which the bee harvests a full pollen load will depend on various factors, such as temperature, the intensity of male harassment, and parasite and predator avoidance, as well as the availability of resources in the environment. Some studies have shown that greatly increased availability of resources has relatively limited effects on foraging rates but does lead to a shift towards production of more females (Kim 1999, Goodell 2003). The time taken for each foraging trip is also affected by the condition in which the nutrients are carried. In most solitary species, pollen is collected dry-packed on the scopae (pollen-carrying brushes), and nectar is carried in the crop. However, in some species, especially Melittidae, the pollen is mixed with either nectar or floral oils, allowing relatively larger external loads to be carried. Among British species, the yellow-face bees *Hylaeus* are alone in lacking scopae for external transport of nutrients. Instead, they ingest and then regurgitate a fluid mix of pollen and nectar. It may be that other species with less well-developed scopae, such as the Little Carpenter Bee *Ceratina cyanea*, also carry a significant proportion of their provisions internally.

The number of foraging trips required to provision a cell is a function of the bee's transport capacity and its conversion ratio: on average, the higher the load it can carry, in relation to its weight, the fewer trips will be needed. Neff recognised that some solitary bees contribute glandular secretions to the food-store, but as the secretions have not been quantified, they are left out of the calculations. Also omitted is the time taken to construct brood cells and complete the nest. Digging burrows is a very laborious activity, and females of many cavity nesters, especially in the Megachilidae, collect materials such as leaf sections, leaf hairs, resin and mud both to line their nests and plug them when complete. These activities are frequently very costly in terms of both time and energy, so any full account of the constraints on cell completion would have to include these activities as well as the collection and delivery of provisions.

These considerations led Neff to conclude that the vast majority of solitary bees can provision only one, or fewer, cells in a day, while a few can manage two, and probably none is able to complete three or more. Exceptionally, NWO observed a Red Mason Bee *Osmia bicornis* female that was foraging from *Cistus*, within 1 m of the nest. This individual completed three cells in one day and began another, but over a longer period of observation, the completion rate (taking into account poor weather) was approximately one per day. Even where it seems that the rate of delivery of provisions to the nests might allow for faster completion of brood cells, the additional time taken to line cells and partitions and to lay an egg generally keeps the rate of cell completion at this surprisingly general rate of one per day. Often females take extended breaks between completing one cell and beginning another, sometimes consuming only nectar for a day or more. It seems likely that this is an indicator of the time taken for each egg to mature, there being a metabolic trade-off between foraging and working on the nest, on the one hand, and investing in egg development, on the other (Neff 2008). It is also suggested that cleptoparasitic bees, lacking this constraint, have a faster rate of egg production.

Sex, nutrition and body size
Large body size in bees is generally thought to confer selective advantages, such as better ability to forage in cool weather, greater longevity, and better chances of success in competition for nest sites or, in males, for mating opportunities. In view of this, we might expect females to adjust their provisioning to produce offspring that approximate the maximum size for their species. Roulston and Cane (2000) analysed data on 31 species of North American bees in order to test likely influences on offspring size. They accept that there are general correlations between the volume of the brood cell, quantity of provision, and size of the adult. However, given that it is generally thought that bees cannot estimate the nutritional content of pollen, and that this varies greatly between plant species, we might expect the offspring of pollen-host generalist species to vary much more in size than those of pollen specialists. Comparing pairs of closely related species with contrasting diet widths, they found, surprisingly, no significant differences in body size variability between the two groups. However, on another source of variability, nesting habit, they did find differences. Comparing ground-nesting and cavity-nesting species, they found a much greater size variation in the latter. In some species, there is evidence that females nesting in narrower cavities make longer cells, but though this increases the cell volume (and therefore the size of the provision mass), there remain significant disparities in volume, and so in offspring size, depending on the diameter of the cavity used. Roulston and Cane argue that these results suggest competition for nest sites may lead cavity-nesting bees to use sub-optimally sized cavities, and so produce sub-optimally sized offspring. By contrast, ground-nesting bees can control the size of their burrows, so they tend to produce more consistently sized offspring.

Roulston and Cane cite a study by Strickler (1982) of a species of *Hoplitis* in which the response to limited availability of pollen did not lead to production of smaller offspring. Rather, foraging effort was increased to yield standard provision sizes. There is also evidence that female bees respond to resource limitations by shifting foraging effort in favour of the (usually) smaller requirements for male offspring.

A plausible alternative interpretation of the results reported by Roulston and Cane is that at least some cavity nesters have an evolved capacity to develop to adulthood on variable nutritional input, as an adaptation to contingent variability of cavities of various sizes. Use of, for example, smaller than optimal cavities may not be a result of competition alone but also of availability. Although a population of variably sized adults will be able to make use of a wider range of cavity sizes, of course, any individual female will be limited by the fit between her own body and the space she has to work within. Usually, cavity nesters have to turn round to switch from nectar provision and work on cell linings to delivery of pollen and egg-laying. In relatively wide cavities, this can be done within the nest, but if the cavity is too narrow, the bee may enter the nest head-first, then back out and re-enter back-end first.

In general, it seems that female solitary bees construct differently sized brood cells for male offspring, and, in the case of cavity nesters with linear nests, these are placed closer to the nest entrance, allowing for protandry, as the males will generally be the first to leave the nest. This implies that sex is determined before each brood cell is constructed. Indeed, a study of the differences between male and female cells of the Sea Aster Bee *Colletes halophilus* showed that by both weight and size, the female cells were as much as 44 per cent larger than those destined for male offspring. The implication is that the 'decision' to lay a male (unfertilised) egg or a female one is made before the brood cell is constructed. Occasionally, female-sized cells were found with male occupants, suggesting that the egg-laying female may have 'changed her mind' after constructing the cell. This might be a response to environmental factors such as shortage of forage. In this species, the females are not under the constraint faced by most cavity nesters to place male-destined cells closer to the exit (Rooijakkers & Sommeijer 2009).

ARCHITECTS AND ARTISANS: WHERE AND HOW TO MAKE A NEST

Next, we will consider some of the extraordinary diversity in the sites bees use for the construction of their nests, the materials they use, and the methods they adopt to protect their future offspring from the physical, chemical and biological hazards that lie ahead. As bees construct their cells and nests in parallel with stocking them with provisions and laying their eggs, we will consider these processes together.

Danforth *et al.* (2019) offer a loose four-fold classification of solitary bee nesting habits. Globally, as well as among British solitary bees, the great majority of species make their nests in underground tunnels, which they dig. Most species of Andrenidae, Colletidae, Melittidae, Halictidae and Apidae are ground nesters. The second major group are the aerial, or cavity, nesters. These species utilise pre-existing crevices or cavities, which are often the hollow stems of woody or herbaceous plants, but also included here are species that make their nests in abandoned holes of wood-boring beetles, gaps and cracks in rocks or masonry, empty snail shells, abandoned plant galls or other convenient holes or crevices provided deliberately or unknowingly by humans. Several genera of Megachilidae belong to this group, as do species in one genus of the Colletidae: *Hylaeus*, the yellow-face bees.

The third group of nest-makers are species whose females make their nests in holes which they excavate in wood or comparable substrates, or others which excavate pith from plant stems. Most of these belong to the carpenter bee group (genus *Xylocopa*, Apidae), of which only one species, the Violet Carpenter Bee *Xylocopa violacea*, is regularly recorded in Britain. The much smaller indigenous Little Carpenter Bee *Ceratina cyanea* clears pith from twigs to make its nests. At least one species of mason bee (*Osmia pilicornis*) and the Fork-tailed Flower Bee *Anthophora furcata* excavate their nests in dead wood. It may be that other cavity nesters are able to modify the internal dimensions of their adopted nest cavities. *Megachile ligniseca* certainly does this. Snail-shell users can choose the diameter they need by selecting the size of shell and also by the positioning of their brood cells within it, since the diameter is tapered.

The fourth category of nest makers distinguished by Danforth *et al.* are ones that make free-standing nests above ground. These may be attached to the underside of a leaf, to a rock surface or to some other solid substrate. They take many different forms, but usually there is a cluster of brood cells enclosed within a structure made of resin, soil or plant materials, bound together with salivary or abdominal secretions. Resin nests of species in the tribe Anthidiini include other materials such as small stones, pith and bark, embedded in the resin, and have a 'spout', allowing exchange of gases. Abdominal secretions are used both to soften and harden the resin (Danforth *et al.* 2019). Internationally, several species belonging to three genera of Megachilidae (*Osmia*, *Hoplitis* and *Megachile*) are included in this group, as well as orchid bees (Euglossini, Apidae), but unfortunately this mode of nest-making is arguably absent among the British species. The two British species closest to this description are the Mountain Mason Bee *Osmia inermis* and the Cliff Mason Bee *Osmia xanthomelana*. Still more unfortunately, both species are very localised and rare in the British Isles. *Osmia inermis* is a very localised species of the Scottish Highlands. Its nests comprise clusters of oval cells made of leaf mastic attached to the underside of rocks, where, when dry, they are said to resemble rabbit droppings (Else & Edwards 2018, Falk & Lewington 2015). Although the British resin bees (*Heriades*) do not construct 'free-standing' nests, their use of resin in nest construction does have parallels with those that do. In particular, their ability both to manipulate resin as a cell lining

FIG 72. Nest of the Mountain Mason Bee *Osmia inermis*. (Courtesy of G. Else)

and to break through thick nest caps of hardened resin when emerging suggests that there may be similar chemical secretions to alter the viscosity of the resin.

Ground nesters

British ground-nesting bees excavate burrows of varying depths, but generally down to 50 cm or less, though Danforth *et al.* cite examples of nests of 5 m or more below ground elsewhere. Some bees show decided preferences for horizontal ground, while others tend to occupy slopes or vertical exposures. Areas of bare ground, such as the edges of paths, dunes or earthen banks are often used, while some species dig their burrows among loose tufts of grass or other vegetation. A commonly used alternative is the exposed root plate of a fallen tree, especially if this is south-facing and exposed to the sun.

Nest density is also very variable among these species. Some species of *Andrena*, *Lasioglossum* and *Colletes*, especially, tend to nest in very large, dense aggregations – sometimes with many thousands of females making their nests in close proximity to one another.

The late David Baldock estimated as many as 10,000 bees at an aggregation of Ashy Mining Bees *Andrena cineraria* on Ham Common in 2001 (Baldock 2008). These aggregations seem to call into question the designation 'solitary', but the term is justified as there is generally no division of labour between individual bees, each female getting on with its business independently of its neighbours. As mentioned above, some species of *Halictus* and *Lasioglossum* do have truly social (eusocial) modes of life (to be discussed in Chapter 4).

ABOVE: **FIG 73.** Smooth-gastered Furrow Bee *Lasioglossum parvulum* nest site in a root plate.

FIG 74. Vast nest aggregation of the Ivy Bee *Colletes hederae*.

Because the activity of nest construction and provisioning takes place underground, opportunities to observe it are rare. What we know about nest architecture and cell construction in the ground-nesting bees is largely derived from excavation of completed nests. From this source, we know that there is considerable variation in nest design among ground nesters. In most cases there is a vertical burrow, with brood cells constructed singly, in clusters, or in linear series on lateral branches. Depending on the friability of the substrate, the bees dig with mandibles or with a combination of mandibles and forelegs. Often, initiating the burrow involves penetrating a hardened soil surface, and in some cases the bees use a period of wet weather to soften the substrate. Disposal of excavated soil or sand is most commonly achieved by a combination of compressing the walls of the burrow (or filling side-branches) and forming a conical mound over the nest entrance. These 'pyramids' (or tumuli) are often quite persistent and are a giveaway to the presence of the nests. In some species (such as the Pantaloon Bee *Dasypoda hirtipes*), the bees use their hind legs to sweep away the excavated materials, and in yet others (such as the Small Sandpit Mining Bee *Andrena argentata*), the sandy substrate simply collapses around the nest entrance – posing a challenge to the returning occupant.

The walls of the part of the burrow allocated for the brood cells are made firm by pressure applied by the bee's abdomen. In most ground-nesting species, there is a flattened area on the dorsal surface of the final abdominal segment (the pygidium), which is used to tamp down the soil comprising the brood cell wall. Some bees (*Andrena* species and some others) 'brace themselves' against the cell walls, using structures (basitibial plates) at the junction of their tibiae and femora (knees). Then a water-resistant lining is applied. This may be to guard against desiccation or inundation and also may include anti-fungal and/or antibacterial compounds. The female bee makes a series of foraging flights, returning with a load of nutrients. These are formed into a pollen-nectar mass, and a single egg is laid – usually attached to the food-store. As just noted, there are relatively few

FIG 75. Small Sandpit Mining Bee *Andrena argentata* female locating her nest.

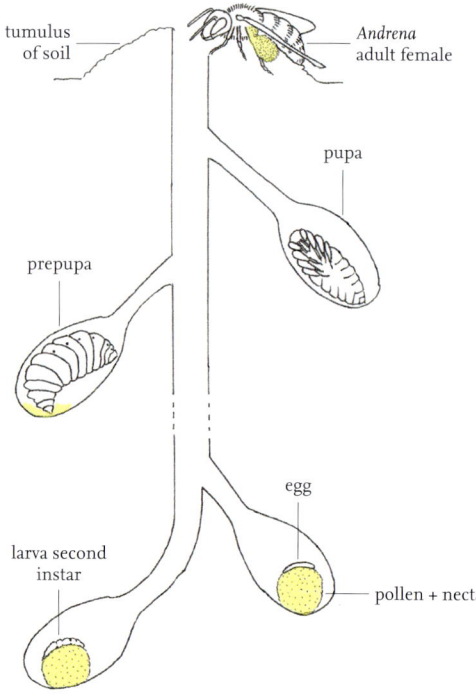

FIG 76. The nest of an *Andrena* bee is typical of the structure of many kinds of mining bees. Soil is excavated using the mandibles and legs to form a narrow vertical shaft. A side chamber is prepared where a ball of pollen and nectar is placed. An egg is then laid on the food-store, and the side-chamber is sealed. A sequence of 6–8 side-chambers is fashioned in turn as the shaft is extended downwards. Eventually the entrance to the whole nest is sealed at the surface, and the parent bee pays it no more attention. In some cases, a second or subsequent nest will be made by the same female.

detailed accounts of the activities of ground-nesting bees within their nests. However, there are some, and they do illustrate the great variability among the different groups of ground-nesting species in their methods of making their nests and constructing and lining the cells.

Melittidae

The British species of the Melittidae are all ground nesters. Relatively little is known about the nesting habits of our four species in the genus *Melitta*, but it seems they nest in small aggregations, with entrances usually concealed among vegetation. By far the rarest species is *Melitta dimidiata*, which is almost entirely confined to Salisbury Plain in the British Isles. Else *et al.* (2020) report the discovery of nest sites along the grassy centre of a wide track in a firing range. The aggregation contained 134 nests, at an average of 13 nests per m². Females were observed returning from foraging trips at a height of about 20 cm, subsequently lowering to 2–3 cm as they flew under fallen grasses. At the nest-approach they slowed and hovered before the nest entrance. After entering and delivering their provisions, they frequently remained at the entrance to the nest, facing out.

Celary (2006) studied the nesting behaviour of a North American population of one of our more common *Melitta* species, the Clover Blunthorn Bee *Melitta leporina*. According to that report, the burrows range from 25–40 cm in depth, with lateral tunnels, each terminating in a single brood cell. The brood cells are lined with a secretion from Dufour's gland, which confers waterproofing and protection against microbes. The female collects a pollen-nectar mix to stock the brood cells, each cell requiring 15 foraging trips, and each trip taking 20–30 minutes. Up to four minutes is taken to deposit each load, and when provisioning is complete, the bee lays an egg and takes a further 20 minutes to close the cell with soil.

The other genera of Melittidae, each with one British species, are more fully studied and are quite distinctive. The very familiar Pantaloon Bee *Dasypoda hirtipes* is so called because of the exceptional development of the scopal hairs on the hind legs of the female. Favoured nesting locations are areas of bare, compacted sandy ground, often south-facing banks. As mentioned above, the females use the dense drapes of hair on their hind tibiae to sweep spoil away as they dig their burrows. These can be very deep, with side-branches each of which terminates in a single brood cell. These are usually at depths from 8–60 cm (Lind 1968) but may be even deeper. This species is considered to be unique among the British bees in that although it flattens the walls of its cells, it does not line them with a protective layer as do other species. The provisions for its future larvae are formed into a relatively solid 'loaf' which is supported by three projections from the brood cell wall, so that it is kept apart from leakage or contamination from surrounding soil. This exception serves to indicate the role of the resistant cell walls in the other species, where the larval food-store is in contact with the cell wall or, in some genera (notably *Hylaeus* and *Colletes*), is quite fluid, having a relatively high nectar content.

FIG 77. Pantaloon Bee *Dasypoda hirtipes* female digging her nest burrow.

A single egg is laid on the surface of the pollen loaf in each cell. The flowers belonging to the Asteraceae, which are most used as pollen sources by this bee, often close by the middle of the day, so it is not common to see the females outside their nests in the afternoon (see Chapter 5).

The other distinctive melittid is the Yellow Loosestrife Bee *Macropis europaea*. This species is an inhabitant of wet, marshy ground, where it specialises in foraging from the flowers of Yellow Loosestrife *Lysimachia vulgaris*. As well as pollen (and nectar which it collects from other flowers), the female bees collect plant oils from minute glands on the surface of the petals and on the bases of the anthers. This agglutination of the pollen with plant oils enables the bee to carry especially large loads of adhered pollen back to the nest, and the oils make a contribution to the nutrition of the bee's larvae. The plant oils are also used in nest construction, providing a waterproof lining to the brood cells. There are very few accounts of the nesting behaviour, but as might be expected from the habitat of the Yellow Loosestrife, nests are reported to be in damp ground, often on clay or peat soils, and in small aggregations. The nest entrances tend to be concealed by vegetation. Excavated nests are reported to comprise four or five cells, with waxy and yellow or greenish oil-based linings (Phipps 1948, Else & Edwards 2018).

FIG 78. Yellow Loosestrife Bee *Macropis europaea* female with full loads of oil-soaked pollen.

FIG 79. Green-eyed Flower Bee *Anthophora bimaculata* female digging and clearing excavated soil from her nest entrance.

Apidae

ANTHOPHORA British flower bee (*Anthophora*) species are mostly ground nesters, including use of soft cliffs, old masonry, vertical mud banks and, in one case (*Anthophora furcata*), dead or rotting wood. The Green-eyed Flower Bee *Anthophora bimaculata* forms large, dense aggregations on bare ground, usually on sandy soil. Each nest is indicated by a loose mound of excavated material, but in this species, the nest entrance is usually at the edge of the mound. The female digs with her forelegs, funnelling material backwards, often with a backward kick of the hind legs or a sideways swipe of both legs, so that the pile of excavated material building up behind her is spread out. Digging activity is frenetic and takes the form of short bursts of as little as 10 or 11 seconds, interspersed with brief flights around the nest entrance. The process is often interrupted by mating attempts on the part of the males which actively patrol the nest sites, emitting a high-pitched buzz. During foraging trips, the entrance hole is left open, but returning bees still seem to require a brief orientation flight over the immediate vicinity of the nest before darting inside. This may be related to the problem of identifying the right nest entrance in the midst of a dense aggregation. When the nest is complete, the female performs an odd 'reverse digging' operation, in which she turns her back to the nest entrance and uses her forelegs to detach sand or soil from the mound of material she had previously excavated in digging her nest burrow. She passes this through under her body, and then flicks it back over the nest entrance with her hind legs. This activity is conducted in short bursts of 4–20 seconds, interspersed with brief flights, until the spoil mound is depleted and the entrance is concealed (TB video).

Andrenidae

ANDRENA With over 60 species in Britain, it is not surprising that this genus includes a great variety of nesting habits, although all species are ground-nesting. Some species, such as the Yellow-legged Mining Bee *Andrena flavipes* and the

FIG 80. *Andrena* (mining bee) females at their nest sites: (a) Grey-backed Mining Bee *Andrena vaga* and (b) Tawny Mining Bee *Andrena fulva*.

Grey-backed Mining Bee *Andrena vaga*, nest in large, dense aggregations, often on bare ground, while others, such as the Tawny Mining Bee *Andrena fulva* and the Ashy Mining Bee *Andrena cineraria*, also may nest in large aggregations, but usually among grasses or other low vegetation. The Black Mining Bee *Andrena pilipes* nests in aggregations on vertical exposures, such as soft cliff faces, while the Cliff Mining Bee *Andrena thoracica* nests in sandy soil, sometimes among patches of heather. The nests of many species are so rarely described that it is difficult to make any generalisation about their preferred substrate. There seem to be considerable differences between species in their ability to penetrate tough surfaces. While species such as *Andrena flavipes* often nest in compacted ground at the edge of paths, both the Sandpit Mining Bee *Andrena barbilabris* and the Small Sandpit Mining Bee *Andrena argentata* favour loose, sandy substrates. *Andrena argentata* females dig directly into the sand with their forelegs, keeping their hind legs close to their abdomen, so that their burrow simply closes in on them

FIG 81. Small Sandpit Mining Bee *Andrena argentata* female with a pollen load, digging to her concealed nest.

as they dig down. On returning from foraging trips, they frequently run about a small area of sand before digging down to the nest. Presumably, within a small area identified visually by landmarks, the precise point that triggers the digging is indicated by scent cues. One of us (TB) watched females of *Andrena barbilabris* at the edge of a well-trodden footpath as they repeatedly attempted but failed to open up nest burrows, digging frantically with front legs for a few seconds before abandoning the attempt and trying again a few centimetres away – all the while being subject to 'hits' from males.

Typically, burrows of *Andrena* species consist of a main shaft, with lateral branches, within which are constructed brood cells. A study of the foraging behaviour of the scarce Large Scabious Mining Bee *Andrena hattorfiana* in southern Sweden (Larson & Franzén 2007) incidentally yielded discoveries of nine nests. These were without an above-ground tumulus of excavated soil and were sparsely distributed (more than 2 m apart), and sometimes with the entrances concealed among leaves. Some nests were excavated and were found to consist of an approximately vertical shaft, with branches each of which terminated in a single brood cell. The branches were filled with soil, but the main shaft was open. The nests were all active when excavated, and although a maximum of four cells were excavated from any nest, the authors supposed that the full complement would have been six cells per nest.

This Swedish study was an attempt to estimate the minimum pollen resource for a sustainable population of the bee, but the researchers were also able to determine the rate at which females were able to fully stock their cells with pollen (from photographic evidence and personal observation, the bees generally imbibe

FIG 82. Large Scabious Mining Bee *Andrena hattorfiana* female foraging for both pollen and nectar.

nectar as they forage for pollen). Eleven foraging trips were required to stock a single cell, and foraging trips took approximately an hour, with between nine and ten minutes spent in the nest after each trip. So, over 12 hours was required to complete each cell and, on the assumption of favourable weather conditions, this amounted to 1.72 days per brood cell (or 10.3 days to complete a nest).

The walls of the brood cells of *Andrena* species are made of compacted soil, and they are lined with a waterproof material secreted by Dufour's gland, which opens at the tip of the female abdomen, close to the base of the sting. Cane (1981) analysed the chemical constituents of the brood cell linings, extracts of pollen balls and of Dufour's gland secretions for several ground-nesting bees, including two *Andrena* species. In both the Orange-tailed Mining Bee *Andrena haemorrhoa* and the Small Scabious Mining Bee *Andrena marginata*, farnesol, isomers of farnesene and farnesyl hexanoate were the dominant compounds. Farnesol is an organic compound that is the starting point for the synthesis of many biologically active compounds. It and its derivatives are associated with floral scents and have been linked to the ability of female bees to recognise their own nest entrances. Farnesol is hydrophobic, and so has an important role in making the nest lining waterproof. It is also an antibacterial agent and a mite pesticide. The dominance of the same compounds in both cell linings and Dufour's gland secretions suggests that the bees use the secretions of this gland when lining their brood cells, and this is confirmed by the great increase in the size of this gland during nest construction. It is much smaller in cuckoo bees. As discussed in Chapter 2, these compounds are volatile and also play a part in sexual signalling. In the Buffish Mining Bee *Andrena nigroaenea*, males can detect farnesyl hexanoate in the 'bouquet' of females, presumably acquired in the course of nest construction. Male sexual response is inhibited by the scent, which indicates that the female has already mated (Scheistl & Ayasse 2000).

On her return from a foraging flight, the bee enters her brood cell and scrapes pollen from her hind leg and propodeal scopae. This is mixed with some nectar, regurgitated from the crop, to form a viscid sphere resting on the cell lining. A single egg is then laid on the food mass before the cell is sealed with compacted soil.

Several studies note that female bees wait for considerable periods of time facing out of their nest entrances before setting off on foraging trips, a habit presumably related to weather conditions in species that fly in early spring. On leaving the nest, some species cover the nest entrance before setting off, presumably a response to the risk of nest parasitism. This is reported in the Grey-backed Mining Bee *Andrena vaga*, and both authors observed this behaviour in a large aggregation at Dungeness. However, it was not universal, and some females returned to open nests. Orientation flights are also often reported as females leave the nest for a foraging trip, though, again, this is by no means universal. Presumably these flights are important in enabling the bee to navigate its return to the nest site, though identification of the individual nest in an aggregation must depend on a mixture of visual and olfactory cues. Reskova *et al.* (2011) observed orientation flights in this species on departure for the first foraging trip of the day, or when the bee had experienced difficulty in re-finding its nest on returning from its previous trip. They also observed females digging 'fake' nest entrances before leaving, presumably as a distraction for would-be nest-parasites. This behaviour is combined with two minutes or more of active 'scuffling' of excavated material to conceal the true nest entrance as the bee leaves on each foraging trip. Both authors have observed this behaviour in Clarke's Mining Bee *Andrena clarkella* (Owens & Benton 2022; see also Chapters 7 & 8).

NWO observed recently emerged females of the Cliff Mining Bee *Andrena thoracica* in Norfolk. After mating and completing a burrow, the bees entered a resting period, rather than seeking pollen immediately. On sunny days, they could be seen basking at their burrow entrances. It has been suggested that this period of quiescence allows time for the maturation of the ovaries (Danforth *et al.* 2019). At this stage there was considerable competition between females still

FIG 83. Nest aggregation of Clarke's Mining Bee *Andrena clarkella* among the roots of a tree.

FIG 84. Clarke's Mining Bee *Andrena clarkella* female covering her nest entrance, with decoy entrances nearby.

FIG 85. Cliff Mining Bee *Andrena thoracica* females basking at their burrow entrances. A would-be usurper is being repelled from the left-hand burrow.

prospecting for a nest site, with some altercations taking place. Some females moved from burrow to burrow, perhaps sometimes displacing the owner. Lone females were sometimes seen several hundred metres from the main aggregation, and these seemed to attract other females to start up burrows nearby.

The diurnal pattern of provisioning behaviour in *Andrena vaga* was studied at a site in the Czech Republic by Reskova *et al.* (*ibid.*). They confirmed the finding of a previous study (Bischoff *et al.* 2003) that pollen and nectar are collected on distinct foraging trips, but in their study, pollen and nectar were collected on different days: two pollen-collecting days to one nectar-collecting day. On pollen days, the bees also had a mixture of pollen and nectar in their crop, but on nectar days, the crop was full of nectar only. Although the bees were active from 8:00 or 9:00 am through to 6:00 pm, they completed surprisingly few provisioning trips in a day. On pollen days, 5 per cent of the bees completed only one provisioning cycle, and only 34 per cent managed two. Smaller numbers managed three or four. The median time taken for a pollen-collecting flight was 1 hour 39 minutes, but when only one pollen delivery was effected in a day, the foraging flight was longer. The bees frequently returned with provisions but left again without entering the nest, only to return again soon afterwards, and variable amounts of time were spent covering the nest entrance, digging their way back into it, expelling surplus material and, depending on weather conditions, basking or waiting at the nest entrance. The rest of the day was spent inside the nest. The

FIG 86. Grey-backed Mining Bee *Andrena vaga* female delivering pollen loads to her nest.

sequence on nectar-collecting days was similar, though the median flight time to collect a full nectar load was longer, and more than one delivery in a day was rare. It is possible that the differences between nectar- and pollen-collecting times might be related to different local availability of male and female willows.

PANURGUS The two British *Panurgus* species are both ground nesters that make their nests in aggregations in sandy ground, often along footpaths. The nesting and associated behaviour of *Panurgus banksianus* was thoroughly studied by Mikael Münster-Swendsen in Denmark during the 1960s. He used some ingenious methods, including making casts of nests, observing foraging behaviour, and partial rearing of immature stages. The bee digs a burrow in sandy soil at roughly 45° from the horizontal, to a depth of 4–5 cm. It then continues up to the surface a few centimetres away from the original entrance. This forms a U-shaped 'bow', with a tumulus of excavated sand over the initial entrance. From now onwards, the bee uses the second entrance hole for working on the nest and delivering provisions. Next, it digs a lateral tunnel (a 'gallery') from the bow. This leads approximately vertically down to a further depth of 8–14 cm, where a brood cell is constructed by firming the walls of the opening and lining the surface with a waxy secretion.

Nest construction takes place mainly in the afternoon, and the following morning the bee forages for pollen from (at this site) Catsear *Hypochaeris radicata*. On average, eight trips are required to fully stock the cell, and this takes approximately 2–2.5 hours in good weather. As the flowers close in the afternoon, the bee is constrained to complete provisioning its cell during the morning. During the foraging period, the pollen is left loose in the cell, and the bee remains inactive in the nest for 1 hour 45 minutes. She then mixes the pollen with regurgitated nectar and forms a spherical mass, onto which she lays an egg. She next closes the cell and digs a new lateral branch from the bow, at the terminus of which she forms a new cell. This sequence takes approximately three hours.

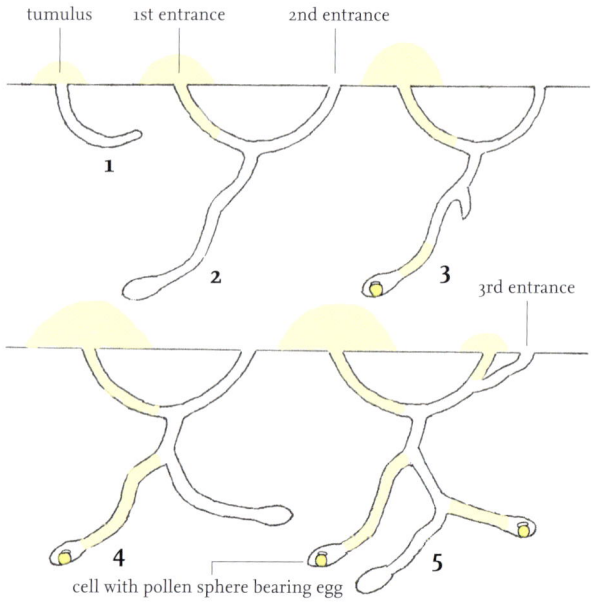

FIG 87. Nest architecture of the Large Shaggy Bee *Panurgus banksianus* (redrawn from Münster-Swendsen 1970).

FIG 88. Large Shaggy Bee *Panurgus banksianus* female returning to her nest with full pollen loads.

On the following day the process is repeated, but after four or five cells have been completed, the excavated material (so far being brushed into tunnels with completed cells) becomes too bulky to be contained in the nest, and a new entrance tunnel is dug from the bow. It was observed that potential nest-parasites showed interest in the tumulus, so it seems possible that it serves as a distraction from the entrance(s) in use, just as in cases of 'decoy' nest entrances used by other species such as *Andrena clarkella* and *Andrena vaga* (see Chapters 7 and 8).

Halictidae

LASIOGLOSSUM The Sharp-collared Furrow Bee *Lasioglossum malachurum* is probably the most thoroughly studied of the halictid bees, partly because its social behaviour potentially sheds light on the evolution of sociality (see Chapter 4). Fertilised females overwinter in underground hibernacula and begin nest-making in spring. Nesting females usually form large, dense aggregations, each nest evidenced by a pyramid of excavated sand or soil. The bee excavates its burrow with mandibles and forelegs, funnelling soil particles back between its middle and hind legs, pressing its dorsal surface against the burrow wall as it does so. The hind legs are then spread wide and rapidly brought together

FIG 89. Nest aggregation of the Sharp-collared Furrow Bee *Lasioglossum malachurum*.

FIG 90. Sharp-collared Furrow Bee *Lasioglossum malachurum* female digging her nest burrow.

again, in a scissor-like motion, to disperse the excavated material upwards to the surface. The sub-group of *Lasioglossum* to which *Lasioglossum malachurum* belongs (species that have in common fine ridges, or carinae, on the rear surface of the propodeum) have a distinctive nest architecture. The entrance burrow ends in a cavity, within which the brood cells are clustered. The cell walls are made of soil or sand, which is compacted and smoothed by the bee, using the pygidium as a trowel. The inner surface of each cell is then lined with a coating of water-resistant material secreted by the bee's Dufour's gland. The compounds used are macrocyclic lactones, and the lining protects against infection by fungi and microorganisms as well as inundation or desiccation during the development of the future larva (Duffield *et al.* 1981, Hefetz *et al.* 1986, Steitz *et al.* 2018).

Colletidae

COLLETES The genus *Colletes* is very widespread globally, and there are nine British species. Though each of these has its distinctive features, there are some important aspects of nesting behaviour that are believed to be held in common by them all. After mating, females either dig burrows or, in some species, use abandoned chambers left from previous generations of the same species. Habitats used may be flat, sloping or even vertical, including substrate such as sandstone cliffs but, more typically among the British species, soft, sandy soil or dune systems. Several British species nest in large, dense aggregations, marked by tumuli of excavated material, but others nest more sparsely. The Early Colletes *Colletes cunicularius* and the three pollen specialists that fly later in the year (the Ivy Bee *Colletes hederae*, the Sea Aster Bee *Colletes halophilus* and the Heather Colletes *Colletes succinctus*) all form large aggregations marked by numerous

FIG 91. Small nest aggregation of the Ivy Bee *Colletes hederae*.

FIG 92. Early Colletes *Colletes cunicularius* female digging her nest burrow.

FIG 93. Sea Aster Bee *Colletes halophilus* males visiting emergence holes of females.

tumuli formed from excavated sand or soil. Females of all four species dig rapidly with mid- and front legs, funnelling excavated material behind them. They retreat at intervals, backing out of the burrow to flick accumulated material to the sides by means of a scissors-like movement of the hind legs. The result of this is that the pyramids of excavated sand commonly have a compacted groove leading from the entrance hole. This is frequently (but not always) used by returning bees as an entrance path.

Within the nest, the entrance burrow runs roughly vertically for most of its length, sprouting side-burrows at or around its terminus. These side-burrows run approximately horizontally, and the females make their cells here. In some *Colletes* species, the cells form short, linear series. This feature is unlike the placing of cells individually or in clusters, characteristic of other ground-nesting bees, and suggests to some authors that the *Colletes* species descended from ancestral cavity-nesting bees. Supporting this suggestion, *Colletes* bees lack the pygidium.

FIG 94. Excavation of part of an Ivy Bee *Colletes hederae* nest aggregation in a suburban garden.

FIG 95. Brood cell of the Ivy Bee *Colletes hederae* at 15 cm depth. (© M. Fremlin)

A small excavation into an aggregation of *Colletes hederae* in a Colchester garden (Benton & Fremlin 2018) found cells at depths from 15–35 cm, in groups of 3–4. Cells varied from 9.2–13.03 mm in length and from 7.25–7.5 mm in diameter, and we estimated their density at some 200 per m². A similar excavation at a large aggregation of the spring-flying *Colletes cunicularius* yielded cells at depths from 15–30 cm, at an estimated density of some 50 per m². Interestingly, several soil-filled 'husks' of a previous season's cells were found, testimony to the resilience of the cell lining material.

A distinctive feature of nesting behaviour of *Colletes* (shared with their cavity-nesting relatives, *Hylaeus*) is the way they line their brood cells with a transparent, cellophane-like substance. It is presumed that this guards against fungal and microorganism infections and is well known for its water-resistant properties. *Colletes halophilus*, as might be expected from its pollen-specialisation on Sea Aster, is a mainly coastal species in Britain, forming dense nesting aggregations which are sometimes below the high tide mark. A large aggregation on a sand-dune nature reserve on the Essex coast suffered prolonged inundation during high tides and storms in the winter of 2013–14, but there was no evident effect on the numbers that emerged the following summer (B. Seago pers. comm.).

Once the burrow is dug, the female begins to construct the brood cells, using her short, bilobed tongue to brush each cell wall with fluids she secretes from her Dufour's gland, hence the popular term 'plasterer bees' given to this genus. This creates the 'cellophane' lining, and several layers are added before the bee emerges from the nest to gather provisions. Though we do not have detailed accounts of the behaviour of the bees within the nest for British species, there

FIG 96. Nest aggregation of the Sea Aster Bee *Colletes halophilus*.

are some remarkable studies published by American observers. Torchio *et al.* (1988) were able to observe females of *Colletes kincaidii* making their nests in glass tubes. In this species, new brood cells are constructed on the cap of a previously completed cell. After initial brushing of a coat of salivary secretion on the walls of the new cell, the bee deposits some droplets of an abdominal secretion (presumably a product of Dufour's gland) close to the cap of the previous cell. She continues to move in both directions along the length of tube, performing half-somersaults as she changes direction, but on reaching the droplets of Dufour's secretions, she appears to mix them directly with salivary secretions, continuing to add more layers of lining to the cell in construction. Torchio *et al.* describe the bees' collection of droplets of Dufour's gland secretion in their mouthparts as they perform their half-somersaults, continuing to add layers to the cell lining with mixtures of salivary and Dufour's gland secretion. Next, a platform is constructed close to the entrance to the new cell, again made of a mixture of secretions from the two sources. This is used to make a rim around the entrance to the cell which later forms the basis for a cell cap. According to their account, these complex operations take from 2–3 hours, and the bee rests for a further two hours. Given their estimate that the bees construct and complete one cell

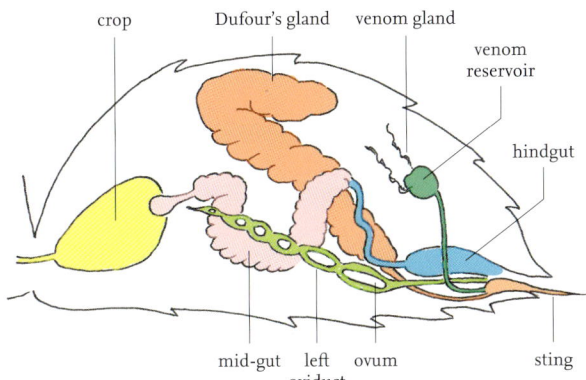

FIG 97. Internal organs of a female *Colletes* showing the large Dufour's gland.

per flight-day, Torchio *et al.*'s observations suggest that at least as much time is taken to construct the cell as to collect and position its provisions. Since stocking of the cell required 11 foraging trips, the implication is of a very high relative investment in cell construction and lining in this species, and possibly in the Colletidae species more generally.

Batra (1980) also described cell construction in *Colletes* (including data on three species). There are some discrepancies between the two accounts, Batra concluding that the bees imbibe the Dufour's gland secretion, so that mixing of secretions from the two sources takes place in the crop. However, the differences are relatively slight and could be due either to differences between the species under study or just the sheer difficulty of making accurate observations of processes occurring with great rapidity. What seems now to be clear, however, is that, of the chemical constituents of the Dufour's gland secretion (hydrocarbons, aldehydes and macrocyclic w-lactones), it is the macrocyclic w-lactones that are the precursors of the distinctive transparent lining material. These are present in a liquid form and are presumed to be polymerised by the action of an enzyme in the salivary secretion as this is brushed onto the inner cell surfaces by the bee (Almeida 2008).

In the spring-flying species *Colletes cunicularius*, pollen is collected mainly from willow (*Salix*) catkins, whose flowering coincides with the flight period of the bee. The bees carry large loads of pollen in scopae on the femora, tibiae and tarsal segments of their hind legs, together with a dusting of pollen grains among the hairs on the ventral surfaces of their body. The late-nesting *Colletes succinctus* and *Colletes hederae* carry pollen to the nest from Ling *Calluna vulgaris* and Ivy *Hedera helix*, respectively. As in *Colletes cunicularius*, large loads are carried on the main segments of the hind legs.

FIG 98. Colletes females returning to their nests with pollen loads: (a) Sea Aster Bee *Colletes halophilus* and (b) Early Colletes *Colletes cunicularius*.

Cavity nesters

The most frequently used alternative to ground-nesting among the British bees is to find an above-ground hollow crevice or cavity in which to make a nest. A great diversity of such cavities may be used, including hollow plant stems, emergence holes of wood-boring beetles, holes in masonry and, for a few species, empty snail shells and vacated plant galls. Luckily, some of these nest sites can be replicated by bee-watchers, who thus attract bees to nest where their antics can be closely observed. Bee hotels in gardens have proved very successful and are now widely used. Even here, though, the behaviour of the bees inside the nest is usually out of sight, unless a specially designed 'observation' nest box is used. One of these, designed by George Pilkington (http://nurturing-nature.co.uk), has semi-cylindrical hollows cut into the sides of wooden inner cores, with transparent plastic covers, through which behaviour can be observed.

Most nest-making British species of the family Megachilidae are cavity nesters, as are the yellow-face bees (*Hylaeus*) from the family Colletidae. Perhaps the best known of all the cavity nesters are the leafcutters (*Megachile*) and some species of mason bees (*Osmia*). Keen gardeners are sometimes disconcerted to find that numerous large symmetrical holes have been cut from the leaves of their carefully tended rose bushes. Usually they (the gardeners) can be placated by the discovery that this is the work of a bee, and even better, that the bee plays a very useful role as a pollinator. Better still, spending time watching the activities of the bee can add immensely to the enjoyment of the garden. Several species of leafcutter bees are common and will readily take to making their nests in bee hotels. Although rose leaves are the most commonly mentioned in reports, the bees seem indifferent to the species of trees, shrubs or herbaceous plants from which they take their leaf cuttings. However, they are clearly not indifferent to the physical or chemical character of the leaves they choose. Females on leaf-collecting trips can be seen to 'try' several leaves before settling on one to cut. One *Megachile* female in TB's garden was observed to attempt cutting into the margin of a holly leaf before giving up and trying something less resistant.

While the bees are searching for a suitable leaf, they are wary and easily disturbed, but once a leaf has been selected, they persist until the job is finished. However, this is a very short time indeed. A female Willughby's Leafcutter Bee *Megachile willughbiella* took between 6 and 12 seconds to cut sections of the leaves of Steeple Bush *Spiraea douglasii* in TB's garden, and females of the Silvery Leafcutter Bee *Megachile leachella* (which, just to challenge our vain attempt to impose order on reality, happens to be a ground-nesting species!) were filmed cutting leaf sections from Rose of Sharon *Hypericum calycinum*, Greater Bindweed *Calystegia sylvatica*, Elder *Sambucus nigra* and a hawkweed (*Hieracium* species). The time taken to cut a leaf section varied between 7 and 10 seconds on Rose of Sharon, 13 and 15 seconds on Bindweed, 9 and 16 seconds on the hawkweed and 20 and 22 seconds on Elder. Occasionally, cutting times were prolonged by the bee's difficulty in cutting through a tough vein, especially on the Elder. In the case of *Megachile leachella*, numerous females converge on a single plant or patch to collect leaf segments and are generally 'harassed' by amorous males as they do so.

The bee rapidly 'saws' with her mandibles, beginning at the leaf margin and continuing to cut on a continuous curve until a symmetrical segment of leaf is cut free. Sometimes the segment is ovoid; sometimes an almost perfectly circular disc. The ovals are used to line the nests, while the discs are used as cell dividers or to plug the completed nest. As she cuts, the bee clasps the part of the leaf to be detached within a cage formed by her legs and body, so as soon as this happens the bee can fly off directly holding her prize. Sometimes the bee performs an orientation flight before flying back to the nest and often repeat visits are made to the same or adjacent leaves for subsequent materials.

Even with an observation nest box it is difficult to see what happens inside the nest. In flight, the bee carries the leaf cutting below her body, caged between her legs and ventral surface and held in her mandibles. On arrival at the nest, the bee clings on with her legs and crawls rapidly along the nest cavity, holding the leaf cutting in her mandibles and sometimes rolling to left or right, depending on the ultimate destiny of her cargo. The pieces of leaf are manipulated and pressed into position by her head, mandibles and forelegs, and where necessary cut into shape by her mandibles. She nibbles the margins of leaf segments to produce a mastic which is used to bind the leaves together to form the distinctive cigar-like cell. Then she licks the inner surface of the cell, applying what appears to be a salivary secretion. This presumably provides the waterproof lining that is needed to hold the nectar and pollen mix which is next brought in from a series of foraging trips. The above account is derived from video clips of *Megachile centuncularis* taken by G. Pilkington, but it seems likely that the behaviour of other leafcutters is similar. Once the cell is fully stocked, the bee lays an egg, and a cell partition is built, using disc-shaped leaf cuttings.

When the cavity is almost full, a gap is left (the 'vestibule') before a final partition is built to form the outer plug of the nest. This final process is more easily observed, and the following account is based on video by TB of *Megachile ligniseca*. The bee makes many return journeys with leaf segments. These are pressed into the nest entrance to create a many-layered plug, as the bee chews fragments of leaf into a 'bed' of wet mastic at the lower edge of the nest entrance. She clings to the surrounding structures with her legs, while she repeatedly and rapidly draws out the mastic with her mandibles and uses it to secure the outer edges and parts of the disc-shaped leaf sections that form the nest plug.

One of us (TB) conducted an 'accidental' experiment that revealed something of the way *Megachile* females re-find their nests. At a coastal site where *Megachile willughbiella* females were nesting in a sandy bank, the bees were leaving and entering their nests without orientation flights. Keen to get photographic images of a female entering her nest with a leaf cutting, TB was defeated by their speed of entry. The next step was to scuff the entrance of one nest. The returning bee flew to within 3–5 cm of the now concealed entrance and retreated. It repeated this twice, before depositing its load and returning to re-open the nest entrance. Still lacking a usable photo, TB tried imposing a plantain leaf in front of the nest. After two retreats, the bee returned and made its way around the obstruction. It seems that the bee relies on visual clues to navigate its way to the nest entrance but is able to respond effectively to small-scale changes in the immediate vicinity. Following these interventions, the bee performed an orientation flight on next leaving the nest (Benton 2017).

Several species of the genus *Osmia*, the mason bees, are widespread and familiar cavity nesters and readily occupy bee nest boxes in gardens. The most well known of the masons is the Red Mason Bee *Osmia bicornis*. In this species, the cells are lined and capped with mud, which the female collects and carries in her mandibles. The bee constructs the mud walls of the first two or three brood cells, using the two forward-projecting 'horns' on her head, before stocking the first (i.e. the one furthest from the entrance) with a quite solid pollen-nectar mix. A female of this species which nested in TB's garden took a full 23 days to complete her nest in an observation nest box. There were 14 cells in sequence along the cavity, with a vestibule the length of three cells towards the entrance

BOX 1 Leafcutter bees at work

A study of both captive and free-flying *Megachile centuncularis* by A. Raw (1988) found the mean number of foraging trips per cell was 18, and this task occupied some 75 per cent of the daily activity period of the female bees. For bees in captivity, foraging trips averaged over 24 minutes, while collecting leaf segments took only 1 minute 2 seconds per trip. Time spent on nest construction was 2 minutes 12 seconds per visit, and it took an average of 3 minutes 34 seconds after each foraging trip to deposit the provisions (presumably including the time taken to lay an egg on completion of

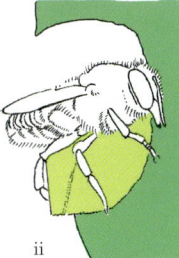
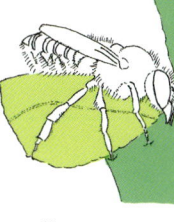
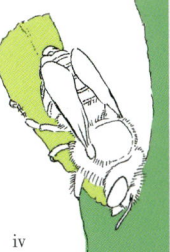

i   ii   iii   iv

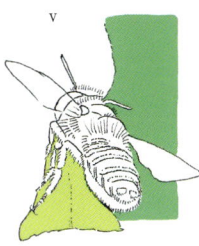

v

the stocking of each cell). More trips were needed to collect leaf sections for each cell (21) than the 18 required to collect provisions for it, though, of course, these trips were much shorter in duration. It seems likely that free-flying bees would have taken longer than the captive bees (at 9 hours 45 minutes) to complete each cell, as they would have had to invest more time in flying and searching. Even so, the average of only one completed cell per day is interestingly the standard across many bee taxa.

FIG 99. (a) Leaves with holes cut by a leafcutter bee. (b) Sequence of leaf cutting by the Silvery Leafcutter Bee *Megachile leachella*. (c) Wood-carving Leafcutter Bee *Megachile ligniseca* female returning to her nest with a leaf cutting. (d) Leafcutter bee nest cap.

hole, and then a very thick wad of mud as a cap to the nest. This feature of a well-armoured cap and an extended vestibule between it and the outermost brood cell is common to many cavity-nesting bees and has presumably evolved as a defence against parasitic wasps with long, piercing ovipositors, such as *Gasteruption jaculator*.

Detailed observations of nest construction and provisioning, using an observation nest (supplied by G. Pilkington, see above), revealed the sequence of events in the *Osmia bicornis* nest in more detail. From the front wall of each completed cell, the bee takes some mud, using its horns, to construct a ring of mud (partition) around the inner surface of the cavity. When filled in, this will form the front wall of the cell under construction. She then goes on a series of (usually nine or ten) foraging trips and returns with pollen loads. To deliver each of these she enters the nest head-first, puts her head through the hole in the partition and pushes the previous pollen load up, sometimes depositing drops of nectar and spreading some of the pollen over it. She then reverses and turns in the cavity (or, sometimes, retreats to the entrance to turn round) so that the tip of her abdomen pokes through the partition. She then uses the fine setae on the inner surface of her hind basitarsi to scrape the current pollen load off into the cell and, following her final provisioning trip for that cell, she backs her abdomen through the partition hole to lay an egg on the mass of provisions. Next, she flies off to gather more mud, returning to plug the hole in the partition to form the front wall of the cell, which is now complete. To form the next cell, she scrapes mud off the outer side of the new partition using the horns on her face and drags it forward to form the ring of mud for the start of the next partition, leaving the hind wall as a smooth inward-facing dome. The process is repeated until a line of completed brood cells extends to 1 cm or 2 cm of the entrance. The nest entrance is then sealed with more mud, leaving an empty vestibule in front of the last brood cell.

Delivery of mud for cell partitions was timed by TB in his urban garden. The time taken for the bee to leave the nest, collect mud and return with its load carried in its mandibles was 1.2–2.5 minutes, and time spent in the nest with each delivery was from 1.5–6 minutes, but usually from 2–3 minutes. Provisioning trips took from 11.5–16.5 minutes, and nine or more trips, taking 2–2.5 hours in all, were required to provision each cell. The full sequence, including mud collection, would have taken significantly longer, and many further mud-collecting trips were required to plug the nest entrance once the full complement of brood cells had been completed. Seidelmann *et al.* (2014) estimate that *Osmia bicornis* females allocate approximately 80 per cent of their time to foraging and 20 per cent to constructing cells and nest plugs.

The Orange-vented Mason Bee *Osmia leaiana* is another familiar species in gardens, with similar nesting habits, though instead of mud to make its cell

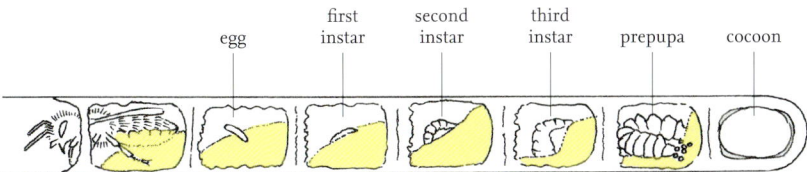

FIG 100. Linear nest of the Red Mason Bee *Osmia bicornis*, showing stages of development.

partitions and nest plug, it uses a mastic formed by chewing up fragments of leaves. Once a cavity is selected, the bee begins by lining a length of it with mastic. Then a series of cells is constructed, the cells being divided from each other by a partition wall, also made of mastic. There are few accounts of the sources of the plant material used by this bee, but several females were observed by TB in his garden cutting leaf fragments from a rose bush. Each bee focused its activities on a small area of the bush, frequently returning to the same leaf on each visit or, if moving on to another, always choosing one close by. As the nesting season advanced, the result was several discreet areas around the bush characterised by leaves damaged by ragged edges and holes, a record of the work of several bees. Video clips of the process reveal that the bee clings to the edge of the leaf with hind legs and mid-legs, rapidly nibbling off parts of the lamina and then chewing and moistening them to form a ball held between head, forelegs and up-curved abdomen. The process takes a minute or more, before the bee flies back to the nest. The full trip, including flight and sometimes repeated 'testing' of leaves before settling on one, took from 1.5–2.5 minutes.

FIG 101. Red Mason Bee *Osmia bicornis* female adding mud to her nest cap.

A video produced by G. Pilkington shows a similar action by the bee on a *Cotoneaster* leaf. The bee works very rapidly, simultaneously 'nibbling' small pieces of leaf, chewing and moistening them with her mandibles. She continues until a ball of masticated material is accumulated. She then flies back to the nest, sometimes settling for a while to continue chewing. Video clips of her activity in the nest show her manipulating the ball of mastic with her forelegs, while she picks off small fragments with her mandibles to complete a cell partition. When the nest is complete, the bee conducts a lengthy operation to provide a strong cap at the entrance. This involves numerous foraging trips for leaf material, which is brought back in the bee's mandibles already partially masticated. Sometimes the bee settles close to the nest and completes the process of mastication. She then flies to the nest entrance, holding on to the surrounding surface with her middle and hind legs, while gripping the ball of mastic with her forelegs. She then repeatedly nibbles pieces of the material with her mandibles and carefully embeds them in the nest cap. Sometimes she may hold on with all six legs and hold the ball of mastic in place by pressing it between her thorax and the nest cap under construction.

FIG 102. Orange-vented Mason Bee *Osmia leaiana* female nibbling leaf fragments.

FIG 103. Orange-vented Mason Bee *Osmia leaiana* female arriving close to her nest with part-chewed material.

FIG 104. Part of a nest of the Orange-vented Mason Bee *Osmia leaiana*, showing brood cells, vestibule and nest cap.

Osmia leaiana females collect pollen in scopae on the ventral surface of the abdomen, and foraging trips timed by TB in the garden averaged 15 minutes. Each cell is provided with a store of pollen and nectar mix, and a single egg is laid on the 'loaf' of provisions.

Perhaps the most remarkable of all the British mason bees is the Red-tailed Mason Bee *Osmia bicolor*. This is one of three British species that make their nests in empty snail shells. The shells of several common land snails (Garden Snail *Cornu aspersum*, Banded Snail *Cepaea nemoralis* and Roman Snail *Helix pomatia*) have spiral cavities into which females of the Red-tailed Mason fit up to four or five cells to make their nests. Then the bee devotes considerable effort to adjusting the position of the shell until the opening faces downwards, either initially or after completion of the nest. Repeated visits with provisions, followed by the laying of an egg, are followed by the completion of a cell by means of a partition made of leaf mastic. Towards the mouth of the shell the bee fills the space with earth, tiny stones and shells and then completes the nest with a plug of chewed leaf mastic. Entomologist Rob Parker, while walking in Kings Forest (in the East Anglian Brecks), noticed a bee flying with a load 'like a witch on a broomstick' – though the piece of dead grass would have been as long as a telegraph pole in proportion. On closer inspection he observed the bee to make repeated trips, bringing back pieces of grass and the previous year's pine needles to make a structure over its snail-shell nest. Eventually 'the construction grew from a trellis to a wigwam, and the snail shell disappeared from view' (Parker 2008). (See Figure 108.)

This behaviour was described as long ago as 1884 by the Victorian entomologist V. R. Perkins (Perkins 1884), and the late David Baldock, in his classic *Bees of Surrey* (Baldock 2008), reported as many as 50 females of this species in an area of 10 m^2 covering their nests with grass stems and beech bud-scales. There is currently a superb video by John Walters on the BWARS website (bwars.com) of a Red-tailed Mason Bee bringing small twigs, grasses and bramble clippings to cover its nest. NWO was able to study the nesting behaviour of this species in great detail at a chalk pit near Cambridge.

Four nests of the Red-tailed Mason Bee *Osmia bicolor* were found within about 2 m of each other in the chalk pit. The nests were made in empty *Cepaea* species

FIG 105. The nesting aggregation was in the area of short grass on the right.

FIG 106. (a) Nest 1: adding mastic to the shell surface and provisioning with pollen (pollen is covering the bee's face). (b) Nest 2: making the final mastic seal, just visible well inside the shell. (c) Nest 2: repositioning the shell with the opening downwards. (d) Nest 2: covering the shell with dry grass stems. During flight, the legs straddle the stem, which is gripped by the mandibles. The stems were manoeuvred into position around the shell. (e) Nest 2: picking up a stem close to the nest. (f) Nest 2: the final covering. It took about 2 hours 30 minutes to completely cover the shell.

FIG 107. (a) Collecting leaf mastic – possibly young elm leaves. (b) After sealing the single brood cell with mastic, small pieces of clay were rapidly collected from bare ground close to the nest, finally filling about 18 mm of the shell's (curved) length. The clay pieces were then sealed in with a second mastic layer, leaving the final part of the shell empty (see also Chapter 8).

snail shells, on a sunny bank with Cowslips, Ground Ivy, Milkwort and shoots of woody plants, as well as patches of straggly Ivy. Several more females were apparently searching for snail shells, all of them flying from place to place and crawling under low vegetation, especially Ivy. The nests were amongst low woody plants with the shells being initially half-hidden. Many males were also present at the site, basking and occasionally dive-bombing females, including onto one female while she carried out an orientation flight as she departed from her nest.

At a third nest, the shell was well covered with stems and out of sight before the final mastic seal was made. The bee continued to fill the shell with particles of dry chalky clay, which were collected and deposited very rapidly, sometimes in seconds. Mastic took longer to collect and the bee also spent much more time depositing it. Having finally sealed the shell, the bee collected more stems to fill the access hole it had used between the stems.

It remains unclear what the function of the nest cover is. It may have to do with temperature control or, alternatively, protection of the shell from predation by birds.

Another well-studied cavity nester is the Wool Carder Bee *Anthidium manicatum*. The females nest in a wide range of cavities, including hollow plant

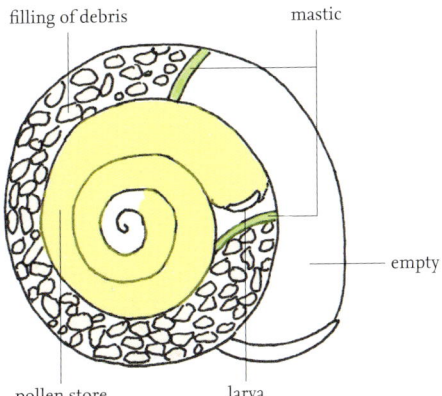

FIG 108. Diagram of completed nest with one brood cell.

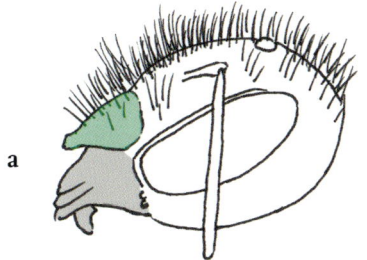
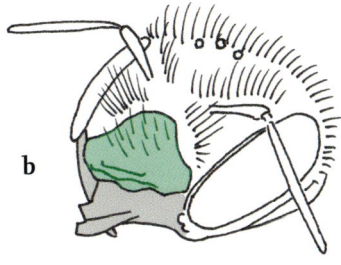

FIG 109. (a) and (b) Red-tailed Mason Bee *Osmia bicolor* heads: the clypeus (green) projects over the mandibles (grey) and has a raised, polished edge. (c) A Red-tailed Mason Bee gripping a stem with its mandibles in preparation for flight. The clypeus appears to be pressed down onto the stem to stabilise it as it is pulled up from the ground and then in flight. The clypeus is not raised in the Gold-fringed Mason Bee *Osmia aurulenta*, which does not carry stems, so perhaps this is a particular adaptation of the Red-tailed Mason Bee.

stems, exit holes of large wood-boring beetles, crevices in old walls, and even metal chair arms (Severinghaus *et al.* 1981). They can also, if rarely, be tempted to occupy garden bee hotels. Having established a nest site, the female bee seeks out plants with hirsute stems or leaves, to provide the lining for its cells. Mullein, Yarrow and plants in the dead-nettle family (Lamiaceae), such as woundworts, are among the wild plants reportedly used, but in gardens, stands of Lamb's Ear very frequently attract a patrolling male of this species. As we noted in Chapter 2, to avoid harassment by the males, females seeking plant hairs for their nests fly low into the patch and settle on the underside of a basal leaf. Alternatively, females may collect plant hairs from a patch of the plant with no patrolling male.

Sometimes the bee will 'test' several leaves before beginning to clip, but more usually they start to clip the plant hairs almost immediately. Tufts of hairs are separated from the leaf by a rapid up and down movement of the head, with

FIG 110. (a), (b) and (c) Wool Carder Bee *Anthidium manicatum* female cutting plant hairs from a stem of Lamb's Ear *Stachys byzantina*.

mandibles spread out in front. The bee clasps the rest of the leaf with mid- and hind legs, while manipulating the cut hairs with the forelegs. Typically, the bee clips a series of swathes of hairs, working backwards from the tip of the leaf, accumulating the resulting load in the cage formed by her mid- and hind legs and up-curved abdomen. This may take from 20–75 seconds and is followed by a second phase in which the bee rears back from the leaf, holding on by her hind legs only, and uses her forelegs and middle legs together with her open mandibles to form the hairs into a symmetrical ball that can readily be carried back to the nest. This second phase takes from 4–7 seconds approximately. Occasionally the bees take hairs from the plant stem, but the process is basically the same. It seems likely the whole process would take much longer on plants less luxuriously hairy than *Stachys*.

Yet another set of nest-building skills within the family Megachilidae is exemplified by the Large-headed Resin Bee *Heriades truncorum*. Until a few years ago, this species was very localised in southeast England, but it is now spreading rapidly and is a familiar occupant of bee hotels. It nests most readily in cavities of 2–2.5 mm in diameter but will use ones of 5 mm or more in diameter, there being very marked size differences among individuals (Early 2020). The cells are approximately 10 mm long and aligned in a series along the cavity, with a vestibule of some 2 cm before the nest cap. Instead of mud or leaf mastic, the cell partitions are composed of resin. The females stock each cell in turn with pollen collected on their abdominal pollen brush, together with regurgitated nectar, and lay one egg in each cell before completing it with a resin plug. As with many other cavity nesters, much work goes into making a secure cap to the nest. The bees first bring malleable pieces of resin, which they laboriously spread across the

FIG 111. Large-headed Resin Bee *Heriades truncorum* (a) sealed nests and (b) female returning to her nest with a pollen load.

entrance to the nest, using their mandibles only. This takes several return flights, after which they bring back tiny pebbles and other material which they embed in the resin. This gives the nest a quite distinctive external appearance – rather like the cases of some caddis larvae.

Having failed to find the sources for the nesting materials used by the numerous *Heriades truncorum* that nested in TB's garden in previous years, it was hoped that the restrictions of Covid-19 year 2020 might make for more successful observations. On 14 July, while sitting in the garden, TB noticed two *Heriades* moving about on the crazy-paving path. On closer inspection, it could be seen that they were rapidly moving over the slabs and along the edges of the path, eventually finding a tiny stone and flying off with it, grasped between the mandibles. The time taken from leaving the nest to returning with a selected

FIG 112. Large-headed Resin Bee *Heriades truncorum* female (a) collecting resin to cap her nest, (b) collecting a tiny stone to complete her nest and (c) completing her nest cap.

stone was from 30 seconds to one minute. Again, the source of the resin the bees collected eluded discovery. However, it must have been close by, as the mean time for resin-collecting trips was only 30 seconds. Eventually TB 'cheated' and placed a cone of Cedar of Lebanon *Cedrus libani* close to the nest box. Immediately this was approached by a female *Heriades*, with a sideways zigzag flight. Subsequently females came to take small pieces of resin from the cone, cutting and pulling with their mandibles until a rough ball of resin was collected under the head and between the mandibles to be carried back to the nest. A study in France found that provisioning of each cell required from 30–50 foraging trips and took from 5–7 hours. Taking into account the time required to collect resin and construct the cell, the author concluded that a female would not complete a cell within a single day (Maciel de Almeida Correia 1981).

Even in artificial nest boxes, where there are many suitable cavities, there is considerable apparently competitive interference between nesting females. In July 2020, conflict between two females was filmed (TB). On this occasion, the pair assumed a '69' formation, apparently grappling one another's legs with their mandibles. This lasted for several seconds, after which one took flight but returned in one second to resume the tussle for a further second before both flew off. One returned some six seconds later.

This conflict could have been a consequence of the closeness of the holes in the bee hotel, but another incident would be hard to interpret other than as either a failed attempt at nest usurpation or as a case of intra-specific cuckoo behaviour. A female with a full pollen load was filmed facing into her nest entrance, with most of her abdomen exposed and rapidly 'pumping'. After six seconds, she slowly backed out of the nest, pulling another female (presumably a would-be usurper) out of the nest, holding its head with her mandibles. The video shows the second bee's head emerging from the nest entrance, whereupon the first bee releases her and the two face each other, mandibles spread wide, in an 18-second 'stand-off'. During this time, the would-be usurper makes small movements with one front leg several times and then retreats back into the nest. She is followed by the first bee, and both are inside the nest for a further 10 seconds, after which the first bee reappears, hind-end first, again dragging out the intruder. However, this time the latter is released immediately and it leaves the nest. They face one another briefly on the surface of the nest box, and then the first bee re-enters its nest, and the intruder moves off to investigate other holes, and preen.

In addition to intra-specific attempts at usurpation, or 'cuckooing', *Heriades* females were observed re-occupying the nests of another species, the Orange-vented Mason Bee *Osmia leaiana* which also nests in the same group of nest boxes. Several *Osmia leaiana* nests from the previous year had presumably failed, as the nest cap remained unbroken (though it may also be that they were destined to emerge after a second year – see Boyes 2020). Despite the availability of numerous unused and apparently suitable cavities, females of *Heriades* cut holes in the *Osmia* nest caps and proceeded to drag out the whole of the residual nesting material before making their own nests in the (unusually large) cavities which they thus opened up. Another instance of cross-species usurpation came from an opened nest of this species in a length of cane. There were 11 cells containing fully developed larvae of *Heriades*, but behind these were two cells of the Common Yellow-face Bee *Hylaeus communis* (Fig. 113) with dead adult bees, presumably trapped by the *Heriades* cells constructed in front of them, preventing their escape. Nest usurpation seems to be particularly frequent in some species, notably the North American mason bee, *Osmia lignaria* (Terpedino & Torchio 1994).

The family Colletidae includes just two genera in Britain. We have already commented on the nesting activity of *Colletes* bees, which are mainly ground-nesting. The other genus, *Hylaeus*, the yellow-face bees, includes mainly cavity-nesting species. The species are rather similar to one another, small and black, usually with some yellow markings on legs and face. In all British species except one, the males have extensive yellow markings on the face, whilst the females usually have smaller, paired yellow patches. G. Pilkington has produced remarkable video clips of the nest-making activities of one of the more common species, *Hylaeus communis* (http://nurturing-nature.co.uk). Once a cavity has been adopted, the bee constructs a series of 'portals' along its length. These are relatively sturdy structures, formed as partial walls but with a hole in the middle, through which the bee can pass. The bee constructs these from drops of a Dufour's gland secretion from her rear end. These are drawn out to form a lattice of strands and then coated with a salivary secretion pasted on with her brush-like tongue. The portals are used as stores for the Dufour's gland secretion and drawn on by the bee to make the cellophane-like cell lining.

The sequence of events in the formation of a new cell are as follows. The bee occupies the space that will eventually form the cell, with her rear end close to the cell wall of the last-completed cell. Her head protrudes through the portal that will become the wall of the cell under construction. She thoroughly brushes a salivary secretion onto the sides of the cavity beyond the one she occupies, while dabbing the tip of her abdomen against the rim of the cell under construction. This appears to be depositing a Dufour's gland secretion. Next, she bends double to reverse her position in the cavity so that she now faces into the part of the cavity that will form the next cell. As she does this, her face passes the tip of her

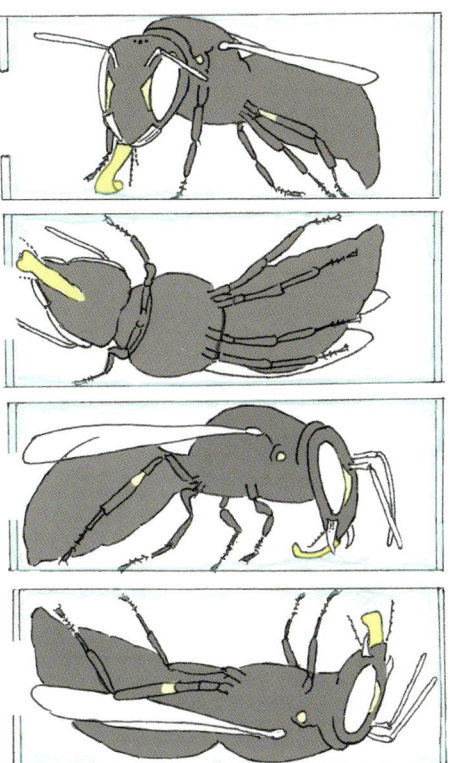

FIG 113. Common Yellow-face Bee *Hylaeus communis* female secreting and spreading the cellophane-like lining to a brood cell (drawings based on video by G. Pilkington).

abdomen so it is possible that more Dufour's gland secretion might have been passed to her mouthparts. She next proceeds to use her tongue to thoroughly brush secretions onto the inner surface of the cavity, including the cell wall of the previously completed cell.

Hylaeus bees do not have any external apparatus for carrying provisions back to the nest, but instead they ingest pollen and nectar and carry it internally. They repeatedly blow what looks like a bubble of the mixture to allow evaporation to condense the fluid, and then suck it back into the gut. However, the mixture is still much more fluid than the stock of provisions left by other solitary bees. When the 'right' concentration has been reached, the female bee returns to the nest and begins stocking the cell just made. After several deliveries, an egg is laid on the surface of the mixture, and the bee completes the cell by closing up the hole in the partition, using a combination of salivary and Dufour's gland secretions.

Another member of this genus is remarkable in its choice of nesting cavities. This is the Reed Yellow-face Bee *Hylaeus pectoralis* (Fig. 116) which nests in evacuated galls caused by the fly *Lipara lucens* and is one of the few British bee species associated with wetland habitat. The fly lays its eggs in the flower stems of the Common Reed *Phragmites australis*. The result is the formation of a cigar-shaped gall at the apex of the stem, and the development of the flower stem is inhibited. The larva of the fly consumes plant tissue in a cavity within the gall and overwinters as a pupa. The adult fly emerges in early summer, leaving a cavity available for occupation by a female *Hylaeus pectoralis*. She enters the gall through the exit hole made by the fly and constructs brood cells with cellophane-like linings, using glandular secretions brushed on with her bilobed tongue (glossa). There are no published accounts of the constituents of the cell lining in this species, but it is likely that these are common to the genus *Hylaeus*. Depending on the size of the gall, there may be from 1–8 cells, and the nest is capped by a plug of leaf fragments at the apex (Else 1995).

In 2020 and 2021, TB carried out a study of this species at a coastal site in north Essex (Benton 2021). Galls containing larvae or pupae of *Lipara lucens* are neatly pointed at the tip, while those containing the immature stages of *Hylaeus pectoralis* have small holes at the top (the exit holes of the previous occupant), usually obscured by a rough tuft of dry leaf-tips. During the late summer and autumn of 2020, TB collected 12 presumed *Lipara*- and 12 presumed *Hylaeus*-occupied galls. Samples of each were cut open (vertically) to determine the internal nest architecture and developmental stage of the occupants. Then, all 12 of each species were stored in separate containers in a shed for the winter and spring. After emergence in June 2021, the resulting adults were returned to the source site, and the remaining vacated galls were dissected. Meanwhile, reed stems with galls left *in situ* were tagged with coloured tape so that the behaviour of the bees could be observed in their natural habitat during the summer of 2021.

TB opened eight of the *Hylaeus*-occupied galls. Of three that were collected on 13 August, two contained fully developed larvae (prepupae), but one contained two cells with eggs or early-stage larvae amid the fluid mix of nectar and pollen characteristic of *Hylaeus*. Both of these went on to complete their metamorphosis and emerged in 2021. Each nest contained a number of brood cells, each one with the characteristic cellophane-like lining, arranged in linear order, either attached to one another or separated by gaps loosely stuffed with clippings from reed leaves. Above the line of cells, towards the apex of the gall, was usually a vestibule, also with leaf clippings, leading to the exit hole made by the fly. Taking the galls opened while still occupied, together with those dissected after the emergence of the bees, the number of cells per gall ranged from 2–7, with 3–4 being the most frequent, and the mean 3.75. One nest with six cells (Fig. 115b) contained two cells closer to the apex than the others, separated by clippings, and significantly smaller than

FIG 114. (a) Gall of the fly *Lipara lucens* and (b) the gall now occupied by larvae of the Reed Yellow-face Bee *Hylaeus pectoralis*.

the others. TB concluded these were distinct male and female cells, the smaller (male) cells being 5 mm long, the larger ones (female) averaging 7–8 mm long.

There was no evidence of parasitic activity in any of the *Hylaeus*-occupied galls. During the spring, the opened nests were regularly inspected, and the first pupa was seen on 1 June 2021. The male emergence began on 10 June and continued until 20 June. The females emerged between 15 and 22 June. This suggests the species is slightly protandrous, with two-thirds of the males emerging before the first female, and male emergence being more drawn out than that of the females. In all, 12 males and 12 females emerged, from what TB estimated to have been a total of 44 cells. Care had been taken to keep the cell linings intact when the nests were opened, and there was no evidence that opening the nests had affected survival rates.

Owing to the density of summer growth, the tagging of galls *in situ* proved to be of little help in finding active nests in 2021. However, nests were discovered in two localities, and bees were observed bringing materials into their nests, and one was filmed plugging the entrance to her nest. She appeared to dab the

FIG 115. Partially opened nests of Reed Yellow-face Bee *Hylaeus pectoralis* showing (a) two brood cells with full food-stores and (b) four brood cells with female prepupae (on the right) and two with (presumed) male prepupae on the left (the nest exit is on the left).

FIG 116. Reed Yellow-face Bee *Hylaeus pectoralis* female completing her nest.

entrance with her abdomen from time to time, presumably depositing a secretion of Dufour's gland, and then pasting the surrounding dried plant tissues with her brush-like labium. This was observed for 1 hour 15 minutes, by which time the nest entrance was filled with a white film. Subsequently, TB collected the nest. It contained six cells, and there was no sign of the white nest plug. None of the previously collected nests had this feature, so it seems to be a relatively short-lived addition, unlike the material used for the cell linings.

The galls occupied by *Lipara lucens* generally enclosed a single larva within an oval cavity. When collected (in autumn 2020), some already contained puparia, and this was the stage in which the occupants spent the winter. The adult flies began emergence on 1 June, though they were preceded by a few days by the emergence of parasitic wasps *Palemochartis liparae* from some galls. Several other inquilines were also found in the *Lipara*-occupied galls, though all would have been vacated by the time of the first emergence of the *Hylaeus*. If these emergence times are consistent with those in the natural habitat, then the *Lipara* galls would be available for occupation by *Hylaeus* in the same year as the flies emerge. The apparent dependence of this species of *Hylaeus* on the galls of *Lipara lucens* poses interesting questions as to the relationship between the populations of the two species and their parasites (especially those of *Lipara*).

A large-scale study of the occupants of *Lipara* galls in reedbeds in the Czech Republic discovered that the galls are used by an astonishing 21 hymenopteran species, including nine solitary bees (seven nest-makers and two cuckoos) (Bogusch *et al.* 2015). One specimen of *Chelostoma campanularum* was reared from a gall, as well as several specimens of two other *Hylaeus* species, three *Heriades rubicola*, and the cuckoos *Stelis breviuscula* and *Stelis ornatula*, though *Hylaeus pectoralis* nests were found relatively frequently. The nest architecture described in this study is consistent with the much smaller number of nests examined on the Essex coast, but the Czech study raises the possibility that other British solitary bees may also use reed galls. Three other species of yellow-face bees (the Common Yellow-face Bee *Hylaeus communis*, the Hairy Yellow-face Bee *Hylaeus hyalinatus*, and the Chalk Yellow-face Bee *Hylaeus dilatatus*) fly together with *Hylaeus pectoralis* at the same site, raising the question whether they too use reed galls, or possibly reed stems.

Wood-borers and pith excavators

The third category in Danforth *et al.*'s classification of nest-building methods (excavation of cavities by the bees) has just a few British representatives – but they are of considerable interest. The best-known carpenter is the Violet Carpenter Bee *Xylocopa violacea*. This bee is not yet regarded as resident in Britain, but its range is moving northwards in Europe and, as mentioned above, it is recorded quite frequently here. The females have powerful jaws which enable them to dig cavities for their nests in sound wood as well as rotten. Their cell partitions are composed of wood fragments.

The Wood-carving Leafcutter Bee *Megachile ligniseca* has some claim to be regarded as a wood-borer. Females of this species can modify a cavity by using their mandibles to cut off shavings from its inner walls, leaving fragments of wood below the nest (TB, G. Pilkington pers. comm.). However, it seems unlikely that this bee is enough of a carpenter to make its nest cavity from scratch. As mentioned above, one British species which is known to cut its nest burrows in dead wood is the Fork-tailed Flower Bee *Anthophora furcata*. Dawson (2018) describes a female of this species making a nest-tunnel over two days, in a cut branch of *Rhododendron*. She alternated between antennating and chewing the wood and clearing away the cuttings with her hind legs. NWO reports this species nesting in a rotting stump of a Rowan *Sorbus aucuparia*. The wood removed was very soft, and the operation left an 'avalanche' of rotten wood below the nest(s).

Recent research on the scarce and declining Fringe-horned Mason Bee *Osmia pilicornis* has shown it to be a clear example of a wood-borer. This small bee inhabits semi-open, usually deciduous woodland on nutrient-rich soils. On the European mainland, its main pollen source is flowers of the genus *Pulmonaria*, to which its mouthparts seem to be closely adapted (see Chapter 6). However, in Britain it collects pollen mainly from Bugle *Ajuga reptans* and also Ground Ivy *Glechoma hederacea*. It is now known that this bee makes its nest in burrows which it cuts in dead wood – usually small fallen branches. Typically, these are in sunny open clearings or the edges of rides but are covered by vegetation. The burrows are 5–6.4 mm wide and 1.7–3.2 cm long. From the entrance, leading into the nest, there is usually (but not always) a vestibule, beyond which the burrow changes direction so that the part containing brood cells runs longitudinally through the wood. The brood cells are few in number – just 1–3 in each nest, arranged in linear order. The bee excavates its burrow using its chisel-shaped mandibles, flying off with wood clippings at intervals to deposit them away from the nest. In their study of the bee in southern Germany, Prosi *et al.* (2016) found that the bee may use the wood of several tree species, whether hard or soft. They report one female as cutting her burrow in the hard, dead wood of Ash *Fraxinus excelsior* from the surface to a depth sufficient to conceal her head and thorax in just 10 minutes.

The brood cells are separated by partitions made of leaf mastic, and the mouth of the nest is sealed by a thick, multi-layered plug of the same material. Prosi *et al.* give Wild Strawberry *Fragaria vesca* as its source for this material in mainland Europe, but in Britain the bee has also been observed collecting leaf

FIG 117. Fringe-horned Mason Bee *Osmia pilicornis* (a) female and (b) nest entrances. (Courtesy of R. Earwaker)

fragments from Tormentil *Potentilla erecta* (R. Earwaker pers. comm.). The bees nibble small pieces of the leaf with their mandibles and fly back to the nest, processing it to form a mastic at the nest. In the study by Prosi *et al.*, the mean time taken to collect the leaf fragments was 27 seconds, and the time taken to process it rather longer, at a mean of 75 seconds.

The Little Carpenter Bee *Ceratina cyanea* is the only British representative of its genus, and its carpentry seems to be limited to excavating pith from plant stems. The females make their nests in dead stems of bramble or rose, or occasionally other plants. They use cut stems, usually ones lying close to the ground and in sunlit situations. They use their mandibles to cut a burrow through the pith in the stem, and their nests can be detected by the 'mounds of fresh, fine pith fragments' directly below the cut ends of a stem (Else 1995). There is a constriction in the width of the burrow close to the entrance, and farther along is a series of brood cells when the nest is complete. Each cell is 7–9 mm long and 3 mm wide, contains a loaf of provisions and an egg, and the cells are separated from one another by partitions made of fragments of pith. Sometimes the outer cell has no partition, and the nest may be guarded by the overwintering female. The life cycle of this bee is quite distinctive, and we will return to it below.

LARVAL DEVELOPMENT

The larvae of the nest-making solitary bees are little more than feeding machines, and they pass through their growth stages, feeding on the provisions left in their brood cell. They develop rapidly, usually completing their development (and exhausting their food supply) within a few weeks. In most species, the larva stays attached to the surface of the food mass, but in the plasterer bees, Colletidae, which have their nutrients in a fluid state, the larva floats sideways and 'swims' through its food.

The head of the larva has a pair of small antennae and minimal mouthparts – mandibles, maxillae and an elementary labium. There are two pairs of respiratory spiracles on the rear segments, and some species have projecting tubercles that assist in movement within the cell. The body is soft, flexible and otherwise without definite external structures. Growth is achieved by repeated shedding of the larval skin (cuticle) as the food mass is consumed. The final larval stage (prepupa) has a more robust cuticle, reported to be both water-resistant and proof against microbial attack (Danforth *et al.* 2019), and this is the stage used by many species for passing the winter months.

The nest of *Hylaeus pectoralis* mentioned above, which was collected on 13 August and opened on 14 August, contained two cells approximately three-quarters full of the fluid pollen-nectar mix, suggesting that the occupants were either at the egg or very early larval stage (see Figure 115 a, above). By 21 August, the larvae in the cells were visible, one of them almost full-grown, with the food mass almost consumed, and the other with the food mass greatly reduced. When checked on 26 November, both of the larvae had reached the prepupal stage. The flight period of this species is from early June to late September, and it is certainly on the wing by mid-June at this site. Clearly, eggs laid early in that long flight period develop to the prepupal stage, and then go into an early resting stage, which continues through the winter. Eggs laid much later (presumably early to mid-August or later) are still able to reach the prepupal stage before winter (see Chapter 9).

The life cycle is managed differently among four of the species of *Colletes* that occur in Britain. These have either very early or late flight periods. As mentioned above, one of these, *Colletes cunicularius*, flies in early spring. At an Essex sandpit site, 13 cells were excavated on 28 November 2019. The cells had opaque brown linings, were quite fragile, and had a strong smell. The bees inside were fully adult, and occasionally buzzed when the cells were touched. Four of them became active during the excavation and escaped their cells. These were subdued by being placed in a fridge, then transferred to narrow glass tubes and left in a cool, dark place. They appeared to settle down and emerged early in 2020 (TB, M. Fremlin pers. obs.). Overwintering as adults enables early emergence, coinciding with the flowering of willow (*Salix* species), the main pollen source.

By contrast, the three late-flying *Colletes* species (*Colletes succinctus*, *Colletes hederae* and *Colletes halophilus*) are pollen specialists on late-flowering plants: Ling, Ivy and Sea Aster, respectively. This association gives little opportunity for

FIG 118. Brood cells of Early Colletes *Colletes cunicularius* containing fully adult bees.

larvae to reach the prepupa stage before winter. Brood cells of *Colletes hederae* were excavated on 11 May 2017. They were mid-brown and contained larvae in various stages of development, some with the yellowish fluid food mass still present. Observation of collected cells, and others left *in situ*, established that the usual season of pupation was the end of July, with the first adults flying by 28 August. Later that year (2 October), M. Fremlin excavated newly constructed nests, finding one with an egg attached to the surface of the food mass and visible through the transparent cell lining. Other cells contained early instar larvae (Benton & Fremlin 2018). These results suggest that this species is very unusual in overwintering in the vulnerable larval stage. It has been argued that the distinctive cell lining of colletid bees, which offers protection from microbial attack, inundation and desiccation, is a preadaptation, which allows these bees to use late-flowering pollen sources while their larvae are protected from mould forming on their food-store and from other hazards during their prolonged period as immature larvae (Müller & Weibel 2020) during the winter months.

A study of the life cycle of the Sea Aster Bee *Colletes halophilus*, another late-flying species, found only young or later-stage larvae from November to April. By May, only 5 per cent of the larvae had reached the prepupal stage (Sommeijer *et al.* 2012), a similar sequence to that observed in *Colletes hederae*. However, a small number of *Colletes halophilus* cells collected by NWO in Norfolk during December were all in the prepupal stage. It seems likely that early-stage diapause in these species is available, but not obligatory. Other *Colletes* species that forage from earlier-flowering pollen hosts probably overwinter as prepupae (e.g. *Colletes daviesanus*; Esser 2004).

In some solitary bees, the full-grown larva spins a cocoon for its protection using secretions from salivary glands which open onto the labium. All species in the family Megachilidae in Britain spin cocoons, and so do some Melittidae, such as *Macropis europaea*. As mentioned above, the nutritional requirements for cocoon construction are very large, so presumably the practice must offer considerable survival benefits for the species that retain this in their life cycle. The larvae of *Heriades truncorum* are particularly well protected as their cells usually have a leathery lining, two layers thick, in addition to the resin cell partitions and nest plug. Nests opened up by TB on 16 March, 24 May, 14 October and 28 November 2020 provided some insights into the life cycle of this species, as both spring and autumn nests contained prepupae in cocoons, confirming that the species overwinters in that stage. The nest opened on 24 May contained a pupa, whose pale colouring suggested it was very recently formed. The first males were noticed on the wing just two days later, suggesting that the pupal stage may be quite brief.

The nest of *Osmia bicornis* in TB's garden, mentioned above, was completed on 23 May. On 8 June, the larvae in the rear-most cells were almost full-grown, having consumed almost all of their provisions, whilst on the outer (most recently constructed) cells, the full food mass appeared to be present, with eggs or larvae not visible. As the first cells had been constructed on about 30 April, it seems that the time taken from egg to completion of consumption of the provisions in each cell would have been between five and six weeks. On 12 September, all cells contained cocoons. A nest of 11 cells was completed the following year, under better weather conditions, in 12 days. The first cell was completed on 4 May, and by 4 June, the first larvae were, again, almost fully grown, with little or no remaining food. The development of the larvae from deposition of the egg in this case seems to have taken between four and five weeks.

FIG 119. Cell of the Large-headed Resin Bee *Heriades truncorum* (a) with prepupa exposed and (b) with pupa.

CHAPTER 4

From Solitary to Social and Back

In Chapter 3, we considered the lives of the nest-making species whose females take on the whole burden of nest-making, foraging and provisioning their nests as 'solitary' individuals. Here, we stretch the scope indicated in our book title to recognise that some of the 'other' bees (i.e. other than bumblebees and honeybees) also display various modes of social behaviour. At the most elementary, some species, most notably some of the mining bees (*Andrena*) and plasterer bees (*Colletes*), make their nests in large, dense aggregations. It seems likely that this is more than just a response to limited available and suitable habitat but involves some form of attraction to nesting in close proximity to other nests (Friedel *et al.* 2017). This might have evolved as a result of some protection against parasites or predators being given, for example.

Another feature, present in the nesting behaviour of a few species, involves the sharing of nest entrances by several (or, in some cases, very many) females of the same species. However, this seems to be the limit of the cooperation, and each female has its own distinct 'compartment'. This very limited mode of social existence is termed 'communalism' and is found in the closely related group of mining bees, the Chocolate Mining Bee *Andrena scotica*, the Big-headed Mining Bee *Andrena bucephala* (Fig. 120) and the very rare Oak Mining Bee *Andrena ferox* (all belonging to the subgenus *Hoplandrena*). Females of more than one species may share a common entrance, as, for example, a nest entrance shared by *Andrena bucephala* and *Andrena scotica* at Cambridge University Botanic Garden (NWO) and a similar association near Oxford, reported by O'Toole and Raw (1999). Worldwide, several approaches to social behaviour have been described in the family Apidae and, especially, in the Halictidae. These are intermediate between the solitary way of life and the highly complex social systems of the honeybees and stingless bees (Meliponini). Some of these intermediate forms can be understood as phases in the development of more fully social modes of life, but in other cases are seen as distinct alternative modes of organisation. O'Toole and Raw distinguish 'quasisocial' and 'semisocial' modes, in which females cohabit and may show some degree of division of labour between nest-making, egg-laying and foraging, from 'subsocial' forms, in which a nest-founding female remains in the nest, giving parental care to the offspring until their adult stage. This is common in bees of the genus *Ceratina*, and in the one British species, the Little Carpenter Bee *Ceratina cyanea* (Fig. 121), the foundress female may remain in the nest after the offspring have reached adulthood, sometimes living for up to 18 months (Else 1995).

FIG 120. Big-headed Mining Bee *Andrena bucephala* female approaching shared nest entrance.

FIG 121. Little Carpenter Bee *Ceratina cyanea* female.

However, none of these arrangements is regarded as fully social – or 'eusocial'. This term applies to the mode of life of species in which females of two or more generations inhabit the same nest, and in which there is a division of labour between an egg-laying female (usually the nest foundress) and a caste of workers who carry out various functions but do not lay eggs. The most commonly recognised eusocial bees are the (non-parasitic) bumblebees and the honeybees (the corbiculate bees), but one group of so-called 'solitary' bees includes species which are eusocial as well as ones which are solitary, and others which can be either solitary or social, depending on geography or climate. This group comprises the Halictidae ('sweat' or 'furrow' bees) and, so far as Britain is concerned, it includes just two nest-making genera: *Lasioglossum* and *Halictus*.

The diversity and relative fluidity of social life among these bees suggests that their sociality is, in evolutionary terms, a relatively recent acquisition. This contrasts with the more long-established and fixed character of the social behaviour of the corbiculate bees and social wasps. It is this relatively recent and labile character of halictid sociality that gives them their special interest as research subjects. The shift to forms of social life is regarded as one of the great evolutionary transitions, and it poses important challenges to evolutionary theory, such as how to explain some individuals in a community sacrificing their own chances of reproduction to provide services to the offspring of another: apparently 'altruistic' behaviour. More specifically, studying the conditions under which these bees adopt or retreat from social modes of life might shed light on the earlier origins of sociality in the more generally recognised social species. In addition, the diversity of patterns of social life among the halictids, and their fluidity in some species, provides insights into the further evolution of sociality once it has been acquired in some lineages.

A widely shared pattern among eusocial halictids (and most bumblebees) consists of an annual life cycle, in which a fertilised female (gyne) overwinters and makes her nest the following spring. She forages to provision a number of brood cells and lays an egg in each, but remains in the nest following the subsequent emergence of her first brood of offspring. These are all or mostly females, and they forage to provision further eggs laid by the foundress female. These female offspring are, then, a worker caste, deferring to the foundress as 'queen'. This may persist for the current brood only, or there may be one or more further broods of workers, before a brood of sexual adults appears: males and fertile females (potential queens). Mated females subsequently overwinter underground, often in the nest from which they originally emerged. The cycle is repeated the following year. This can be thought of as a two-phase cycle: an initial solitary phase, from nest construction to the emergence of the first workers, followed by a social phase, in which workers contribute to the nurturing of the queen's offspring until the new reproductive adults complete the cycle. As we shall see, there are numerous variations on this theme. The number of broods between the emergence of the first brood of workers and the shift to production of sexuals is one such variation, as is the number of workers reared in each brood. In some species, the separation of solitary and social phases is not so clear-cut, with some of the first-brood females mating and going into their winter resting stage without serving as workers, while others become workers, helping to raise the offspring of the foundress queen. The division between queens and workers is also highly variable, with significant size differences in some species but not in others, and there are differences between species in the mechanisms by which the switch from production of workers to production of the sexual generation occurs and the means by which queens maintain their dominance.

Most of the research on halictid sociality has been conducted in Europe and North America and has focused on a small number of species. Fortunately, these include several species that also occur in Britain: the Orange-legged Furrow Bee *Halictus rubicundus* (Fig. 122) and the Common, Bloomed, Lobe-spurred and Sharp-collared Furrow Bees (*Lasioglossum calceatum*, *Lasioglossum albipes*, *Lasioglossum pauxillum* and *Lasioglossum malachurum*). Of these, the first three are regarded as relatively flexible in their social behaviour, retaining an ability to revert to solitary lifestyles, while for *Lasioglossum malachurum* and *Lasioglossum pauxillum*, the social mode of life is obligatory, though still variable.

Of the species that have been thoroughly studied, three (*Lasioglossum calceatum*, *Lasioglossum albipes* and *Halictus rubicundus*) can be either eusocial or solitary. In Europe, more northerly populations, and those at high altitudes, tend to be solitary in habit, while those in warmer, generally southern and lowland localities are social. In northern, solitary populations, mated females emerge from their winter resting places in late spring, establish a nest and rear a brood of adults, including both males and females in roughly balanced ratio. These go on to mate, the males dying off and the females going into diapause until the following spring. In the more southerly, eusocial populations, females emerge earlier, establish a nest and produce a brood of mainly female adults. The foundress female continues to lay eggs in the same nest, while the daughters (workers) forage to stock the subsequent brood cells containing the queen's eggs and larvae. These subsequently emerge as fertile females and males. These mate, the fertilised females overwintering, while the males die off before winter.

There remains considerable controversy as to whether the two alternative 'behavioural phenotypes' represent an innate plasticity, such that environmental conditions provide the trigger for either pathway, or whether there is an underlying genetic difference between the populations of the same species expressing eusocial or solitary lifestyles. Field *et al.* (2010) studied populations of the Orange-legged Furrow Bee *Halictus rubicundus* at different localities in the British Isles and succeeded in transplanting mated female bees from a social (southerly) population to a more northerly one, where the local bees were solitary. The newcomers directly adopted the solitary mode of life. The results of the reverse transplant were less clear, though almost half of solitary-sourced populations did adopt the eusocial mode. Key evidence for this was that reproduction was strongly 'skewed' in favour of a single female (queen) in each nest, implying that the first-brood females had sacrificed their own direct reproductive interests in favour of the queen's offspring. In some nests, the foundress died before the emergence of first-brood adults, in which case one of the first-brood females replaced her. These results show that both phenotypically solitary and social populations retain the capacity to adopt the alternative mode, given relevant environmental triggers, though it still remains uncertain whether some genetic differences are involved in the ability to adopt the social mode of life.

HALICTUS RUBICUNDUS: A 'FACULTATIVELY' EUSOCIAL BEE

Soro *et al.* (2010) used a combination of genetic methods to investigate the relatedness of separate populations of *Halictus rubicundus* across a range of sites in the British Isles. Their results showed that the degree of genetic differentiation between the populations was a function of geography: differentiation correlated with geographical distance, but with evidence that the Irish Sea constituted a significant barrier to gene flow. Genetic differentiation did not correlate with the social–solitary division, and there was continuing gene flow between solitary and social populations.

This is consistent with a purely environmental trigger determining the two life cycle options but is in sharp contrast with findings reported in Soucy and Danforth (2002) concerning North American populations of the same species. Their study used mitochondrial DNA and markers of genetic relatedness and revealed that social and solitary populations of *Halictus rubicundus* belonged to distinct lineages, with barriers to gene flow between them. This suggests a genetic underpinning of the social–solitary transition in North American populations. This difference between North American and British (European) populations of the same species might be explained in terms of the different histories of the spread of populations of the species into Europe and North America, together with features of the British populations (notably, high levels of gene flow between them) that have so far resisted local adaptation to one or other of the two life cycles. In the case of another 'socially polymorphic' species, *Lasioglossum albipes*, a laboratory study has shown that bees sourced from solitary and social populations retained their original mode of life even when subjected to reversed environmental conditions (Plateau-Quénu *et al.* 2000), suggesting a genetic basis to the differentiation between social and solitary modes of life in that species.

FIG 122. Orange-legged Furrow Bee *Halictus rubicundus* female.

A six-year intensive observational study of a population of *Halictus rubicundus* in North America (Yanega 1988, 1989, 1993) yielded still more complexities. In this population, both social and solitary cycles coexisted, even in the same nest. Typically, the earlier (and usually smaller) females in the first brood became workers, but later-emerging females tended to leave the nest within a few days and, after mating, enter into diapause without engaging in any foraging for the maternal nest. In this population, on average 25 per cent of first-brood offspring were males. Meanwhile, the initial group of workers continued to rear a second brood, this time of fertile females and males, emerging later in the year. Yanega was able to show that approximately half of the females establishing nests in the following spring were ones from the first maternal brood that had overwintered without serving as workers in the previous year's maternal nest. Although these tended to be smaller than the products of the social phase, the two groups overlapped in size. Yanega hypothesised that the key trigger that induced some first-brood females to opt for the solitary route was mating. More generally, some environmental trigger correlated with the extent of future suitable conditions for foraging (such as day length) may be involved in the transition to the social mode of life, given the more extended flight period required for rearing a worker brood and then a subsequent brood of sexuals. Since overwintered females in 'solitary' locations emerge from diapause later than those in 'social' locations, it could be that climatic conditions in early spring play an important role in determining which pathway a colony will adopt.

LASIOGLOSSUM MALACHURUM: AN 'OBLIGATELY' EUSOCIAL BEE

The common and widespread Sharp-collared Furrow Bee *Lasioglossum malachurum* has been subject to extensive study by researchers with an interest in the evolution of sociality. Unlike *Halictus rubicundus*, this species shows eusocial behaviour wherever it occurs – this is 'obligate' eusociality. The initial phase of nest-building and foraging for a first batch of brood cells is, as in the case of *Halictus*, carried out by each fertilised and overwintered female. Usually the nests are constructed in very large aggregations, but each female works independently. When the first batch of brood cells has been provisioned, the foundress remains in the nest and closes it until the first brood of adult bees emerges. The first brood are (almost) exclusively female, and these act as workers, carrying out various functions in the nest and foraging for subsequent brood(s). In many

FIG 123. Sharp-collared Furrow Bee *Lasioglossum malachurum* overwintered female.

FIG 124. Open nests of Sharp-collared Furrow Bees.

FIG 125. Sharp-collared Furrow Bee worker meeting a nestmate on returning to the nest.

FIG 126. Sharp-collared Furrow Bee male.

lowland and southerly British locations, one or more further broods of workers will be raised, culminating in a final brood of males and fertile females. These mate, and the fertilised females go on to overwinter, often in their natal nests. At cooler localities, there is usually just one small brood of workers prior to the production of the sexual brood.

One intensively studied population (in the area around Tübingen in Germany) exhibits the 'northern' colony cycle of one worker brood prior to the production of males and gynes (fertile females). Paxton *et al.* (2002) used highly sensitive genetic markers to explore the kinship relationships among nestmates in this population. They excavated 18 nests and collected the occupants at different stages in the colony cycle. All nests were 'monogynous': that is, had a single dominant female (queen). In the majority of nests, the workers were full sisters, but genetic analysis showed that some workers in the same nest were products of more than one maternal mating – up to three. There were also 'alien' workers in 33 per cent of nests, and, contrary to expectations, worker reproduction contributed as much as 19 per cent of the final, sexually fertile, brood of females. In this context, 'alien' workers were defined as nestmates that were not daughters of the queen. Paxton *et al.* considered two possible sources of this mixing of genetic lines in some nests. One of these was that the nest had been usurped by another female during the phase of worker-production, so that the 'alien' current workers were the offspring of the displaced foundress. The other possibility was 'drift' of adult workers from adjacent nests. In the case of one nest, alien workers were excavated as pupae, ruling out the latter possibility, and Paxton *et al.* favoured nest usurpation as the main explanation of mixed-lineage nests.

Their analyses also suggested that alien workers were the most likely source of worker-produced gynes. It seemed that where queens survived up to the end of the provisioning phase of the sexual brood, they were able to suppress the reproduction of their own daughter-workers but did not suppress that of 'alien' nestmates. This raises interesting questions about what, if any, might be the selective advantages to workers drifting to a non-natal nest, and what might favour their acceptance by the host nest, as well as how queens achieve control over their own daughters, while offering 'concessions' to alien workers.

This pattern differs considerably from a population in southern Greece, studied by Richards *et al.* (2005). Here, under milder climatic conditions and a longer season, there are up to three successive broods of workers prior to the emergence of the sexual brood, resulting in much larger numbers of workers per nest (as many as 35) than is the case in northern populations. Alien workers were found much less frequently (in 17 per cent of nests). On the hypothesis that alien workers derive from usurpation, this might be expected, as the offspring of the displaced foundress would have died off before the subsequent broods emerged. The Greek population also revealed that alien workers made little or no genetic contribution to the sexual generation, implying full control of worker reproduction by the incumbent queen. Interestingly, this runs contrary to the expectation that queen control should decline as numbers of workers rise, and it raises the question whether there might be underlying genetic differences between these widely separated populations. The authors of this study calculated that the eusocial mode of life would have produced approximately 10 times as many fertile offspring than would have been produced if each female had reproduced as a solitary bee, but it was still probably in the reproductive interest of the workers to have nested in the solitary mode had they been able.

Further complexity in the sociogenetic structure of *Lasioglossum* nests was discovered by another study – this time carried out with a population at an 'intermediate' geographical location (Eichkogel, Austria). Soro *et al.* (2009) found a high incidence of queen-less colonies (48 per cent) when the sexual brood was being produced. This might have been the result of matricide, as workers assert their own direct reproductive interests towards the end of the colony cycle. At this site, alien workers were found in approximately 25 per cent of nests but had probably arrived there by drifting from nearby nests, rather than being offspring of a usurper of the foundress. This was because they were generally found as singletons in the nests and also were genetically aligned with other local nests. Finally, multiple matings (up to three times) were found to be relatively common among queens at this site, as at Tübingen, in contrast to the population studied in southern Greece, where only 3 per cent of queens mated twice (Richards *et al.* 2005).

The general practice of nest foundresses sealing their nests for the period(s) between completion of foraging for each brood and the emergence of that cohort of adults has been difficult to interpret. In the case of production of the first brood of workers, it has been suggested that nest closure is a strategy used by incumbent gynes to prevent usurpation by rival females. However, this seems unlikely. In early spring, during the phase of nest formation, there is often intense competition for nest sites between overwintered females, as some establish nests while others 'float' through the aggregation, investigating nest entrances. Sometimes this results in antagonistic interactions between floaters and residents and, rather uncommonly, in actual usurpation. Marion *et al.* (2007) found that it tended to be the larger females that established nests and fought off potential rivals, but also that the rivalry for nest sites diminished as foraging for the first brood advanced, so that there was little or no risk of usurpation by the time incumbent females sealed their nests. It seems more likely that nest closure may protect the brood from invasion by nest-parasites, while the foundress gyne reduces wear and tear and avoids the risks associated with activity outside the nest.

Nests of *Lasioglossum* species, including the social species, are prone to invasion by *Sphecodes* cuckoos, and this raises a number of questions. How are the life cycles of the cuckoos of the social species aligned with the life cycles of their hosts, and what is the impact of nest parasitism on the hosts' modes of social life? In the case of *Lasioglossum malachurum*, it seems that the number of workers may be limited by nest parasitism (see Chapter 8). The Lobe-spurred Furrow Bee *Lasioglossum pauxillum* is an obligate eusocial species that is host to the Swollen-thighed Blood Bee *Sphecodes crassus* (Fig. 127). In a small population of the *Lasioglossum* observed by TB, *Sphecodes* females were observed entering and leaving the 'turret' nest entrances of the furrow bee in late May and early June 2021. At these dates, the *Lasioglossum* nests already had four or more workers, seen

FIG 127. Swollen-thighed Blood Bee *Sphecodes crassus* female entering a nest of the Lobe-spurred Furrow Bee *Lasioglossum pauxillum*.

entering with pollen loads. However, *Sphecodes crassus* females were present at the site as early as 22 March 2022. It may be that they wait until the first brood of workers has become adult before entering the *Lasioglossum* nests, to lay their eggs in brood cells being stocked by first-brood workers. If the occupants of those brood cells are destined to become the sexual generation adults, the *Sphecodes* may benefit from greater nutrient content. However, at this site, *Lasioglossum* nests with workers bringing in pollen loads were seen as late as 28 July 2021, with males of both the *Sphecodes* and *Lasioglossum* not seen until early September. This suggests that at least in some years, *Lasioglossum pauxillum* nests raise more than one brood of workers (or, possibly, there is a continuous production of workers over as much as two or three months) before the sexual generation appears. Female *Sphecodes* appear to be present throughout, but males are not produced until late summer, raising the question as to what triggers the transition to sexuals – or, at least, to males – in *Sphecodes*.

PHYLOGENY AND THE EUSOCIAL TRANSITION

Studies of these and other examples of variable sociality in the Halictidae have both informed and challenged theoretical approaches to the evolutionary origins of social life and the conditions that sustain it. Current evidence bearing on the evolutionary history of the eusocial insects confirms the view that the insect groups with the most complex and stable forms of sociality (the termites, ants and corbiculate bees) adopted their social modes of life in the late Cretaceous or earlier, more than 65 million years ago, much earlier than groups within the family Halictidae. By contrast, a combination of fossil and genetic evidence puts the period of emergence of halictid eusociality as between 20 and 22 million years ago (Danforth 2002, Brady *et al.* 2006) – relatively recent by geological standards.

In a series of studies, Packer and co-authors mapped the diversity in social behaviour in a subgroup of Halictidae onto a reconstruction of their evolutionary history (phylogeny). Phylogenies were reconstructed by

FIG 128. *Halictus petrefactus*, an early Miocene species, preserved in laminated mudstones from Rubielos de Mora, Spain. (© M. Engel)

a method involving electrical separation of the enzymes (allozymes) coded by alleles at selected loci in the genomes of bees belonging to species with known behavioural repertoires. Packer (1991) distinguished nine behavioural characters subject to variation within one especially variable halictid group. This group comprises species belonging to two sub-genera within the genus *Lasioglossum* (*Dialictus* and *Evylaeus*). *Lasioglossum (Evylaeus)* includes the group of species whose females have fine ridges bounding the rear face of the propodeum – known by British writers as carinate species and including *Lasioglossum calceatum*, *Lasioglossum albipes*, *Lasioglossum malachurum*, *Lasioglossum pauxillum* and *Lasioglossum fulvicorne*, among others. The common Green Furrow Bee *Lasioglossum morio* was the sole British member of the sub-genus *Dialictus*, which was included in the study as an 'outgroup'.

The behavioural characters mapped by Packer included: annual or perennial cycles, single foundresses or multiple, number of broods per year, differentiation between worker caste and queens, workers mating, features of nest architecture, and whether brood cells were open during juvenile development. One North American species (*Lasioglossum marginatum*) establishes perennial nests, but this was considered to be derived from an ancestral annual cycle. Packer's results suggested that the eusocial condition was ancestral in the sub-genus *Evylaeus* and possibly ancestral to both that and the sister group, *Dialictus*. Both of these sub-genera

FIG 129. 'Turret' nest entrance of the Lobe-spurred Furrow Bee *Lasioglossum pauxillum*.

FIG 130. Lobe-spurred Furrow Bee worker leaving the nest for a foraging trip.

include species that are currently social as well as species that are currently solitary. Packer's inference is that eusociality evolved more than once in the tribe Halictini and that a reversion from social to solitary life cycles has happened several times.

The behavioural repertoires listed form a suite, some combinations representing a more firmly integrated and fixed eusocial lifestyle, with others representing more openly flexible and environmentally variable states. Of the British species, *Lasioglossum malachurum* offers the most fully integrated condition in the subgenus *Evylaeus*, indicated by three key features: a single queen, the ability to produce more than one worker generation in a year and non-overlapping size differences between workers and queens. *Lasioglossum albipes* and *Lasioglossum calceatum* are close relatives but produce only one worker brood in the year, have smaller size differences between queens and workers and, like *Halictus rubicundus*, revert to solitary life cycles in more northerly, inhospitable geographical locations. Among the British species included in the study, the very small *Lasioglossum pauxillum* approaches *Lasioglossum malachurum* in its evolutionary commitment to sociality. It is argued that for these species, most notably where workers and queens are highly differentiated, reversion to solitary status is unlikely.

A further dimension of the variability among the halictids is nest architecture. Most members of the *Evylaeus* group construct a cavity, either at the terminus of the main nest burrow or on a lateral branch from it. The brood cells are located as a cluster within the cavity. Packer *et al.* (1989) suggest the cavity may function as a protection from waterlogging, though there are some grounds for doubt about this. An alternative possibility is that the clustering of brood cells within a wider space would favour simultaneous and (in the social phase) cooperative stocking of the cells, compared with the more usual pattern among ground-nesting bees of

FIG 131. Lobe-spurred Furrow Bee worker returning with yellow pollen (probably daisy).

FIG 132. Lobe-spurred Furrow Bee worker returning with white pollen (probably Small-flowered Cranesbill *Geranium pusillum*), suggesting division of foraging labour among the workers.

a linear series of cells, with sequential mass provisioning. It is also reported that cells are opened for inspection in some *Evylaeus*, and they are apparently kept open in *Lasioglossum malachurum*. This could be relevant to caste determination, if there is some parental care associated with the open cells. Finally, one respect in which *Lasioglossum malachurum* does not easily fit into the more evolved pattern of sociality has to do with the contribution of worker reproduction to the sexual brood. Packer's analysis assumes this at less than 1 per cent, but the studies reported above put this at very much higher in the more northerly populations, and workers with developed ovaries and full spermathecae are found in all populations. As this is a key issue for understanding the evolution of social life in these lineages, we will return to it a little later.

Research by Danforth and his colleagues, using a combination of analysis of mitochondrial and nuclear DNA, morphology and fossil evidence, produced more finely differentiated phylogenies and dating of the evolutionary transitions to eusociality in the halictids. As mentioned above, this put the emergence of eusociality in this group as occurring between 20 and 22 million years ago and identified three independent origins in the sub-family Halictinae. One of these is in the sub-family Augochlorini, which includes North American species, commonly with brightly coloured metallic cuticles. The other two are ancestral to the species discussed above, which are widely distributed in Britain and Europe. The common ancestor of the species of *Halictus* was eusocial, with from 4–6 subsequent reversals to solitary living. In the large and complex genus *Lasioglossum*, there was a single transition to eusociality in the common ancestor of the three sub-genera *Evylaeus*, *Dialictus* and *Sphecodogastra*, again with subsequent reversals to solitary nesting. In all, the research concludes that there were three independent transitions to eusociality in the Halictinae, followed by 10–12 reversals, and that the transitions took place within a single, 'relatively narrow' time frame (Brady *et al.* 2006).

WHAT CONDITIONS MIGHT HAVE FAVOURED THE TRANSITION TO EUSOCIALITY IN THE HALICTINAE?

These reconstructions of the ancestral lineages of the social halictids, together with observations of the variability of their social forms today, yield some hypotheses about the conditions favouring the evolutionary emergence of eusociality and its subsequent persistence. First, for the nest foundress to produce one or more broods of workers, followed by a further brood of sexuals to continue into the following year, a prolonged warm season with conditions suitable for foraging is needed. As Brady *et al.* (2006) point out, the period within which the three halictid transitions are thought to have taken place was one characterised by global warming. Any tendency to bivoltinism or eusociality would have been intensified as the climate warmed, especially for the northern hemisphere *Halictus* and *Lasioglossum* species. This is consistent with the geographical variation in both obligate and facultative eusocial behaviour within these groups today, eusociality being strongly favoured within the most thoroughly studied Halictidae in warmer, more southerly (and lowland) locations. Relatedly, the length of the required period of foraging activity would preclude narrow pollen diets, linked to brief flowering periods of pollen hosts. The eusocial halictids are, indeed, like the corbiculate bees, polylectic, and Danforth (2002) points out that two of the proposed reversals from social to solitary lifestyles were correlated with shifts from polylecty to oligolecty.

Another condition favouring transition to social modes of life is the degree of genetic relatedness of the social group. The generally accepted theoretic account of 'altruistic' and cooperative behaviour locates the gene as the level at which natural selection operates, so individual organisms might sacrifice their own reproductive interests in favour of a close relative who shares many of their genes, thus enhancing their own inclusive fitness. On these assumptions, the hymenopteran reproductive system (haplodiploidy), in which males are the product of unfertilised eggs, should be generally favourable to sociality since daughters of once-mated females are more closely related to one another than in other versions of sexual reproduction (sharing 75 per cent of their genes). On these assumptions, eusociality, in which a single, monandrous female founds a nest and sets her daughters to work, would be favoured. However, there is some deviation from this state in the case of *Lasioglossum malachurum*. Presence of alien workers appears to be widespread, especially in more northerly populations, and multiple matings, with more than one patriline present in the same nest, also seem to be relatively frequent.

This raises questions about how order is maintained in the nest. Is some form of coercion involved in maintaining the subservience of the workers and preventing them from developing their own direct reproductive potential? And how effective is it? Queens in the population studied by Paxton *et al.* (2002) near Tübingen seemed capable of suppressing the reproductive development of their own offspring but conceded a share of the reproductive output of the nest to alien workers. In southern Greece (Richards *et al.* 2005), where alien worker presence in nests was much less frequent, workers did not contribute genetically to the sexual generation. How and whether queens of *Lasioglossum malachurum* can discriminate between kin and non-kin among their nestmates is unclear. It seems that the Tübingen queens were able to, as worker reproduction was suppressed among the daughters of the queen, while alien workers provided 19 per cent of the output of sexuals. In contrast, a study of the chemistry of the cuticular bouquet and Dufour's gland secretions of workers taken from the Eichkogel site showed strong commonality among nestmates and differentiation by caste but not by kin (Soro *et al.* 2011).

The size difference between queens and workers in *Lasioglossum malachurum* is such that queens could achieve control by physical coercion, but in general, coordination of activity within the nest is likely to be achieved by pheromonal communication. This, indeed, seems to be a further faculty closely related to the transition to and further development of eusociality. The differing functions in the division of labour between nestmates in the social phase of the life cycle of eusocial bees would involve communication of reproductive status versus various worker tasks, such as foraging for different foods, brood cell construction and nest-guarding. It is to be expected, then, that there will be greater investment in chemical communication in eusocial species as compared with solitary ones (LeConte & Hefetz 2008, Kocher & Paxton 2014, Leonhardt et al. 2016).

Wittwer et al. (2017) devised a study to investigate the relationship between communication systems and social behaviour in halictids by grouping a set of 36 species according to assessments of their evolutionary history. The phylogenies of the species studied were constructed by the authors, and their results coincided closely with those of Danforth and colleagues, discussed above. They used measures of sensilla density in the antennae of females as an indicator of differences in communication systems and divided the species into three groups in terms of their different evolutionary histories. The three groups were: eusocial species, species considered to be ancestrally solitary species, and species currently solitary but considered to have formerly been eusocial. Despite some outliers, there was a significant overall pattern: sensilla density was greater in the eusocial species compared with the formerly eusocial species that had become solitary. The authors take this to support their view that there is a close relation between social behaviour and communication systems, and that eusocial species in general invest more in communication systems than solitary. However, the results of the comparison between eusocial species and ancestrally solitary ones were anomalous, as sensilla density across these groups was similar.

This study also included a complementary analysis of differences between the solitary and eusocial states in a facultatively eusocial species, the Bloomed Furrow Bee *Lasioglossum albipes*. Using specimens sourced from French populations, the authors were able to combine their measure of chemical sensitivity with assays of the chemical signals detected in the cuticles of females from eusocial and solitary populations of this species. As expected, sensilla density was greater in females from eusocial populations, and this was correlated with differences in Dufour's gland and cuticular chemical profiles. In the eusocial populations, there were marked differences between queen and worker chemical profiles, indicating the significance of chemical signalling in the social division of labour.

In a further study (Steitz et al. 2018), obligate and facultatively eusocial species were compared with both secondarily and ancestrally solitary ones, with respect to the odour profiles of females at different phases of the colony cycle. Specimens were drawn from several European and North American populations, with one ancestrally solitary species from Australia. The chemical analyses focused on two groups of compounds (n-alkanes and macrocyclic lactones), which earlier studies have shown to be indicative of the reproductive status of female halictids. Their results confirmed those of Wittwer et al. in revealing a higher differentiation between workers and breeding females (queens) in obligate eusocial species (including *Lasioglossum malachurum* and *Lasioglossum pauxillum*).

In facultatively eusocial species (including *Halictus rubicundus*), they also found marked differentiation between worker and queen odour profiles, but these were related to the reproductive status of workers. Steitz *et al.* interpret their results as illustrating greater investment in chemical communication associated with the transition to social modes of life, but conclude that this takes the form, primarily, of heightened differentiation between castes. In the obligate eusocial species, the presence of odour profiles differentiating queens and workers, independently of their reproductive status, indicates the evolution of a fertility signal, or 'queen pheromone', in those species.

Finally, it seems likely that the distinctive life cycle of the halictid subfamily Halictinae may have played a role in the solitary-to-social transition. This very large group of species is subdivided into two tribes, the Augochlorini and the Halictini. The latter tribe includes five genera represented in the historic British bee fauna, but due to likely extinctions, it may now include only three: *Halictus*, *Lasioglossum* and the nest-parasitic *Sphecodes*. As we noted above, the genera *Lasioglossum* and *Halictus* include species that are eusocial as well as both ancestrally and secondarily solitary. The significant feature of the life cycle, common to both ancestrally solitary and eusocial members of the Halictinae, is the overwintering of adult fertilised females. In Europe, both males and fertile females emerge during the summer, mating takes place, and females dig a hibernaculum (often using their natal nests), where they spend the winter. In spring they emerge, and after a few days construct a nest, forage for provisions and begin egg-laying. As they already have their complement of stored sperm, they are able to get an early start, which may be an advantage if the season is long enough to allow the production of more than one generation.

Getting an early start may, in itself, favour transition to sociality, but more significant is the ability of the females to control the sex of their offspring – by laying either fertilised or unfertilised eggs. Studies of the life cycles of obligate eusocial species have shown that all, or almost all, of the first (worker) brood are female. In the absence of potential mates, females of that brood would have no reproductive option but to cooperate in rearing the next brood of their mother's offspring. It seems that this facility on the part of female halictids may have served as a preadaptation for the transition to eusociality that took place repeatedly in the group: 'Perhaps the fact that there are no males available in the early spring, when the female's first-brood offspring emerge, gives her a unique opportunity to "enslave" her daughters as workers' (Danforth *et al.* 2019). The non-parasitic bumblebees, though making their transition to eusociality much earlier in geological time, also have annual life cycles, with overwintering fertilised females (gynes).

CHAPTER 5

Bees and Flowers Part I

'All the adaptations, the intricate beauty of flowers and the details of all the interactions of pollination...are really about maintaining at least the possibility of cross-fertilisation.'

—Proctor, Yeo and Lack, New Naturalist,
The Natural History of Pollination (1996)

'Plants cannot know or see which insects are in the vicinity, nor can they accept or reject a visitor. They can only offer their commodities to a diverse army of potential visitors from where they stand, and try their luck.'

—E. L. Clare et al. (2013)

It is largely to insects, especially bees, that we owe the colours and beauty of flowers, evolved through millions of years of competition for the attentions of pollinators. The intimate relationship between insects and flowers has intrigued and enchanted many human generations, yet the biological significance of insects as pollinators was not fully appreciated until the late 18th century, largely through the insights of the German naturalist and teacher Christian Konrad Sprengel, who developed a fascination for flowers. He was the first fully to appreciate that the colours, scents and structures of flowers serve to attract insects, whose role it is to carry pollen from one flower to another, thereby fertilising them and allowing fruits and seeds to develop. In 1793, he published his classic work *The secret of Nature revealed in the construction and fertilisation of flowers*, describing the flowers of 461 plant species. He viewed flowers as the work of the creator.

Sprengel comments that 'nature seems to intend that no flower should be fertilized by its own pollen' (cited Müller 1883) and it is now well understood that cross-fertilisation (outbreeding) is critically important in reducing or preventing inbreeding depression, characterised by poor growth and development or failure to survive. Sprengel's interpretation of pollination by insects provided a foundation for all subsequent studies of floral ecology, pioneered in England by Charles Darwin and in Germany by Hermann Müller in the 19th century. In his autobiography, Darwin notes 'my interest [in the cross-fertilisation of flowers] was greatly enhanced by having procured and read in November 1841, a copy of C. K. Sprengel's wonderful book'.

Despite their significance as pollinators, solitary bees have received relatively little attention in pollination research compared with honeybees and bumblebees. The focus of this chapter is to review what is known about the interactions between solitary bees and flowers, from the plant's perspective. We will examine how flowers influence and manipulate the behaviour of bees to the plant's advantage, tempting bees to visit them but controlling the pollen and nectar rewards they offer. To what extent have flowers achieved control over their solitary bee visitors? Do solitary bees sometimes evade flower mechanisms either through evolutionary change or by learning to take short cuts to reach the rewards? On the other hand, how do some plants trick bees into visiting them without giving any reward at all?

On a broader scale, we will consider how solitary bees 'decide' which flowers to visit amongst the variety on offer in flower-rich habitats. Do they tend to stick with one flower type or move more randomly? Is it possible to recognise simple categories of flower types which are patronised by some bee families more than others? We will consider whether any generalisations are possible regarding patterns of interaction across different flower forms and whole plant-pollinator communities. Some personal observations and studies of solitary bees are woven into the chapter. The following chapter (Chapter 6) switches emphasis to the means by which bees find and collect floral resources and transport them to their nests. First it is necessary to review the structure of a representative 'typical' flower and look in some detail at flowering plant breeding systems, examining the services that flowers might require from pollinators such as bees.

FLOWER STRUCTURE AND FERTILISATION

Bees are totally dependent on flowering plants, since they feed themselves and their larvae entirely on plant material, in the form of pollen and nectar and in some cases plant oils. Rare exceptions include some tropical bees which feed their larvae partly or entirely on freshly dead carrion (Roubik 1989). Bees are central place foragers, harvesting food for their offspring from their surroundings and placing it in a suitable nest cavity. This requires female bees to expend considerable time and energy in flight, limiting the resources that can be passed to their offspring. Successful reproduction by solitary bees can occur only where suitable flowering plants are present in abundance, close to potential nesting sites.

Flowers come in a great variety of forms. Each plant family has its own characteristic configuration of sepals, petals, anthers and ovaries, but there is considerable variation even within families. Figure 133 shows a vertical section through a flower of Herb Robert *Geranium robertianum*, which has a relatively unspecialised structure. *Geranium* was the genus which first caught the attention of Sprengel in 1787 when he noticed drops of nectar protected by small hairs, which he realised 'is secreted for the sake of insects' (Müller 1883).

All flower parts are attached to the receptacle, the swollen end of the flower stalk. The most obvious flower structures are the petals, which together make up the corolla. The sepals surround the corolla to form the calyx, enclosing the flower when in bud. At the base of each petal or stamen there is often a nectary, secreting sugar-rich nectar which, like the petals, can be scented. The flower's male reproductive organs, the stamens, are encircled by the petals or attached to their base, each stamen consisting of a filament (stalk) surmounted by a four-chambered anther, which produces pollen. The female reproductive organs are situated at the centre of the flower, comprising an ovary (or ovaries), which is a hollow structure containing one or more ovules. Pollination happens when pollen becomes attached to a stigma, while fertilisation, the fusion of male and female gametes, happens somewhat later, within the ovary.

In flowering plants, male gametes do not 'swim' to the female gamete, using a tail, as do those of bryophytes (liverworts and mosses), pteridophytes (ferns)

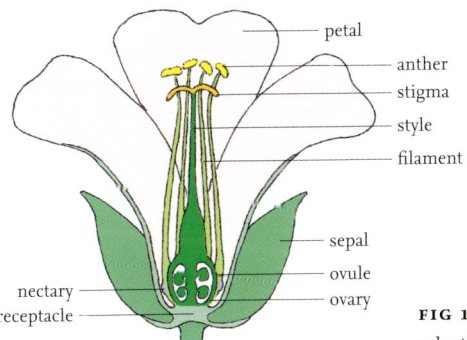

FIG 133. Structure of Herb Robert *Geranium robertianum* flower in vertical section.

and the majority of animals. Instead, the male gamete, in the form of a nucleus, travels down a narrow tube which grows from the pollen grain down through the style until it reaches an ovule inside the ovary. The tip of the pollen tube enters a small pore in the ovule. On entering the pore, the male gamete (male nucleus) meets the female gamete (egg), and the two sex cells fuse to become a zygote: a fertilised egg. The fertilised ovule, containing the zygote and a food-store, is now an immature seed. The zygote develops into an embryo plant within the seed before the seed is dispersed. The pollen grains themselves contain stored nutrients which contribute to the growth of the pollen tube, if they are not diverted for the nourishment of animals such as bees. Pollen sticks readily to a bee's hair as well as to a plant's stigma (Fig. 134).

Pollen grains vary greatly in size and shape between plant species. Minute projections (papillae) and secretions on the stigma, as well as substances on the surface of the pollen grain itself, help pollen to stick on and germinate to form pollen tubes. There is competition between pollen grains for their pollen tubes to reach the ovules first, but the number of pollen grains must at least match the number of ovules in the ovary for each ovule to become a seed. Foreign pollen from other plant species on the stigma can cause interference or even hybridisation.

The protein-rich nutrients contained in pollen are the main component of the food provision of bee larvae. Nectar is also secreted by most plants and is a vital means of attracting pollinators. Nectar complements the nutrients in pollen

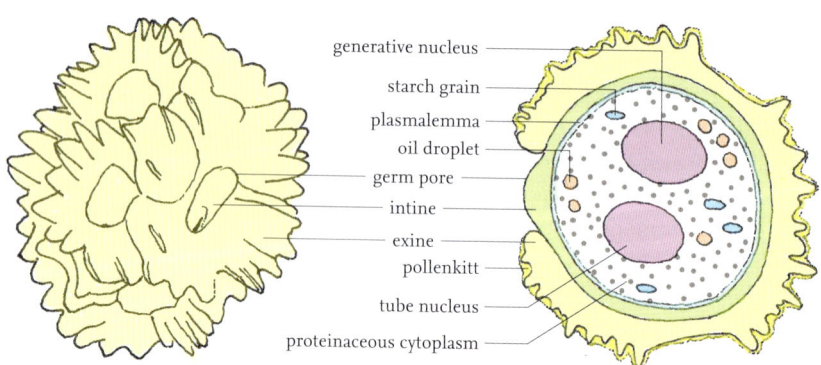

FIG 134. Structure of a pollen grain of an Asteraceae species. Most insect-pollinated plants produce slightly asymmetrial pollen, often with a spiky surface. Lipids known as pollenkitt cover the pollen surface and assist with adhesion and species recognition and also provide colour. The pollen grain has openings (germ pores) from which the pollen tube can emerge. The tube nucleus controls the growth of the pollen tube. The generative nucleus divides into two nuclei which travel down the pollen tube and fertilise the ovule.

FIG 135. Black Mining Bee *Andrena pilipes* female on Sea Kale *Crambe maritima*. Pollen is covering her face and thorax and also gathered into her scopae.

by providing a solution of sugars which can be metabolised by bees for growth, movement and thermogenesis. Nectar also provides water, amino acids, lipids, vitamins and, as with pollen, sometimes includes toxins. However, about 6 per cent of flowering plants provide no nectar (Russell *et al.* 2016). Such plants are often bee specialists, offering copious pollen. As explained below, bees are able to learn to associate floral signals with the presence of both nectar and pollen (Muth *et al.* 2015).

THE IMPORTANCE OF SOLITARY BEES TO POLLINATION

In temperate zones, around 78 per cent of flowering plants are estimated to be animal-pollinated, the remainder being pollinated by wind, water or other means. As many as one in ten terrestrial animal species worldwide is thought to be involved in pollen transfer. Butterflies and moths (Lepidoptera) are the most diverse group, with some 140,000 species possessing the long, tubular mouthparts needed to obtain nectar from flowers. Next come beetles (Coleoptera), with an estimated 77,000 pollinating species, and Hymenoptera (bees, wasps, ants and close relatives), with 70,000. The fourth major insect group visiting flowers is flies (Diptera), of which an estimated 55,000 species pollinate flowers. Here it should be borne in mind that the majority of the world's insect species, including some bees and many Lepidoptera and Diptera, have yet to be described. Amongst vertebrates, over 1,000 species of birds are known to be pollinators, while bats, rodents and lizards also play a very significant role (Ollerton *et al.* 2011, Ollerton 2017, 2021, Wardhaugh 2015). These figures do not take any account of the abundance, distribution or effectiveness of the species concerned; they are simply provisional (but useful) estimates of the numbers of species involved.

How important are bees, and solitary bees in particular, to pollination, compared to other insects? Some perspective comes from a wide-ranging study by Willmer *et al.* (2017), who compared pollen transport by bees with that of all other insects (non-bees). Plant communities were observed in Kenyan savannah, Mediterranean garrigue (scrub), British gardens and British heathland, noting pollen placement on newly exposed stigmas of 76 plant species from 30 families. Flowers were temporarily covered with bags then exposed to offer their virgin stigmas to insects. In all four localities, bees were found to be more effective at pollen deposition on stigmas than were non-bees, and visit lengths were shorter in bees than in non-bees, allowing bees to visit more flowers in a given time. In British heathland and in Mediterranean garrigue, significantly less heterospecific pollen (pollen of a different plant species) was deposited by bees than by non-bees. Across the whole study, both social and solitary bees were found to be the most frequent flower visitors and the most effective pollen vectors and to show higher fidelity to particular plants than non-bees.

Solitary bees usually carry pollen which is dry and loose, in contrast to bumblebees and honeybees which carry pollen moistened with nectar, compacted into pollen baskets (corbiculae). Compacted, wetted pollen is less likely to be transferred to stigmas than loosely held dry pollen. There is also evidence that pollen transferred from a corbicula is less viable than pollen carried by the scopa of solitary bees (Parker *et al.* 2015). Non-corbiculate bees (solitary bees) make up around 98.5 per cent of the 20,000 bee species worldwide and about 90 per cent of British species. Studies by British naturalists over many decades have resulted in an inventory of the flowers visited by Britain's bee species, including details of the plants from which pollen is gathered, and much of this information has recently been brought together in the species accounts of Else and Edwards (2018).

CROSS-POLLINATION OR SELF-POLLINATION? PLANT BREEDING SYSTEMS

How do flowers work? The challenge for the floral ecologist is to explain the forms and functions of flowers in all their bewildering variety. The breeding systems of flowering plants are very diverse and need a fairly detailed introduction in order to understand their strategies and how they might affect solitary bees. The colourful petals and abundant nectar of flowers can be interpreted as evolved traits which promote cross-pollination: the transfer of pollen from the anther of one plant to the stigma of another plant of the same species. In the vast majority of plants with colourful flowers, this is carried out by vectors such as insects, bats, birds and lizards, which travel from one plant to another seeking pollen, nectar or other resources and in doing so inadvertently deposit pollen on the flowers they visit.

Charles Darwin was a pioneer of experimental studies into the effects of cross- and self-pollination of flowers, demonstrating that self-pollination leads to reduced height, weight, vigour and fertility of plant offspring as well as to reduced variation in features such as flower colour (Darwin 1862, 1876, reviewed by Barrett 2010). Darwin gave credit to the Herefordshire horticulturalist Thomas Knight who, long before Darwin, realised from his experiments with pea plants that crosses between different varieties resulted in much more vigorous offspring (Knight 1799, 1841). In a similar vein, William Herbert also preceded Darwin in emphasising the importance of cross-breeding in plants as well as domestic animals (Herbert 1837).

The twin advantages of cross-fertilisation – avoiding inbreeding depression and enhancing the variety and adaptability of offspring – are reflected in the wide variety of plant strategies which promote cross-pollination. About 10 per cent of flowering plants have male and female flowers on different plants. In these dioecious plants, there is clearly no possibilty of self-pollination. British examples include White Bryony *Bryonia dioica*, Red Campion *Silene dioica*, willows (*Salix* species) and Holly *Ilex aquifolium*.

However, the majority of flowering plants have flowers of both sexes on the same plant and are therefore potentially able to fertilise themselves. Some species are monoecious, having separate male and female flowers on the same plant, but the great majority of insect-pollinated plants have hermaphrodite flowers, containing both male and female sex organs, like Herb Robert described above. Despite this arrangement, self-pollination is generally avoided or reduced

FIG 136. Chalk Furrow Bee *Lasioglossum fulvicorne* male seeking nectar on Rosebay Willowherb *Chamaenerion angustifolium*. The flower is hermaphrodite, protandrous and in the female stage: the anthers have withered, and the stigma is receptive.

through the anthers maturing before the stigmas (protandry – the usual form in insect-pollinated flowers) or vice versa (protogyny). The positioning of male and female parts within the flower can also reduce the chances of self-pollination, if anthers are well separated from the stigmas.

Mirror-image flowers, which are known across 10 flower families, possess either right- or left-bending flower structures, such that bees visiting the two flower types pick up pollen on opposite sides of their bodies, thus favouring cross-pollination (Barrett 2010). Most members of the primrose family (Primulaceae) and a wide range of plants across 27 other plant families exhibit the phenomenon of heterostyly, in which each individual plant has flowers of one of two (in one case three) alternative forms, with complementary arrangements of the 'vertical' positions of anthers and stigma. A well-studied example is the Common Primrose *Primula vulgaris* first investigated experimentally by Darwin (1884). All the flowers on a plant are either the 'pin' form or the 'thrum' form. Pin flowers have a long style, with the stigma above the anthers and thrum the reverse. There are also differences in pollen size and stigma structure between pin and thrum (Fig. 137). Heterostyly favours cross-pollination, carried out by long-tongued pollinators, such as the Hairy-footed Flower Bee *Anthophora plumipes*. Differences in pollen and style morphology also reduce the extent of pollen wastage (Barrett & Shore 2008). A study of captive male *Anthophora plumipes* (Keller *et al.* 2014) confirmed that Primrose and Oxlip *Primula elatior* pollen was differentially distributed on the face and tongue of the bee, as predicted by Darwin (more thrum pollen on the face and upper tongue and more pin pollen towards the tip of the tongue), leading to most pollen being transferred from pin to thrum or vice versa – though only male bees were used in this study (Box 2). There are other plant species, such as Thrift *Armeria maritima*, which also have two forms of pollen and stigma but do not show heterostyly.

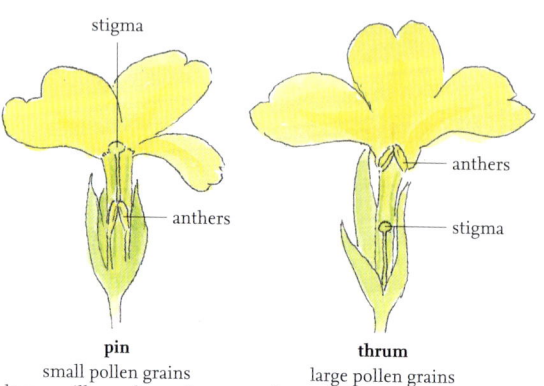

FIG 137. Common Primrose *Primula vulgaris* flowers with the front two petals removed, showing the reciprocal arrangement of anthers and stigma, promoting cross-pollination.

BOX 2 Which insects pollinate Primroses?

Charles Darwin experimentally protected Primrose plants from insect visitors and found that these produced only a very small amount of seed. He concluded that they are self-incompatible but that some tiny insects had entered the Primrose covers, causing a few seeds to develop. When NWO repeated this test, covered plants (both pin and thrum) produced no seed at all except on flowers on the same plants that were exposed to insects before the covers were put on. These produced fully developed fruits containing many seeds. In a long series of investigations, Darwin

FIG 138. Gwynne's Mining Bee *Andrena bicolor* female collecting pollen from a thrum Primrose.

(1884) found that full fertility and seedling vigour occurred only when a pin pollinated a thrum or vice versa. However, Darwin commented 'it is surprising how rarely during the day insects can be seen visiting the flowers [of Primroses] but I have occasionally seen small kinds of bees at work'. This is somewhat surprising as Primroses do receive a wide range of visitors. The bees Darwin refers to may well have been Gwynne's Mining Bees *Andrena bicolor* (Fig. 138) which regularly gather pollen from Primroses at many sites. However, this bee cannot reach the pollen in a pin Primrose or the nectar of either form. It lands on pin and thrum flowers about equally but quickly leaves pins. On thrum flowers, however, it often remains for a minute or more gathering large quantities of pollen into its scopae. It seems that it can carry pollen from thrum to pin but not vice versa. Hairy-footed Flower Bees *Anthophora plumipes* have a long tongue, eminently suited to foraging from Primroses. Females of this species were often seen foraging systematically along a Primrose-covered bank, each time generally moving to the nearest Primrose plant in a fairly linear fashion, with no obvious discrimination between pin and thrum. Two hundred or more Primrose flowers were typically visited in a sequence as the bees gathered both nectar and pollen. Bee-flies are often seen on Primroses and have a long tongue. However, specimens collected as they departed from Primroses had pollen only on the prominent hairs on the front of the head and not on the tongue, which is very smooth. It seems again that they would transfer pollen only from thrum to pin. The long-tongued hoverfly *Rhingia campestris* may do the same, but some solitary bees, such as *Lasioglossum morio*, are small enough to crawl down the corolla tube of a Primrose and could serve as pollen vectors in either direction. Long-tongued Garden Bumblebees *Bombus hortorum*, usually recently emerged queens, are often the only bees visiting Primroses in northern England and Scotland. They take only nectar and can serve as effective pollinators.

FIG 139. Buffish Mining Bee *Andrena nigroaenea* female taking nectar from Thrift *Armeria maritima*. The pollen on the bee's head will not fertilise the plant it is on, since Thrift pollen comes in two forms, which are compatible with alternative types of stigma, differing in their surface sculpturing (Proctor *et al.* 1966).

Many hermaphrodite plants have evolved a system of avoiding self-fertilisation by being self-sterile: their pollen will not fertilise their own ovules, even when it is placed on the stigma. In these 'self-incompatible' plants, the processes leading to fertilisation are arrested at differing stages, depending on the plant species. The flowers of such plants typically provide a striking visual display. Examples include Common Poppy *Papaver rhoeas* (Box 3), Meadow Cranesbill *Geranium pratense* and clovers (*Trifolium* species). Self-incompatibility is under genetic control: recognition genes of the two parent plants must be different in order for cross-fertilisation to happen. These barriers can, however, sometimes fail, allowing a degree of self-fertilisation when pollen vectors are lacking (Briggs & Walters 1997). A somewhat different system of self-incompatibility occurs in Common Birds-foot-trefoil *Lotus corniculatus*, in which embryos arising from self-pollination die within the seed, thus avoiding wastage of resources on poor-quality seeds. Birds-foot-trefoil is highly reliant on mobile, long-tongued bees such as bumblebees, mason bees and leafcutter bees, which move rapidly from plant to plant.

Despite the variety of flower forms that promote cross-pollination, an estimated two-thirds of all plant species in temperate zones remain capable of self-fertilisation, providing a fall-back (reproductive assurance) when pollen vectors are lacking or if the weather is unfavourable. Self-fertilisation ensures that at least some seeds will be produced, even though they are likely to be of poorer quality. Any degree of self-fertilisation reduces a plant's dependence on long-distance pollen vectors and the need for energetically expensive attractants in the form of large colourful flowers with copious pollen and nectar. Plant life cycles involving self-pollination also tend to be quicker, and the more uniform offspring that arise will be better suited to the local conditions in which

BOX 3 A day in the life of a Common Poppy flower

At 5:00 am on a sunny June day in my garden (NWO), each Common Poppy *Papaver rhoeas* plant is likely to bear two or three newly opened flower buds, with their petals quickly unfolding. The first bees to arrive are likely to be bumblebees, which make loud buzzing movements with their wings to shake pollen from the anthers, but they are soon accompanied by solitary bees. The flowers attract a wide variety of species, both long-tongued bees such as *Osmia bicornis* and *Osmia caerulescens*, which shuffle around the ring of anthers, gathering pollen in the scopa under the abdomen, and small, short-tongued bees such as *Lasioglossum pauxillum* and *Lasioglossum leucozonium*, which scrabble amongst the anthers and load pollen onto the scopae on the hind tibiae. Other visitors may include *Megachile centuncularis*, a leafcutter bee, and tiny parasitic conopid flies which jump on the small halictids to lay eggs (see Chapter 7). By 8:00 am the pollen is rapidly diminishing. At this stage, hoverflies sometimes eat pollen already placed on the stigmas, and *Andrena* bees glean fallen pollen from the petal surfaces. By 9:00 am the stigmas are fully loaded with pollen: the stamens and petals fall off, and bees move on to other types of flowers until the following morning.

FIG 140. (a) White-zoned Furrow Bee *Lasioglossum leucozonium* harvesting pollen from Common Poppy *Papaver rhoeas*. (b) Red Mason Bee *Osmia bicornis* exploring a flower. (c) Buff-tailed Bumblebee *Bombus terrestris* worker buzz-pollinating a flower with two White-zoned Furrow Bees, apparently harvesting the liberated pollen.

Meadow Cranesbill
Geranium pratense
flower large
first exclusively male then exclusively female
incapable of self-fertilisation

FIG 141. Four species of British Geranium showing the relationship between flower size and breeding system (after Lubbock 1875).

Hedgerow Cranesbill *Geranium pyrenaicum*
flower small
first exclusively male, then hermaphrodite
generally fertilised by insects

Dove's-foot Cranesbill *Geranium molle*
flower smaller
first exclusively male, then hermaphrodite
often self-fertilised

Small-flowered Cranesbill *Geranium pusillum*
flower smallest
first exclusively male, soon hermaphrodite
generally self-fertilised

FIG 142. Meadow Cranesbill *Geranium pratense* flower. Nectar guides lead to nectaries at the base of the stamens between the petals. Glandular hairs deter nectar robbers from crawling up from below.

the parent plant is growing. Self-fertilisation is especially characteristic of short-lived plants in frequently disturbed or marginal habitats, such as arable land or riverbanks, where the plants are able to colonise bare ground rapidly. Plants which are predominantly self-pollinating tend to have smaller flowers, with their anthers and stigmas in closer proximity, and each flower generally produces less pollen in relation to the number of ovules. Such plants are often visited by small solitary bees such as *Lasioglossum* and *Hylaeus* species and small Andrenas. Self-pollination results in some (limited) variation among the seeds from one parent plant, but nearly all flowering plants do carry out some degree of cross-pollination sooner or later (Proctor *et al.* 1996, Cruden 2000, Willmer 2006).

FIG 143. Cliff Mining Bee *Andrena thoracica* female on Bramble *Rubus ulmifolius*, one of over 300 recognised bramble microspecies in Britain (Bull 2015).

A further aspect of plant breeding systems is that, rather strangely, many plants develop seeds from ovules without any need for fertilisation. This phenomenon, known as apomixis, occurs widely in members of three plant genera which are very important to solitary bees and other insects, namely *Rubus* (brambles), *Hieracium* (hawkweeds) and *Taraxacum* (dandelions). Species belonging to these genera are difficult to classify since they have developed many very similar 'microspecies' by random genetic change. Despite not needing fertilisation, many have large showy flowers, probably because pollen may still be needed to trigger embryo development, even though male and female gametes never meet.

Pollen vectors, such as bees, clearly play an essential role in the reproduction of self-sterile plants by flying from one plant to another, carrying pollen. But self-fertile plants can be just as dependent on insect vectors for carrying pollen from the anthers of flowers in the male stage to the stigmas of flowers in the female stage. This can result in self-pollination or cross-pollination, but from the plant's point of view, pollen from another plant is more valuable than pollen from the same plant. Bees may need strong encouragement to move from a bountiful flowering plant where they are finding copious nectar and pollen, but visits to multiple flowers in one sequence are not in the best interests of the plant. Self-fertile plants therefore need highly attractive advertisements to tempt pollinators away from competing displays but benefit by not detaining each pollinator for too long, as this risks more self-pollination. Encouraging insect visitors such as solitary bees in just the right way is a balancing act for flowering plants. A compromise is for plants to have highly conspicuous flowers with relatively few open at any one time, as happens for example in members of the Rosaceae (rose family), such as Tormentil *Potentilla erecta* and Rock Rose *Helianthemum nummularium*.

Natural plant breeding systems are continuously adapting, with species able to switch from one system to another (self-incompatible to compatible, monoecious to dioecious, heterostylous to homostylous) and back again through evolutionary time. For example, Thrift populations in the Arctic are self-compatible, perhaps because relatively few insect vectors are present (Baker 1948). Bees are able to adapt to the flowers that are present in their surroundings both through evolutionary change (such as by developing a longer tongue) or by individual learning (see Chapter 6).

FLOWER MORPHOLOGY AND SOLITARY BEES

Bees visit flowers seeking pollen, nectar and other rewards, advertised by a combination of colour, shape, texture, pattern and scent. Bees associate the floral display with a reward, either through innate behaviour or through learning. Large floral displays and scent gradients can attract bees from afar, with whole communities of plants combining their effects. At close range the characteristics of individual flowers come into play. Flower signals first influence the bee's orientation and landing, which is followed by the extension of the bee's antennae and mouthparts and the movements involved in placing pollen on the scopa. Plants compete for the attentions of bees both within and between plant species but, as we have seen, once a bee has arrived, it may be to a plant's advantage not to delay it for too long. Flowering plants also face the dilemma of attracting pollinators without encouraging pollen thieves or other kinds of florivores (flower-eaters) and they cannot afford to be too 'generous' even to authentic pollinators. This is particularly true in the case of female bee visitors, since they are capable of transferring large amounts of pollen to their nests rather than to the stigmas of other flowers. Pollen and nectar are often hidden within the flower structure and rationed out in small portions, necessitating close inspection and frequent visits by potential pollinators. This, in turn, has led to the evolution of flowers which are deceptive, enticing bees to visit when little or no reward is present. Flowers are not simply passive plates of food; they react and adapt in complex ways to the insects which visit them. For example, floral signals, especially colours and scent, can change soon after pollination, directing bees to flowers yet to be visited. The following examples illustrate some of the ways in which flowers influence the behaviour of solitary bees.

FIG 144. Green Furrow Bee *Lasioglossum morio* female taking nectar from Germander Speedwell *Veronica chamaedrys*. The bee's middle leg has grasped the filament, pulling the anther into contact with her body. The stigma matures at the same time as the anthers, allowing some self-pollination.

FIG 145. Tormentil Mining Bee *Andrena tarsata* female on Tormentil *Potentilla erecta*. While the bee reaches for nectar with her tongue, her underside is dusted with pollen, which is collected on the hind legs. (© Liam Olds)

Small symmetrical flowers such as Germander Speedwell *Veronica chamaedrys* and Tormentil *Potentilla erecta* tend to attract small *Andrena*, *Nomada* and *Lasioglossum* species, as well as diptera such as hoverflies (Syrphidae). Germander Speedwell advertises the location of nectar by radial nectar guides and a contrasting central ring. The pollen, too, is easily accessible to any small visitor. The anthers are arranged such that an insect seeking nectar tends to grasp the base of the filament, drawing the anther towards its body (Müller 1883). Female Red-girdled Mining Bees *Andrena labiata* take pollen very largely from this plant, while males of this bee and its cuckoo (the Short-spined Nomada Bee *Nomada guttulata*) also favour it. Males and both sexes of cuckoos seek only nectar but can nevertheless act as pollen vectors, partly offsetting the large quantities of pollen that females remove to stock their brood cells. The Tormentil Mining Bee *Andrena tarsata* forages almost entirely from Tormentil. It has a slow, delicate flight, enabling it to fly from flower to flower amongst taller vegetation. The bee can be found across many parts of Britain where soil conditions are acidic and moist, as far north as Orkney and the Outer Hebrides. Tormentil also frequently attracts the Impunctate Mini-miner *Andrena subopaca*, *Lasioglossum* species and syrphids.

Members of the bellflower family Campanulaceae have a somewhat more exclusive range of visitors. Part of their exclusivity arises because bees are more adept than many other insects at entering hanging, bell-shaped flowers, which largely exclude diptera (Proctor & Yeo 1973). In Britain, the Bellflower Blunthorn Bee *Melitta haemorrhoidalis* and the Small Scissor Bee *Chelostoma campanularum* have long been recognised as Campanulaceae specialists (Kirby 1802, cited Curtis 1823), with the males of these two species often seeking nectar and mates in the same flowers. The second generation of Gwynne's Mining Bee *Andrena bicolor* also often uses *Campanula* pollen. In Harebells *Campanula rotundifolia* and most other bellflowers, pollen is protected within the petals by being placed on the hairy surface of the style, where the pollen adheres. This secondary pollen presentation (SPP) was first noted by Sprengel in the 18th century and is now recognised in 16 plant families (Howell *et al.* 1993, Leins & Urbar 2006; see also Figure 146 a). SPP involves the placement of pollen from the anthers onto a different part of the flower, sometimes before the flower opens. SPP is thought to protect pollen from hazards such as ultraviolet light, consumption by florivores or loss to poor vectors. It can also make pollen rationing very precise.

It is to a plant's benefit if each visit by an insect takes only a small proportion of the pollen output of each flower. Each flower will then receive a succession of insect visitors, providing a more effective pollination service. Increasing portion size beyond a certain point does not provide a proportional increase in pollen dispersal and ovule fertilisation – there are diminishing returns. Retaining pollen

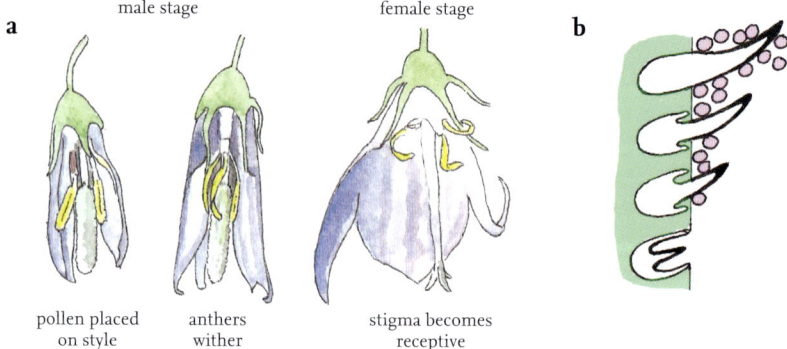

FIG 146. (a) Harebell *Campanula rotundifolia* flower at three stages (nearest petals removed). Left: Before the bud opens, pollen is placed on the style by the surrounding anthers. Centre: Pollen covers the style surfaces, and bees come into contact with pollen while taking nectar from the base of the flower. Right: The anthers have withered, and all pollen has been removed by visiting insects. The three stigmas separate and become receptive. (b) Detail of the surface of a Harebell's style. The hairs withdraw in sequence by invagination, allowing successive zones of pollen to be released and available for harvesting by bees (redrawn from Leins & Erbar 2006).

FIG 147. Gold-fringed Mason Bee *Osmia aurulenta* male using his long tongue to seek nectar from a brassica flower.

for too long can, however, result in a decline in pollen viability and also reduces the probability of pollen reaching stigmas before the pollen of a competing plant (Harder & Wilson 1994).

Campanulaceae ration their pollen by the phased withdrawal of the hairs on the style, which retract inwards, releasing bands of pollen in sequence (Brongniart 1831, Erbar & Leins 2006). Despite this rationing, the costs to a *Campanula* plant of pollen loss to bees can be high, with only a small proportion of pollen found to reach the stigmas of some Campanulaceae species (Schlindwein *et al.* 2004). Florivores, including a specialised pollen-eating weevil (*Miarus campanulae*), can also take a toll.

Deep tubular flowers

Deeper, tubular flowers, such as some members of the cabbage family (Brassicaceae), release nectar at the base of the corolla tube, and this attracts insect visitors with a long tongue, such as *Osmia* bee species, though short-tongued bees may still take pollen.

Thus far we have considered radially symmetrical (actinomorphic) flowers. Long-tongued solitary bees are particularly attracted to flowers with bilateral symmetry (zygomorphic flowers) and a more complex morphology, especially members of the pea family (Fabaceae) and the mint family (Lamiaceae). Fabaceae flowers place pollen on the underside of a bee, whereas Lamiaceae usually place pollen on the top. Precise placement onto a bee's body increases the precision of pollen transfer onto the stigmas of the flowers they subsequently visit. Zygomorphic flowers generally possess a deep corolla tube with hidden pollen and often a long, hollow spur containing nectar. There is typically a wide, flat petal forming a landing stage, advertised by coloured spots or lines. The rewards in zygomorphic flowers are usually accessible only to a subset of specialised insect species, reducing wastage to poor pollen vectors and directing more pollen to stigmas of their own plant species. Complex flower structures may also help to reduce water evaporation from nectar (Corbet 1978, Giurfa *et al.* 1996, Neal *et al.* 1998).

Fabaceae

Common Bird's-foot-trefoil *Lotus corniculatus* (Fabaceae) is a key plant for many long-tongued solitary bee species. Each plant displays a sequence of flowers over many weeks, and several rare *Osmia* species in northern Britain rely on it. The authors visited Gait Barrows National Nature Reserve in Lancashire to track down one of these, the elusive Wall Mason Bee *Osmia parietina* (Fig. 148 a). After we had watched Bird's-foot-trefoil at a likely spot on limestone pavement for three days of indifferent weather, the bee finally appeared, and some video was

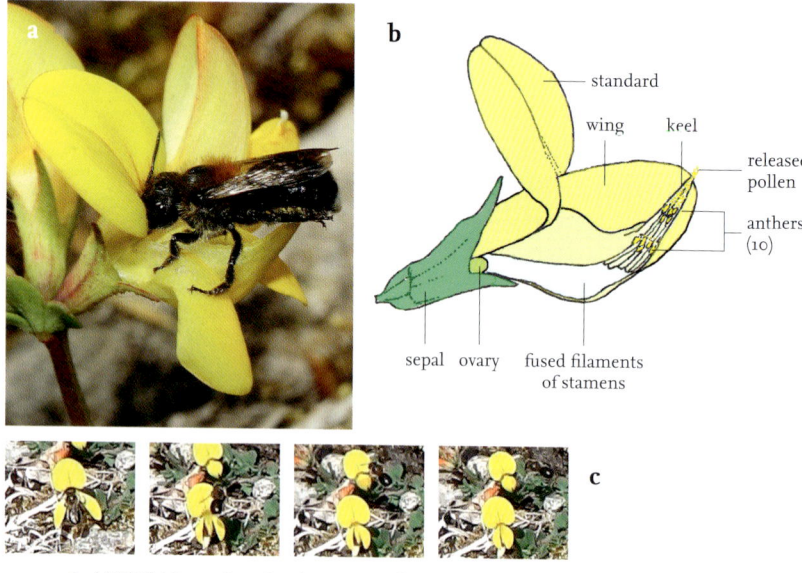

FIG 148. (a) Wall Mason Bee *Osmia parietina* female taking nectar from Common Bird's-foot-trefoil *Lotus corniculatus*. The tip of the pointed keel is applying pollen to the bee's scopa. Note the flexible splayed legs of the bee, which spread and depress the wing petals. (b) Anatomy of a Common Bird's-foot-trefoil flower, with near-side wing petal and keel petal removed. (c) Successive video frames of a Wall Mason Bee female visiting Common Bird's-foot-trefoil, showing the scissor-like movements of the wing petals, which snap shut after each insect visit.

obtained. This revealed the speed at which the bee and the flower mechanism work. The bee's front and middle legs spread out widely as it lands on the wing petals, which separate to reveal the keel petals in which the anthers and stigma are enclosed. As the bee seeks nectar, pressure on the keel causes the anthers to push a plug of pollen out of the pointed tip, applying a small portion of pollen onto the scopa on the underside of the bee's abdomen, assisted by the up-and-down pedalling movements of the hind legs (see Chapter 6). As the bee moves to the next flower, the wing petals spring back into shape in a fraction of a second, ready for another bee to arrive. Fabaceae flowers come in many sizes, suited to the weight of different sized bees, each Fabaceae species showing slight variations on this theme. Some of the larger species such as Gorse *Ulex europea* and Broom *Cytisus scoparius* have explosive flowers from which the anthers burst out of the keel petals, spreading pollen more liberally (Owens 2020b).

Lamiaceae

Flowers of the mint family (Lamiaceae), such as White Dead-nettle *Lamium album* and Red Dead-nettle *Lamium purpureum*, work very differently from Fabaceae flowers. The four anthers and forked stigma are situated under the hood petal. A bee is enticed to land on the spotted platform and to take nectar from the deep corolla tube. Pollen is dusted onto the bee's upper surface, or, at a later stage, the forked stigma picks up pollen from the bee's back, the pollen originating from a flower in the male stage, potentially from a different plant. Pollen interference between species is reduced by the pollen of each Lamiaceae species being placed on a different part of a bee's body.

FIG 149. (a) White Dead-nettle *Lamium album* and (b) Red Dead-nettle *Lamium purpureum* (Lamiaceae) both attract the Hairy-footed Flower Bee *Anthophora plumipes*, but the anthers (tucked under the hood) place pollen on the thorax or on the face, respectively.

FIG 150. (a) A Lamiaceae shrub places pollen on the upper surface of a Hairy-footed Flower Bee *Anthophora plumipes* female, while (b) the white pollen of Ground Ivy *Glechoma hederacea* is scraped off by the setae on the upper part of the tongue and (c) Red Dead-nettle *Lamium purpureum* places orange pollen on the face of a male.

Boraginaceae

Many members of the borage family (Boraginaceae) also have zygomorphic flowers and provide nectar and pollen for a wide range of solitary bees. The anthers and stigma of Viper's Bugloss *Echium vulgare* reach well beyond the petals and a bee will intercept them as it approaches the flower to take nectar from the deep corolla. The pollen is physically difficult to gather as it is held at the tips of long, protruding anthers, ensuring that pollen is dusted on many parts of the approaching bee. The arrangement is not proof against small bees, which can act as pollen thieves, taking pollen without touching the stigma. The Viper's Bugloss Mason Bee *Hoplitis adunca*, newly established in Britain, specialises on this plant (see Chapters 6 and 9).

Clustered flowers

A wide variety of plants create a conspicuous display by placing many small flowers together, all supported by a single stem, as in members of the parsley family (Apiaceae), such as Hogweed *Heracleum sphondylium* and Wild Carrot *Daucus carota*. Species in the daisy family (Asteraceae), such as Daisy *Bellis perennis* and dandelions (*Taraxacum* species), and the teazel family (Dipsacaceae), such as

FIG 151. Green-eyed Flower Bee *Anthophora bimaculata* female and Viper's Bugloss *Echium vulgare*.

Field Scabious *Knautia arvensis*, have even more dense 'flower' heads known as capitula. Each capitulum consists of hundreds of individual flowers, termed florets, massed together in a disc. The floral layout creates a combined signal, offering a large quantity of accessible nectar within a small area, which is nevertheless difficult for insects to obtain without brushing against the protruding anthers and stigmas. Capitula may be especially attractive to bees (and other insects) because they can visit several flowers without taking flight, reducing their foraging costs and increasing their ability to make an energetic profit on small flowers. The closely packed florets also deter florivores from stealing nectar from the base of a flower. The outer (ray) florets typically differ from the inner (disc) florets in having a single,

FIG 152. Common Yellow-face Bee *Hylaeus communis* female swallowing pollen from Viper's Bugloss *Echium vulgare* without touching the stigma.

FIG 153. Wood-carving Leafcutter Bee *Megachile ligniseca* female taking nectar from a cultivated aster. The 'flower' (capitulum) is composed of many yellow disc florets surrounded by a ring of mauve ray florets.

outward-facing flag petal, as in the white fringe of a common Daisy. In some cases, the ray florets are sterile and simply for display. The combined florets provide a target-like 'flower' which is highly visible (Neal et al. 1998).

New rings of florets open in succession, starting at the perimeter and moving inwards. Bees tend to orient themselves to the ring of open florets and receive pollen on their head and underside. In the Asteraceae, pollen is rationed out steadily as the floret matures. Pollen is presented to pollen vectors on the surface of the style before this becomes receptive, reducing the likelihood of self-pollination, though this can happen as a fall-back. In the subfamily Cichorioideae, the hairy style collects a coating of pollen as it pushes up between the anthers (Fig. 154), whereas in the subfamily Asteroideae, the tip of the style behaves more like a piston, pushing up a plug of pollen ahead of it. In both cases the process is gradual, dispensing pollen in small quantities (Leins & Erbar 2006). A similar system operates in thistles and knapweeds (Carduoideae).

In a wide range of Asteraceae species, the filaments of the anthers are able to contract in response to being touched by an insect's tongue, causing pollen to be pushed out in response, as first noted by Joseph Köhlreuter in the 18th century (cited Müller 1883). In *Centaurea* species, the filaments contract several millimetres when stimulated, causing the anther tube to slide down, exposing a plug of pollen (Proctor & Yeo 1973). In the absence of stimulation, the piston-like immature stigma pushes pollen out gradually from the long, tubular florets.

Populations of Sea Aster *Tripolium pannonicum* show a mosaic of rayed, partially rayed and rayless capitula. One of the main pollinators, the oligolectic

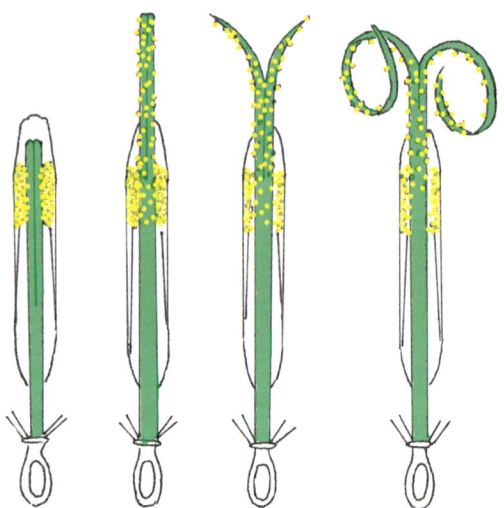

FIG 154. Time sequence of pollen presentation in Asteraceae florets. The rough outer surface of the stigma brushes pollen from the surrounding anthers as the style elongates. The stigma opens out but is not receptive to incoming pollen until most of its own pollen has been dispersed. Self-pollination is, however, possible if no other pollen is received.

FIG 155. (a) Common Knapweed *Centaurea nigra* and a Green-eyed Flower Bee *Anthophora bimaculata* female taking nectar and pollen. (b) In the absence of insects, the white pollen gradually accumulates. The capitulum had been covered by a mesh bag for seven days. Broad ray florets (lacking in this example) occur in Common Knapweed populations largely where there is competition for pollen vectors from Greater Knapweed *Centaurea scabiosa* (Lack 1976).

Sea Aster Bee *Colletes halophilus*, readily moves between the various forms and appears to show no selectivity. This variation may relate to competition between or within plant species for pollinators but has not been fully investigated.

The wide diversity of Asteraceae and Apiaceae species provides a sequence of foraging opportunities for solitary bees throughout the season, but members of the Dipsacaceae, such as scabious, generally flower from mid-summer. In all of these, the capitulum floral structure promotes rapid pollination of

FIG 156. Sea Aster Bee *Colletes halophilus* female collecting pollen from (a) a rayless and (b) a partially rayed form of Sea Aster *Tripolium pannonicum*.

multiple flowers (florets) and protects them from nectar thieves, but they remain vulnerable to pollen thieves such as beetles and thrips. Some Asteraceae defend themselves by closing their flowers early in the day, as explained below. A comparison of insect visitors to 10 species of Asteraceae and 10 of Apiaceae in Germany showed that bees comprised 43 per cent of Asteraceae visitors but only 11 per cent of Apiaceae visitors (Müller 1883, Corbet 1970). In Britain, Asteraceae have 17 oligolectic bee species but Apiaceae only two. Asteraceae and Apiaceae seemingly cater for a somewhat different spectrum of visitors.

THE FATE OF POLLEN

Rationing pollen is just one of many evolved mechanisms by which plants reduce pollen wastage. Many plants produce toxic or indigestible pollen, limiting the range of bees which consume their pollen to those able to tolerate it. Pollen of the plants concerned is therefore more likely to be transported to a member of their own species (Rivest & Forrest 2020; see also Chapters 6 & 7). The physical properties of pollen can also limit the range of bees which collect it. Studies of bumblebees and honeybees show that corbiculate bees have difficulty in collecting very large spiny pollen grains, which is a characteristic of species in the mallow family (Malvaceae), such as Common Mallow *Malva sylvestris*, Hollyhock *Alcea rosea* and cotton (*Gossypium* species). The large pollen grains do not pack easily into the pollen basket, and this has been attributed to the size and spikiness of the pollen and/or to properties of the waxy pollenkitt covering the pollen's surface (Konzmann *et al.* 2019, Vaissière & Vinson 1993). The spines of *Alcea rosea* pollen lack pollenkitt and inhibit clumping of the pollen (Lunau *et al.* 2015). Hao *et al.* (2020) studied 64 wild plant species pollinated largely by bees and found that pollen grain diameter was significantly larger in plant species from which bumblebees took nectar but did not groom the pollen adhering to their hair into their pollen baskets – though pollen grooming by solitary bees was not recorded.

Three members of the Dipsacaceae (Field Scabious *Knautia arvensis*, Small Scabious *Scabiosa columbaria* and Devil's-bit Scabious *Succisa pratensis*) have very large pollen grains, over 0.1 mm in diameter. Bumblebees and honeybees rarely harvest the pollen, but the pollen is, however, gathered by two oligolectic solitary bee species, the Small Scabious Mining Bee *Andrena marginata* and the Large Scabious Mining Bee *Andrena hattorfiana*, which are capable of holding the large pollen grains amongst their loose, feathery scopa hairs (Owens 2017, 2020b). An attempt was made to get some idea of the impact of these two oligolectic bees on their host plant, relative to other scabious visitors (Box 4).

BOX 4 Scabious bees: beneficial or harmful?

Knautia arvensis and *Scabiosa columbaria* flowers were watched for a combined total of eight hours at three sites in Norfolk, where both *Andrena marginata* and *Andrena hattorfiana* were present. Of 429 insect visits observed, 47 per cent were by bees, largely comprising mining bees (*Andrena* species) and bumblebees (*Bombus* species). All but one of the 33 *Andrena* visits were by *Andrena hattorfiana* or *Andrena marginata*. Pollen was collected almost entirely by these two bees, with much smaller amounts collected by occasional visits by three furrow bees and one leafcutter bee species. Some pollen was also probably consumed by hoverflies (Syrphidae). None of the workers of the seven (non-cuckoo) bumblebee species or the honeybees collected any pollen, and 72 per cent of bumblebee visits were by males. Amongst non-bees, Lepidoptera, especially Burnet Moths, and hoverflies were the most numerous visitors. Thirteen hoverfly species were observed but they made up only 7 per cent of insect visits. In total, 65 insect species visited the scabious flowers.

The two scabious bees may be particularly significant pollinators when there is strong competition between different plant species in a community, since they will not readily be diverted from scabious to other flowers. Scabious is protandrous and the two *Andrena* specialists continued to visit flowers in the female stage for nectar when no pollen remained (Fig. 157). Moreover, some *Knautia arvensis* plants have flowers which are entirely female, making this behaviour even more significant. However, the host plants thrive in many places where these specialist Andrenas are absent and their flowering continues well beyond the flight period of the bees. It is possible that scabious pollen also has toxic properties, since other members of the Dipsacaceae produce toxins in their pollen (Tang *et al.* 2019). In either case, the two scabious bees are able to specialise on a pollen source which is used by relatively few other insect species, at least at the sites observed. It remains an open question whether these two scabious bees should be regarded as beneficial or harmful to scabious.

FIG 157. Small Scabious Mining Bee *Andrena marginata* female taking nectar from a flower of Small Scabious *Scabiosa columbaria* in the female stage (stigmas protruding), with *Scabiosa columbaria* pollen (white) and Field Scabious *Knautia arvensis* pollen (pink) in her scopa, gleaned from other plants.

There are numerous examples of solitary bees appearing to evade the adaptations of flowering plants. Very small bees such as *Hylaeus* and *Lasioglossum* can take pollen from anthers with little probability of touching the stigma, while Orange-legged Furrow Bees *Halictus rubicundus* in Sutherland were seen to take nectar from between the bases of the wing petals of Common Bird's-foot-trefoil *Lotus corniculatus* flowers and bypassing the pollination mechanism. Similarly, *Nomada* species sometimes use the rob-holes made by bumblebees in comfrey (*Symphytum*) flowers (NWO pers. obs.).

COLOUR CHANGE

The foraging activities of solitary bees are dictated by the daily and seasonal rhythms of flowers as well as by their structure. In a very wide range of plants, flowers change colour after pollination, as seen especially in the Boraginaceae, whose flowers generally change from red to blue as they age. Flowers in the red phase are the most rewarding, and visiting flower bees (*Anthophora* species) and bumblebees learn to select them, though blue flowers continue to provide long-range attraction (Oberrath & Böhning-Gaese 1999). Flower colour changes are an honest signal to bees about where to find the best floral rewards and they also benefit the plant by directing pollinators to unfertilised flowers. The mechanism of colour change is not fully understood, but in some plants, there is evidence that cross-pollination acts as a trigger, operating through the growth of pollen tubes down the style (Weiss 1991, Nuttman & Willmer 2003, Nuttman *et al.* 2006).

FIG 158. Orange-legged Furrow Bee *Halictus rubicundus* female probing the base of a Common Bird's-foot-trefoil *Lotus corniculatus* flower.

THE TIMING OF FLOWERING

An extended flowering season may compensate for the vagaries of the weather and the unpredictability of

FIG 159. Red Mason Bee *Osmia bicornis* male on Wood Forget-me-not *Myosotis sylvatica*. The yellow centre of the flower changes to white after pollination. Yellow-centred flowers offer more pollen and nectar and are visited preferentially by hoverflies (Nuttman & Willmer 2008) and also by solitary bees including *Nomada* species and *Osmia bicornis* males (NWO pers. obs.).

pollen vectors. Flowers are often borne on branching stems, with a sequence of flowers appearing over several weeks or months. The anthers of individual flowers and capitula also typically dehisce in sequence over several days as, for example, in Devil's-bit Scabious *Succisa pratensis* (Dipsacaceae) – see Figure 160. Viper's Bugloss flowers last for 3–4 days, with the anthers ripening first then withering just as the stigmas become receptive. There are hourly changes in nectar quantity and quality resulting from interactions between the flowers, insect visitors and the

FIG 160. Staggered pollen release by Devil's-bit Scabious *Succisa pratensis*. The pink anthers have not yet opened (dehisced).

FIG 161. Tawny Mining Bee *Andrena fulva* female on *Prunus* species. The style bends sharply and touches the anthers in response to wet and windy weather (NWO pers. obs.).

weather, leading to changes in the spectrum of flower visitors (Corbet 1978). Poor weather or damage by herbivores can stimulate increased flower production (McCall & Irwin 2006, Wise *et al.* 2008), and some flowers apparently respond to extreme weather by hastening self-pollination (Fig. 161).

Some flowers open and close in response to changes in light intensity, temperature or internal cycles (Doorn & Meeteren 2003). Celandines *Ficaria verna* open when the air temperature reaches 10–15 °C, but direct sunlight induces much more rapid opening, even when the shade air temperature is lower than this threshold; temperatures close to the ground in full April sunshine quickly reach 25–30 °C (NWO pers. obs.), and temperatures in a pale flower such as this are higher than the ambient temperature (S. Corbet pers. comm.). Prokop and Fedor (2015) used wire loops to hold Celandine flowers open at night and a spray to mimic rain during daylight hours. Neither treatment had an effect on flowering time, which averaged four days, or on seed production, but flowers held open overnight in field experiments received significantly more damage from slugs and deer.

Some flowers close from about midday during sunny weather, perhaps as a means of protection against florivores. Daytime closing is particularly marked in yellow Asteraceae in the subfamily Cichorioideae, which typically close the capitulum around midday. Fründ *et al.* (2011) measured the speed of closure of flowers of Smooth Hawksbeard *Crepis capillaris* sown in experimental plots amongst other plant species. When insects were excluded, Smooth Hawksbeard flowers remained open for most of the day and closed at night. Adding an assortment of wild bees to the caged flowers caused the flowers to close much earlier in the day, and this was found to be a direct response to pollination rather

FIG 162. (a) A Field Sowthistle *Sonchus arvensis* flower (capitulum) was visited several times by a Patchwork Leafcutter Bee *Megachile centuncularis* female between 11:00 and 11:15 am. (b) and (c) The flower became closed completely within an hour of the bee's first visit. (© M. Fogden)

FIG 163. On a sunny day, exposed flowers of Beaked Hawksbeard *Crepis vesicaria* started to close at 11:30 am, while flowers enclosed in bags remained open.

than simply to physical disturbance by an insect. Artificial self-pollination did not cause closure, however; closure necessitated pollen from a different plant. Observations in a garden (NWO) showed that Field Sowthistle *Sonchus arvensis* flowers closed from around noon on sunny days. When insects were excluded using permeable cloth bags, the flowers remained open for an average of five hours longer per day than exposed flowers. All flowers closed before nightfall whether bagged or not, even when the flowers were picked and placed in a vase. Similar dramatic responses to enclosure happened in Rough Hawkbit *Leontodon hispidus* and Beaked Hawksbeard *Crepis vesicaria*. It is possible that differences in humidity or other abiotic factors within the bags caused this difference, but the effects were so dramatic and long-lasting that it was considered most likely to be the effects of cross-pollination as described by Fründ *et al*. Furthermore, the flowers opened inside the bags on cue the following day (Owens 2018).

Solitary bee species which take pollen exclusively from yellow Asteraceae, such as the Pantaloon Bee *Dasypoda hirtipes*, the Buff-tailed Mining Bee *Andrena*

FIG 164. (a) Pantaloon Bee *Dasypoda hirtipes* female, (b) Large Shaggy Bee *Panurgus banksianus* female and (c) Buff-tailed Mining Bee *Andrena humilis* female with large pollen loads from Asteraceae.

humilis, the Hawksbeard Mining Bee *Andrena fulvago* and shaggy bees (*Panurgus* species), share the characteristic of having extremely long pollen-collecting hairs which are able to carry large volumes of pollen. At a nesting aggregation of *Dasypoda hirtipes* in Norfolk, pollen was gathered rapidly in the mornings (on fine days), but most nest burrows remained sealed in the afternoons with the bees inside, as their main pollen source, Field Sowthistle, was no longer available (Owens *loc. cit.*). In the mornings, female *Dasypoda hirtipes* tended to follow a regular beat between stands of its food plant, seemingly making efficient use of time. These daily rhythms can change the outcome of field studies of natural pollination networks, which produce different outcomes depending on the time of day they are carried out (Fründ *et al. loc. cit.*).

Carl von Linné, the celebrated creator of the binomial taxonomic system, was well aware of the diurnal patterns of flower opening times and created a 'floral clock'. He proposed that the time of day could be discovered by examining which flowers were open (Linné 1783). However, he did not entertain the notion that the daily closure of flowers was a response to bees! Twenty-seven out of the 44 flowers represented in Linné's floral clock are members of the Cichorioideae (Fründ *et al. loc. cit.*).

OTHER SELECTION PRESSURES ON FLOWER FORM

Solitary bees often choose to rest in bowl-shaped or tubular flowers, which tend to be warmer than their surroundings. On a grassy common in Norfolk, the authors observed male and female Red-girdled Mining Bees *Andrena labiata* (Fig. 165) repeatedly retreating into buttercup (*Ranunculus*) flowers when clouds drifted across the sun, remaining to bask for a few seconds when the sun returned, before resuming flight. Similarly, in mountainous regions in southeast Spain, female Gwynne's Mining Bees *Andrena bicolor* use the corolla of the local wild daffodil *Narcissus longispathus* (a close relative of Britain's wild daffodil) to warm up. The daffodil corollas reached 8 °C above the ambient temperature and were warmest around the anthers. This relationship benefits both the plant and the bee, which is the daffodil's main pollinator (Herrera & Medrano 2017). Cultivated daffodils can be at least 2–3 °C warmer inside the corolla than outside and are sometimes occupied by *Anthophora plumipes* females on cool spring mornings (NWO pers. obs.). Sandy substrates warm up quickly and cool down quickly. On sand dunes, Silvery Leafcutter and Coast Leafcutter Bees (*Megachile leachella* and *Megachile maritima*) continue to bask on warm sand for some time after clouds cover the sun.

FIG 165. Red-girdled Mining Bee *Andrena labiata* female basking on a buttercup flower.

MIMICRY AND DECEPTION

Plants sharing a habitat often compete for pollinators, and it can be an advantage for flowers to evolve signals (colours and scents) which are distinctively different from the flowers of other nearby species. However, a less-rewarding plant can sometimes gain an advantage by resembling a more rewarding species. At a grassland site in Gloucestershire, several Ashy Mining Bees *Andrena cineraria* were observed collecting pollen exclusively from Rock Rose *Helianthemum nummularium* flowers, but one was also seen to land momentarily on two Bulbous Buttercup *Ranunculus bulbosus* flowers in succession, before reverting to Rock Rose. This was perhaps 'by mistake', since these two flowers appear very similar, at least to human eyes. The buttercup had use of a pollination service with little cost to itself in terms of rewards since the bee did not collect any pollen or nectar from the buttercup but potentially pollinated the second flower it visited. There are well-documented cases of floral mimicry in many habitats, though careful research is needed before drawing firm conclusions. To confirm this type of mimicry (Batesian mimicry), it is necessary to demonstrate that the mimic does better in the presence of the model than without it. An alternative interpretation is that a variety of rewarding flowers growing together all benefit by converging in appearance and pooling their floral signals (Müllerian mimicry). Since plant species inevitably differ in the quantity and quality of their rewards, these alternative types of mimicry are almost impossible to differentiate (Benitez-Vieyra *et al.* 2007, Willmer 2011).

Even within one species, individual plants growing in proximity vary in the amount of reward they produce. Individuals producing less nectar may benefit

FIG 166. (a) Bulbous Buttercup *Ranunculus bulbosus* (Ranunculaceae) and (b) Rock Rose *Helianthemum nummularium* (Rosaceae), a possible case of floral mimicry.

FIG 167. (a) Butterwort *Pinguicula vulgaris* and (b) Dog Violet *Viola canina* sharing a moorland habitat. The colours and complex structure of the flowers are similar.

from the more prolific individuals. Competition by deception within and between plant species works best if pollinators cannot easily detect the rewards available without landing and exploring the flower. Complex (zygomorphic) flowers with a deep corolla are well suited to such deception, as pollen and nectar are hidden within. Some complex flowers provide no reward at all and simply exploit their pollinators. Some bees, such as *Anthophora* species, are able to hover close to flowers without landing and are perhaps quicker at gauging the rewards on offer than some other solitary bee species, such as Andrenas. An evolutionary arms race may be expected to exist whereby pollinators get better at identifying the location and size of rewards, while deception plants improve their mimicry, enslaving pollinators without giving too much away. Conversely, rewarding plant species tend to evolve complex flower signals, involving a combination of colour, shape and scent, which reduce uncertainty in the receiver, assisting pollinators to distinguish between a model and its mimic and benefiting both the pollinator and the plant (Leonard *et al.* 2012, Kooi *et al.* 2020).

Deception was first noted by Sprengel, though doubted by Darwin. Orchids offer some prime examples. About one-third of European orchid species are now believed to give no reward at all to insect visitors (van der Pijl & Dodson 1966, cited Claessens & Kleynen 2011). Most British members of the genus *Orchis* are non-rewarding and deceive bees into visiting them by looking similar to the rewarding flowers of other plant species. Some are pollinated by bumblebees, but large solitary bees such as leafcutters also play a part. In many orchids, the mass of pollen is transferred in the form of a compact, pear-shaped lump, known as a pollinium. Each pollinium becomes firmly attached to the bee by a stalk (caudicle) which gradually bends after attachment, in such a way that the pollinium makes contact with the stigma of subsequent flowers visited by the bee. Small sub-units of the pollinium (maculae) detach and adhere to the stigma, resulting in pollination. Since there is no reward within the flowers, bees tend to move fairly quickly to another flower. Darwin proposed that the caudicle bends at a rate that makes any pollen transfer unlikely before the bee has moved to another plant, thus favouring cross-pollination. Since Darwin's time, measurements of

FIG 168. Coast Leafcutter Bee *Megachile maritima* male probing a nectarless flower of Southern Marsh Orchid *Dactlyorhiza paetermissa*. The bee emerges from a flower with a single pollinium attached to his face.

the speed of caudicle bending in a wide range of orchid species give support to Darwin's contention (Claessens & Kleynen *loc. cit.*). Experiments in which nectar is added to deception orchid flowers show that bees remain longer on each plant and the self-pollination rate increases, indicating that secreting nectar is likely to be disadvantageous to the plant rather than beneficial (Johnson *et al.* 2003).

Bee orchids (*Ophrys* species) have received a lot of attention from European ecologists. A very wide range of species occurs in Mediterranean habitats, though there are just four (very interesting) species in Britain. Bee orchids are so called because their flowers mimic a female bee, both in appearance and in scent, which serves to attract male bees, usually of just one species. Flowers of the Bee Orchid *Ophrys apifera* look rather like a bumblebee, but in Britain the orchid is entirely self-pollinating. This is possibly because their particular pollinator (said to be a *Eucera* or *Tetralonia* species) is absent from Britain, though some overseas *Ophrys apifera* flowers do also self-pollinate (NWO pers. obs.). The pollinia dangle down and are blown by the wind onto the stigma (Darwin 1862, Claessens & Kleynen *loc. cit.*).

Early Spider Orchids *Ophrys sphegodes* behave in a more typical *Ophrys* fashion. Their flowers mimic a female Buffish Mining Bee *Andrena nigroaenea* and release a scent which closely mimics the female pheromone (see Chapter 2). Males are attracted to the flowers, and when they attempt to mate (pseudocopulation), pollinia become attached to their head. Observations suggest that the male bee lands on the flower for a couple of seconds, at most (K. Wilson pers. comm., based on 12-plus observations). Each plant differs slightly in the composition of the mimic pheromone and in appearance, and this may tempt males to try more Early Spider Orchid plants rather than giving up (Ayasse *et al.* 2000). After pollination, each flower reduces its level of attractants but increases the amount of farnesyl hexanoate, which acts as an antiaphrodisiac in female Buffish Mining Bees. The substance deters further visits to a previously visited flower, preventing disturbance of any deposited pollen and encouraging visits to new flowers, sometimes on the same plant (Schiestl & Ayasse 2001). Male Buffish Mining Bees emerge before the females, and this gives the orchid an opportunity to attract males with little competition from 'real' females. Seed set is highly variable (Claessens & Kleynen 2011; NWO pers. obs.), but there is evidence that climate change is advancing the emergence of the bee slightly more than the flowering of the orchid. This could reduce the number of males still flying by the time the orchid flowers open and could also give rise to more competition between the orchid and female bees for males (Robbirt 2012).

Red Helleborine Orchids *Cephalanthera rubra* are also entirely dependent on solitary bees for pollination but work in a different way. The flowers appear rose-coloured to humans but in fact reflect light across very similar wavelengths

FIG 169. (a) Flower of Early Spider Orchid *Ophrys sphegodes*. It has 'eyes' and shiny 'wings' with a suggestion of a hairy body. There is much variation within and between populations. (© R. Laurence) (b) Buffish Mining Bee *Andrena nigroaenea* male pseudocopulating with an Early Spider Orchid. Pollinia are attached to his head from flowers he has visited previously.

to (blue) bellflowers (Nilsson 1983). In Britain, this serves to attract Small Scissor Bees *Chelostoma campanularum*, a species which, as we have seen, is a *Campanula* specialist. Male scissor bees swarm on Red Helleborine flowers in mistake for *Campanula* flowers and pick up pollinia on their back. The orchid is very rare in Britain and often fails to set seed. The chief pollinator in mainland Europe is *Chelostoma fuliginosum*, which is absent from Britain, and this may partly explain the orchid's lack of success in its Buckinghamshire and Gloucestershire localities. *Chelostoma campanularum* is present but may be too small to act as a reliable vector (Harvey 2006).

Perhaps the most elegant British example of bee manipulation by flowers occurs in the Lady's Slipper Orchid *Cypripedium calceolus*, a plant which is entirely dependent on the services of solitary bees. In the 19th century, the orchid was collected almost to extinction in Britain. A famous survivor in the Yorkshire Dales was carefully guarded, human visitors even being required to be blindfolded during their approach and departure. British Lady's Slipper Orchids have now been successfully propagated and reintroduced. One of us (NWO) was able to watch some perfect reintroduced specimens in Lancashire on a sunny May morning. On one plant, the large, yellow flowers attracted several female Rufous-footed Furrow Bees *Lasioglossum rufitarse* (Fig. 170). Some flew directly into

FIG 170. Rufous-footed Furrow Bee *Lasioglossum rufitarse* female escaping from Lady's Slipper Orchid *Cypripedium calceolus* flower.

the slipper via the main opening, while others landed on the edge and slipped inside. The flower has an oily covering, inside and out. A few landed on the purple-spotted staminode (a modified anther), which works like a playground slide. Once inside, the bees crawled about and buzzed their wings, but the opening was too constricted to allow them to fly out. Within 60–90 seconds, the bees moved towards the transparent 'windows' at the back of the flower. The windows produce a light gradient, guiding bees towards two exits, each partially guarded by a stigma and anther. A bee needs to squeeze past the stigma and then the anther to escape and turns sideways to do so. If the bee is of suitable size, pollen, resembling yellow toothpaste, is squeezed out onto its back or underside. The bees we saw were small and received very little pollen. Fortunately, another observer had previously taken photographs of a female Orange-tailed Mining Bee *Andrena haemorrhoa* (Fig. 171) visiting the orchid. This solitary bee is just the right size and occurs widely in northern regions of Britain where the Lady's Slipper naturally grows. When a pollen-loaded bee passes through another flower, pollen is scraped off by the papillae on the stigma, thereby pollinating it (Claessens & Kleynen 2013).

A rather subtle (non-orchid) example of within-species mimicry seems to occur in Red Campion flowers, which, as we have seen, are either male or female (the species is dioecious). Male flowers can be a good pollen source to

FIG 171. Orange-tailed Mining Bee *Andrena haemorrhoa* female escaping from a Lady's Slipper Orchid flower with pollen paste on its face and back. (© H. Harrison)

FIG 172. (a) Red Campion *Silene dioica* male flowers. (b) The white scales mimic the stigmas of female flowers.

solitary bees and have white scales around the flower entrance which resemble the stigmas of female flowers. Female flowers provide more dilute nectar but in greater quantity than in male flowers. However, some female bees apparently visit female flowers in mistake for male flowers, which they have learnt to recognise as a pollen source (Kay *et al.* 1984).

FIG 173. Black-headed Mining Bee *Andrena nigriceps* female collecting pollen from Red Campion male. The tongue is too short to reach the nectar, but she could potentially pollinate a female flower.

STAY OR GO? FLOWER CONSTANCY

Foraging bees often go from flower to flower of one species of plant for long periods of time and are seemingly unresponsive to possible alternatives in their vicinity. It is in a plant's interest for pollinators to show this kind of fidelity, since pollen is taken to stigmas of its own species and not wasted elsewhere, and stigmas are not contaminated with heterospecific pollen. There is evidence that bees are less likely to move between flowers whose signals are distinct from one another, so in a stable plant community, we might expect flower characteristics, such as petal colour and shape, to diverge over evolutionary time. However, as already mentioned, there can be selection in the other direction, since poorly rewarding plants can benefit by having flowers which resemble those that provide more nectar and pollen. Bees will then visit the poorly rewarding species 'in mistake' for the more rewarding ones. Groups of rewarding species may 'cooperate' by sharing flower characteristics, creating a more attractive overall display. But despite these caveats, it is thought likely that many aspects of floral diversity do arise from distinctive flower adaptations, which promote fidelity by bees (Chittka *et al.* 1999).

When a female solitary bee first emerges from the nest burrow or from hibernation, it begins to make decisions about which flowers to visit. At first the bee will seek food only for herself, but she will soon start collecting pollen and nectar to stock her brood cells. The bee may have some inbuilt genetic preferences for particular types of flowers, confining its visits to flowers of a particular plant family or genus. Preferences may also derive from the food the bee was provided with as a larva. These oligolectic bees probably recognise their preferred flower type on emergence and will largely ignore other plant species throughout their adult lives (see Chapter 6). More surprisingly, generalist (polylectic) bees very often restrict their flower visits for several minutes to one or a few particular plant species, typically foraging from one type of flower for long sequences and bypassing alternative flowers which appear equally rewarding. For example, when Primroses *Primula vulgaris* and Cowslips *Primula veris* grow close to each other, an *Anthophora plumipes* female might move from one Primrose flower to the next and ignore the Cowslips, while another bee of the same species foraging alongside it might visit only Cowslips for a while. From time to time, each bee could switch to the other flower type.

This kind of transient fidelity was first noted in honeybees by Aristotle (350 BCE; cited Bennett 1883) and is usually referred to as flower constancy. The *Anthophora plumipes* example just described can be considered 'true' flower constancy since bees are (temporarily) being faithful to one flower type and bypassing equally good or better alternatives. Each individual bee shows constancy for a while to one flower type before moving to another, and at any one time, individuals of a given bee species may be visiting different plant species. There has been much debate about the nature and causes of flower constancy and whether the behaviour benefits bees (Chittka *et al.* 1999). The topic has considerable relevance to crop pollination as well as to the pollination of wild plants.

Here, for simplicity, we use the term 'flower fidelity' for any tendency of a bee to show a preference for a restricted range of flowers, with flower constancy being a special case of flower fidelity. The simplest explanation for flower fidelity is that flowers often grow in patches or clusters and bees (or other insects) inevitably forage within each patch for a while before moving on. This happens in monocultures such as orchards and Oilseed Rape *Brassica napus* crops and also in some specialised plant communities such as a salt marsh rich in Sea Aster *Tripolium pannonicum*. These cases cannot be interpreted as flower constancy since bees are not bypassing more rewarding flowers.

Waser (1986) distinguished between the 'fixed preference' seen in oligolectic bee species and the 'labile preference' exhibited by polylectic bees, whose flower selection is more flexible. The discussion that follows refers only to the labile preference of polylectic bees and their choice of flowers (Table 5). Optimal Foraging Theory (OFT) proposes the common-sense idea that a bee (or any animal) seeks to maximise its food intake with the least expenditure of time and energy (maximising its rate of net energy gain). This would be achieved by a bee visiting only the most abundant and rewarding types of flowers (but not carrying excessively heavy loads, as this reduces lifespan (Wolf & Schmid-Hempel 1989)). Waser restricts 'true' flower constancy to instances where foraging polylectic bees apparently go against the predictions of OFT and confine their visits to a single plant species regardless of cost, sometimes travelling greater distances to do so and ignoring equally or more rewarding alternatives. At first sight this would not seem to make much sense for the bee. A widely accepted explanation for flower constancy, so defined, is based on supposed limitations to a bee's ability to adapt its behaviour to a different floral morphology. Switching flower type is thought to slow down foraging for a while as the bee learns to tackle the different floral layout. In Waser's words, 'skipping other flowers is the best strategy if reduced handling time more than offsets increased travel costs'. One prediction would be that bees are more reluctant to transfer to species with dissimilar morphology than to similar ones. Waser carried out experiments using pairs of similar or dissimilar Asteraceae flowers placed in arrays in the Arizona desert, visited by wild *Anthophora* species. As predicted, the bees showed more fidelity when one plant species had ray florets (like a common Daisy flower) and the other did not, than when both had ray florets. Similar results were obtained for bumblebees in Costa Rica, using more contrasting flower forms.

TABLE 5. Possible reasons for flower fidelity in bees. The labile preference categories are not all mutually exclusive.

Fidelity cause	Explanation of flower fidelity (feeding for a long sequence on one flower type)
Oligolectic bees (fixed preference)	Genetic factors or early imprinting lead to fidelity to a small range of flower species.
Polylectic bees (labile preference)	
Flower abundance	Superabundance of one plant species: there is no alternative choice nearby.
Quantity of reward	Bees select flowers offering the greatest rewards and remain with them.
Quality of reward	Bees select flowers offering the highest nectar concentration or the best pollen quality.
Optimal foraging	Bees select flowers which offer the highest rate of net energy gain.
Flower constancy	Bees select flowers which are not necessarily the most rewarding and bypass others, various possible explanations.

Although Waser's observations were consistent with the expectations of floral constancy, it is very difficult to know in a natural situation whether some of the bypassed flowers are in reality equally or more rewarding. Rewards are influenced by the density of individual flowers and the quantity and quality of reward each one holds, and also by the bee's comparative skills at collecting the pollen or nectar from the alternatives. Bees may reject an alternative pollen- and nectar-loaded flower because the bee is not well adapted to foraging from that particular flower structure, perhaps by virtue of its tongue length. The quantity of reward may also be affected by recent visits of other foragers. Some bees, such as *Anthophora*, *Anthidium* and *Eucera* species, are able to hover in front of flowers while assessing them, while others, such as *Andrena* species, need to land to check the rewards present. Such differences in behaviour will influence measurements of flower constancy. A further complication is that bees may select dilute nectar when they are dehydrated rather than always going for the richest nectar (Prys-Jones & Corbet 2011).

LEARNING AND CONSTANCY

Chittka *et al.* (1999) reviewed the subject of flower constancy, based largely on studies of the learning abilities of bumblebees and honeybees. The authors distinguish between sensory and motor learning by bees. Sensory learning refers to remembering flower shape, colour and scent, while motor learning involves remembering how to move the body appropriately when accessing a flower, as noted by Darwin (1876). When a new flower is encountered, familiarisation takes some time, and learning may be more effective when it involves a wide range of sensory and motor features. Insects appear to have short-term memory (STM) and long-term memory (LTM), similar to that in mammals. STM is a 'working memory' and limits to its capacity may partly explain flower constancy. Working memory of a particular flower can be lost when a bee shifts its attention to a different flower type. STM also retains a running tally of floral rewards during foraging, with bees moving away from unrewarding patches but turning more sharply when rewards are high, thereby staying within the same patch. Such decisions are more complex when flowers of different types grow together.

Experiments with honeybees and bumblebees show that bees can store knowledge of multiple flower types in LTM. When returning to a familiar flower, bees need to access their LTM and bring the information to their STM, causing some delay. Bees therefore tend to bypass familiar flowers which are not in an uploaded state in their brain. When dealing with new flowers, bees can learn more quickly by visiting sequences of one new flower type rather than alternating between two new flower types, though the latter strategy may help a bee to acquire a better association between sensory and motor tasks. These considerations suggest that bees have reason to show constancy both to familiar flowers and to new ones, but that more complex patterns of learning, involving alternating between flower types, can sometimes be advantageous.

How quickly can bees learn? Honeybees typically extend their proboscis by reflex action when sugar (sucrose) solution, such as nectar, is detected. Experiments show that just one trial, pairing sucrose with a particular scent, is sufficient to condition the reflex such that the tongue is immediately extended in the presence of the scent alone, with the response persisting as a long-term memory (Hammer & Menzel 1995). If honeybees are typical, it would seem that bees can achieve associative sensory learning (learning scents, colours and shapes) very quickly when a reward is provided. In various experiments with captive bees, including the solitary bee *Osmia bicornis*, it was found that bees in general can learn the sensory features of a flower with just 1–20 rewarded trials and that the resulting LTM can last for several weeks, roughly matching the typical lifetime of an adult bee. Learning motor skills generally takes longer, however, requiring visits to around 30–100 flowers, so it is likely to be the rate that bees learn or recall motor skills, rather than sensory skills, that selects for flower constancy in bees. Each time a bee diverts to a different flower, the working memory of the previous one can be lost, so hopping between plant species with contrasting flower structures potentially slows down overall foraging speed (Chittka 1986, Chittka *et al.* 1992).

REALITY CHECK: OBSERVING BEES IN A SUMMER PASTURE

Field observations of solitary bees can provide some clues about the nature and possible causes of flower constancy in natural conditions. Bees of 12 species were observed foraging together on flower-rich limestone grassland in Gloucestershire during June (NWO pers. obs.). Individual female bees were watched for several minutes in turn and their flower choices noted, including whether they were taking nectar, pollen or both. All the bee species observed showed considerable flower fidelity. Buffish Mining Bees *Andrena nigroaenea* and Ashy Mining Bees *Andrena cineraria* (Fig. 1/4) both took pollen and nectar exclusively from Rock Rose *Helianthemum nummularium* and ignored other flowers. *Andrena nigroaenea* typically visited 50 or more Rock Rose flowers in succession, despite the presence amongst them of two buttercup species and three Fabaceae species, all of them yellow (to human eyes). *Andrena cineraria* showed similar strong fidelity to Rock

FIG 174. Ashy Mining Bee *Andrena cineraria* female collecting pollen from Rock Rose *Helianthemum nummularium*.

Rose. Hawksbeard Mining Bees *Andrena fulvago* favoured Mouse-ear Hawkweed *Pilosella officinalis* flowers, as expected in an Asteraceae specialist (Elsc & Edwards 2018). White-zoned Furrow Bees *Lasioglossum leucozonium*, another oligolege of Asteraceae, was seen only on Rough Hawkbit *Leontodon hispidus*, while individuals of the Chalk Furrow Bee *Lasioglossum fulvicorne*, a polylectic species, foraged on both Rough Hawkbit and Meadow Buttercup *Ranunculus acris*. Four species of bumblebees used Rock Rose as a pollen source but regularly diverted to nectar-providing flowers, perhaps because nectar is required to adhere pollen grains in their pollen baskets. This requirement, which is not typical of solitary bees, resulted in solitary bees displaying much greater flower fidelity than bumblebees during these (limited) observations.

The flower choices of the solitary bee species appeared not to differ between individuals, but honeybees foraged in a strikingly different manner. When taking nectar, each honeybee displayed its own preference, with individuals feeding on Thyme, Common Bird's-foot-trefoil, or Horseshoe Vetch for several minutes, with no honeybee observed switching between these species. Bird's-foot-trefoil and Horseshoe Vetch are members of the family Fabaceae and physically very similar, but individual honeybees nevertheless bypassed the species they were not using, perhaps distinguishing them by scent. Other individual honeybees focused on pollen-gathering from Rock Rose. Honeybees overall visited a wider range of flowers than any of the other 11 bee species observed, but each honeybee showed complete flower fidelity (this was probably the fidelity noticed by Aristotle).

This type of individuality within a bee species was not seen in any of the solitary bees or bumblebees present but is consistent with the findings of experimental studies on honeybees. Wells & Wells (1983) tested honeybees with artificial flower patches of mixed colours and rewards and found that individual bees remained constant to one flower colour or morph, even when the flower chosen was less rewarding in terms of quantity or quality; in other words, they showed true flower constancy. Similarly, Hill *et al.* (1997) presented honeybees with patches of yellow and blue artificial flowers. The bees remained constant to the first colour they landed on, some remaining constant to yellow flowers and some to blue. Only 17 out of 3,017 visits were to the flower colour that was not initially landed on. Lubbock (1882) found that individual honeybees trained to go to blue or green colours retained their preference from one day to the next. This pattern of behaviour is thought to assist resource partitioning for the honeybee colony, reducing competition by establishing a society of individual specialists. It is not known whether eusocial halictid bees behave in any comparable way, but interesting evidence from *Lasioglossum pauxillum* (see Chapter 4) suggests that this is likely.

A somewhat different line of reasoning suggests that minimising risk is important and that bees remain with a reasonably good food source rather than take the risk of transferring to one which might be worse (minimal uncertainty hypothesis). Patterns of foraging by honeybees may reflect flower constancy, optimal foraging and minimising uncertainty (Wells & Wells 1986). In natural communities of plants, nectar rewards can be lower and the range of flower choices much greater than in experimental arrangements. Assessing all the options therefore takes more time, making flower constancy the best option (costly information hypothesis) (Chittka *et al.* 1999).

POLLINATION SYNDROMES AND SPECIALISATION

Insects, including bees and a wide range of other animal visitors, provide a 'pollination service' to an estimated 308,000 species of flowering plants worldwide (Ollerton *et al.* 2011, Christenhusz & Byng 2016), and multiple attempts have been made to recognise patterns (syndromes) of plant-pollinator relationships amongst them, with the ultimate aim of understanding how pollinators have contributed to floral evolution. The pollination syndrome concept proposes that flowers of different plant species, often from different families, have converged towards sharing similar traits (such as colours, scents and structural forms) in response to natural selection by their pollinators, which in turn have evolved complementary adaptations in size, structure, physiology and behaviour (Benton 2017). For example, flowers of the families Fabaceae, Scrophulariaceae and Lamiaceae are typically zygomorphic and provide copious nectar hidden within the flower. The conventional explanation for their similarity would be that the flowers of the three families are pollinated by insects with a long tongue, and flowers and pollinators have become mutually adapted. Each pollination syndrome is considered to attract a different suite of pollinators, these being the pollinators which are the most effective pollen vectors for that syndrome and therefore provide the strongest selection pressure on flower form.

Joseph Kölreuter was one of the first to recognise that animals could act as pollinators, but he realised that pollinators and plants could have somewhat different interests rather than being entirely mutually supporting. He also noticed that a plant could be visited by many insect species rather than just a few specialists and, conversely, that some insects visited a wide range of plants, suggesting that there was not usually a close one-to-one fit between pollinators and flowering plants (Kölreuter 1761, cited Waser 2006).

The Italian Federico Delpino, a correspondent of Darwin, assigned flowers to functional groups according to their main pollinators. Following Delpino and others, Vogel (1954) proposed distinctive flower adaptations attracting either bees, butterflies, moths, flies, birds or bats (Table 6). These different pollination styles are now generally referred to as pollination syndromes. Additional syndromes have been recognised, such as a beetle syndrome, while flies and moths can both be divided into sub-categories (Willmer 2011). Vogel's bee syndrome characteristics seem to apply largely to long-tongued bees and the complex flowers they generally visit, but not to less specialised bee and plant species. More recently, Faegri & Van de Pijl (1979) descibed six main pollination syndromes (dish- and bowl-shaped, bell- or funnel-shaped, head- or brush-shaped, tube- and gullet-shaped) and also recognised numerous different syndromes within some flower families. They considered there to be 'harmony between visitor and blossom', somewhat overlooking the point first made by Kölreuter that plants and their visitors can have different interests. Succeeding authors have classified flowers into a number of pollination categories, variously referred to as structural blossom classes, spectra or syndromes, with orchids sometimes being placed in a group of their own. Apart from orchids, the categories are only loosely related to taxonomy.

The pollination syndrome concept has provided useful insights into the mutual adaptations of flowers and bees, especially the significance of concealed nectar and tongue length, but it is not easy to pigeon-hole all characteristics of bee anatomy and flower structures into particular syndromes. The syndromes themselves have not been consistently defined, and the term has been applied both to the flower visitors (birds, bees, etc.) and to the flowers themselves. More

TABLE 6. Pollination syndromes based on Vogel (1954), cited Waser (2006). Examples of plants representing each pollination syndrome have been added.

Syndrome	Colour and pattern as perceived by humans	General shape	Details	Nectar	Scent as perceived by humans	Opening and closing times	Other features	Examples
Bees	Blue, violet, purple, yellow, white: nectar guides	Flag, gullet, tubular, brush (nototribic and sternotribic): pollen on back or underside	Landing stage, narrow tube, thin peduncles	Hidden up to 15 mm deep	Strong, honey-like	Day	Silken or velvety sheen: robust	Red Dead-nettle, Common Bird's-foot-trefoil
Butterflies	Scarlet, purple, blue, yellow, white: nectar guides	Long tube or spur (salverform): pollen on wings, head, proboscis	Disc-shaped with long tube, anthers pendulous	Hidden up to 40 mm deep	Pleasant, honey-like	Day	Delicate	Valerian, Buddleia
Nocturnal moths	White, cream, violet, yellow-green: no nectar guides	Salverform paintbrush type: pollen on wings, head, proboscis	Star-shaped, narrow tube, anthers pendulous	Hidden up to 200 mm deep	Pleasant, intoxicating	Evening and night	Waxy surface, delicate	Honeysuckle, Tobacco flower

continued overleaf

TABLE 6. continued

Syndrome	Colour and pattern as perceived by humans	General shape	Details	Nectar	Scent as perceived by humans	Opening and closing times	Other features	Examples
Flies	Brown, brown-red, dirty yellow, green-white	Basin or saucer: pollen on proboscis or legs	Flat, clear patches, wrinkles	Exposed, accessible	Nauseating	Day	Reflective or dull, warty, ciliated	Ivy, Hogweed
Birds	Scarlet, red-orange, yellow-green, pure blue, pure white, dark violet: nectar guides	Tubular, salverform, paintbrush type: pollen on throat, forehead, beak	Expanded opening, pendulous, anthers and stigma bent upwards together	Varying depth, plentiful, dilute, slimy	None	Day	Robust	No British species *Heliconia*, *Hibiscus*
Bats	White, cream: no nectar guides	Wide tube			Fruity, beet-like	Night	Fleshy sepals and petals	No British species take nectar

significantly, pollination syndromes imply close specialisations (harmony) between pollinator and plant and this assumption has been seriously challenged. As we have seen, very close mutual relationships between flowers and bees are not the norm. Close mutualisms amongst solitary bees and flowers in Britain are limited to rare cases where plants provide special substances such as floral oils, as in the case of Yellow Loosestrife *Lysimachia vulgaris* and the Yellow Loosestrife Bee *Macropis europaea* (see Chapter 6), or in plants not providing any reward at all, as in *Ophrys* Orchid species. A review by Waser *et al.* (1996) noted that 'pollination systems are much more generalised and dynamic than traditions of closely co-evolved pollination syndromes have suggested'. There is now considerable support for this less-rigid view from research on plant communities. One example is a four-year study of 30 h of scrub near Athens (Petanidou *et al.* 2008), where plant-pollinator interactions involving 262 bee species varied widely from year to year. Plant species which appeared as specialists in one year tended to be generalists or to interact with different insect species in other years. Generalist bees visited some plants which had few other visitors, and specialist bees generally visited plants which had many other visitors. There were no one-to-one species specialisations, though many oligolectic bees were present.

Studies of individual British plant species and their solitary bee visitors are limited. One example is a study of Sheep's-bit *Jasione montana*, an atypical member of the Campanulaceae, with flowers displayed in a capitulum. Forty bee species have been recorded taking pollen from the plant across its European range, only five of them being oligoleges of Campanulaceae, with a further 91 bee species noted as visiting the flowers for nectar only. Other visitors included 76 species of wasps (Early 2020). The plant is clearly a generalist and highlights the degree of selectivity achieved by other members of the Campanulaceae. Another case is Wild Mignonette *Reseda lutea* (Resedaceae) which, at a site in Surrey, is visited by 24 solitary bee species, only one of which, the Large Yellow-face Bee *Hylaeus signatus*, is a *Reseda* specialist (J. Early pers. comm.).

The absence of a close tie between many pollinators and flower structures poses the problem of how the more generalist flower types have evolved. Have flower forms been moulded by the actions of just the most effective pollen vectors of each plant, as suggested by Stebbins (1970), or by a wide spectrum of reasonably good ones? Specialist (oligolectic) bees are not necessarily the best pollen vectors, since they take large quantities of pollen for their nests, though this can be offset by their strong flower fidelity, as we have seen in scabious species. Oligolectic bees generally specialise on plants which also attract a very wide range of other pollen vectors, and plants are not usually dependent on the specialists themselves (Minckley & Roulston 2011).

TABLE 7. Broad associations between solitary bee genera and flower types.

Genus	Bowl-shaped	Brush	Densely clustered	Trumpets and bells	Floral tubes	Zygomorphic nototribic*	Zygomorphic sternotribic**	Composite	Orchids
Andrena	✓	✓	✓	✓	✓		✓	✓	✓
Anthidium						✓			
Anthophora	✓	✓		✓	✓	✓		✓	
Apis	✓	✓	✓	✓		✓		✓	✓
Bombus	✓	✓	✓	✓	✓	✓	✓	✓	✓
Ceratina	✓		✓	✓	✓	✓		✓	
Chelostoma	✓			✓					✓
Colletes		✓	✓					✓	
Dasypoda								✓	
Eucera							✓		
Halictus	✓		✓	✓	✓	✓		✓	
Heriades								✓	
Hoplitis	✓		✓			✓		✓	
Hylaeus	✓		✓	✓	✓	✓		✓	
Lasioglossum	✓		✓	✓	✓			✓	✓
Macropis	✓								
Megachile	✓		✓				✓	✓	✓
Melitta				✓			✓		
Osmia	✓		✓		✓	✓	✓	✓	
Panurgus								✓	
Xylocopa	✓				✓				

* Pollen placed on upper surface of a bee's body.
** Pollen placed on underside of the body.

The apparent lack of close specialisation in British flowering plants may relate partly to pollinators of past plant generations rather than to current ones. The ranges of flowering plants often extend beyond the range of the pollinator community in which they evolved (Herrera 1996). Britain has around 1,300 species of native flowering plants which colonised Britain in post-glacial times and 150 'archaeophytes': plants which arrived in Britain by human agency before the year 1500, mostly since the advent of farming in Neolithic times. Since 1500, we have gained many further established aliens ('neophytes') and frequent 'casuals' which together number about as many species as our natives (Preston *et al.* 2002; C. Preston pers. comm.). Most British solitary bees are no better adapted to native species than to new arrivals. Some inborn preferences for flower signals may be present in newly emerged solitary bees, but they also quickly learn to deal with the plants they encounter in their surroundings. These considerations suggest that we cannot expect that a solitary bee visiting a flower will be closely adapted to that particular plant species. Nevertheless, we do see patterns of association between solitary bee genera and flower type, as would be expected on the basis of taxonomic differences in tongue length, scopa type, oligolecty and other characteristics of solitary bees (see Chapter 1). Table 7 attempts to highlight the floral types (syndromes) generally visited for pollen by each British bee genus, based on personal experience and published reports of pollen sources (Else & Edwards 2018).

A fuller understanding of flower-pollinator interactions requires careful study of individual plant species and their known visitors in a wide range of habitats, ideally over several seasons. This can potentially enable functional groups of flowers to be matched with functional groups (rather than taxonomic groups)

FIG 175. Long-horned Bee *Eucera longicornis* female visiting Bush Vetch *Vicia sepium*, which has sternotribic zygomorphic flowers.

of pollinators (Corbet 2006). A simple version of this approach with bees in a single garden (NWO pers. obs.) did show some clear patterns. The garden offers a spring display of wild and cultivated flowers, competing with each other for the attentions of pollinators, giving an opportunity to begin identifying links between the morphology of bees and the flowers they visit for pollen and/or nectar. The garden is bordered by a small stream adjacent to open farmland. The same floral menu was available to a variety of bees, but each species made somewhat different choices. Table 8 shows the flowers visited by six frequently seen bee species inhabiting the garden and stream bank during early April. Individual bees were followed for as long as possible, and the relative frequency of their visits to each flower type was determined.

There was a clear link between tongue length and the category of flowers chosen: bowl-shaped, densely clustered and composite flowers were used very largely by short-tongued bees, whereas tubular and zygomorphic flowers were much more likely to be visited by long-tongued species. There were some anomalies: *Anthophora plumipes* females readily visited domestic pear (*Pyrus*) blossom, immediately changing their foraging style from zygomorphic to bowl-shaped flowers and remaining for long sequences on pear without apparent difficulty. Honeybees spent a lot of their time taking nectar from Ground Ivy, a zygomorphic flower but small enough for a short(ish)-tongued bee to tackle. Ground Ivy was also visited by *Andrena bicolor*, which collected both nectar and pollen, sometimes by adopting an inverted position on the flower. *Andrena bicolor* also collected pollen from (thrum) Primroses and nectar and pollen from Celandines. It was perhaps necessary for the bee to visit Celandines from time to time to obtain nectar, returning to Primroses periodically, where pollen collection may have been quicker and/or the quality better (the bee's tongue is too short to reach the nectar of a Primrose). Bell-shaped flowers were visited about equally by short-tongued and long-tongued bees.

Pollinators can differ markedly between sites, so this snapshot will be particular to this locality. For example, Marsh Marigolds and Dandelions at Dunnet Bay in Caithness are visited largely by Diptera, especially hoverflies such as *Cheilosia* species (NWO pers. obs.). Only four solitary bee species are known at this latitude in Britain (see Chapter 9). Solitary bees cannot function in cool, damp conditions since pollen must be dry and powdery and perhaps electrostatically charged to allow its collection. In cool spring weather and in shady habitats, Diptera are the main visitors in southern counties, too, diluting the selection pressures by bees on flower structure. Also, flower features that make them attractive to bees will be less advantageous if they also attract herbivores. The effects of herbivores may become clear only when taking account

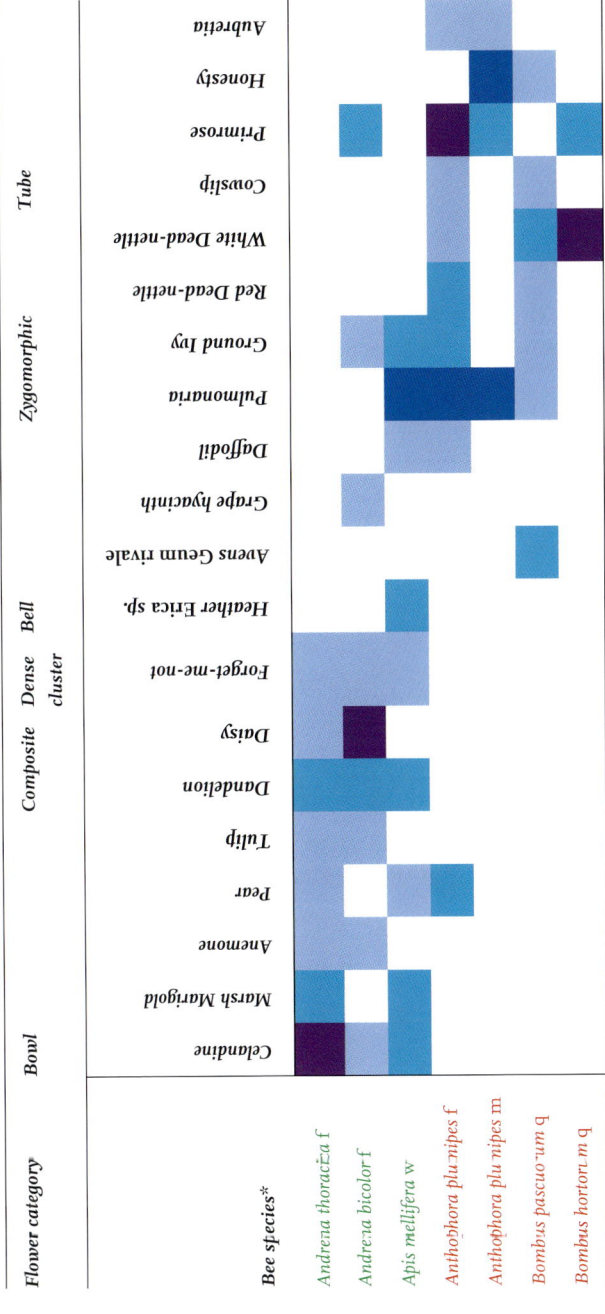

TABLE 8. Bee visitors to garden flowers in early spring. The species in green are short-tongued; those in red are long-tongued. The dark-shaded boxes indicate that the bees visited regularly, the lighter-shaded boxes that they visited often, and the lightest-shaded boxes that they visited occasionally.

* f = female, w = worker, m = male; q = queen.

of a plant's whole life history. In the Crucifer *Erysimum mediohispanicum,* long stalks and high flower numbers increased pollinator visiting rate, but plants with these features were also preferred by Spanish Ibex as a source of food. High flower numbers led to the germination of more seeds only when ibex were excluded from the plants (Gómez & Zamora 2006). Similarly, the hoverfly *Cheilosia albitarsis* is attracted to Creeping Buttercup *Ranunculus repens* flowers, along with potential bee pollen vectors such as *Chelostoma florisomne,* but the hoverfly larvae cause damage by feeding on the buttercup roots. Such factors influence the advertisements, rewards and opening times of flowers and can favour generalisation rather than specialisation.

Over evolutionary time, flowers can change in either direction between specialisation and generalisation according to the 'pollinator climate' (Corbet 2011). The evolution of syndromes can potentially be investigated from a genomic perspective. The genomes of plant species representing different syndromes (such as birds, bees and moths) are now available, and the genetic architecture of floral traits can be compared (Clare *et al.* 2013). Useful reviews on these issues include Corbet 2006, Fenster *et al.* 2004, Ollerton *et al.* 2009, Ollerton 2021, Rosas-Guerrero *et al.* 2014, Waser 2006, Waser *et al.* 1996 and Willmer 2011.

POLLINATION NETWORKS

A flower-rich grassland on a sunny summer's day is alive with bees, butterflies, moths, beetles, flies, wasps and many other insects, mostly seeking nectar, pollen or other rewards from the plants. A recent survey of 644 h of ancient grassland in the Breckland of East Anglia yielded 167 species of flowering plants and 877 species of invertebrates, of which 44 species were bees (Hawkes *et al.* 2020). Making sense of such complexity, even for bees alone, is a major challenge. One approach is to attempt to link all flower visitors with the plant species present, creating a pollination network (or pollination web) which can be analysed visually or using specialist software. Pollination networks (webs) are comparable to food webs and can give clues about whole community processes as well as patterns of interactions between individuals. (Unlike food webs, a pollination network is bimodal: pollinators connect to plants but not to other pollinators, whereas in a food web consumers can also be eaten.) Pollination networks can reveal patterns of interactions between pollinators and flowers which are not apparent when looking at individual species.

Figure 176 shows a relatively simple pollination network from an Essex meadow, displaying the interactions between 28 bee species and ten plant species.

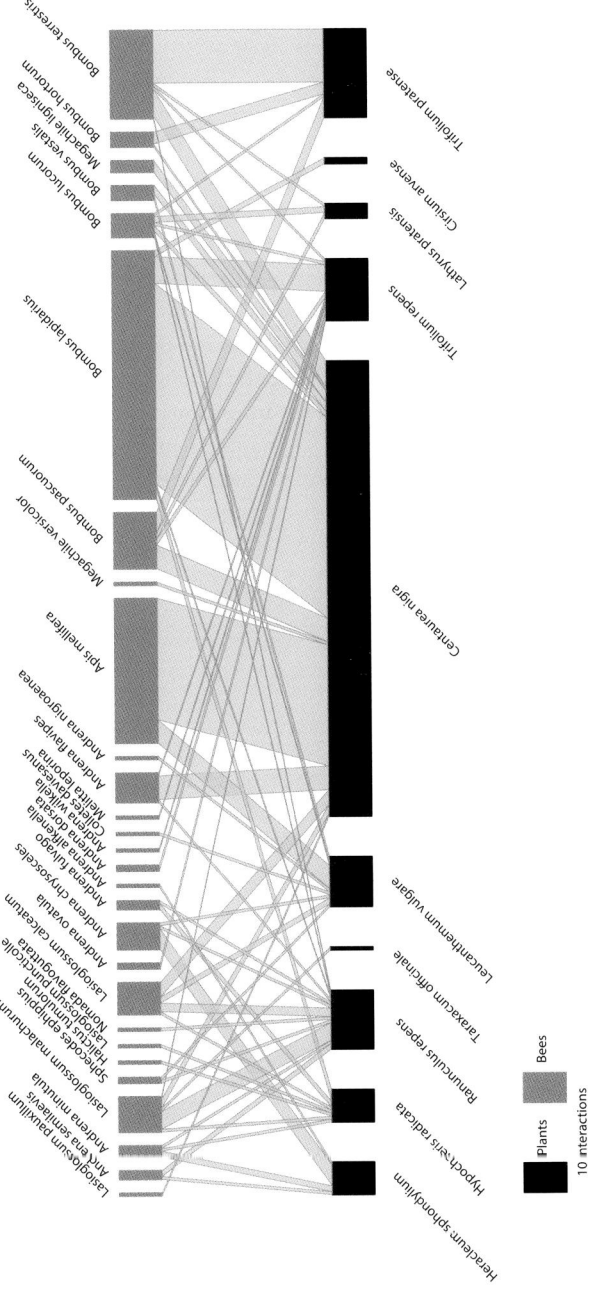

FIG 176. Pollination network for an Essex nature reserve.

The bees are arranged in increasing size order along the top, and the plants are along the base. The width of the rectangles (termed 'nodes') is proportional to the number of interactions of each node, and the width of the connecting lines ('edges') is proportional to the number of interactions (visitor frequency) between each bee node and plant node (Rumeu *et al.* 2018).

Plant-bee networks are often characterised by a small number of 'core generalist' plant species which occupy the centre of the network, interacting with a wide range of generalist as well as some oligolectic solitary bees. Such networks are said to be 'nested': if plants are arranged in decreasing order of the number of visitor species they receive, the visitors can be arranged in a sequence which gives a symmetrical pattern (Fig. 177). Natural pollination networks usually show some resemblance to such a pattern, with a majority of connections displaced towards the top left corner of the matrix, showing that they are far from random (Jordano *et al.* 2006). Pollination networks in artificial habitats such as gardens can show nestedness in a similar fashion to those in more natural habitats (Erenler 2013).

Core generalists in British grassland plant communities are represented by abundant species, such as Field Scabious, Knapweeds, thistles (*Cirsium* species) and yellow Asteraceae. In scrub and hedgerows, core generalists include plum *Prunus* species, Hawthorn *Crataegus monogyna*, Dogwood *Cornus sanguinea* and Apiaceae such as Hogweed *Heracleum sphondylium*, while late in the season, Ivy *Hedera helix* attracts a wide range of visitors, including the specialist Ivy Bee *Colletes hederae*. In woodlands, willows (*Salix*) and Sycamore *Acer pseudoplatanus* attract many bee species. Pollination networks are sometimes compartmentalised (subgroups interacting largely with each other), often on a phenological basis but also to some extent according to syndrome groupings (Dicks *et al.* 2002). The garden study above, though on a small scale, clearly shows separate long-tongued and short-tongued bee compartments but with some overlap, especially in honeybees.

To identify all the insects and plants at any one site is very challenging and time-consuming and requires the cooperation of a large number of taxonomic specialists. Inevitably, network studies have frequently classified insects into broad taxonomic groups, which limits their explanatory power. Network studies which identify insects and plants to species are much more likely to reveal valid patterns of interaction. Even within one insect species there can be individual specialisation; in bees there is considerable variation in size within a species through differences in sex, caste and larval food provision. Body size is closely linked to tongue length across all bee species and strongly influences flower selection. In the study of an Essex meadow described above (Rumeu *et al. loc. cit.*), it was found that a trait-based network – based on each individual

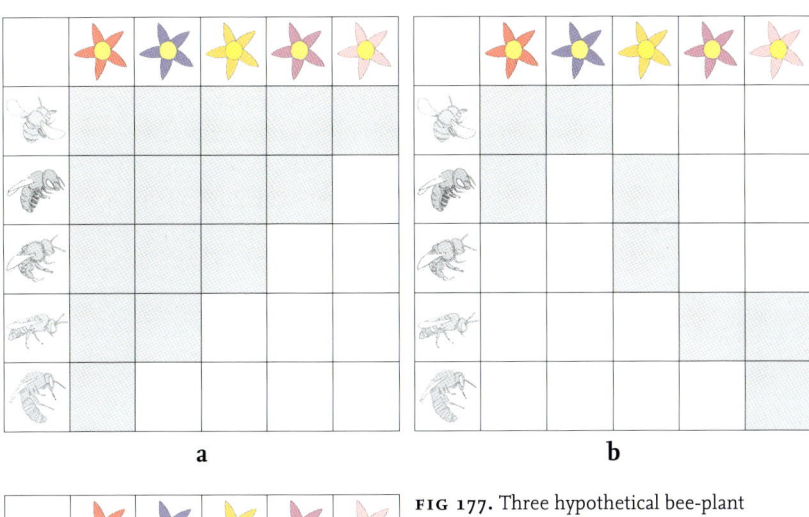

FIG 177. Three hypothetical bee-plant networks, each displayed as a matrix, showing nested, compartmented and random patterns for five species of solitary bees visiting five plant species, represented by different colours (based on Willmer 2011).

bee's body size and the depth of the flower's corolla it was visiting – provided a clearer view of community structure than a species-based network. A somewhat different approach was taken in a study of individual differences in pollen loads of bees foraging at high altitude on the island of Mallorca, where the bee and plant community is relatively simple compared with lowland habitats (Tur *et al.* 2014). Individual bees tended to specialise in their pollen choices and most had narrower niches (a narrower range of pollen types used) than the species as a whole; in other words, generalist species were composed of specialist individuals, especially when the bee species was abundant. These findings offer a challenge to the view that the majority of solitary bees behave as generalists.

'Analysis of pollination networks now dominates significant parts of the pollination literature' (Willmer 2011). Research funding is often aimed at assessing the possible effects of habitat degradation and species loss on pollination. The high inter-connectedness of many pollination networks implies that pollinators will show some resistance to change, since they appear often to use several alternative food sources, while plants have a large number of alternative pollinators. This assumption has been strongly challenged, however, since pollination networks generally rely on records of insect visitation without providing evidence of whether the insects concerned actually behave as (good) pollen vectors (King *et al.* 2013). In cases where detailed studies have been carried out, only a small proportion of insect visitors have been found to transfer pollen to stigmas, as for example in milkworts, *Polygala* species (Castro *et al.* 2013). The degree of flower-pollinator specialisation and the consequent effects of the loss of an insect or plant species from a community can therefore be greatly underestimated, giving false reassurance concerning the 'pollination crisis' (reviewed by Willmer 2011; see also Fogden 2022). Aside from the intrinsic fascination of bees and flowers, these are powerful reasons for attempting a better understanding of the ways in which bees and other insects interact with the flowers they visit. This is the subject of the following chapter.

CHAPTER 6

Bees and Flowers Part II

'Over millions of years of evolution the plants and animals have developed intimate relations that are among the most enthralling subjects in biology.'
—Barth (1991)

In Chapter 5, we explored the close relationships between bees and flowers, with the emphasis on the floral side of the relationship. Here, we focus more on the bees. For the plants whose flowers are visited by bees, it is the contribution the bees might make to pollination that is the key benefit. For the bees, the floral rewards, mostly pollen and nectar, are the main source of nutrition both for themselves and for their offspring. For some bees, other plant products are also used, especially in nest construction, and there is now evidence that flowers are an important hub for the transfer of both beneficial and pathological microorganisms (Voulgari-Kokota *et al.* 2019). Nectar, too, may contain toxins as well as the sugars which are its main nutritional reward (Pain 2015). A secondary function of flowers in the lives of bees is, as we saw in Chapter 2, to serve as 'rendezvous' sites, where males can make a play for foraging females. Also, flowers may just be used simply as physical structures – as perches for basking or as refuges for shelter during inclement weather.

All British bees rely on flowers to meet their nutritional needs, and so all can incidentally play some role in pollination, but our focus here will be on the nest-making species, leaving the cuckoos for another chapter (Chapter 8). For the nest-making species, the floral rewards, in the form of pollen, nectar and/or oils, provide energy to sustain their general metabolism, maintain body temperature above the ambient in adverse conditions, and power their flight. As we saw in

Chapter 2, males of most species maintain remarkably persistent patrolling flights in search of receptive females and a few actively defend territories. This intense activity is maintained by regular 'breaks' to sip nectar from a nearby flower.

Females, too, need to imbibe sugar-rich nectar to sustain their activities, but for them the burden is much greater. The nest-making species are described as 'central place foragers', as they have to fly back and forth many times between their food sources and the nest, and so need sufficient energy to power their flights, as well as to sustain their foraging activities. Once the female bees have established a nest, they are no longer foraging just for their own activities but also to meet the requirements for the full development of their offspring. This involves the construction and provisioning of each brood cell, and, for those species whose larvae spin cocoons, the stock of food placed in the cell must be sufficient for both the growth and development of the larva, and for its construction of the cocoon and exit from the nest. Internally, too, the formation of the eggs and the synthesis of secretions used in the cell linings in many species also add to the female's demand for the goods the flowers provide. As we noted in Chapter 3, a female bee may have to collect four or five times her own weight in food to provision a single brood cell.

Added to these demands are the sensory and cognitive skills required to recognise relevant flower species and harvest their rewards, which are often difficult to access. Then, for those foods not needed for immediate consumption, skills are needed to gather them together, 'pack' them, and transport them back to the nest. As we saw in Chapter 3, the round trip may be repeated many times. According to Neff (2008), provisioning of a single brood cell may involve as many as 40 round trips. Sometimes, to make matters more complicated, there are detours, caused by obstructions, parasite avoidance or changes of flower source, so that considerable navigational abilities will also be required. Most species of bees take nectar from a wide range of different plant species, but females vary greatly in how selective they are in the collection of pollen. Species that have narrow, specialist pollen diets are referred to as 'oligolectic', whilst those that collect pollen from flowers from widely different plant families are termed 'polylectic'. The most extreme specialism – collection of pollen from one plant species only – is termed 'monolecty' but is rare. There is some terminological confusion on this topic, as pollen specialism is a matter of degree, with no clear dividing line between the generalists and specialists. Some authors have introduced terms that allow a more fine-grained analysis, recognising, for example, 'narrow' and 'broad' oligolecty, or 'mesolecty'. Later in the chapter we will discuss these distinctions in more detail, consider attempts at explaining the various pollen-foraging strategies employed by different bees, and give detailed accounts of the foraging behaviour of a selection of particularly interesting species.

For now, it is worth noting that this feature of foraging behaviour imposes a further set of demands on the bees with narrower pollen diets. Since suitable nest sites may be found in quite restricted locations, and the appropriate plant species may also be quite localised, the forager faces a strong constraint in the form of the distance between nest and flower patch, and it may also need to spend time, initially at least, searching for the 'right' plants. The perceptual, navigational and memory requirements for this must be considerable (Menzel *et al.* 1997; see also Chapter 5).

Finally, as nests are very attractive to a great range of parasites and parasitoids (see Chapters 7 and 8), and the bees themselves are vulnerable to predation on their foraging trips, they are under strong time constraints: the longer the nest is left unguarded, the more likely it is that it will be invaded. In some species, the nest entrance is covered while its owner is away, but this is not foolproof and can make it difficult for the bee herself to find the entrance when she returns. Some species even make 'false' or 'accessory' nest entrances which are thought to confuse their enemies (Evans 1996, Rezkova *et al.* 2012, Owens & Benton 2022). Some parasites trail their hosts from foraging sites back to their nests, and females are also subject to frequent mating attempts from males during foraging trips. Females may need to take evasive action or forage at more inaccessible flowers, adding further complexity to the process (Stone 1995; see also Chapter 2).

On leaving the nest, usually for the first foraging flight of the day, but also if there is a change in the surroundings of the nest, foragers perform an 'orientation flight'. This takes the form of a zigzag or spiral with a gradually increasing amplitude as the bee flies up and away from the nest. It seems the bee is able to store a visual map of the landmarks in the vicinity of the nest, which it uses to navigate on its return. At close quarters, scent also plays a part, especially where bees are nesting in dense aggregations. As suitable sources of nectar and pollen may be at a considerable distance from the nest, the bee will need to find her way back and forth over that distance. Though honeybees may forage as much as 14 km away from their hives, it is believed that solitary bees generally have shorter foraging ranges, usually less than 500 m, with smaller species generally having much shorter ranges than that (Gathman & Tcharntke 2002, Danforth *et al.* 2019). Individual *Anthophora plumipes* females at a nesting aggregation in Cornwall were, however, found to travel quite long distances: two marked females were seen more than 900 m from their nest site and one at 1.9 km, foraging on Bugle *Ajuga reptans* (P. Saunders pers. comm.).

Since the bees may need to follow more than one route from the nest to sources of nectar or pollen or, in some species, to gather nesting material, they must have the learning and memory abilities to identify and steer among landmarks in a local landscape. For example, females of the Large-headed Resin Bee *Heriades truncorum*, nesting in artificial nest boxes in a Colchester garden, flew to the west over a garden fence to collect resin, and over a tall hedge to the north to collect small stones from a path, and to nearby flowers to forage for nectar and pollen. Initially foraging for pollen from *Geranium x oxonianum*, several of them switched to their more usual pollen source, Ragwort *Senecio jacobaea*, as soon as it came into flower (TB pers. obs.).

At a distance, bees recognise flowers by visual cues of colour and shape. The compound eyes of bees are sensitive to the wavelengths we see as yellow and blue, as well as ultraviolet light, so their visual world is somewhat different from ours. In particular, the ultraviolet hues contrast strongly with green backgrounds. Although bee eyes are not sensitive to the red end of the spectrum, flowers that appear red to humans can still be seen by bees, as they are sensitive to other wavelengths reflected by them (Chittka & Waser 1997).

At close quarters, bees can follow a scent-gradient to the flower. Many flowers have visual guide-markings that help to direct the bees to floral rewards, regulating their access to them in ways most likely to facilitate pollination. These guide-marks are frequently complemented by scent trails within the floral structure. The floral scents belong to several classes of compounds, including aromatics (such as benzaldehyde), monoterpenes and sesquiterpenes, and these are detected by receptors in the bees' antennae. The combined visual and chemical cues offered by flowers differ from species to species of plants, and the foraging strategies used by bees are shaped by both innate and learned responses to them. These responses are more limited in oligolectic species than in polylectic bees (Dobson & Bergstrom 2000, Dobson & Ayasse 2012, Milet-Pinheiro *et al.* 2012, Milet-Pinheiro *et al.* 2013, Milet-Pinheiro *et al.* 2016).

HARVESTING THE FLORAL REWARDS

1. Collecting nectar: the bee 'tongue'

What happens next depends on whether the bee's visit is to take nectar, pollen, or, more rarely, plant oils. Commonly, female bees collect both nectar and pollen on each foraging trip, but others make separate foraging trips for each resource. As mentioned in Chapter 3, for example, the mining bee *Andrena vaga* has distinct pollen- and nectar-collecting days. On pollen-collecting days, the bees make from one to four trips, returning to the nest with pollen loads as well as nectar, but on 'nectar days', they return with clear nectar in the crop and no pollen loads (Rezkova *et al.* 2011). It seems that the nectar-only foraging trip is the final one for each cell, before an egg is laid and the cell closed.

To collect nectar, the bee has to access floral nectaries. In open, disc- or bowl-shaped flowers with superficial nectaries, the bee can generally use the corolla as a platform while it sucks or laps nectar through its extended mouthparts. More complex floral structures may involve the bee in learning how to access hidden nectaries, which may be at the base of a tubular corolla or within gullet-type blossoms, such as those in the dead-nettle (Lamiaceae) family, or the flag-type structures of the pea and vetch family (Fabaceae). These provide a platform for the bee, from which it can probe down and access the nectaries, meanwhile brushing against the (usually) more exposed anthers and stigmas. In the Fabaceae, the bee frequently has to force open the flower to access the nectar, and in doing so it exposes the reproductive parts of the flower, which are usually enclosed within the keel formed by the lower petals.

It is likely that an electrostatic charge plays a role in the attachment of pollen to a bee's hair (Corbet *et al.* 1982). The hair often acquires a positive electrostatic charge, whereas a negative charge often exists on flowers, causing pollen grains to move from the anthers to the bee. Charge differences can even stimulate petunia flowers to release more scent in response to approaches by bumblebees (Montgomery *et al.* 2021). Other recent research provides further evidence that plants are not mere passive recipients of nectar-seeking insects. An Israeli study (Viets *et al.* 2019) measured the responses of flowers of a species of Evening Primrose *Oenothera drummondii* to the sounds of the wingbeats of approaching insects (including a carpenter bee), using both playbacks of recorded natural sounds and artificial sounds of various frequencies. At low frequencies close to wingbeat sounds, the flowers responded with vibrations and within three minutes increased the sugar concentration of the nectar they secreted. Field observation of patterns of insect visitors to the flowers showed that the three-minute interval was consistent with the plants' increasing their attractiveness to pollinators. It remains to be seen if this phenomenon is more widespread among plants and their visitors.

The main apparatus used by bees to suck, lap or sip nectar is a complex structure formed by a combination of two elements: the maxilla and labium. These are themselves quite complex structures. The maxillae, a pair of elongated outer structures, can form a sheath around the inner structure, the labium. In external view, the main segments making up each maxilla are the stipes (plural 'stipites') and the (distal) galea (plural 'galeae'). A segmented organ that arises from the tip of each stipes is called the maxillary palpus. The labium is also a segmented organ that lies between the maxillae, the more exposed segments being termed the mentum, the prementum and the glossa, which is tipped by minute hairs and is the organ that directly probes the nectaries. Either side of the glossa are two further segmented organs – the labial palpi. The labial and maxillary palpi bear sensory cells that pick up chemical and tactile stimuli as the bees probe flowers.

This complex organ, the tongue, is articulated basally and supplied with powerful muscles that allow it to be stowed away under the head when it is not in use. When about to be used to collect nectar, the whole apparatus is brought forward, so that it projects in front of the head. For this to happen, the bee's mandibles have to be spread wide, and the flap under the bee's face (the labrum) that shields the rest of the mouthparts when they are not in use has to be raised. Often the mouthparts are opened up in this way while the bee is still in flight and approaching the flower. In our description of foraging behaviour, we will usually refer to the whole structure, loosely, as the 'tongue'.

While the above description covers the main features that bee mouthparts have in common, there are important variations. There is much discussion in the literature about the significance of a distinction between long- and short-tongued bees, especially in terms of variations in their ability to probe flowers

FIG 178. Four-banded Flower Bee *Anthophora quadrimaculata* female with tongue extended.

FIG 179. Tongue of the Hairy-footed Flower Bee *Anthophora plumipes* female. Long-tongued bees such as this are able to slide the glossa in and out of a tube formed by the stiffened, overlapping pair of galeae (part of the maxillae) and the upper four segments of the two long, flexible labial palpi which lie beneath the labium. The two final segments of the labial palpi are exposed (labelled in the diagram) and have a sensory function. In this species, the galeae are covered in stiff setae (bristles) which dislodge pollen from flowers such as Red Deadnettle *Lamium purpureum* and Primrose *Primula vulgaris*.

FIG 180. The end of the glossa of a Hairy-footed Flower Bee female. The hairs soak up nectar, and the glossa has a tiny spoon-shaped tip.

with deep corollae. However, the distinction has a technical meaning that is rather confusing as it does not always correspond to observed differences in the tongue lengths of bees. Strictly, the long-tongued bees are ones in which the basal two segments of the labial palpi are elongated, the short-tongued ones having even-sized palpi segments. Of the bee families that occur in Britain, the Melittidae, Colletidae, Halictidae and Andrenidae are technically short-tongued, while the Apidae and Megachilidae are long-tongued. The method of nectar-collecting differs between these two groups, with the extra length of the first two segments of the labial palpi in the long-tongued group allowing the formation of an enclosed tube around the glossa. This can slide up and down within its sheath and enables the larger long-tongued species to probe deeper corollae and suck up more nectar than the comparable short-tongued species (Michener 2007, Danforth *et al.* 2019).

FIG 181. Tongue of the Shaggy Furrow Bee *Lasioglossum villosulum* female. Short-tongued bees such as this extend their tongue by unfolding it. Some short-tongued bees have quite long tongues, but they are not as versatile as those of long-tongued bees. When not in use, the tongue of all bees folds into a cavity under the head, protected in front by the labrum.

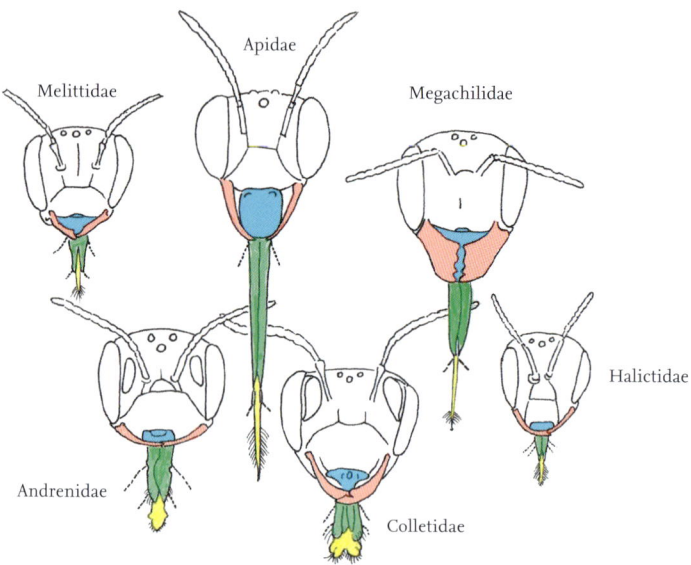

FIG 182. Mouthpart structure of representatives of each of the six bee families. Blue = labrum, pink = mandibles, green = galea (part of the maxilla), and yellow = glossa (part of the labium). Dotted lines show the maxillary palpi (upper) and labial palpi (lower).

In fact, many 'short-tongued' species have extended mouthparts or other features that enable them to handle deeper flowers, and there are few floral structures that can be accessed in only one way, so there is no direct correlation between (technical) tongue length and ability to access different sorts of flower structures. However, there are related differences in mouthpart structures among the bee families that certainly do have consequences for foraging behaviour. The Colletidae (in Britain the genera *Colletes* and *Hylaeus*) are short-tongued bees whose mouthparts have evolved as 'brushes' for applying secretions to their brood cell walls (see Chapter 3). In these species, the glossa is spread out at the tip in the form of two lobes, an adaptation which limits its role in nectar-collecting. Species in these genera tend to collect both nectar and pollen from shallow flowers with exposed nectaries, including umbellifers (Apiaceae), willows, brambles, Ivy and heathers.

The Ivy Bee *Colletes hederae* (Fig. 183) is a very clear example. The flowers of Ivy are carried on dome-shaped umbels. Each flower bears five small yellowish petals and five stamens, surrounding a conical ovary and a single central style. The upper surface of the ovary is green, turning brown later. This is minutely ridged and studded with tiny pores in microscopic view. Nectar is secreted directly onto this surface, its complex micro-structure helping to limit evaporation. Glucose is the main nutritional component of the Ivy nectar, but it also contains sucrose and fructose. Both male and female Ivy Bees collect nectar from the Ivy flowers, walking from flower to flower on each umbel with their mouthparts projecting down and dabbing their tongues onto the surface of the ovaries as they go. Females sometimes collect nectar on umbels even after the anthers are spent, but more often they can be seen simultaneously taking nectar and pollen as they

FIG 183. Ivy Bee *Colletes hederae* female harvesting both nectar and pollen from Ivy flowers.

move from flower to flower, rapidly 'flicking' their forelegs over the anthers to dislodge pollen.

TB timed foraging females at Ivy in an Essex hedgerow in early October, when males were no longer present. Of 17 visits recorded, the mean time spent at each umbel was 22 seconds, but some visits were as short as six seconds and two were of a minute's duration. The bees rarely foraged from more than five flowers on any umbel, sometimes brushing pollen from their faces with their forelegs as they departed. At intervals, females settled on a leaf or dead flower head to scrape pollen from their body hairs onto their hind legs, a process taking an average of 38 seconds. As usual, other species of insects were present – the hoverfly *Eristalis tenax*, common wasps *Vespula vulgaris*, honeybees *Apis mellifera* and a single worker bumblebee *Bombus lucorum*. The *Colletes* greatly outnumbered the other species, and only one brief instance of 'interference competition' was observed (between *Colletes* and an *Eristalis* hoverfly). These observations make interesting comparison with the study of Ivy pollination by Jacobs *et al.* (2009). That study suggested that *Vespula* wasps were the most effective pollinators compared with other flower visitors, including honeybees, bumblebees, large hoverflies and bristly flies. That study was conducted before the rapid range expansion of *Colletes hederae* and, as the authors anticipate, this 'may have implications for fruit availability and pollination in these areas'.

2. Harvesting pollen
This section describes the main methods of collecting pollen among different groups of bees, sometimes in relation to different flower structures.
As in the case of the Ivy Bee, both nectar and pollen may be collected in the same trip, but pollen is frequently the main objective, and we now turn to the work of the bees in collecting and transporting this crucial floral resource. The branched or 'feathery' hairs forming the bees' coats seem to be an adaptation for collecting pollen (as well as for temperature regulation), and they have a positive electrical charge, which also attracts pollen onto their bodies. This incidental collection of pollen may be important for the role of the bee as pollinator, especially where pollen is deposited on a part of the body which is difficult for the bee to reach with its legs when grooming (often the head and parts of the dorsal surface). However, the magnitude of the bee's demand for pollen, up against the plant's evolved mechanisms for restricting access to this highly costly and biologically important product, dictates more than mere incidental deposition. Many groups of flowering plants have evolved complex floral structures, to which bees have responded with either abilities to learn appropriate handling skills or innate behaviours, sometimes complemented by highly specific anatomical adaptations.

These latter are, of course, more likely to be found in the narrowly oligolectic species with a long history of association with a particular range of flower species.

For many short-tongued bees foraging from flowers with exposed anthers, their methods of releasing pollen are described variously as 'scrabbling' (as in the example of *Colletes hederae*, just mentioned) or 'drumming' and 'tapping' (Cane 2017). Where pollen is dislodged in this way, it is dispersed among the body hairs, and the bee usually needs to groom it onto the specialised structures (scopae) used to transport it. However, in the members of the family Megachilidae, the scopae take the form of areas of long stiff hairs on the underside of the abdomen. In many of these species, the abdomen is brushed or tapped against the anthers of host flowers so that pollen accumulates directly in the scopae.

More complex floral structures demand still greater handling skills and, sometimes, anatomical adaptations. Frequently, the reward, in terms of the nutritional quality of the pollen, is correspondingly greater in these more complex flowers. In the 'gullet-type' flowers of Scrophulariaceae (figworts) and Lamiaceae (dead-nettles, woundworts and their relatives), a platform is provided for the forager, with nectaries in a narrow tube below, while the reproductive parts are located above, shrouded in a lobe-shaped petal (a 'bilabiate' flower structure; Classen-Bockhoff 2007). As the bee reaches down for nectar, she brushes against anthers and stigmas, with pollen deposited on the head and dorsal surfaces that are difficult for her to groom (termed 'nototribic' deposition). If the bee already has pollen grains on her back, and the flower is in a female phase, the pollen is likely to be deposited on a stigma, so effecting pollination. Bees that specialise in, or frequently forage from, flowers in this group commonly have specially modified hairs, or setae, on the face, mouthparts or forelegs that are used to detach and capture pollen. The Fork-tailed Flower

FIG 184. Spined Mason Bee *Osmia spinulosa* female harvesting pollen from Common Fleabane *Pulicaria dysenterica*.

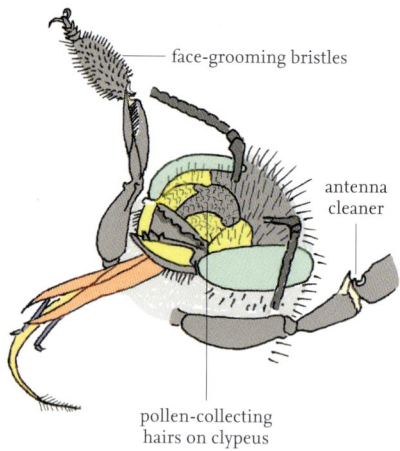

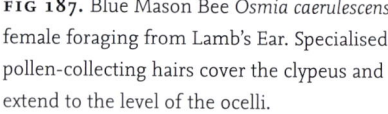

FIG 185. Face and forelegs of a Wool Carder Bee *Anthidium manicatum* female. Adaptive, curved, pollen-collecting hairs cover the (partly yellow) clypeus. Other parts of the face have long straight hairs.

FIG 186. Wool Carder Bee female on Lamb's Ear *Stachys byzantina* showing pollen-collecting hairs on the clypeus.

FIG 187. Blue Mason Bee *Osmia caerulescens* female foraging from Lamb's Ear. Specialised pollen-collecting hairs cover the clypeus and extend to the level of the ocelli.

Bee *Anthophora furcata* (Fig. 188) is a rather localised British species that is strongly associated with flowers in the Lamiaceae, and it has erect hairs on the clypeus that are curved towards the tip and used in pollen collection. Wool Carder Bee *Anthidium manicatum* females (Figs 185 & 186) collect pollen from a range of flowers but commonly use Lamb's Ear *Stachys byzantina* (also used for nesting materials), Black Horehound *Ballota nigra* and other Lamiaceae. They also have a covering of forward-projecting hairs on the face, used for pollen harvesting from these plants, and the inner surface of the fore tarsi is clothed with a dense pile of short stiff bristles. These are used for combing pollen from the face (Müller 1996).

FIG 188. Fork-tailed Flower Bee *Anthophora furcata* female collecting nectar and pollen from Lamb's Ear.

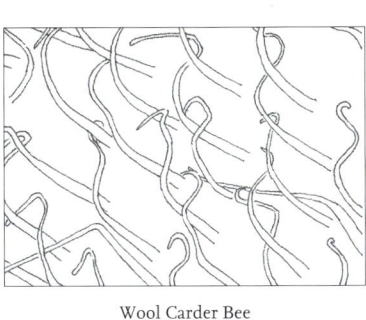

Wool Carder Bee
Anthidium manicatum

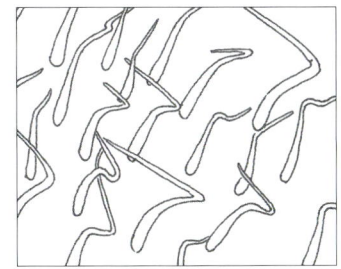

Blue Mason Bee
Osmia caerulescens

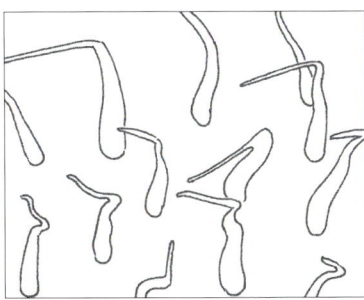

Gold-fringed Mason Bee
Osmia aurulenta

Fork-tailed Flower Bee
Anthophora furcata

FIG 189. Detail of specialised pollen-collecting clypeus hairs in four solitary bee species. The hairs have a broad base and usually a sharp bend. They are not universal in the genera concerned and are thought to have evolved independently by convergent evolution (drawn from Müller 1996).

FIG 190. Fringe-horned Mason Bee *Osmia pilicornis* female collecting pollen on its face from Bugle *Ajuga reptans*.

The plant family Boraginaceae includes some flowers such as the lungworts (*Pulmonaria* species) which have their anthers concealed within narrow corolla tubes. Bees that specialise in these flowers also have modified hairs on their face or front legs that are used to detach pollen (Müller 1995). One example is the Fringe-horned Mason Bee *Osmia pilicornis* (Fig. 190), which has stiff black setae on the galeae (outer sections of the tongue). Although regarded as polylectic (Westrich 2019), this bee forages preferentially on *Pulmonaria* flowers in central Europe, but in Britain it collects pollen mainly from Bugle *Ajuga reptans* (Prosi *et al.* 2016; TB & NWO pers. obs.). Presumably the adaptations for collecting pollen from *Pulmonaria* also work for Lamiaceae.

One of the most discussed methods of collecting pollen is termed 'sonication' – or, more familiarly, 'buzz pollination'. This is a specialised interaction between some bee and plant taxa, most familiarly members of the Solanaceae (e.g. nightshades). The plants involved often have tubular anthers with slits or pores at the top ('poricidal' anthers). The pollen grains are typically small with relatively smooth surfaces and are especially rich in proteins. To gain this reward, the bee grasps the anthers with her legs or mandibles, curves her abdomen below the flower, and vibrates her wing muscles. Because of the communicated kinetic energy, centripetal forces and possibly electrical charges, pollen grains are expelled from the anther. Sound vibrations themselves are now thought not to be sufficient to dislodge pollen, and the bees also use vibrations produced by physical movements of legs and mandibles (Vallejo-Marín 2019). Buzz pollination is widespread among plants, with 22,000 known species from 72 families

BOX 5 Why do bees buzz?

Buzz pollination in bumblebees, as described above, involves quite complex behaviour, but some groups of solitary bees (notably species of *Megachile*, *Anthophora* and *Melitta*) buzz loudly when flying among flowers as they collect nectar or pollen. It may be that, as in the case of the response of flowers to wingbeat sounds, these sounds, too, produce a response from the flowers. This may be, as in the case of Evening Primrose (discussed above), to increase the sugar content of its nectar or to render pollen more accessible. The Silvery Leafcutter Bee *Megachile leachella* and both sexes of the Hairy-footed Flower Bee *Anthophora plumipes* (Fig. 191), for example, buzz as they move from one flower to another while foraging. However, these species also buzz at their nest sites and, in the case of male *Anthophora*, also while flying along patrol routes. There is much more to be learned about how many, and which, of our solitary bees are capable of buzz pollination, and which plants they use.

FIG 191. Hairy-footed Flower Bee *Anthophora plumipes* female 'buzzing' a Cowslip *Primula veris* flower.

Buzz pollination appears to be rare in British solitary bee species, though it is perhaps overlooked. It has been recorded in *Anthophora furcata* in which buzzing assists the release of pollen from Lamiaceae flowers onto the specialised facial hairs (Müller 1966). In NWO's experience, it is used on *Stachys byzantina* and *Linaria purpurea*. The buzzing is very quiet and not made on every flower visited, perhaps being reserved for flowers with anthers at a suitable stage. Buzz pollination was also observed in *Anthophora plumipes* visiting White Dead-nettle *Lamium album*, Cowslip *Primula veris* and Red Poppy *Papaver rhoeas*. In this species, buzzing is loud and made on successive flowers, with the posture of the bee resembling that of a buzz-pollinating bumblebee.

worldwide. The ability to access concealed pollen by buzz pollination is similarly widespread among bees, possessed by species of at least 74 genera, but does not occur in honeybees (Cardinal *et al.* 2018). As the plant families dependent on sonication include several important agricultural crops (tomatoes, peppers, kiwi fruit, blueberries and cranberries), there has been extensive research on the topic. However, this has generally been confined to bumblebees, as they are the main pollinators used in the pollination of commercially significant crops.

As we saw in Chapter 5, the flag-type flowers of the Fabaceae (peas, vetches and their allies) are bilaterally symmetrical, with petals differentiated into an upper 'standard', a 'wing' at each side, and a boat-shaped keel below. Basally, the petals are fused to form a corolla tube containing nectaries. The sexual parts of the flower are typically concealed within the keel until a flower visitor settles on the wing petals to access nectar by reaching down into the corolla tube. If the weight of the insect is sufficient, the keel is depressed and the anthers and stigma are exposed, brushing against the ventral surface of its abdomen (or in a few cases, such as broom, also on the dorsal surface of the bee), so either depositing pollen onto the bee or receiving pollen onto the stigma. The Silvery Leafcutter Bee *Megachile leachella* (Fig. 192 a) forages for nectar and pollen from a wide range of flowers, but at one of its coastal sites in Essex, it was observed foraging

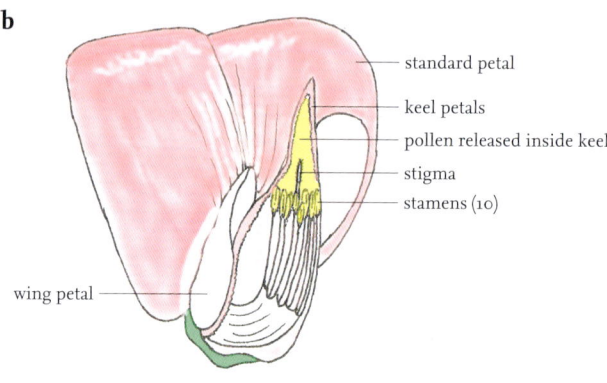

FIG 192. (a) Silvery Leafcutter Bee *Megachile leachella* female foraging from Spiny Restharrow *Ononis spinosa*. (b) Flower structure of Spiny Restharrow, with one keel petal removed.

for pollen mainly from Spiny Restharrow *Ononis spinosa* and Bird's-foot-trefoil *Lotus corniculatus*. On flowers of both plants, the bee typically lands on the wing petals and reaches forward to access the opening of the corolla tube with her mouthparts, while holding on to the wing petals with forelegs and mid-legs. To obtain pollen, the bee thrusts downwards with her abdomen, while sweeping her hind legs backwards against the exposed anthers and simultaneously brushing them against her abdominal scopa. The thrusting movement is repeated up to five times, the whole sequence taking some seven seconds, and the bee flies to the next flower, where she may repeat the process.

In the same habitat, the small Wilke's Mining Bee *Andrena wilkella* (Fig. 193) also forages for pollen and nectar from Common Bird's-foot-trefoil. The foraging behaviour of this bee is quite different. She settles on the keel to access the

FIG 193. (a) Wilke's Mining Bee *Andrena wilkella* female taking nectar from a flower of Common Bird's-foot-trefoil *Lotus corniculatus*. (b) She turns round to force open the keel with her abdomen before (c) 'scrambling' with her legs to remove pollen

opening of the corolla tube for nectar and then turns on the keel to face away from the corolla tube, forcing the keel open with the rear segments of her abdomen. With rapid scrambling movements of her legs, she detaches pollen from the anthers and then raises her head, while brushing pollen from her face with her forelegs. As she does this, she turns round on the keel again to take another sip of nectar. The bee may then fly to the next flower or settle on nearby vegetation to scrape pollen from her body onto her hind-leg scopae. During the period of observation, several attempts to force open the keel failed, and the bee then moved on to another flower (TB pers. obs. & video).

TRANSPORTING THE PROVISIONS

As we have seen, female bees forage from flowers both to meet their own nutritional needs and to take provisions back to the nest to provide a store for their future offspring. As well as imbibing nectar to power their foraging activity, maintain body temperature in cool weather, and sustain their flights between nest and foraging site, female bees consume pollen as a source of the protein needed for egg production. Usually this is consumed directly or brushed forwards to the mouth by the forelegs.

Honeybees and bumblebees and some solitary bees take a proportion of the nectar that they have harvested for their brood cells back to their nests in

FIG 194. (a) Long-horned Bee *Eucera longicornis* female with full loads of a pollen-nectar mix. (b) Hairy Yellow-face Bee *Hylaeus hyalinatus* female with regurgitated pollen-nectar mix.

a mixture with pollen on the corbiculae or scopae. This is typical of *Melitta* species, as well as the Long-horned Bee *Eucera longicornis* (Fig. 194 a). In the Yellow Loosestrife Bee *Macropis europaea*, plant oils form an additional part of the mix. It may be that the greater adhesion of the mixture allows the bee to collect a larger mass of provisions in each foraging trip, though, from the standpoint of the plant, it may reduce the likelihood of pollination as pollen grains are much less detachable. In most species of solitary bees, nectar that is not used for direct consumption is stored internally, in the crop, and regurgitated when the bees return to their nest. The nectar is then mixed together with pollen in the brood cell, to form a more or less solid 'loaf'. In the genus *Hylaeus* (the yellow-face bees), both pollen and nectar are carried in the crop. The bee adjusts the consistency of the mixture by partial regurgitation of droplets which are exposed externally on the face to allow evaporation of surplus water.

In honeybees and bumblebees, pollen (with or without nectar) is carried back to the nest packed into the corbiculae, whereas in solitary bees, as we have seen, pollen is gathered into the scopa or scopae. In nest-making females of the solitary bee family Megachilidae, the scopa is beneath the abdomen (metasoma), whereas in the other five solitary bee families, the scopae are on the hind legs, which are supplemented in some genera by flocci on the hind femora and/or by hairs at the sides of the propodeum (see below).

Before the pollen can be carried back to the nest in these structures, it first has to be groomed from the various places on the bee's body or head where it has caught into the bee's body hairs, and then packed into the scopae and other structures adapted for pollen transport.

Bees of most groups are equipped with dense pads of stiff setae on the inner surfaces of their tarsi (feet) or lower leg segments. These are used while the bee forages, or often during pauses between sequences of flower visits, to brush pollen from the face or body hairs to the scopae. This process is similar across the bee families that carry pollen on the hind legs. A female Grey-patched Mining Bee *Andrena nitida* (Fig. 195 a), filmed by TB as it moved from foraging on flowers of Alexanders *Smyrnium olusatrum* to settle on a leaf, is a typical example. The bee brushed its forelegs several times downwards over its face, then brushed the forelegs against the mid-legs, transferring pollen to them. The mid-legs were then rubbed upwards along the hind legs, thus depositing (and packing) the pollen grains onto the scopae. The process continued, with repeated use of the mid-legs to brush pollen from body hairs on the thorax onto the scopae, and back-and-forth movements of the hind legs against the sides and ventral surface of the abdomen, brushing pollen from those areas (the latter phase sometimes involved flexing the mid-legs to raise the abdomen). The flower bees (*Anthophora*

FIG 195. (a) Grey-patched Mining Bee *Andrena nitida* female scraping pollen onto her hind-leg scopae. (b) Hairy-footed Flower Bee *Anthophora plumipes* female holding on with her mandibles while scraping pollen from her face and body onto her hind-leg scopae.

species) also have hind-leg scopae, and females of the Hairy-footed Flower Bee *Anthophora plumipes* (Fig. 195 b) usually settle on an inconspicuous perch to wipe pollen from their copious body hairs onto their scopae, though they are still often seen and harassed by males as they do so. All three pairs of legs are used to brush pollen from the head, tongue and the rest of the body, and, as in *Andrena*, pollen is passed back to the hind legs. Often the action is so energetic, even frenetic, that the bee hangs on by only one leg. A female Four-banded Flower Bee *Anthophora quadrimaculata* was observed to hang on by its mandibles alone while pollen-grooming and seemed at imminent risk of falling (TB).

There is considerable diversity among the species and genera that have scopae on their hind legs, not only in where the scopae are placed but also in the fine structure of the hairs, their density and their inclinations from the body surface. In some species that collect pollen from a narrow range of plants, the apparatus for carrying the pollen shows features that can be seen as specialised adaptations. We will consider some examples later in this chapter, but here we'll outline the main variations in pollen-carrying equipment across the families of bees represented in Britain.

BOX 6 Brushes, baskets and combs

Female bees belonging to the nest-making genera are equipped with a variety of arrangements for carrying plant products back to their nests. These are generally similar among the species belonging to each family or genus, but where there are close evolved relationships between a species and a specific group of flowers, there are sometimes very distinctive features. Species belonging to the family **Melittidae** are all pollen specialists and so their pollen-transporting structures vary from species to species. These will be described later in this chapter.

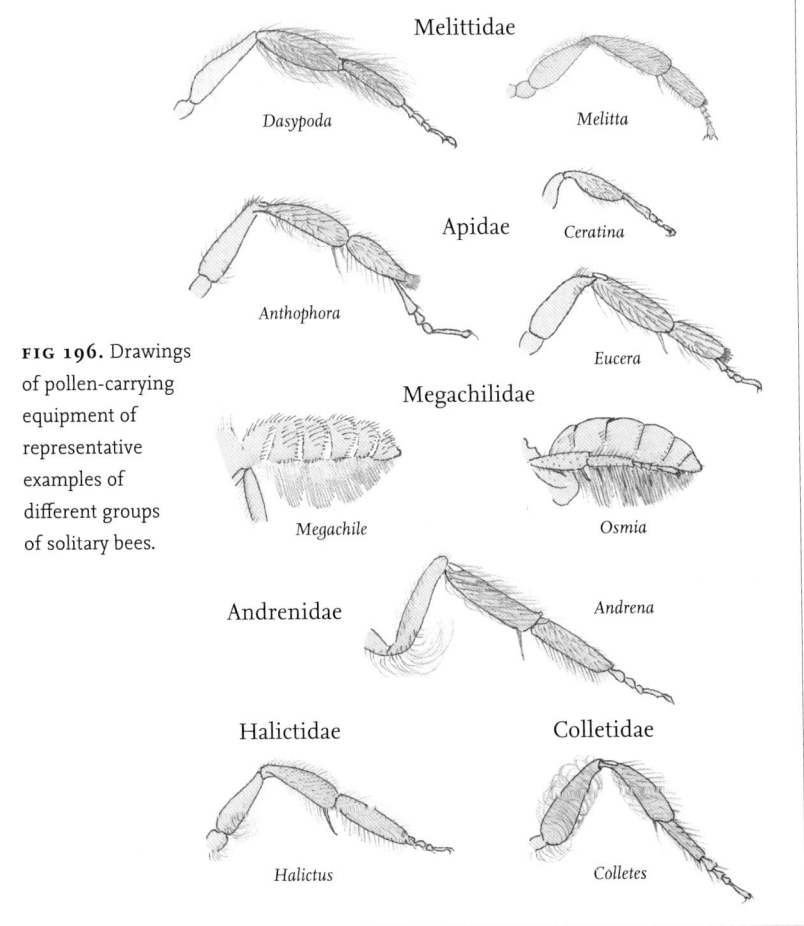

FIG 196. Drawings of pollen-carrying equipment of representative examples of different groups of solitary bees.

continued overleaf

BOX 6 Brushes, baskets and combs *continued*

The family **Apidae** includes the honeybee *Apis mellifera* and the bumblebees, which (except for the cuckoo species) carry their provisions back to the nest packed into hind-leg corbiculae.

The flower bees of the genus *Anthophora* have a dense covering of long hairs on the outer surface of the hind tibiae and basitarsi and a fan-like dense brush of hairs at the tip of the hind basitarsi. The inner surface of each of the basitarsi is clothed with a dense mat of oblique sturdy setae, which are used to detach pollen from body hairs and pass it back to the hind-leg scopae. The scopae of *Eucera* are very similar, but there are some plumose hairs among the mainly simple ones comprising the hind-leg tibial scopa. Our one species of *Ceratina* has very little development of external means of carrying pollen. There is a scattering of fine short hairs on the body, and specimens usually have pollen grains attached haphazardly among these. There are longer fine hairs on the hind tibiae, and some of those close to the leading edge are plumose. These could be taken to form loose scopae, but images of the bees foraging do not show them as carrying pollen loads. It is possible that *Ceratina cyanea* females gather pollen from their body hairs when they return to the nest, but it seems more likely that they have come to follow the habit of the *Hylaeus* species and carry provisions back internally.

The scopae of the family **Megachilidae** consist of dense piles of hairs projecting downwards and backwards from the ventral surface of the abdominal segments. Pollen is collected either by direct contact with floral anthers or by being transferred from other body hairs by pads of stiff backward-slanting setae on the inner surface of the

FIG 197. Wood-carving Leafcutter Bee *Megachile ligniseca* female, showing scopa on ventral surface of the abdomen.

tarsi. Bees in the genus *Megachile* have a distinctive posture when foraging – with the abdomen kept in line with head and thorax, or slightly upturned, so enlarging the volume available to pollen loads.

Despite the similarity of the pollen-transport structures of *Osmia*, their posture when foraging is very different from that of *Megachile*. Typically, they arch their bodies around usually centrally placed anthers, to give a more compact impression as they forage. This results in direct capture of pollen on the scopae, but *Osmia* frequently pause while foraging and use all three pairs of legs to sweep pollen from their body hairs onto their scopae. A substantial minority of *Osmia* species is oligolectic and, as discussed below, some have anatomical adaptations for harvesting pollen from their host plants.

The megachilids comprise several other genera with representatives in Britain. Two of these genera (*Coelioxys* and *Stelis*) include only nest-parasitic species, and we will be discussing them in Chapter 9. The other genera in the Megachilidae have few British species, and all share the pattern of scopal hairs on the underside of the abdomen. *Chelostoma* (two species), *Hoplitis* (two species), and *Heriades* (one species fully established) are mostly oligolectic and show variations in their pollen-transporting apparatuses, some of which will be discussed below. The single species of *Anthidium* is regarded as polylectic but has a close association with flowers in the families Lamiaceae and Fabaceae.

The family **Andrenidae** has only two genera that are represented in Britain – the large genus *Andrena* and the much smaller *Panurgus*, with just two species. In Britain, the majority of *Andrena* species are polylectic, and the differences in pollen-carrying equipment among the species are found mainly in the oligolectic species, such as *Andrena hattorfiana* and *Andrena marginata*.

Andrena females have a conspicuous covering of long hairs on the outer surface of the hind-leg tibiae. These carry a large proportion of the pollen load, but there are also 'baskets' formed by flat, bare areas on each side of the propodeum (rear of the thorax), with associated drapes of long curved hairs. In addition, the hind femora have a bare area on the ventral surface, subtended by a row of long curved hairs from the dorsal edge and overhung by a long fan of usually plumose curved hairs (known as the floccus) that arise from the ventral surface of the trochanter. All three of these structures can be packed with pollen during a pollen flight.

Both our *Panurgus* species are pollen specialists on Asteraceae. The scopae cover the outer surface of the hind tibiae and basitarsi and are made up of densely packed long hairs, plus other specialised features, to be discussed later in this chapter.

continued overleaf

BOX 6 Brushes, baskets and combs *continued*

The species belonging to the family **Halictidae** are almost all polylectic, and the pollen-collecting and transporting equipment is quite uniform through the group. There are three main structures for pollen transport, not unlike *Andrena* species, but there is additional capacity on the ventral surface of the abdomen in some species. There are brushes of long hairs on the outer surfaces of the hind tibiae and basitarsi, but these are complemented by highly developed pollen baskets on the sides and undersides of the hind femora. These are formed by long plumose hairs that arise from the basal area of the femora (resembling the flocci of *Andrena* as described above), combined with curved

FIG 198. Ventral view of a Bronze Furrow Bee *Halictus tumulorum* female, showing the quantity of pollen collected among hairs on the sternites in addition to the scopae.

and plumose hairs arising from the sides of the femora and arching over the ventral surface, so creating an extensive enclosed space below each femur. A large proportion of the full pollen loads is carried in these femoral baskets. There are propodeal pollen baskets, but these are less well developed than those of *Andrena* species. Finally, some species have broad bands of long hairs on the underside segments (sternites) of the abdomen, and though they are much less well developed than the ventral scopae of the megachilids, they do still accumulate significant pollen loads.

The family **Colletidae** is represented in Britain by just two genera, *Colletes* and *Hylaeus*. As mentioned above, the yellow-face bees (*Hylaeus* species) are very unusual in having no external structures for conveying pollen and nectar back to their nest. The pollen-transporting equipment of members of the genus *Colletes* is similar to that of the *Andrena* species, with scopae on the hind tibiae and pollen baskets on the hind femora and at the sides of the propodeum. The hind tibiae have long hairs on the inner face as well as the outer, and fully laden bees have most pollen packed onto the inner surface of the tibiae and the large femoral pollen basket. There is some variation among the species, with the propodeal pollen basket undeveloped in the spring-flying *Colletes cunicularius*.

SPECIALISTS AND GENERALISTS: OLIGOLEGES AND POLYLEGES

So far, we have mentioned the differences in the pollen diet of different species of solitary bees, some of them narrowly dependent on pollen from one or a small number of plants, while others forage widely for pollen from a large number of different plant taxa. For those bees that have a narrow pollen diet, this adds an extra constraint on their foraging strategy and limits their options for nesting habitat. It would seem that specialising in this way could be a handicap, imposing extra time and energy searching for the 'right' flowers and also possibly imposing extra flight times between nest site and foraging locations. There is a risk, too, of a mismatch between the flight period of the bee and the flowering of its favoured pollen host. In addition, specialisation implies giving up on the chance to obtain resources from all the other flowers that might be available to a non-specialist. On the other hand, a long history of association with a narrow range of pollen hosts allows for the evolution of specialised anatomical and behavioural adaptations and associated modifications on the part of the 'chosen' flowers.

One difficulty in writing about this aspect of bee lives is that there is some (understandable) terminological confusion. First, it is important to recognise the difference between the restricted pollen diets of some bee species and the more variable behaviour pattern of individual bees, referred to as 'floral constancy' or 'fidelity' (see Chapter 5). This is seen when an individual bee flies from one plant to another of the same species during a single foraging session, sometimes for several subsequent foraging trips, only occasionally switching from one plant species to another. Other populations of the same bee species may be constant to other flower species, as these are available in different locations or at different times.

The form of specialisation we discuss here is quite different. It concerns pollen collection only and is an inherited trait, pervasive throughout the populations of a species, across different locations, and from one year to the next. We have already briefly introduced the terms 'monolecty', 'oligolecty' and 'polylecty' to classify the different pollen-collecting strategies. The bees that use these strategies are referred to as 'monoleges', 'oligoleges' and 'polyleges', respectively. By tradition, the term 'monolecty' refers to the most specialised pollen diet – where females of a bee species collect pollen from flowers of one plant species only. At the opposite extreme are bees that forage for pollen from a very wide range of flowers – belonging to more than 40 families in the case of some populations of the honeybees that have been studied and 15 families or more for some solitary bees (Cane & Sipes 2007). These are referred to as polyleges (and their dietary pattern as polylectic). Between these extremes are bees whose pollen diets are restricted to some degree. These are the oligoleges.

This threefold classification is useful for some purposes, but Cane and Sipes have pointed out some of the problems it poses for researchers and have suggested new terms. One difficulty is how to establish what the pollen diet of a bee species is. We could rely on observing flower-visiting on many different occasions and in many places, but it is often difficult to tell whether a bee is taking nectar or collecting pollen on any specific visit to a flower, and there is a problem of 'observer bias' when we direct our observations according to preconceptions. Some studies analyse pollen in the scopae of museum specimens, while others take pollen loads from bees caught when foraging or when carrying their pollen loads back to the nest. Analysis of the contents of the brood cells is another method. A further difficulty to overcome is that not all pollens are readily identifiable down to the species of flowers from which they were taken. However, there is enough evidence, derived from a combination of all these approaches, to show that species-specific differences in pollen-foraging strategy are real and that there is a basis for a more refined classification of them.

Cane and Sipes point out a problem with the definition of 'monolecty': some species so designated are obligately monolectic, in the sense that they simply do not forage if they are denied access to the appropriate flowers. Others (facultatively monolectic) will take pollen from other flowers if they have no access to their usual pollen-host species. More challenging is the problem that a species that appears to be fixated on one plant species within a particular geographical zone may forage from a closely related plant species that occurs elsewhere. In other words, apparent monolecty may be (and, it turns out, usually is) merely a function of the geographical distribution of a bee's potential hosts. An example from Britain is the Bryony Mining Bee *Andrena florea*, which collects pollen only from White Bryony *Bryonia dioica* here but also uses *Bryonia alba* where it occurs in mainland Europe. In a later discussion, Cane (2020) used these considerations to argue for the definition of 'monolecty' to be widened to include species which confine their pollen diet to flowers of plants belonging to one genus.

To introduce more precision into distinguishing different degrees of pollen specialisation, Cane and Sipes suggest retaining the term 'oligolecty' to characterise bees that forage from plants from 1–4 genera within a family, and they introduce the term 'mesolecty' for bees that use plants from more than four genera from 1–3 different families (or tribes within very species-rich families – such as the Asteraceae). Foraging from plants from 4–25 per cent of the available plant families would count as 'polylectic', with a further category of 'broad polylecty' to characterise the strategies of bees that collect pollen from more than

25 per cent of the available families. Clearly, these distinctions are to some extent arbitrary, but the effort at classification draws attention to the great diversity of bee pollen-diets and poses interesting questions for research.

As we mentioned above, bees that are limited in the range of plant species from which they collect pollen appear to be handicapped, in that they necessarily pass up on resources that are potentially available from other plants. The demands of finding a narrow range of forage plants are likely to limit options for nest sites, as well as impose longer and more demanding flights between the nest and the appropriate floral host. Specialist bees also run the risks that follow from being dependent on a small number of flower species – these may become rare or locally extinct or may be subject to climate-induced vagaries of flowering period. Of course, the bees cannot calculate these risks, and the evolution of pollen specialism will generally depend on the balance of selective advantage at specific times and places. Only subsequently may the risks associated with dependency have their effects on potential decline, or even extinction, of an oligolectic lineage.

Despite the apparent disadvantages of specialisation, the proportion of (nest-making) bees that are to some degree specialist pollen collectors is high. It varies greatly geographically, with the highest proportions of specialists found in hot, dry habitats, such as deserts. As many as 50–60 per cent of species in such habitats are oligolectic, compared with roughly 20–30 per cent in temperate zones, such as in central Europe (Scheuchl & Willner 2016). Minkley and Roulston (2006) suggest that generalist bees may lose out in harsh environments because of their dependence on less well-adapted plant species. Westrich (1989) estimates that some 30 per cent of the bees of northern and central Europe are oligolectic. As many as 45 per cent of the British species are reported to show definite preferences for some flower types over others, though in the absence of further research, it is not clear how far this might simply be a consequence of the abundance or availability of those flowers in the habitats of the bees concerned. For example, Milet-Pinheiro *et al.* (2016) carried out an experimental study of the foraging behaviour of the common Gwynne's Mining Bee *Andrena bicolor*. This species is bivoltine and was believed to be polylectic in the spring generation and oligolectic on *Campanula* flowers in the summer generation. The researchers tested female bees' innate (i.e. 'flower-naïve') responses to both colour and scent cues from *Campanula* and a common spring pollen source (Dandelion *Taraxacum officinale*) and found no significant differences. Presumably, the apparent specialisation of summer generation bees on *Campanula* was a learned response to the abundance or availability of campanulas in that part of the bees' range and at that time of year.

However, there is evidence that at least some oligolectic species are innately attracted to particular flowers or related groups (Praz *et al.* 2008, Sedivy *et al.* 2008), but it is also possible that other, non-genetic mechanisms might predispose bees to prefer a narrow range of pollen sources. One suggestion is that developing or newly adult bees in the brood cell might become conditioned to, or 'imprinted' on, the chemical aroma of the pollen that they have consumed (Linsley 1958). This, rather than genetic inheritance, might explain the pattern of initial preferences of 'naïve' bees. Dobson *et al.* (2012) set out to test this suggestion at two sites in Vienna, using the widespread and common polylectic bee, the Red Mason Bee *Osmia bicornis*. In distinct phases of the experiment, they reared the bees exclusively on each of two flower pollens and confronted the resulting adults with a range of foraging options when they first emerged. Although one group did show a greater preference for the flower used for the pollen upon which it developed, compared with the control group of wild-resourced bees, the difference was not statistically significant. Instead, the pattern of observed preferences suggested that the bees – whether experimental or controls – shared a strong initial preference for one of the flowers whose pollen had been used (Oilseed Rape *Brassica napus*), but also for one of the other flowers whose pollen had not been used to rear the larvae (buttercups, *Ranunculus* species). While discounting the 'imprinting' hypothesis, at least for this species, the experiment showed evidence of innate foraging preferences even in a broadly polylectic species.

Recent research using fossil evidence, morphological comparison and genetic analyses has allowed the (often tentative) reconstructions of the evolutionary history of different bee families and genera (phylogenies). On this basis, it has been suggested that some families, including the Melittidae, Megachilidae, Halictidae and Andrenidae, had basal (i.e. original) members that were oligolectic. If these studies are right, it seems that although oligolecty may seem to be the foraging strategy most in need of explanation, often it will be a shift to polylecty that requires evolutionary explanation. Switches from oligolecty to polylecty have occurred in the ancestry of some particular groups and species, as well as switches in the opposite direction. Among oligolectic lineages, switches from one host plant to another have also been noted, sometimes associated with the emergence of new species of bees (Larkin *et al.* 2008). So, although there is considerable stability through time in host-plant specialism, it is also the case that shifts from one host plant to another can occur, as well as changes in dietary breadth, from more to less specialised and the reverse.

This poses the question: what selective advantages might bees gain from either pollen specialisation or generalisation? An initial hypothesis has been that specialisation reduces competition for resources, as each species corners its own supply, a process known as 'resource partitioning'. Unfortunately for this approach, there are many cases where it does not seem to apply. It is quite rare to find a tight one-to-one fit between bee and flower species. It is common for flower species with one or more 'dedicated' pollen specialists to be at the same time used as a pollen source by numbers of generalist species (Minckley & Roulston 2006). In the case of the Bryony Mining Bee *Andrena florea*, for example, foraging females have to share the pollen resources of the Bryony with honeybees, other *Andrena* species (such as *Andrena bicolor*) and various species of *Lasioglossum* (not to mention beetles, wasps, hoverflies and sundry other visitors). An alternative hypothesis is that specialists might forage more efficiently from their pollen source than generalists do. This is a difficult comparison to make experimentally, though Strickler (1979) did compare an *Echium* specialist (*Hoplitis anthocopoides*) with generalist species foraging from the same plants. Though the generalists were able to collect as much pollen, the *Hoplitis* females moved more rapidly from plant to plant and so were held to have foraged more efficiently.

Some specialist bees have definite morphological features that seem to be adaptations to the demands of pollen collection from their favoured hosts, but not all of them do. The species without physical adaptations may forage more efficiently as a result of behavioural adaptations. An inherited disposition to collect pollen from a particular type of flower might confer savings in terms of learning and remembering how to access and collect pollen from it. However, not all the flowers that have dedicated specialists are especially demanding in that respect. In fact, many are not and are attractive to generalist species, too. Also, of course, for specialisation to be an advantageous dietary strategy, any increased efficiency in pollen collection from the favoured host would have to compensate for the loss of the (generalist) opportunity to gather pollen from other available sources. This invites comparison with flower constancy as a feature of individual bee behaviour. In that context, it is sometimes argued that switching between different floral structures in a single pollen-collecting trip is likely to make demands on the bee's information processing and handling ability that would reduce its overall foraging efficiency. The cognitive and motor abilities of solitary bees are much less well known than those of honeybees and bumblebees but are likely to be comparable. Menzel (2001) provides grounds for thinking that bees are cognitively up to the tasks of accessing different flowers without significant loss of handling times. If that is right, then it seems behavioural specialists would be unlikely to have a selective advantage over generalists, at least in virtue of the costs to the latter of shifting between different flower types (see Chapter 5).

Another hypothesised advantage of pollen specialisation is that the host plants of the oligolectic species may have especially nutritious pollen. Recent research in the USA provides important evidence that the nutritional content of pollen, especially the ratio of protein to lipids, is an important driver of foraging strategies in honeybees and bumblebees, with the implication that foragers are able to adjust their flower visits according to the nutritional requirements of their brood. So far, one species of solitary bee, *Osmia cornifrons*, has been included in the study. Its foraging preferences result in high protein to lipid ratios in the provisions delivered to its brood cells (Vaudo *et al.* 2015, 2016, 2020). However, it remains unclear whether individual bees are able to adjust their foraging according to an appropriate nutritional mix, or whether inherited floral preferences secure this result indirectly. For oligolectic species, it seems likely that securing an appropriate nutritional reward is an outcome of an evolved relation to a limited range of host plants. As we have seen (see Chapter 3), pollen contains a range of nutrients, but most relevant research has focused on protein as the main one. This varies from 12–60 per cent of mass, and there is evidence that higher protein content favours faster larval development (Roulston & Cane 2002). However, although some oligolectic bees do forage from plants with high protein content in their pollen, this is not true of all of them (Roulston *et al.* 2000).

Another possible condition favouring pollen specialism also has to do with the chemistry of the pollen. Many plants produce toxins as a defence against herbivores, and it is now known that these toxins are often also present in the pollen. Plants in several families, including Boraginaceae, Asteraceae, Fabaceae and Ranunculaceae, have pollen which is either toxic or difficult for bee larvae to digest (Roulston & Cane 2000). The hypothesis is that bees whose larvae have evolved the ability to feed safely on the pollen of these host species may benefit from the relative absence of competition from generalists which avoid it. For example, studies of bee survival on diets of *Ranunculus* pollen have shown that it is toxic for many bee larvae but that some have physiological adaptations that enable them to survive (Haider *et al.* 2013, Sedivy *et al.* 2017).

Most research on this topic focuses on the plant family Asteraceae. A study of 60 species of the genus *Colletes* that occur in the Western Palaearctic (Müller & Kuhlmann 2008) compared pollen-specialist species with generalist ones. Fourteen pollen specialists collected pollen predominantly or exclusively from one sub-family of Asteraceae (the species-rich Asteroideae), which was very little used by most of the generalists. This suggests that an inherited ability of larvae to digest pollen that is toxic to other species allows pollen specialists to corner the rewards offered by this group of flowers. In other words, this ability on the part of the larvae enables resource partitioning as a benefit of specialisation. However, it seems that not all generalist species are excluded from using Asteroideae pollen. Asteraceae, including species with pollen known to be toxic, seem to be used extensively by both specialists and generalists belonging to several bee genera in Britain, including many species in the large genus *Andrena* (Wood & Roberts 2017, 2018).

If oligolectic bees can benefit directly from larval immunity to plant toxins, there may be a further 'layer' of protection that they gain. If their nest-parasites or pathogens are not similarly tolerant, then the plant toxins may serve to eliminate, or at least reduce their vulnerability to, attacks from these enemies. This idea is explored by Danforth *et al.* (2019), citing a study of American species of the genus *Osmia* that are parasitised by a *Sapyga* wasp (*Sapyga pumila*). Two of the *Osmia* species were pollen generalists, one was a legume specialist and three were specialists on Asteraceae pollen. None of the nests of the Asteraceae specialists was parasitised by the wasp, contrasting with high levels of parasitism in the other three. Independent attempts to rear *Sapyga* larvae showed much lower survival rates on Asteraceae pollen (Spear *et al.* 2016). In the British Isles, *Sapyga quinquepunctata* parasitises several *Osmia* species, at least one of which (*Osmia leaiana*) does collect pollen primarily from Asteraceae.

There are two further hypotheses as to selective pressures favouring or conserving pollen specialism. First, abundant flowers with readily available resources would be expected to attract specialisation. Minckley and Roulston (2006) list a number of 'superabundant' pollen sources among North American flower groups which also host large numbers of oligolectic species. However, as these flowers usually attract other, generalist, species, it is not clear what advantage would accrue to obligate pollen specialists, unless they have adaptations which allow them to forage more efficiently from the shared pollen source than do the generalists.

Finally, it has been argued that pollen specialism could arise from a coincidence between the flight period of the bee and the flowering of the host plant. This synchrony is particularly noted in hot, arid habitats, where flowering periods may offer very brief 'windows' for pollinating insects. There is evidence that for pollen specialists, those bees whose flight period most exactly matches the flowering period of the host have more reproductive success than those that are less well synchronised. However, this would be a selective pressure towards closer synchrony only for bees that already specialise in the relevant host. It seems likely that synchronised phenology would be a consequence of coevolved mutualism or of adaptation of the oligolege to the phenology of its host, rather than an independent cause of the association.

The requirement for pollen specialists that their flight period coincides with the flowering period of their host also, of course, carries risks. If climatic disturbances cause either partner to depart from the usual cycle, there could be problems for the foraging success of the bee and for the flower, too, if it has relied on its specialist for pollination. But this is perhaps just a special case of the wider risk incurred by an evolved pollen specialist: it becomes dependent on its host species. So, set against the selective advantages that might accrue to an oligolectic species, there are also potential disadvantages. The synchrony between bee and host plant might go awry, the host plant may decline in abundance or distribution, and the bee will have abandoned the option of accessing other available pollen sources. This may be why only a minority of surviving bee species (approximately 30 per cent of the pollen-collecting species in Britain) are pollen specialists. However, given the potential disadvantages and risks associated with specialisation, this could be seen as a surprisingly large minority. Of course, bees cannot foresee the potential risks they face, and the disadvantages may not befall them for a very long time.

So far, we have considered the various hypothetical selection pressures that might tell for or against pollen specialism as if they were offered as general explanations. Of course, in each evolving lineage, these may be acting in concert, or crosscutting one another, with different outcomes at different times. An alternative approach might be to focus on particular genera and species and consider how far the hypotheses we have considered might go in explaining their different foraging strategies. It needs to be remembered, however, that pollen specialism is now thought to have been a basal trait in the evolution of bees and that it generally tends to be conserved (Larkin *et al. op. cit.*). In many cases, what stands in need of explanation is the shift to polylecty rather than conservation of oligolecty in a lineage.

There are inherent traits in some lineages that resist shifts to alternative pollen hosts, or from pollen specialism to generalisation, even if the environmental conditions might favour it (e.g. if there are abundant and rewarding flowers other than the host taxa available in the habitat). These traits might include the outcomes of long periods of coevolution between bee and host plant. Bees make very large demands for pollen, and since the production of pollen is very costly to plants, they have developed ways of balancing the need to attract pollinators with constraints either to limit pollen harvesting or, at least, to manipulate the behaviour of visitors in favour of pollination. These strategies on the part of plants include the evolution of flower shapes that conceal the anthers, making perceptual and cognitive demands on insect visitors, delivery of pollen onto a part of the insect body that is relatively inaccessible (e.g. nototribic release), slow release of pollen, requiring repeated visits, poricidal anthers that only release pollen in response to sonication by bees, and the use of chemical defences or other devices that make digestion of the pollen difficult (see Chapter 5).

Until recently, it has been assumed that pollen represents a generally usable source of protein and other nutrients for bees and their larvae. However, experimental attempts to rear bee larvae on pollen drawn from different sources have called this into question. In one important experiment, the larval development of four species of pollen-specialist bees on non-host pollen was studied (Praz *et al.* 2008). In each case, comparison was made with a control group, fed on their usual host pollen. The results for larvae fed on non-host pollen were very varied. The Asteraceae specialist (*Heriades truncorum*) fared well on non-host pollen except for that of buttercup (*Ranunculus*), while the other species failed to develop fully on Asteraceae pollen, and another failed on *Ranunculus*. There was a high survival rate in the control groups, suggesting that the experimental manipulation itself was not the explanation for these failures to develop.

It seems that these varying responses of bee larvae to pollen from different sources have to do with the properties of the pollen. There are several possibilities: variations in the nutritional content of the pollen (e.g. the concentration of protein or the presence or absence of other essential nutrients), presence of plant toxins, or difficulties in the way of digesting pollen contents. The experimental design used by Praz *et al.*, together with results of other studies, make low protein content in the pollen an unlikely explanation of poor larval development in these examples. However, lack of essential nutrients, such as certain amino acids, in Asteraceae pollen could explain the failure of non-specialists on this pollen.

The toxicity of some pollens is another explanation. Pollen grains are commonly coated with an oily layer called the pollenkitt, which contains volatile molecules that act as cues to visiting insects, either attracting or warning them (Detzel & Wink 1993, Dobson & Bergström 2000). This layer in *Ranunculus* pollen contains the compound protoanemonin, which is toxic to bees and is especially abundant in some Ranunculaceae taxa that are visited by bees. The pollen of Viper's Bugloss *Echium vulgare* was also tested in the study by Praz *et al*. It is the pollen host of *Hoplitis adunca*, whose larvae were reared successfully on it, but it proved fatal for larvae of the *Campanula* specialist *Chelostoma rapunculi*. Praz and colleagues note that this pollen contains high concentrations of toxic pyrrolizidine alkaloids which can presumably be successfully metabolised by *Hoplitis* (and also by *Heriades truncorum*, an Asteraceae specialist).

Independently of the presence of toxic compounds, some pollens have structural features which may make digestion difficult for some bee larvae. Pollen grains have a protective outer wall (exine) which is not digested, but within this is a further layer, the intine. This must be digested so that the protein in the pollen protoplasm can be extracted through pores in the exine. It is suggested that the failure of some bees that are not Asteraceae specialists to develop on Asteraceae pollen may be a consequence of their lack of ability to degrade the intine wall (see Chapter 5, Fig. 134).

These very different abilities of bee species to develop on pollen from various sources are not confined to pollen specialists (Sedivy *et al.* 2011), but they do offer some indications concerning the pressures that might bind a bee lineage to a particular group of plants and, perhaps, set limits to the range of possible pollen-host switches. Asteraceae, for example, are abundant and offer readily accessible floral rewards, so that a bee species that evolves the ability to digest pollen from this family of plants has an advantage over other potential harvesters

FIG 199. (a) A pollen specialist, the Large Scissor Bee *Chelostoma florisomne* female collecting pollen from Creeping Buttercup *Ranunculus repens*. (b) A pollen generalist, the Common Furrow Bee *Lasioglossum calceatum* also collecting pollen from Creeping Buttercup.

and is liable to become oligolectic on that family. *Ranunculus*, too, encourages pollen specialisation to those (rather few) species that can deal with its toxins and is rarely visited for pollen by non-specialists. *Echium* pollen is rich in protein, so it, too, offers large rewards to a bee that can overcome its highly toxic pollen. Interestingly, the *Campanula* specialist in the study by Praz and colleagues failed to thrive on the other pollen sources tried, but *Campanula* pollen did not appear to be toxic to specialists on other plant taxa.

HOW FAR IS DIETARY BREADTH AN INHERITED TRAIT?

A superficial overview of British species in terms of their dietary breadth reveals striking differences among the families and genera of nest-making species. All British species in the Melittidae are regarded as oligolectic. Among the (solitary) Apidae, all *Anthophora* and our one *Ceratina* species are polylectic, while *Eucera*, with just one currently extant species, is oligolectic. Among the Megachilidae, the one established species of *Heriades* is regarded as oligolectic, both species of *Chelostoma* are oligolectic, one species of *Hoplitis* is oligolectic, while roughly half of the *Osmia* species are polylectic, the single species of *Anthidium* is polylectic, and all but one *Megachile* species are polylectic.

The species of the family Andrenidae show mixed pollen-collecting strategies. Just under half of the species of *Andrena* whose foraging preferences are well studied are oligolectic (approximately 32 species), while a small number of those are rigidly and narrowly so. Both of the British species of *Panurgus* are oligolectic (on Asteraceae). Currently, the British bee fauna includes only two genera of nest-making species in the family Halictidae: *Halictus* and *Lasioglossum*. All but one of the British species of *Halictus* are regarded as polylectic, and all but one of the very large genus of *Lasioglossum* are also considered to be polylectic. Finally, the Colletidae comprise two genera with British representatives. Almost all species in the genus *Hylaeus* are regarded as polylectic (10 out of 11) while most British species of *Colletes* are considered to be oligolectic (8 out of 9). At the level of genus, there are striking differences. Some genera have all or almost all oligolectic species, others are wholly or predominantly polylectic, while others have more or less balanced mixtures.

As the British bee fauna is relatively impoverished, with some genera represented by only one or two species, comparison with the well-studied, overlapping but more extensive bee fauna of Germany is illuminating. We'll take Paul Westrich's *Die Wildbienen Deutschlands* (Westrich 2019) as the authoritative source for the comparison. A few species considered oligolectic in Britain are treated as polylectic by Westrich, possibly because of the broader range of pollen sources available there, but also to some extent reflecting the lack of clearly agreed definitions of the terms. Apart from this, the pattern of distribution of oligolectic species among the different families and genera is very similar in the two countries (Table 9). Germany has five more species in the Melittidae, all oligolectic. Among the solitary Apidae, there are 11 species of *Anthophora* and three of *Ceratina*, all of them polylectic, while fully eight species of *Eucera* are, like the one British species, oligolectic. Among the Megachilidae, *Anthidium*, with just one British species, which is polylectic, has 10 species in Germany, four of which are listed as oligolectic. *Heriades* and *Chelostoma* have only one and two established British species, respectively, all oligolectic. The German list adds a further four species, all oligolectic. *Megachile*, with just one oligolectic species (out of eight) in Britain, has just three oligoleges out of 22 species in Germany. In Germany, as in Britain, *Osmia* has roughly half its species in each category. Among the Andrenidae, the very species-rich genus *Andrena* has, as in Britain, roughly half its species (approximately 55 out of 116) listed as oligolectic, whilst *Panurgus*, with two oligolectic species in Britain, has a further one, also oligolectic, in Germany. Both halictid genera have more species reported from Germany, at 18 for *Halictus* and 70 for *Lasioglossum*. All of these species are listed as polylectic except for one. Finally, Colletidae in Germany includes *Colletes*, at 15 species, of which 10 are oligolectic, and *Hylaeus*, of which only four out of 37 species are considered oligolectic. Although the proportions are slightly different, in both countries, the majority of *Colletes* are oligolectic, while the majority of *Hylaeus* are polylectic.

Confirmation of these broad patterns in both countries does suggest that some families and genera are much more strongly constrained in their foraging strategies than others. The three melittid genera stand out as uniformly oligolectic, as do *Eucera*, *Heriades*, *Chelostoma* and *Panurgus* from other families. The most extreme contrast is with the two halictid genera: both of them being species rich and both almost exclusively considered polylectic. *Anthophora* and *Hylaeus*, too, are overwhelmingly polylectic. *Osmia*, *Anthidium* and *Andrena* each have roughly equal numbers of oligolectic and polylectic species. Whilst this pattern suggests that dietary breadth is strongly linked to taxonomic grouping, an interesting feature is that in some genera which are strongly oligolectic, the specialisms of individual species may be very different. For example, the six species of *Melitta* listed for Britain and Germany variously take Fabaceae, *Odontites*, Campanulaceae, *Onobrychis* and *Lythrum* as their host flowers. Insofar as host-plant choice is linked to speciation, it seems that in some lineages, pollen-host specialisation is constant, though switches to a different plant remain open. This raises interesting questions about what enables or constrains pollen-host switches in each lineage.

TABLE 9. Proportions of polylectic and oligolectic species, by genus, Britain and Germany compared. *Note*: Allocations to foraging strategies based on information from Else and Edwards (2018) for Britain and Westrich (2019) for Germany, omitting species thought to be currently extinct in Britain and those with incompletely known foraging strategies. Criteria for allocating species to either category are not always explicit, so the figures given should be treated as provisional and approximate.

Family	Genus/country	Polylectic species	Oligolectic species
Melittidae	*Dasypoda*		
	Britain		1
	Germany		3
	Macropis		
	Britain		1
	Germany		2
	Melitta		
	Britain		4
	Germany		6
Apidae	*Anthophora*		
	Britain	5	
	Germany	11	
	Ceratina		
	Britain	1	
	Germany	3	
	Eucera		
	Britain		1
	Germany		8
Megachilidae	*Anthidium*		
	Britain	1	
	Germany	6	4
	Chelostoma		
	Britain		2
	Germany		5

continued overleaf

TABLE 9. continued

Family	Genus/country	Polylectic species	Oligolectic species
Megachilidae continued	Heriades		
	Britain		1
	Germany		3
	Megachile		
	Britain	7	0
	Germany	19	3
	Osmia*		
	Britain	9	4
	Germany	19	18
Andrenidae	Andrena		
	Britain	42	17
	Germany	61	55
	Panurgus		
	Britain		2
	Germany		3
Halictidae	Halictus		
	Britain	4	
	Germany	18	
	Lasioglossum		
	Britain	28	2
	Germany	69	1
Colletidae	Colletes		
	Britain	2	7
	Germany	5	10
	Hylaeus		
	Britain	9	1
	Germany	33	4

*Hoplitis is included here in Osmia, following Westrich (2019).

POLLEN SPECIALISTS, POLLEN GENERALISTS AND THEIR HOST PLANTS: SOME EXAMPLES

Melittidae

The Melittidae are represented in Britain by only six species, belonging to three genera (*Dasypoda, Macropis* and *Melitta*). Germany adds a further 11 species, and all species in both countries are regarded as oligolectic. This strongly suggests that pollen specialism is a fixed trait in the family, and this is confirmed by a comprehensive study of the phylogeny and pollen hosts of species belonging to five genera in this family, distributed across the world (Michez et al. 2008, 2009).

Dasypoda

This genus has only one British representative, the Pantaloon Bee *Dasypoda hirtipes*. Michez et al. (2008) include this species, along with 15 others, in the subgenus *Dasypoda s. str.*, all of which share a common descent and are oligolectic on plants in the family Asteraceae. The Asteraceae are common plants, with accessible nectar and pollen, and it seems likely that the lineage to which *Dasypoda hirtipes* belongs was ancestrally oligolectic on this family of plants. As we have noted above, the pollen of some sub-families of Asteraceae is toxic or, at least, presents problems for digestion, so it may be that the persistence of *Dasypoda hirtipes* and its relatives in their specialisation is enabled by an adaptation on the part of their larvae to digest Asteraceae pollen. When females forage for both nectar and pollen from Asteraceae flowers (frequently those of Bristly Oxtongue *Helminthotheca echioides*, in Britain), they move rapidly with a circular motion around the platform of the inflorescence. As they do so, pollen grains are detached from the projecting anthers onto the long plumose hairs that arise from the ventral surfaces of the bees' thorax, face, and mid- and hind femora, and onto bands of shorter plumose hairs on the margins of the abdominal sternites. Periodically, the bees cease foraging and raise their heads, rapidly scraping the face and associated ventral surfaces with the forelegs and passing the collected pollen back to the hind-leg scopae, which are clothed with very long (twice the width of the tibia) drapes of slightly plumose hairs. There are dense brushes of shorter stiff setae on the inner surfaces of mid- and fore basitarsi which are used to transfer the pollen back. The especially long scopal hairs (which justify the vernacular name Pantaloon Bee) are generally reported to aid brushing away excavated soil during nest-making. However, since the flowers visited by this bee generally close after noon, NWO argues that the enlarged scopae could be an adaptation which enables rapid collection of large pollen loads. Other bee species, especially *Colletes*, make very similar movements when excavating nests but do not have enlarged scopae.

FIG 200. Pantaloon Bee *Dasypoda hirtipes* female (a) digging her nest burrow, (b) scraping pollen from her face with her forelegs and (c) with full pollen loads.

Macropis

Michez *et al.* (2008) list 10 species of *Macropis*, all of which are oligolectic on the pollen of plants in the genus *Lysimachia*. This bee genus is represented across Europe and Asia as well as North America (where they forage on a different subgenus of *Lysimachia*). The bees collect plant oils as well as pollen from the host plant, and the generality of this host-plant specialisation across the genus suggests that the connection is ancestral, as does the presence of distinct morphological adaptations to collecting oils and the role of the oils in both nest construction and food provisioning for the larvae.

The Yellow Loosestrife Bee *Macropis europaea* (Fig. 201) is unique among British bees in collecting plant oils. Females collect both pollen and plant oils from Yellow Loosestrife *Lysimachia vulgaris*, but as this plant does not secrete nectar, they need to visit other flowers (as do the males). Females collect oils from epithelial oil glands, situated on the filaments of the anthers and lower areas on the petals. Each gland is a three-celled structure which the bees appear to puncture with stiff setae present in dense pads on the inner surface of the fore- and mid-basitarsi. As Michez *et al.* point out, it is hard to see what function these pads on the fore- and mid-basitarsi could have other than collecting oils, suggesting a definite morphological adaptation to collecting plant oils from this or a closely related plant.

Other morphological features can also be linked to pollen specialisation to this host plant. The females have a covering of fine, pale plumose hairs on the sides and ventral surface of the thorax and bands of plumose hairs on sternites three to five. The outer surfaces of the fore and mid-tibiae and tarsi have a very dense covering of stiff black setae, and on the inner surface of the tarsi are dense pads of setae used to detach oils. The hind-leg scopae have complex arrays of hairs and setae. The outer surface of the hind tibia is clothed with dense, long white hairs, especially to the front, but these give way in the rear to a dense bed of finely plumose white hairs, the whole scopa reaching out to the sides, especially to the rear, to obscure the outlines of the underlying tibia. The inner surface of the hind tibia has a dense covering of short yellowish hair. The hind basitarsi are wide and flattened, with a dense covering of firm black setae on both surfaces.

Observation of the foraging behaviour of the bee reveals how these features are used. When engaged in brood cell construction, the females forage for oils only, but when foraging to provision their brood cells, they accumulate large loads of oil-soaked pollen on the hind legs. Oil is collected mainly from the oil glands on the filaments of the stamens, directly onto the fore- and mid-tarsi, and then transferred to the 'bed' of finely plumose hairs on the hind tibiae. When the bee is foraging to provision a brood cell, she moves her middle and forelegs rapidly over the ripe anthers, often 'cradling' the central cluster of stamens between mid- and hind legs, thus removing pollen from the open, outwardly facing anthers. Pollen is detached onto the hairs on her ventral surface and the hairs on the inner surface of her legs. From there it is transferred to the scopae on the hind tibiae and basitarsi. Here, the oil already collected on the hind tibiae seeps into the pollen to form a large congealed mass on both inner and outer surfaces of the hind tibiae and basitarsi. As the bee builds up this load, she holds both hind legs out behind her, projecting beyond the corolla of the flower, and leaves with a scissor-like movement. This behaviour may aid her movement in the flower and also may possibly serve as a 'keep off' signal to males, which actively patrol the plants.

FIG 201. Yellow Loosestrife Bee *Macropis europaea* female (a) using pads on her feet to soak oils from glands on the anther filaments, (b) with oil on her hind tibiae, beginning to collect pollen, and (c) with large loads of oil-soaked pollen.

Melitta

Michez *et al.* (2008) give analyses of the pollen loads of 43 *Melitta* species, drawn from Europe, Asia and North America. By their criterion (90 per cent or more of pollen from one family of plants), 36 of these are oligolectic, but across the genus, the host plants used are very diverse. Twelve host-plant families are listed: Brassicaceae, Boraginaceae, Asteraceae, Campanulaceae, Ericaceae, Fabaceae, Lamiaceae, Lythraceae, Malvaceae, Resedaceae, Scrophulariaceae and Zygophyllaceae. All four of the British species are closely associated with one or other of these families, though, on the strict criterion used by Michez *et al.*, one of them, the Bellflower Blunthorn Bee *Melitta haemorrhoidalis*, is regarded as mesolectic. Both Else and Edwards, as well as Westrich, treat this as an oligolectic species, and at 88 per cent, the pollen analysis given by Michez *et al.* is extremely close to their threshold for oligolecty (90 per cent).

Of the British species, the Sainfoin Bee *Melitta dimidiata* is extremely local and is oligolectic on Sainfoin *Onobrychis viciifolia* (other species of *Onobrychis* elsewhere in Europe), the Bellflower Blunthorn Bee *Melitta haemorroidalis* is oligolectic on Campanulaceae (almost exclusively on *Campanula*), the Clover Blunthorn Bee *Melitta leporina* is more broadly oligolectic on various Fabaceae, and the Red Bartsia Blunthorn Bee *Melitta tricincta* is oligolectic on the genus *Odontites* in Europe and confined to Red Bartsia *Odontites vernus* in Britain.

The females of the Red Bartsia Blunthorn Bee (Fig. 202) forage for pollen from the small inconspicuous red or pink flowers of the semi-parasitic plant Red Bartsia. The flowers are arranged along one side of the flowering stems and come into flower and shed pollen in sequence from base to apex of the inflorescence. The individual flowers are zygomorphic, with fused petals which open to form a lobed lower lip, and an upper hood-like canopy that partially encloses the four anthers, and the style, which projects beyond the corolla. Both male and female bees crawl rapidly between the open flowers on each spike and generally move to adjacent flower spikes with each flight. When collecting pollen, females spend no more than one or two seconds at each flower and usually forage from only one or two flowers on each flowering stem, buzzing loudly as they do so. Presumably the bees are able to detect the maturity of the anthers in each flower, as they almost always grasp a flower of the 'correct' stage, using forelegs to grip and manipulate it, while clinging on to lower (later-stage) flowers with mid- and hind legs.

The flowers of Red Bartsia are nototribic (that is, the anthers are above the bee as it probes the flower), an arrangement that aids pollination as the pollen lands on upper parts of the flower visitor that are difficult to reach for grooming. However, the bees reach into the flowers with tongue projecting and face uppermost, collecting pollen on the face and tongue. There are long fine hairs on the flattened galeae (outer sections of the tongue) and plumose hairs on the face, which brush against the anthers. At intervals the bees pause, usually while gripping an inflorescence, and perform a repeated, rapid scraping motion over the head and face with the front legs. There are dense brushes of yellow-orange hairs on the inner side of the fore basitarsi which are used to clear pollen from the face while mixing it with nectar. The mixture is passed back immediately to the scopae on the hind tibiae and basitarsi. These are formed of relatively sparse, long, gently curved pale hairs (approximately equal in length to the width of the tibia at its widest point). Presumably, this structure is sufficient to hold the sticky pollen-nectar mix in place, as the bees carry large, robust pollen loads. It is unclear whether the loud buzzing of these bees as they forage plays any part in releasing

FIG 202. Red Bartsia Blunthorn Bee *Melitta tricincta* female (a) with pollen among hairs on her face and (b) using her forelegs to clear pollen from her face.

pollen from the anthers, but there seem to be sufficient anatomical and behavioural modifications to suggest a long history of adaptation of the bee to its host plant.

Bellflower Blunthorn Bee *Melitta haemorrhoidalis*

As we reported above, the Bellflower Blunthorn Bee (Fig. 203) is treated as mesolectic by Michez *at al.* but is regarded as oligolectic on *Campanula* in Britain (Else & Edwards 2018). In places such as the East Anglian Brecks, where they fly in flower-rich habitats, the females forage exclusively from Harebell *Campanula rotundifolia*, but they also use other species of both wild and cultivated bellflowers, notably on chalk downland but also in gardens.

Campanulas have a distinctive pollination mechanism that requires some handling skills on the part of insect visitors. The reproductive cycle of the flowers has two phases, starting with a male phase, during which the anthers deposit the pollen onto the developing central style. This frequently takes place before the flower opens. As the style grows, it carries a coating of pollen grains among minute hairs along its length. When pollen is collected by insects, it is thought to stimulate the opening of the stigmas at the apex of the style. Now the second, female, stage begins, and the stigmas are receptive to pollen incidentally deposited from the body hairs of insects, now visiting for nectar only (see Chapter 5).

When harvesting pollen from Harebells, the females fly rapidly from flower to flower, entering directly into the mouth of the bell, seeming to 'scrabble' around while concealed within. Each flower visit takes from 3–4 seconds, and intermittently the bees settle among flowers to scrape pollen onto the scopae on their hind legs. Presumably, nectar is regurgitated and mixed with the pollen during this process. On larger-flowered campanulas, the bell is often wide enough to enable inspection of the bees' activity within (or, as the authors attempted in Cambridge University Botanical Gardens, strips of corolla tube can be pulled away without any apparent effect on the bees' activity). The bees enter directly and reach for the nectaries at the base of the corolla, rubbing against the pollen-bearing central style as they do so. They then rotate to access each nectary in turn, clearing pollen from the outer faces of the style as they move. There are long plumose hairs on the head and ventral surface of the bees' abdomen and on some of the upper leg segments, and pollen is captured on these as the bees forage. There is a dense covering of stiff, dark, downward-pointed bristles on the inner surfaces of the basitarsi, and these have numerous pollen grains caught in them as the bees forage. It seems likely that these brushes are used in transferring the nectar and pollen mix from body hairs to the scopae, which are situated on the hind tibiae and basitarsi.

This suggests some anatomical adaptation to foraging from campanulas, but it may also be that there are inherited sensory biases in favour of these flowers. Recent research on other bee species that specialise on *Campanula* flowers for pollen has identified both olfactory and visual cues that attract freshly emerged females to flowers of this genus. *Chelostoma rapunculi* (which does not occur in Britain) was shown by Milet-Pinheiro and colleagues (2013) to respond both behaviourally and by independent testing of its antennal responses to a small subset of the chemical compounds in the floral 'bouquet' of most *Campanula* flowers. These compounds, spiroacetals, while widespread in many organisms are rare components of floral scents. Inexperienced bees responded to these compounds, but with more foraging experience they acquired a broader search image, inclusive of other, more abundant and more common, volatile compounds in the scent. It seems that, for this species at least, there is an innate response to the 'signature' components of the *Campanula* scent, but once the bee has foraged from its host plant, it can use the wider array of scents to assess potential floral rewards from particular plants and perhaps to identify its host species at a greater distance.

In another study (Milet-Pinheiro *et al.* 2015), the authors showed that the same bee species shows an innate preference for the UV-blue/blue bee colour-space that corresponds to the colour of light reflected by *Campanula* flowers. Brandt *et al.* (2017) further showed that three bee species that are oligolectic on campanulas were more sensitive to the spiroacetal 'signature' compounds than were comparable pollen generalists. The species that figured in the study by Brandt *et al.* included the British species the Small Scissor Bee *Chelostoma campanularum* (Fig. 203 b) which can often be found visiting the same flowers at the same time as *Melitta haemorrhoidalis*.

FIG 203. Bellflower Blunthorn Bee *Melitta haemorrhoidalis* female (a) collecting pollen from a *Campanula* flower and (b) visiting the same flower as a Small Scissor Bee *Chelostoma campanularum* female.

Although it is not clear whether *Melitta haemorrhoidalis* has similar innate visual and olfactory responses to its host plant, these findings for other *Campanula* specialists are suggestive and, combined with the anatomical modifications mentioned above, give reason to think *Melitta haemorrhoidalis* may have an evolved adaptation to its host genus. It is known that *Campanula* pollen is high in protein, and the most widespread species in Britain (Harebell *Campanula rotundifolia*) also has a prolonged flowering period. These features, together with the relative accessibility of pollen and nectar, render this an attractive plant genus for pollen generalists as well as specialists. The innate ability to recognise the plant and adaptations to forage efficiently from it might well confer selective advantages where the plants are abundant and would tell in favour of persisting with pollen specialisation to it.

Apidae
Anthophora

All British species of *Anthophora* are generally regarded as polylectic but three of them show a distinct preference for flowers of Lamiaceae (dead-nettles) and Scrophulariaceae. Structures used to carry pollen loads are similar across the British species (see Box 6), but an interesting study of some central European species by Müller (1996) argues that 13 species from six bee genera and one honey wasp all have specialised facial structures that are used in harvesting pollen from nototribic flowers in these two plant families. Like the flowers of Red Bartsia, discussed above, the typical mode of access by bees results in pollen being deposited dorsally, often on the head. Nototriby is usually regarded as an adaptation that both allows but also restricts the harvesting of pollen, in the interests of the plant. Although the bees under discussion are all polylectic, Müller uses analyses of pollen loads and samples of pollen from brood cells to show that pollen from nototribic Lamiaceae and Scrophulariaceae flowers forms a high proportion of the nutritional intake of the larvae in these species (from approximately 30–100 per cent in the samples they analysed).

This suggests that, for the bees concerned, there is an advantage to having specialised morphological structures and associated behaviours to aid pollen harvesting. Müller's study confirmed earlier research that found distinctive facial hairs in bees from the genera *Rophites* and *Celonites*, and found comparable adaptations in other species, including several that also occur in Britain. The adaptation in the Fork-tailed Flower Bee *Anthophora furcata* is of particular interest as it is a species closely associated with dead-nettles (Lamiaceae) in Britain. The clypeus and labrum are clothed with modified hairs, which are sturdy at the base but end in fine 'tails' which are angled to produce a more or less enclosed space between them and the cuticle (see Figure 189). The bees may visit the flowers for pollen or nectar or both, depending on whether the flowers are in their male or female phases.

When collecting pollen, the bee alights on the lower lip of the flower, presses its head against the anthers present in the upper lip, and emits a buzzing sound which may function to release pollen. If nectar is also required, the bee moves forwards to probe the nectaries at the base of the corolla tube. When the bee leaves the flower, she brushes the pollen from her face with rapid movements of the forelegs. *Anthophora quadrimaculata*, which lacks the modified hairs, also collects pollen from Lamiaceae but uses a quite different method, settling on the upper lip and reaching down to the anthers. Müller does not mention our most common flower bee, the Hairy-footed Flower Bee *Anthophora plumipes* (Fig. 204), but this species, too, is strongly associated with Lamiaceae,

FIG 204. Hairy-footed Flower Bee *Anthophora plumipes* female with pollen on her tongue from Ground Ivy *Glechoma hederacea*.

notably with Red Dead-nettle *Lamium purpureum* and various cultivated members of the same family, during its spring flight period. In this species, pollen is harvested on fine hairs on the labrum and face.

Eucera

This genus is strongly oligolectic. The Long-horned Bee *Eucera longicornis* (Fig. 205) is now probably the only British species, as *Eucera nigrescens* is believed to be extinct in Britain. Both are pollen specialists on Fabaceae, and a further six *Eucera* species that occur in Germany are all oligolectic. Two of these are also associated with Fabaceae, another two are Asteraceae specialists, and the final two are specialists on Lythraceae and Malvaceae, respectively. This pattern is consistent with the view that narrow diet specialism tends to be conserved in some lineages, even when switches to alternative host plants occur. Interestingly, such switches are not necessarily to taxonomically close relatives of the initial host plant. In this case, Fabaceae seem to be the more widely favoured host (four out of eight species

FIG 205. Long-horned Bee *Eucera longicornis* male on a break from patrolling to take nectar from Meadow Vetchling *Lathyrus pratensis*.

among the ones included here), but the three other plant families have quite different colour patterns and structures, as well as being taxonomically distant.

Although Westrich gives a list of several different flowers in the Fabaceae as pollen hosts for *Eucera longicornis*, and there are records of different primary pollen hosts associated with different habitats in Britain (Falk & Lewington 2015, Hennessy et al. 2020), observations at coastal sites in Essex revealed persistent preference for Meadow Vetchling *Lathyrus pratensis* on the part of females collecting pollen among arrays of other Fabaceae, including the visually similar Common Bird's-foot-trefoil *Lotus corniculatus*. Mate-searching males, too, patrol patches of *Lathyrus pratensis*, ignoring adjacent patches of *Lotus corniculatus*. It would be interesting to know what sensory cues the bees use in making this discrimination, and to what extent this local fine-tuning of host specialisation is a result of locally learned foraging experience.

Females of this species forage for nectar and pollen at the same time, and the pollen loads are mixed with nectar to form a congealed mass on the hind tibiae. The basitarsi of the forelegs have a sweep of long hairs on the inner surface and have an inward-directed projection at the anterior tip, which bears a comb of robust curved spines. These number from 10–12 and a similar, less-sturdy pattern is repeated on the following three tarsal segments. There is also a dense pile of plumose hairs on the labrum. It seems likely that this arrangement is associated with wetting pollen with regurgitated nectar before passing it back to the scopae (see Figure 194 b, above).

Megachilidae
Megachile
Although the established British species are regarded as polylectic, the larger species, such as Willughby's Leafcutter Bee *Megachile willughbiella* and the Wood-carving Leafcutter Bee *Megachile ligniseca* (Fig. 206), forage extensively on larger thistles, such as Spear Thistle *Cirsium vulgare*, and other Asteraceae, such as Greater Burdock *Arctium lappa*, Common Ragwort *Senecio jacobaea* and Common Knapweed *Centaurea nigra*. On these flowers, they work their way around the capitulum, taking nectar from deep nectaries while 'roughing up' the anthers with rapid scrabbling movements of their legs. Pollen is then passed onto the abdominal scopa, sometimes accompanied by up-down movements of the abdomen, rather resembling those of *Heriades*. When the bees forage for nectar, the abdomen is held in alignment with the rest of the body or curved upwards.

Müller and Bansac (2004) describe an arrangement of brushes of hair on the hind trochanter and femur of certain Western Palaearctic *Megachile* species, which are used in pollen harvesting from thistles and knapweeds. The bees were

FIG 206. Wood-carving Leafcutter Bee *Megachile ligniseca* female, showing hind tarsi used to harvest pollen.

observed to squeeze adjacent flowers between femur and tibia of the hind leg and then draw the leg upwards to comb off the pollen. The subsequent transfer of pollen to the scopae was not observed but assumed to be achieved during flight. Müller and Bansac link the possession of this feature with species that are oligolectic on these plants, as they did not find it on polylectic species. We could not find this feature on any of the British *Megachile* species, but there are dense pads of stiff setae on the tarsi that play a similar role in detaching pollen from the exposed anthers of this group of composite flowers. These setae are also used to draw pollen from the keel when the bees are foraging from plants in the Fabaceae (NWO video).

The Silvery Leafcutter Bee *Megachile leachella* is a smaller species which collects pollen from a variety of flowers. On bramble, for example, they move around the flower, probing for nectar while 'scuffling' the prominent anthers with their legs and simultaneously pressing their abdominal scopa against them. They also emit a high-pitched, whining 'buzz' as they do so, though it is not clear whether this has any role in shaking out pollen. They seem equally at home collecting pollen from flowers in the Fabaceae, such as Spiny Restharrow *Ononis spinosa* and bird's-foot-trefoils *Lotus* species (see Figure 192 a, above).

Osmia

As we saw in earlier discussion, this genus holds both oligolectic and polylectic species. The Blue Mason Bee *Osmia caerulescens* is regarded as polylectic, but

FIG 207. Red Mason Bee *Osmia bicornis* female, a pollen-generalist *Osmia* species.

it figures in Müller's study of adaptations to nototribic flowers. His electron microscope images show modified hairs on the clypeus (as in *Anthophora furcata*). These are wide at the base with fine 'tails' at right angles to the main stem. These are just visible with light microscopic examination but are frequently worn away in older specimens. Similar structures are also present in *Osmia caerulescens* (NWO pers. obs.). The Fringe-horned Mason Bee *Osmia pilicornis* is an oligoletic species that in Britain collects pollen mainly from Bugle (see Figure 190, above).

Hoplitis

The Viper's Bugloss Mason Bee *Hoplitis adunca* (Fig. 208) is widespread in mainland Europe but new to Britain and still highly localised. The bee is a pollen specialist on Viper's Bugloss *Echium vulgare* in Britain. The bee recognises its host plant at a distance by visual cues but at close quarters by its distinctive scent. A study by Filella

FIG 208. Viper's Bugloss Mason Bee *Hoplitis adunca* female encountering anthers and stigma of a Viper's Bugloss *Echium vulgare* flower as she approaches for nectar.

et al. (2011) showed that the bee discriminated by scent between Viper's Bugloss and a closely related plant species, probably detecting different ratios of certain compounds in the scent of the pollen. In Britain, this ability is not required, as *Echium vulgare* is our only *Echium* species. Here, the foraging females fly rapidly from flower to flower of *Echium*, using their long tongue to access nectar. During the pollen-producing phase of the flower, they hold on to the corolla with forelegs and sometimes mid-legs, whilst 'scrabbling' pollen from the projecting anthers with their hind legs as their abdominal scopa brushes against them (TB pers. obs. & video).

Chelostoma

Most, if not all, species in this genus worldwide are oligolectic (Michener 2007), but their specialisms vary widely, including plants in families Hydrophyllaceae and Saxifragaceae, as well as Campanulaceae and Ranunculaceae, in Britain. Most European species have plants in the Campanulaceae as their pollen hosts, and this is true of one of the two British species – the Small Scissor Bee *Chelostoma campanularum* (Figs 209 & 210), which is narrowly oligolectic on flowers in the genus *Campanula*. As described above in relation to another *Campanula* specialist, *Melitta haemorrhoidalis*, *Campanula* flowers have a distinctive pollination mechanism. *Chelostoma campanularum* is much smaller than the *Melitta* and has its own approach to foraging. Trailing Bellflower *Campanula poscharskyana* is a common plant in urban gardens and is extensively used for pollen and nectar by this bee. Females forage actively from the flowers, gripping the central style with forelegs and mid-legs or sometimes holding on with their unusually long curved

FIG 209. Small Scissor Bee *Chelostoma campanularum* female, harvesting pollen from the style of a *Campanula* flower.

mandibles only. Facing towards the base of the flower, they rapidly shuffle their legs against the style, passing pollen back to the abdominal scopa while dabbing the scopa itself onto the pollen grains along the style. Periodically, they move back up the style, raise the abdomen and use their back legs to transfer pollen onto the scopa. Sometimes this is repeated against the other faces of the style. When the flowers have passed to the female phase, the bees continue to visit but solely to take nectar from the surface of the ovaries at the base of the flowers (TB pers. obs.).

Other bee species were observed foraging from this *Campanula* species in TB's urban garden. Females of the polylectic Gwynne's Mining Bee *Andrena bicolor* collect pollen by reaching down towards the base of the flower, brushing their ventral surfaces against the pollen-laden style and brushing the hind legs in a backwards motion against it. Before flying off, they comb pollen from the face with the forelegs, while perching on the lower petals with hind legs and mid-legs. Unlike many other *Campanula* species, in the Trailing Bellflower, the five petals soon open out to form a flattish platform that is widely accessible to a range of generalist foragers. *Hylaeus communis, Lasioglossum morio, Lasioglossum leucopus, Megachile willughbiella, Megachile centuncularis* and *Megachile ligniseca* were all observed collecting pollen from Trailing Bellflower in the same garden, and the flowers were very rapidly denuded of pollen. An apparent response to this intense competition for pollen on the part of the specialist *Chelostoma campanularum* was for the females of this species to enter the flowers while they were only beginning

FIG 210. Small Scissor Bee *Chelostoma campanularum* female, using her mandibles to access a *Campanula* flower before it is fully open.

to open and the styles were still replete with pollen. The small size and narrow body shape are an advantage for this manoeuvre. As noted above, *Chelostoma campanularum* also frequently forages together with *Melitta haemorroidalis* on the larger, bell-shaped flowers of other *Campanula* species (see Figure 203 b, above).

It seems likely that oligolecty is a strongly conserved feature of the *Chelostoma* lineage, and as *Campanula* species are common and have rewarding flowers – first 'admirably described and explained' by Sprengel (Müller 1883) – there might have been no strong pressure to switch to an alternative pollen host. However, the other British species of *Chelostoma* – the Large Scissor Bee *Chelostoma florisomne* (Fig. 211) – is oligolectic on *Ranunculus* pollen. *Ranunculus* (buttercup) flowers are abundant, with easily accessible nectaries and dense central clusters of anthers. They frequently predominate in their grassland habitats, especially when these are grazed, since domesticated animals tend to avoid them. When foraging for pollen, the bees move very swiftly from one flower to the next, spending only a few seconds on each (see Figure 199, above). Pollen appears to be gathered on the ventral scopa by direct contact, and still photos show the elongated mandibles clasping one or more anthers. This may be a means of steadying the bee while it sweeps the rest of the anthers, or the bees may bite into the anthers for pollen. That the mandibles have some function specific to the females is suggested by their noticeably larger relative size in the females of this genus.

As noted above, there is evidence that *Ranunculus* pollen is toxic to the larvae of many bee species (Sedilvy *et al.* 2007), and very few species regularly collect pollen from buttercup flowers, although they are frequently visited by many insects for nectar. It seems likely that the larvae of *Chelostoma florisomne* have

FIG 211. Large Scissor Bee *Chelostoma florisomne* female, showing large mandibles.

evolved some physiological tolerance of *Ranunculus* toxins, thereby securing almost exclusive access to buttercup pollen. This seems to be an example of resource partitioning as a benefit of oligolecty.

Heriades

Currently there is just one species of this genus established in Britain. The Large-headed Resin Bee *Heriades truncorum* (Fig. 212) was formerly highly localised in southern Britain but is now spreading rapidly. It is broadly oligolectic, collecting pollen mainly from the composite flowers of the family Asteraceae. When females forage for pollen on ragworts (such as *Jacobaea vulgaris*) or Common Fleabane *Pulicaria dysenterica*, which have both disc and ray florets, they move rapidly in a circular route around the capitulum, probing each of the open disc florets in turn for nectar. As they do so, the abdomen is vibrated rapidly up and down, dislodging pollen directly onto the abdominal scopa. Meanwhile, the hind legs flick up and down in synchrony with the abdomen, patting down the pollen in the scopa (Konzmann *et al.* 2020).

However, on other yellow Asteraceae, such as Fox and Cubs *Pilosella aurantiaca*, which have only ray florets and no disc florets and tall, filamentous stamens, the method is quite different. In these flowers, the female reaches down for nectar while grappling the stamens with her legs, cupping her abdomen around the anthers and ruffling them. Pollen is also collected from other groups of flowers, such as *Geranium* species. Here, the female uses a 'scrambling' technique, pressing her dorsal surface against the inner surface of the cup-shaped corolla and moving round so as to make contact with the anthers with her scopa (TB pers. obs.).

FIG 212. Large-headed Resin Bee *Heriades truncorum* female foraging from Ragwort *Jacobaea vulgaris*.

Andrenidae
Andrena

The genus *Andrena* includes a roughly even division between polylectic and oligolectic species. A study of North American *Andrena* species conducted by Larkin *et al.* (2008) used mitochondrial and nuclear data to trace the phylogeny of selected polylectic species, as well as that of a group of closely related oligolectic species. Their study concluded that the ancestral Andrenas were oligolectic, with subsequent shifts to more generalised pollen diets, as well as some reversals. Their research included a study of an oligolectic clade (a group of species with a common ancestor), which provided further evidence of shifts in dietary breadth. All the species in the group were oligolectic on Asteraceae pollen, some being defined as narrowly so (i.e. confined to one tribe within the Asteraceae) and others as broadly oligolectic (i.e. taking pollen from flowers belonging to more than one tribe). Although there were numerous shifts of host plant, these were mainly within the same tribe of Asteraceae, suggesting an inherent resistance to host-shifts to less closely related plants. Such shifts to more distantly related host plants did, however, occasionally occur and gave rise to increased adaptive radiation, suggesting a close link between changes of pollen host and speciation, at least in this lineage. Another finding of interest is that, while most North American *Andrena* species fly in spring, there have been several shifts to autumnal emergence. These have all been species oligolectic on Asteraceae pollen, which enabled them to use late-flowering Asteraceae and to extend their range northwards.

This study concludes that the genus *Andrena* has a history of relatively frequent changes of dietary breadth and, for the more specialised species, of shifts from one pollen host to another. This makes the genus a potentially informative one for understanding the causes and consequences of changes in dietary patterns. Another study (in Michigan) of North American species of *Andrena* (Wood & Roberts 2018) used a more differentiated classification of degrees of dietary specialisation (including distinctions between narrow and broad oligolecty, mesolecty, polylecty with strong preference, and polylecty). The species studied could be divided almost equally between specialists and generalists, with specialists favouring plants drawn from four families. Specialists on Asteraceae were a focus for the study, following from earlier work (Müller & Kuhlmann 2008; see also below) that drew attention to an 'Asteraceae paradox'. This is that for at least one genus of bees (*Colletes*), while specialists gained all or almost all of their pollen from plants in the Asteraceae, generalist bees in the same genus more or less completely excluded it. The proposed explanation of the paradox is that Asteraceae pollen is either toxic or for some other reason poses challenges to the bee's digestive system. For bees that have evolved the physiological ability to digest this pollen, there are considerable advantages to becoming specialised on it, as other bees cannot compete for the same resource. There is also a suggestion that physiological adaptation for a specialist diet may reduce the ability of a species to utilise an alternative one (Scriber 2005, Rasmann & Agrawal 2011), so reinforcing dependence on the specialised pollen source.

The results of the study carried out by Wood and Roberts confirmed the presence of the 'Asteraceae paradox' in the foraging patterns of the species of *Andrena* studied in the Michigan area. The seven Asteraceae oligoleges collected pollen almost exclusively from Asteraceae, but the pollen generalists almost completely avoided it. Interestingly, the same pattern was observed for the other three plant families that hosted pollen-specialist bees, but unfortunately relatively little is known about their pollen chemistry. However, to add a new layer of paradox, the study also included a comparison with the dietary patterns of a selection of British *Andrena* species, and this did not correspond to the expected Asteraceae paradox. Generalist bees were capable of collecting pollen from all of the plant families that hosted specialists, and this also applied to specialists on Asteraceae.

Wood and Roberts consider various possible explanations for this difference between British and North American foraging patterns among the Andrenas. The Asteraceae is a very large and diverse plant family, with rather different flora in the two regions. It could be that the pollens differ nutritionally or in digestibility, or that the British *Andrena* generalists have more recently acquired the ability to use Asteraceae pollen. Another possibility is the effect of different climatic

FIG 213. Buff-tailed Mining Bee *Andrena humilis* female, an Asteraceae specialist, foraging from Common Catsear *Hypochaeris radicata*.

conditions. The climate in Britain is arguably less sharply seasonal than that around Michigan, so that flowering periods of host plants may be more extensive, loosening the pressure for strict specialisation and allowing for changes in flight periods. Notably, the proportion of bivoltine species of *Andrena* in Britain is much higher than in North America. These conditions would favour a shift from narrow to broader pollen diets. Finally, it seems that 'challenging' pollen can be tolerated as a component in mixed pollen provisions, even where larvae do not develop when it is supplied as the sole dietary source.

Among the (approximately) half of the British species of *Andrena* that are oligolectic, several are specialists on Asteraceae pollen, several use Fabaceae flowers, while others take Ericaceae as their pollen hosts. Many of the flowers in these families are abundant and are often predominant in their habitats (heather heath, flower-rich grassland). A few species are very narrowly oligolectic, and their relations to their pollen hosts are of considerable interest.

The Small Scabious Mining Bee *Andrena marginata* (Fig. 214) is another pollen specialist, collecting pollen almost exclusively from Devil's Bit, Field Scabious or Small Scabious flowers. In the East Anglian Brecks, this species is common along wide rides, often with its scarcer relative, the Large Scabious Mining Bee *Andrena hattorfiana*. Both species are pollen specialists, *Andrena hattorfiana* foraging mainly from Field Scabious (see Chapter 3, Fig. 82), while *Andrena marginata* generally forages from Small Scabious where both scabious species are present. When foraging for pollen, the females of this species alight on inflorescences of the plant and walk rapidly from floret to floret, probing each one for nectar with tongue extended, while brushing against the prominent anthers as they do so

FIG 214. Small Scabious Mining Bee *Andrena marginata* female foraging from Small Scabious *Scabiosa columbaria*.

with their legs and ventral surface of their body. These movements detach pollen, which is brushed onto the hind-leg scopae by movements of the forelegs and mid-legs as the bees forage. There are large 'bushy' scopae on the hind tibiae, with long feathery hairs on the inner surface and stiffer dark hairs on the outer surface. Pollen-carrying capacity is particularly highly developed on the femora, with long feathery hairs projecting from the sides and a similarly feathery floccus forming a cage for pollen below and to each side of the femur. The propodeal pollen baskets, too, are highly developed. The much-branched hairs form a web to capture the unusually large scabious pollen grains, and fully laden bees give the impression of carrying large sacks at the sides of their body as well as on the hind femora and tibiae (see also the discussion in Chapter 5).

White Bryony *Bryonia dioica* is the sole pollen source in Britain for the Bryony Mining Bee *Andrena florea* (Figs 215, 216 & 217). Male and female flowers are carried on separate plants, and females collect pollen and nectar from the male flowers, also visiting female flowers for nectar. The short, wide stamens have an apical wavy ridge from which the pollen is produced. As the female bee alights on the flower, she brushes her ventral surface and the scopae on the hind legs against the ridge of pollen as she probes down beside the anthers to sip nectar, seeming to 'cradle' the anthers between her legs and body. Males, when foraging for nectar, have a quite different, erect posture on the flowers. The female flowers have a central style which opens out into three irregularly shaped stigmas. Both male and female bees visit these flowers for nectar only, incidentally brushing pollen onto the stigmas.

This pollen specialist, like many others, shares its favoured flower species with many other visitors, including honeybees, other solitary bees (including the mining bees *Andrena bicornis*, *Andrena thoracica*, *Andrena nigroaenea*, and *Andrena bimaculata* and several furrow bees, including *Lasioglossum sextrigatum*, *Lasioglossum sexnotatum* and *Lasioglossum morio*) and hoverflies. Elsewhere in Europe, this bee collects pollen from another bryony species (*Bryonia alba*), and its distribution across Europe corresponds closely with the combined distribution of the two plants. It seems clear that the bee is dependent on bryony flowers for the nutrition of its larvae, but the closeness of its association raises interesting questions concerning its possible role in pollination. Although the dioecious habit of the plant secures it against the risk of self-pollination, there must also be a converse risk of failing pollination altogether. Assuming the plant requires cross-pollination for fruit and seed production, there must be some likelihood that insect vectors visiting male flowers will go on to visit female ones. It seems unlikely that this can be left to chance, as male and female plants are often quite widely separated (and the female flowers are much less conspicuous than the male ones).

As we have noted, many other insect species visit the bryony flowers (often just for nectar), but it is unclear how many show 'fidelity' to the plant, extending to visiting both male and female flowers. Both male and female *Andrena florea* visit both sexes of the plant, with males actively patrolling and nectaring from the flowers from the moment of their first opening (TB, pers. obs.). It could be that chemical cues bind the bee to the bryony flowers for both pollen and nectar and that their mating system consolidates a role in pollination.

FIG 215. Bryony Mining Bee *Andrena florea* female collecting pollen from a White Bryony *Bryonia dioica* flower.

FIG 216. The pollen on the scopae of a Bryony Mining Bee female making contact with the stigmas of a White Bryony flower.

FIG 217. Bryony Mining Bee female transferring pollen from her body hairs to her hind-leg scopae.

A small number of species collect pollen exclusively, or almost so, from heather, especially Ling *Calluna vulgaris*. One of the most familiar of these is the Heather Mining Bee *Andrena fuscipes* (Fig. 218). The flowers of Ling are small and clustered around the tips of flowering stems. Bees foraging for pollen and nectar cling to the group of flowers, probing each one for nectar with the extended tongue. Pollen is brushed from the ripe anthers onto a dense covering of finely branched hairs on the face of the bee, especially on the clypeus (lower part of the face) and the base of the outer segments of the tongue (galeae), which have a rough surface and

FIG 218. Heather Mining Bee *Andrena fuscipes* female (a) foraging from Ling *Calluna vulgaris* and (b) grooming pollen from her face and onto her scopa.

rows of fine hairs. From time to time, the fine pollen grains are brushed back to the scopae and gathered together with other grains caught among the bands of long plumose hairs on the sternites (underside of the abdomen). The scopae on the hind tibiae consist of densely packed curved hairs, angled close to the surfaces of the tibiae, some of them with hooked tips to hold in the packed pollen. The scopae are complemented with highly developed propodeal pollen baskets and pollen baskets on the hind femora enclosed by the long feathery hairs of the flocci.

Panurgus

There are just two British species, and only three in Germany. All are oligolectic on Asteraceae and have a very distinctive method of foraging from these plants. The female collects pollen and nectar simultaneously, moving rapidly in a circular (or spiral) mode around the capitulum, with her extended tongue probing the nectaries of florets as she moves around the flower head. The thorax and abdomen are held sideways (lateral surface of one side of the bee against the capitulum), while the bee thrusts her way around, rapidly moving her legs and performing a repeated 'scissor' movement. In this the abdomen is briefly folded against the ventral surface of the thorax, 'ruffling' exposed anthers as she goes, and then quickly opened out again. This action brushes pollen onto bands of erect plumose hairs on the ventral surface of the abdomen, from where it is brushed onto the scopae. The hairs constituting the hind leg scopae have a distinctive 'corkscrew' structure and are bent over towards the tip, enabling more efficient capture of pollen grains. These features, together with their distinctive behaviour, suggest

FIG 219. Large Shaggy Bee *Panurgus banksianus* female foraging for pollen, 'looping' around the capitulum of the flower head on her side.

a history of association with Asteraceae in the *Panurgus* lineage, perhaps with an inherited ability on the part of the larvae to thrive on Asteraceae pollen.

Halictidae

The family Halictidae includes just two nest-making genera currently present in Britain – *Halictus* and *Lasioglossum*. As discussed in Chapter 4, recent research on the phylogeny of this family has provided evidence that eusocial modes of life emerged independently at least three times, but also that reversions from social to solitary states have occurred as many as 12 times. Compared with the more complex forms of sociality exhibited by the bumblebees and honeybees as well as ants and termites, which emerged as long ago as the Cretaceous, the eusocial Halictid clades are thought to have evolved relatively recently, approximately 20 to 22 million years ago (Brady *et al.* 2006, Danforth 2002). Descendants of two of those eusocial clades among the British fauna include all the species in the genus *Halictus* and almost all of the British *Lasioglossum* species. Of the four established *Halictus* species, three are regarded as eusocial, or capable of existing as social or solitary, and one is believed to be solitary. Among the British *Lasioglossum* species, approximately 10 out of 30 species have eusocial modes of life in at least part of their range (see Chapter 4).

What all of these Halictids share is a distinctive feature of their life cycle: females mate in late summer or autumn and then enter diapause for the winter (usually in a burrow associated with the nest in which they emerged). Only females survive the winter, and in spring they make their nests, provisioning and laying an egg in each cell. In solitary species, both males and females emerge in summer, mate, and the females go into hibernation, repeating the cycle the following spring. In eusocial species, or ones capable of sociality where conditions permit, one or more broods of females are reared and contribute to the provisioning of the larvae destined to emerge later in the year as reproductive males and females.

The evolution of sociality in this group of species is of interest in its own right, but its relevance here is that the distinctive life cycle not only favours the evolution of sociality where the summer season is long enough and the habitat sufficiently rich in forage sources, but it also makes oligolecty very unlikely. This is because although they have an annual life cycle, there are two flight periods in the year: in spring, when overwintered females establish their nests and provision their brood cells, and then later in the year, when the offspring emerge. This generation may simply forage for their own nutrition and mate before going into hibernation. Alternatively, in the social species, there may be one or more intermediate broods of workers. In this respect, halictids resemble bumblebees in requiring availability of a sequence of forage sources through spring and summer, as well as the ability to forage effectively from them. Of the British species, all *Halictus* are polylectic, and only one *Lasioglossum* species is regarded as oligolectic.

In line with these expectations, halictid bees are very versatile in their ability to access pollen from a wide variety of flower structures. Their small size is another factor enabling them to access 'hidden' pollen.

Hot-lips Sage *Salvia microphylla* is a popular horticulturally modified lamiate with an enlarged platform, a narrow opening leading down to the nectaries, and a boat-shaped apical section of the corolla which encloses the anthers, with the bifid stigma projecting from the top. Bumblebees such as *Bombus hortorum* and *Bombus lucorum* are able to forage 'legitimately' from these flowers, probing down to the nectaries. However, the flowers are also favoured by small Halictids, such as the Green Furrow Bee *Lasioglossum morio* (Fig. 220 a) and the Bronze Furrow Bee *Halictus tumulorum* (Fig. 220 b). Females of these species forage from these

FIG 220. (a) Green Furrow Bee *Lasioglossum morio* female using her mandibles to access part of a flower of Hot-lips Sage *Salvia microphylla* (Fabaceae). (b) Bronze Furrow Bee *Halictus tumulorum* female collecting pollen from Common Fleabane *Pulicaria dysenterica* (Asteraceae). (c) White-footed Furrow Bee *Lasioglossum leucopus* female collecting pollen from a *Campanula* flower (Campanulaceae).

flowers for pollen only, and they access it by clinging on to the apical section of the corolla with mid- and hind legs and forcing apart the edges of the folded petal with their mandibles and head. This may be an unusual case of pollen larceny, as distinct from the more usual nectar larceny, but it also illustrates the ability of bees to 'learn new tricks'. Though complex floral structures seem to be adapted to attracting specific groups of pollinators, and guiding their behaviour to favour pollination, other species often find unexpected ways of accessing the rewards they need.

Colletidae
Colletes

The species in this genus are roughly equally divided between pollen specialists and generalists. The study by Müller and Kuhlmann (2008), described above, found that, of 60 Western Palaearctic species of *Colletes*, 14 were pollen specialists on the problematic family Asteraceae (notably the sub-family Asteroideae), while the other, generalist, species used very little or no pollen from these plants. These observations seem consistent with the dietary patterns of the *Colletes* species that occur in Britain. Four widespread species (*Colletes halophilus, Colletes similis, Colletes fodiens* and *Colletes daviesanus*) are oligolectic on Asteraceae, predominantly collecting pollen from the sub-family Asteroideae. Other species (*Colletes marginatus, Colletes floralis, Colletes hederae, Colletes cunicularius* and *Colletes*

FIG 221. Hairy-saddled Colletes *Colletes fodiens* female, an Asteraceae specialist.

FIG 222. (a) Heather Colletes *Colletes succinctus* female collecting pollen from Ling *Calluna vulgaris*. (b) Sea Aster Bee *Colletes halophilus* female collecting pollen from a rayless form of Sea Aster, *Aster tripolium*.

succinctus) are either polylectic or oligolectic on plants from other families and use little or no Asteraceae pollen.

However, the pollen diets of three late-flying 'sister' species are of particular interest. Morphologically, the Heather Colletes *Colletes succinctus* (Fig. 222 a), the Ivy Bee *Colletes hederae* and the Sea Aster Bee *Colletes halophilus* (Fig. 222 b) are very similar, and it is supposed that their separation into distinct species occurred recently. According to the study by Müller and Kuhlmann, both *Colletes succinctus* and *Colletes hederae* are polylectic, though they show a very strong preference for heather (Ericaceae) and Ivy pollen, respectively. The third species, *Colletes halophilus*, is regarded as 'broadly oligolectic', with almost 90 per cent of its pollen loads from Asteroideae (in Britain, the bee forages almost exclusively on Sea Aster *Aster tripolium*). Both *Colletes succinctus* and *Colletes hederae* have been regarded as oligolectic, but Müller and Kuhlmann's data include pollen loads of each species with substantial contents of the other's host pollen. Subsequent field reports confirm their view that both species are more flexible in their dietary choices than was initially supposed. Müller and Kuhlmann introduce another division to the classification of types of dietary breadth to deal with this: 'polylectic with strong preference'. It seems likely that this is an example of speciation associated with switches to alternative host plants, together with possible future development of oligolecty in both *Colletes succinctus* and *Colletes hederae*. It is notable that the 'strong preferences' of both species relate to plants which flower in late summer, coinciding with their flight period, predominate in the relevant habitats (woodland edge, hedgerow and heather heath), and are rewarding and easily accessible. Similar considerations apply to the host plant of the Sea Aster Bee *Colletes halophilus* in its coastal habitats.

Hylaeus
Hylaeus species (yellow-face bees), having short, bilobed glossae, forage from shallow or open flowers with accessible nectaries. For example, the Reed Yellow-face Bee *Hylaeus pectoralis* on an Essex coast site forages from Wild Carrot *Daucus carota*, Bramble (*Rubus fruticosus* agg.), Creeping Thistle *Cirsium arvense* and Creeping Cinquefoil *Potentilla reptans*. Females were observed gripping the anthers of the last-named flower with their mandibles, presumably to extract pollen. *Hylaeus hyalinatus*, a common species in gardens, forages from *Geranium x oxonianum* by walking down the corolla to probe the nectaries, moving round to each one and then preening its face with forelegs before flying to the next flower. The bees appear to take both nectar and pollen directly into their mouth but also can be seen brushing pollen forward from their ventral surfaces, using dense brushes of sturdy setae on the inner surfaces of their basitarsi. Getting the mixture of pollen and nectar which they gather in their crop to the 'right' consistency sometimes requires them to expel a globule of the mix to allow excess water to evaporate (see Figure 194 b, above). This is held between the labrum and the open mandibles and is eventually sucked back into the crop. A pollen specialist in this genus, *Hylaeus signatus*, forages mainly from Weld *Reseda luteola* and Mignonette *Reseda lutea*. Foraging females scurry around the flowering spikes of the former plant, sometimes probing under the upper, lobed petal of each flower for nectar, sometimes seeming to bite anthers with their mandibles.

FIG 223. Large Yellow-face Bee *Hylaeus signatus* female, a pollen specialist, collecting pollen from Weld *Reseda luteola*.

CONCLUSION

Bees are almost wholly dependent for their own nutrition, and that of their offspring, upon rewards provided by flowers. For most bee species, these rewards are nectar and pollen, with plant oils used in addition by just one British species. The mouthparts of bees are adapted for 'sucking' or 'licking' nectar, and most bees are able to take nectar from a wide range of flowers. Pollen is a more costly reward for plants, and insect-pollinated plants have evolved ways of combining effective pollination with conservation of pollen, usually by restricting pollinators' access to the resource. Bees have evolved reciprocal adaptations (sometimes morphological, sometimes only behavioural) that enable them to access sufficient pollen for their nutritional requirements. Bees differ greatly in how closely these adaptations are aligned with specific groups of flower species. Some species visit a very wide range of flower types, and their larvae thrive on provisions comprising mixtures of pollen from various sources. Species that use such a wide range of pollen sources are said to be polylectic. Other species (a substantial minority among the British species) have a narrower diet, drawn from one family, one sub-family of a large plant family or, in a few cases, just one plant genus. These species are described as oligolectic. Provisional research into the evolutionary history of bees suggests that they were originally oligolectic, but as they became differentiated into distinct families, genera and species, some groups shifted to broad pollen diets and, in some cases, subsequently reverted to oligolecty. Some lineages, most notably those comprising the family Melittidae, have retained narrow pollen specialism as an inherited trait. Even where shifts have occurred to new pollen hosts, the bees have continued with narrow host-plant preferences. Other bee families and genera show various proportions of oligolectic and polylectic species, and several hypotheses have been advanced to explain how dietary breadth might have changed in either direction. Specialisation carries evolutionary risks associated with dependence on a small range of plants, as well as the disadvantage of inflexibility when more rewarding flowers are available. It has been suggested that specialists might gain by evolving a relationship with a plant species such that other potential visitors are excluded (resource partitioning), or they might gain by superior efficiency in achieving access to more complex floral structures. Oligolecty might also be favoured by the availability of an abundant source of nectar and pollen coinciding with the flight period of the bee. By contrast, polylecty would be required by doubly brooded bees or ones with prolonged flight periods (notably social species) that exceed the flowering period of particular groups of plants.

CHAPTER 7

Parasites and Predators

'With the industrious folk who go quietly about their business, the labourers, masons, foragers, warehousers, mingles the parasitic tribe, the prowlers hurrying from one home to the next, lying in wait at the doors, watching for a favourable opportunity to settle their family at the expense of others.'
—Jean-Henri Fabre

'True parasites are a motley crowd of mostly rather unpleasant creatures, although many of them have highly interesting biographies.'
—A. D. Imms, New Naturalist, Insect Natural History (1947)

The female solitary bee illustrated overleaf is gathering pollen, so we can conclude that she has a nest somewhere nearby, which she has temporarily deserted to collect more supplies. Pollen and nectar are scarce resources, scattered in small quantities around the bee's environment, and each foraging trip could last for five minutes or perhaps for as long as two hours. The bee needs to work quickly, leaving her nest for as little time as possible. As the nest matures, there will be eggs and larvae as well as food-stores. Each successive excursion she makes leaves a greater potential prize for opportunist thieves and parasites.

A highly diverse range of animals and microorganisms exploits the food-stores and young stages of bees, or preys on bee larvae or adult bees (Table 10). Many of these are classed as parasites. A parasite lives in or on another organism (called the host) and does it harm but does not immediately kill it. Examples include fleas, which consume blood through the skin, and tapeworms, which

FIG 224. Yellow-legged Mining Bee *Andrena flavipes* female collecting pollen from bramble.

live in the gut and absorb some of the food eaten by the host. The parasite lives on income (blood and food in the gut are continually replenished by the living host), whereas predators, such as sparrow hawks, live on capital (they kill and eat most or all of their prey). A parasite's life is a compromise between taking as much nourishment from the host as it can but not to an extent that reduces the host's vitality excessively, or kills it (Elton 1927).

In reality, there are no hard and fast distinctions between parasites and predators. For example, some animals live as an internal parasite as the host grows before finally killing and eating it. Some animals have a parasitic larva but are free-living as an adult. Parasites are generally much smaller than their hosts, while carnivores are usually bigger, but there are exceptions. Cleptoparasites (stealing parasites) take the food collected by a different species; for example, a Black-headed Gull *Chroicocephalus ridibundus* might steal a worm caught by a Lapwing *Vanellus vanellus*, which is similar in size. Bee cleptoparasites (cuckoo bees) lay their eggs on the food-stores of another bee species and can be larger than the host bee. Bee parasites and predators come in many guises and often defy easy classification.

A developing bee larva and its food-store make a tempting target, being enclosed in a brood cell and having little means of self-defence. Many parasites feed first on the living body of the bee larva in the nest but ultimately eat the entire larva. These are known as parasitoids and can be ectoparasitoids, living on the larva's surface, or endoparasitoids, living within the larva's body. Most parasitoids have a free-living adult stage, often a winged insect which seeks new hosts in which to lay eggs. Once again, distinctions become blurred in the frequent cases where a parasite consumes a bee's food-store as well as its egg or larva; it is both a cleptoparasite and a parasitoid. The terms can nevertheless be useful as a shorthand.

It is easy to be judgemental about the behaviour of parasites, as in the above quotations, in view of the potential damage they can cause to 'honest, hard-working bees', but of course parasites do not share our moral concerns. Terms such as 'aggressive' or 'sneaky' imply bad motives, but in reality, bee parasites (and their hosts) are responding to stimuli in a more or less automatic way. Parasites

TABLE 10. Solitary bee parasites described in this chapter.

Order	Common name	Group	Genus
DIPTERA (flies)	Shadow flies (aka satellite flies)	Anthomyiidae	*Leucophora*
		Sarcophagidae	*Miltogramma*
	Bee-flies	Bombyliidae	*Bombylius*
	Fat-headed flies	Conopidae	*Myopa*
			Physocephala
			Thecophora
			Zodion
		Drosophilidae	*Cacoxenus*
STREPSIPTERA	Twisted wing parasites		*Stylops*
			Halictoxenus
COLEOPTERA (beetles)	Oil beetles	Meloidae	*Meloe*
			Sitaris
HYMENOPTERA	Aculeate wasps	Sapygidae	*Sapyga*
			Monosapyga
		Mutillidae	*Smicromyrme*
			Myrmosa
		Chrysididae	*Chrysura*
			Trichrysis
	Evanioidea	Gasteruptiidae	*Gasteruption*
	Chalcids	Chalcidoidea	*Aximopsis*
			Calosota
			Melittobia
			Monodontomerus
			Pteromalus
			Torymus
ACARINA	Mites	Chaetodactylidae	*Chaetodactylus*

are of great interest in their own right and equally in need of conservation and protection, forming more than half of all biodiversity, though often overlooked. Their behaviour and life cycles are remarkable and provide insights into the processes of evolution and population regulation. Some of the associations are highly species-specific, reflecting close coevolution between the parasite and its host, while other parasites are less choosy. Here, we review the range of parasites and other associates of solitary bees, looking especially at the strategies used by parasites to exploit their hosts and the means by which bees attempt to defend themselves. Predators of bees are described later in this chapter, while parasitic bees – the cuckoo bees – appear in Chapter 8.

FLIES (DIPTERA)

Shadow flies

A large nesting aggregation of mining bees appears at first sight to be a chaotic buzzing swarm. Gradually their behaviour starts to make sense: each female arriving with pollen is circling and searching, picking up landmarks and homing in on her own nest entrance, while others are basking, grooming or departing to forage, sometimes making widening circular orientation flights before leaving. Males are attempting to intercept unmated females or even bombarding females with established nests. It takes a little longer to notice that you are not the only spectator; small flies are waiting motionless on twigs and stones oriented towards the nest burrows. As a female bee arrives with pollen, they dart out and tail her every move, just a few centimetres behind. The bee may try evasion, but the fly is often able to follow her successfully to her nest, and here the fly waits motionless for the bee to unload her cargo of pollen and depart. Now the fly enters, deposits her eggs and returns to a lookout to await the next chance. The fly's eggs hatch into larvae which destroy some or all of the eggs or larvae of the bee. They feed either on the larva's food-store, as a cleptoparasite, and/or on the bee larva itself, as a parasitoid (Paxton *et al.* 1966). Ultimately, adult flies will emerge in time for the next bee-nesting season, usually in the following year. They are known as shadow flies or satellite flies, but the various species have no common names. Many belong to the genus *Leucophora* (family Anthomyiidae), which usually targets *Andrena* species, though not exclusively. Shadow flies belonging to the genus *Miltogramma* (family Sarcophagidae) often specialise on plasterer bees in the genus *Colletes*.

Leucophora and *Miltogramma* females have independently developed similar behaviour in tracking pollen-laden bees, but once at the nest, their behaviour is somewhat different. *Leucophora* females often wait for the host to leave before entering to lay their eggs and remain inside for some time. *Miltogramma* females deposit small larvae rather than eggs and can do so in seconds, allowing them to enter nests when a bee is still inside, and then quickly escape. *Miltogramma* eggs hatch inside the parent female's body: she is said to be viviparous (giving birth to live young). The *Miltogramma* larvae may not need careful placement since they are able to crawl into a brood cell. At least six *Leucophora* species are cleptoparasites of bees in Britain, and knowledge about which bee species are associated with particular hosts is gradually being established (Owens & Welch 2020 a & b, Welch & Owens 2022). *Leucophora* have also been observed entering the nests of ground-nesting hunting wasps.

At spring nesting aggregations of the Cliff Mining Bee *Andrena thoracica*, both *Leucophora obtusa* and *Leucophora personata* females can be present. Stones and sprigs of vegetation close to clusters of nests are used as vantage points. When the flies spot a moving insect nearby, they immediately give chase. Sometimes a nomad bee or even a spider-hunting wasp is briefly followed, but in these cases the fly quickly returns to her lookout. Female *Andrena thoracica* arriving at the nest aggregation are pursued more persistently, back to their nests. The *Andrena* sometimes crash-lands with its heavy pollen load, then pauses for a while before flying directly into the nest entrance, which remains open while she is away. Sometimes the shadow fly loses interest while the *Andrena* pauses, and at other

FIG 225. Cliff Mining Bee *Andrena thoracica* female challenging a *Leucophora* female at the nest entrance.

times the fly fails to follow the full distance back to the nest. When a bee is followed to its nest, it sometimes turns to face the fly rather than entering.

Leucophora obtusa females can be induced to orient themselves beside an artificial depression in the ground, made by pushing a pencil into a small mound of earth. The artificial nest burrow alone is apparently attractive, since the fly will sometimes enter and inspect it.

When *Andrena thoracica* has been shadowed successfully to her nest burrow, the *Leucophora obtusa* typically stops a little short of the entrance while the bee enters. It then generally crawls to the rim of the burrow entrance and partially enters before backing out and moving to a lookout a few centimetres away. One *Leucophora obtusa* female was observed waiting motionless for 35 minutes on a sprig of heather for a female *Andrena thoracica* to re-emerge from her nest. As soon as the host had departed, the fly explored the ground around the nest entrance, then partially entered, head-first, before quickly rotating and backing into the nest burrow. She then remained out of sight for 12 minutes, presumably laying eggs. On re-emerging, she moved about 20 cm away and rested, before

FIG 226. After following a Cliff Mining Bee *Andrena thoracica* to her nest, a *Leucophora obtusa* female (a) partially enters the nest, then backs out and (b) waits just above the nest entrance on a twig for 35 minutes until the host emerges and leaves, then reverses in and disappears to lay eggs, before (c) emerging from the nest 12 minutes later.

taking flight. A *Leucophora personata* female at the nest of *Andrena flavipes* behaved slightly differently, inspecting the nest entrance after the host entered, then turning and backing in while the host bee was in the nest burrow. The shadow fly stayed inside for about two minutes and left before the host re-emerged. The strategy used by *Leucophora* may vary depending on the position of the host bee inside the nest and the architecture of the nest itself, but behaviour within the nest is not easy to observe. *Leucophora grisella* and *Leucophora sponsa* generally lay eggs just inside the bee's nest entrance (Huie 1916, Wolton 2022), and this is also reported in *Leucophora obtusa* parasitising *Halictus farinosa* in North America, by Nye (1980), who notes that in this partnership, the first instar larva of the fly can be carried into the host nest on the hair of the bee.

Some *Andrena* species, including *Andrena thoracica*, are parasitised by both *Leucophora obtusa* and *Leucophora personata*. These two shadow fly species are active from early spring, but *Leucophora grisella* appears later in the year and targets bee species that emerge in the summer months, as well as weevil-predating wasps in the genus *Cerceris*. The rare *Leucophora sericea* emerges in late July and is associated with the second generation of the Yellow-legged Mining Bee *Andrena flavipes*. We now know that this bivoltine bee is attacked by two *Leucophora* species in the first generation and by a third species in the second generation. An extended 'cat and mouse' sequence was observed in which a second-generation *Andrena flavipes* entered and exited a sequence of nest burrows and crevices for more than 30 minutes, pursued all the while by one or sometimes two *Leucophora sericea*. The bee appeared to be diverting the fly away from her own nest burrow into others nearby (Welch & Owens 2022).

Knowledge of the hosts of *Leucophora* is still very incomplete, especially for the rarer species. Host specificity is at least partly determined by the emergence times of host and shadow fly, and it is possible that each *Leucophora* population is locally adapted to its chosen host. There is also uncertainty about the feeding behaviour of the *Leucophora* larva within the nest. There is evidence that a *Leucophora grisella* larva uses only the food-store provided by one known host bee, the Tormentil Mining Bee *Andrena tarsata* (Huie 1916), but it is possible that with other host species, the *Leucophora* larva behaves as a parasitoid, ultimately consuming the bee larva.

Leucophora can potentially have a major impact on a bee population. The reproductive potential of a bee is much more limited than that of the fly since the bee provides each egg with a large food-store, which takes time and energy to collect. It is therefore not surprising that bees have evolved defensive and evasive tactics against shadow flies. Bees are faced with difficult 'decisions' when targeted by a shadow fly. If they wait too long to fend it off, they sacrifice foraging time, and it may be more profitable to take a chance and seek more food, even if a shadow fly

TABLE 11. Recorded associations between *Leucophora* species and hosts in the British Isles.

	Leucophora grisella	*Leucophora obtusa*	*Leucophora personata*	*Leucophora sericea*	*Leucophora cinerea*	*Leucophora sponsa*
Bees						
Andrena argentata	✓					
Andrena cineraria		✓	✓			
Andrena clarkella		✓				
Andrena ferox		✓				
Andrena flavipes		✓	✓	✓		
Andrena fulva		✓				
Andrena fuscipes	✓					
Andrena haemorrhoa		✓				
Andrena humilis		✓				
Andrena labialis			✓			
Andrena nigroaenea			✓			
Andrena nitida		✓				
Andrena scotica		✓				
Andrena tarsata	✓					
Andrena thoracica		✓	✓			
Andrena trimmerana			✓			
Colletes cunicularius			✓			
Colletes halophilus	✓					
Colletes succinctus	✓					
Lasioglossum nitidiusculum					✓	
Lasioglossum parvulum						✓
Panurgus calcaratus		✓				

	Leucophora grisella	*Leucophora obtusa*	*Leucophora personata*	*Leucophora sericea*	*Leucophora cinerea*	*Leucophora sponsa*
Wasps						
Cerceris rybyensis	✓					
Cerceris arenaria	✓					
Diodontus tristis					✓	

Data from Welch & Owens (2022).

is nearby. Some bees, such as Clarke's Mining Bee *Andrena clarkella*, cover their nest entrance before departing to forage, sometimes leaving a decoy nest entrance. Even so, *Leucophora obtusa* females are able to dig into the nest after the bee's departure (Owens & Benton 2022). Very large aggregations of the Heather Colletes *Colletes succinctus* on the banks of the River Wharfe in Yorkshire are targeted by large numbers of *Leucophora grisella*. The bees typically land some distance from their nest entrance, then walk to it through the grass, often with a *Leucophora* attempting to follow on foot (S. Saxton pers. comm.). *Colletes succinctus* aggregations in north Norfolk appear not to be subject to parasitism by *Leucophora*, however.

It is likely that nesting in aggregations rather than singly provides some protection from cleptoparasitic flies. Paxton and Pohl (1999) investigated a Tawny Mining Bee *Andrena fulva* nesting aggregation near Cardiff, using emergence traps to measure rates of parasitism by *Leucophora obtusa* and other parasites. Small cages made of fine netting were placed over individual nest entrances before the bees' spring emergence (Fig. 227). Bees and parasites emerged into the traps, allowing a daily tally to be taken. A total of 631 insects was collected, of which 502 were *Andrena fulva* and 15 were *Leucophora obtusa*, the remainder being other Diptera species (19) and the cuckoo bee Panzer's Nomad Bee *Nomada panzeri* (95). This gave a maximum rate of parasitism by *Leucophora obtusa* of 3 per cent, assuming that each emerging fly replaced one bee. In an earlier study, Paxton et al. (1996) found just two *Leucophora personata* caught in emergence traps at Chocolate Mining Bee *Andrena scotica* nest aggregations in Sweden, from which

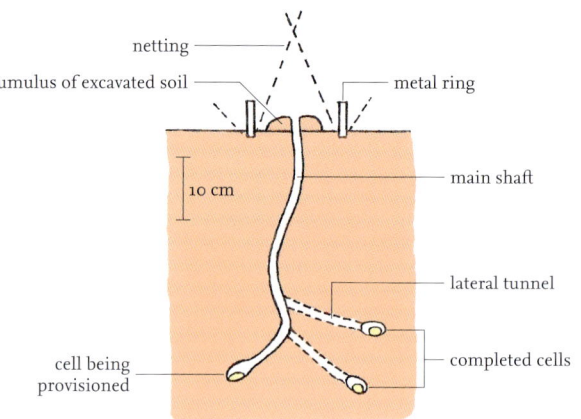

FIG 227. Emergence trap used by Paxton and Pohl (1999), redrawn.

8,900 adult bees emerged. This low level of parasitism was thought to relate to *Andrena scotica* often having communal nests in which several females share a common nest entrance. A study of the Violet-winged Mining Bee *Andrena agilissima* in Italy showed that *Leucophora personata* females often followed their host into the communal nest entrance but remained for a shorter time if a second female bee entered after them, again suggesting a protective effect of a shared entrance (Polidori *et al.* 2005).

Knerer and Atwood (1967) observed *Leucophora* species at nesting aggregations of social halictid bees in southern Ontario. The flies were seen backing into nest entrances and remaining inside for 20–80 seconds while ovipositing. Up to four fly larvae were noted in the same pollen ball. Twenty per cent of the nests that were inspected had been completely destroyed by *Leucophora* by the end of the nesting season. The authors ascribe this high parasitic level to the vulnerability of social bee species, which often remain at one locality for many years, allowing parasite numbers to build up.

The two British species of *Miltogramma* shadow flies do not appear until about June. *Miltogramma germari* parasitises the Pantaloon Bee *Dasypoda hirtipes* and probably also the Silvery Leafcutter *Megachile leachella* and the Green-eyed Flower Bee *Anthophora bimaculata*, all three being mining bees which nest in loose sandy ground (Welch & Owens 2019). *Miltogramma punctata* parasitises at least five *Colletes* species in the British Isles (Table 12). Sea Aster Bees *Colletes halophilus* make closely packed nests in dunes or gravelly beaches near the salt marshes where its main food plant (*Tripolium pannonicum*) grows. A large aggregation in north Norfolk was the subject of recent observations of interactions between host and parasite (Owens & Welch 2020a). When a bee was shadowed and had entered its nest, the

TABLE 12. Host associations of *Miltogramma* recorded in Britain.

		Miltogramma punctata	*Miltogramma germari*
Pantaloon Bee	*Dasypoda hirtipes*		✓
Davies' Colletes	*Colletes daviesanus*	✓	
Hairy-saddled Colletes	*Colletes fodiens*	✓	
Sea Aster Bee	*Colletes halophilus*	✓	
Bare-saddled Colletes	*Colletes similis*	✓	
Heather Colletes	*Colletes succinctus*	✓	
Silvery Leafcutter Bee	*Megachile leachella*		✓
Green-eyed Flower Bee	*Anthophora bimaculata*		✓
Large Shaggy Bee	*Panurgus banksianus*	✓	
Bee Wolf (wasp)	*Philanthus triangulum*	✓	

fly typically moved right up to the edge of the nest entrance, then waited for a few seconds before going in to larviposit, without waiting for the bee to come out. The fly was inside the nest for between 2 and 20 seconds, averaging just 7 seconds (18 observations). It was not possible to see how many larvae were deposited, but possibly just one in this short time. The speed of larva deposition probably reduced the chances of any retaliation by the bee. Sometimes a fly backed out quickly before re-entering, presumably because the bee was still close to the nest entrance.

Dissection of several female *Miltogramma punctata* females showed that over 50 larvae were commonly present in the ovary, packed in rows rather like bottles in a crate, each one ready to be ejected when its turn came. This implies that each fly could potentially destroy about 50 bee larvae, assuming that each fly larva leads to the demise of one bee larva. This kind of evidence is a first step towards working out the possible impact of shadow flies on bee populations. We also need to know the relative numbers of flies and bees at a nest site. The aggregation in north Norfolk was estimated at 1,000 nest burrows with up to 10 *Miltogramma*

FIG 228. (a) and (b) *Miltogramma punctata* female waits then approaches as a Sea Aster Bee *Colletes halophilus* female enters her nest with a load of Sea Aster *Tripolium pannonicum* pollen. When the bee had fully entered, the fly quickly followed to deposit a larva.

punctata females observed at one time. A nest of *Colletes halophilus* usually contains 5–6 cells (O'Toole & Raw 1991). At a very rough approximation, the nest aggregation could contain 5,000 bee cells, with *Miltogramma punctata* depositing 10 x 50 larvae = 500 cells, which is 10 per cent of the brood cells. Despite their high reproductive potential, *Miltogramma* are not usually abundant at bee nesting aggregations, though there might be a sequence of flies emerging during the season. An Australian study found that predation of adult *Miltogramma* by birds and wasps was high, especially when flies were newly emerged and expanding their wings. Five per cent of the host bee (*Amegilla dawsoni*) nests were estimated to be parasitised (Alcock 2000).

As with *Leucophora*, host bees use a variety of tactics against *Miltogramma*. At a *Colletes succinctus* nest aggregation in soft cliffs, *Miltogramma punctata* often waited until a bee had left its nest before entering, rather than darting in and out while the bee was inside. A departing bee was observed to block its nest entrance for 15 minutes when a *Miltogramma* was detected nearby, but the fly did succeed in entering the nest when the bee finally left (Owens & Welch 2017). By waiting for a bee to depart, a fly loses a lot of time but avoids any chance of the bee repelling it inside the nest. Similarly, for the bee, there is a trade-off between deterring shadow flies and gathering more food. Not far from the *Colletes succinctus* aggregation, *Miltogramma punctata* was seen shadowing female Bare-saddled Colletes *Colletes similis* while the bees were collecting pollen from Common Tansy *Tanacetum vulgare*. Similar behaviour was seen at another site (V. Bartlett pers. comm.). The shadow fly is perhaps able to follow a bee back to its nest burrow, a strategy which would be suited to the more scattered nests of this particular host.

It is likely that nesting aggregations are easier for parasites to find than individual nests (O'Toole & Raw 1991). The *Colletes halophilus* aggregation just described occupies the highest part of a sandy ridge, which is rarely covered by high tides but is close to the bee's Sea Aster pollen source. There may be a conflict for bees between remaining at an ideally placed nesting site and moving to a new site with a lower parasite load. However, more isolated nests can still be at risk from a shadow fly visit: one fly marked with a spot of paint at the main nest site was observed 125 m away, close to a small pioneer cluster of nests. Other potential parasites observed at the *Colletes halophilus* site included the cuckoo bee *Epeolus variegatus* and the shadow fly *Leucophora grisella*. Possible predators of

FIG 229. *Miltogramma punctata* shadowing a Bare-saddled Colletes *Colletes similis* on Sneezewort *Achillea ptarmica*. The bee was seen to hide from the fly under flowers. (© V. Bartlett)

FIG 230. (a) and (b) A Heather Colletes *Colletes succinctus* female blocked her nest entrance for 15 minutes before flying off. (c) The *Miltogramma punctata* female then immediately entered to larviposit.

FIG 231. *Miltogramma germari* female shadowing a pollen-laden Pantaloon Bee *Dasypoda hirtipes* female as she enters her nest burrow.

both shadow fly and host were present in the form of the German Wasp *Vespula germanica* and the insectivorous bird species Pied Wagtail *Motacilla alba* and Meadow Pipit *Anthus pratensis*.

Differences in the behaviour and phenology of *Miltogramma punctata* suggest that there could be slightly different forms parasitising each *Colletes* species, but this remains unproven. The much scarcer *Miltogramma germari* has recently been found to be strongly associated with the Pantaloon Bee *Dasypoda hirtipes*, which is not targeted by *Miltogramma punctata* (Welch & Owens 2020). *Dasypoda hirtipes* is quite large, but *Miltogramma germari* is smaller than *Miltogramma punctata*. This mismatch in size may indicate that the shadow fly deposits more than one larva in each *Dasypoda hirtipes* nest cell. *Dasypoda hirtipes* females generally leave their nests open when they forage, but one was observed repeatedly re-covering its nest entrance when disturbed by a *Miltogramma germari* female (Wood 2015).

Bee-flies

The dancing flight of bee-flies in the spring sunshine is an enchanting sight. Their family name, Bombyliidae, comes from their resemblance to hairy bees, as mentioned in the Introduction. The long tongue is used to probe flowers while the fly hovers, rather like a minute hummingbird, often resting the tips of its legs on the petals as it does so. Baldock & Early (2015) list 36 species of spring flowers visited by this bee-fly in Surrey. The presence of a bee-fly is an indication that solitary bees are nesting nearby, since its larvae are parasites of mining bees, especially *Andrena* species.

The agile flight of bee-flies is an essential part of their egg-laying behaviour, the eggs being scattered at bee nesting sites, sometimes directly into a nest entrance. The resulting bee-fly larvae crawl into bee nests and live as parasitoids; the bee-fly larva feeds on the bee larva itself rather than on the food-store. There are four British species of Bombyliidae which behave in this way: *Bombylius canescens*, *Bombylius discolor*, *Bombylius major* and *Bombylius minor*. In 2016, a further species, *Anthrax anthrax*, was reported from a bee hotel in a Cambridge garden. This bee-fly targets stem-nesting bees and is able to flick its eggs into a horizontal nest entrance.

The tip of the bee-fly's abdomen bears a pocket which the fly fills with sand. This is scooped up from the ground as the fly pauses, tail-down, with its wings whirring. Sand grains surround and stick to the eggs as they are deposited into the pocket. The plug of sand protrudes from the sand chamber, allowing the fly to flick out egg-covered sand grains as it hovers over a bee's nesting site. From video sequences of *Bombylius major* laying eggs at an *Andrena* nest aggregation, it was clear that the female was orienting to positions just above and behind open

FIG 232. Egg-laying behaviour of the bee-fly *Bombilius major*: (a) collecting sand, (b) hovering with sand chamber full and (c) flicking eggs into a nest entrance of a Tawny Mining Bee *Andrena fulva*.

nest entrances of Cliff Mining Bees *Andrena thoracica* and Gwynne's Mining Bees *Andrena bicolor*. The female *Bombylius* hovered over each nest entrance in turn, making small bobbing movements and fine adjustments to her position before making a final deeper dip with a louder buzz as (presumably) the eggs were flung out, usually then rising higher and moving to another nest. At other times, she distributed eggs while hovering over low-growing heather where the target was less obvious. When artificial nest entrances were made in damp soil, using a pencil, females readily laid eggs into them, suggesting that it is largely the visual stimulus of a nest entrance rather than scent that triggers the behaviour. Holes without a tumulus of soil around them were also used.

The eyes of female *Bombylius* bulge forwards and downwards, as would be needed for visual aiming. Boesi *et al.* (2009) found that bee-flies in Italy deposited more eggs and spent more time hovering close to bee nest entrances than elsewhere, though their discrimination was not perfect. Some members of the bee-fly family can lay 1,000 eggs in a day, presumably compensating for the low chances of a larva successfully reaching a bee's nest. The first instar larva, called a planidium, bears long setae (bristles) and pseudopods (false legs), enabling it to seek out suitable nests. Subsequent larval stages are less mobile and feed externally on the host larva, killing it only in the final stages of development of the bee-fly larva (Yeates & Greathead 1997). The sites used by bee-flies often harbour several bee species. For example, a nesting aggregation of bees in a sandy bank at Dungeness, Kent, contained at least six species of *Andrena*, with *Bombylius major* females ovipositing widely across the site. It is likely that planidium larvae are able to develop on a variety of hosts and that the imprecise method of egg dispersal limits the degree of host specialisation, in contrast to shadow flies and cuckoo bees, which place their eggs directly into a host's nest.

Bombylius major is common in most parts of lowland Britain across a wide range of habitats and uses many of the spring-emerging bee species as hosts. Its northerly range limit has now expanded well into Scotland. All of its hosts are ground-nesting. It can be difficult to establish exactly which hosts are used in mixed nesting aggregations, since the adult bee-fly emerges from the pupa some distance from the host bee's original nest burrow and usually takes two years to mature (Paxton & Pohl 1999). Bee-fly pupae have spikes on their anterior end, allowing them to break through adjacent nest cells when they dig their way out of a nest, and this can destroy many of the remaining bee larvae (Kunić & Stanisavljević 2006). However, they are not generally numerous at nesting aggregations. At a spring aggregation of about 200 nest burrows of *Andrena thoracica*, the maximum count of *Bombylius major* females during the spring brood was eight (NWO pers. obs.), and in a study of the Chocolate Mining Bee in Sweden, the average rate of parasitism by *Bombylius major* was found to be 6 per cent (Paxton *et al.* 1966).

One advantage of the bee-fly egg-laying strategy is that hosts have little defence. Amongst the *Andrena* species observed nesting together at Dungeness, some, such as the Grey-backed Mining Bee *Andrena vaga*, sometimes closed their

FIG 233. *Bombylius major* pupa from an *Andrena* nest aggregation on heathland. The head is to the right.

nest entrance before leaving to forage, whereas others, such as the Yellow-legged Mining Bee *Andrena flavipes*, did not. Closing the nest entrance may give some protection from searching planidium larvae, but it takes the bee time and energy to cover and then re-excavate its nest entrance, perhaps making it unprofitable in many circumstances. A covered nest entrance is also likely to be more difficult for a returning bee to find. In some instances, bee-flies appeared to follow *Andrena vaga* as they arrived with pollen loads.

The hovering flight of the bee-fly is very energy-demanding, compelling the fly to spend a considerable time seeking nectar. Boesi *et al.* (*loc. cit.*) found that bee-flies in Italy spent about half their day taking nectar, mostly in the early morning and late afternoon, visiting bee nest sites in between. Bee-flies are at the northern edge of their range in Britain. *Bombylius major* is reported to be unable to fly when the temperature is less than 17 °C, though it can be active in lower temperatures during sunny weather, making good use of microclimates and by whirring its wings before take-off (Stubbs & Drake 2014). Table 13 gives some details about other species of bee-flies found in the British Isles. However, there is a much greater diversity of Bombyliidae in mainland Europe, especially in the Mediterranean region, and there are over 270 species worldwide (Boesi *et al.* 2009).

TABLE 13. Bee-flies (Bombyliidae) occurring in the British Isles.

Scientific name	*Common name*	*Host*	*Distribution*
Bombylius major	Dark-edged Bee-fly	Several *Andrena* species	Widespread in England, Scotland, Wales, with a few records in Ireland
Bombylius discolor	Dotted Bee-fly	*Andrena flavipes*, *Andrena cineraria*	Southern England as far north as Birmingham and south Wales
Bombylius minor	Heath Bee-fly	*Colletes succinctus*	Scarce: largely confined to coastal counties of SW England
Bombylius canescens	Western Bee-fly	Halictidae	SW England and Wales, parts of Ireland and Scotland
Anthrax anthrax	Anthracite Bee-fly	*Osmia* species	First record Cambridge 2016; recent records from Kent

FIG 234. A Heath Bee-fly *Bombylius minor* hovering in front of a probable Heather Colletes *Colletes succinctus* nest burrow. (© I. Cross)

FIG 235. The Dotted Bee-fly *Bombylius discolor*. The spotted wings are diagnostic.

BELOW: FIG 236. The first Anthracite Bee-fly *Anthrax anthrax* recorded in the British Isles, at a bee hotel. (© R. Mills)

The impact of bee-flies on their host populations may be considerable and some of their effects may be indirect through competition for nectar. Boesi *et al.* (ibid.) found that bee-flies mostly used flowers close to bee nesting sites and sometimes used the same flower species as their host. Bischoff (2003) reported a decline of a population of *Andrena vaga* in Germany, which appeared to be partly a result of parasitism by *Bombylius major*. About half of the *Andrena* females moved to a new nest site each season, which was thought to be a response to high levels of parasitism.

Little is known about how bee-fly larvae compete with other parasitoids which share the same hosts. A bee-fly larva is reported as eating the larva of an *Epeolus* cuckoo bee (Nielsen 1903, cited Stubbs & Drake 2014). A *Bombylius major* that was egg-laying into an *Andrena* nest entrance was seen to crash against a *Leucophora* female waiting on adjacent vegetation, causing it to depart. The overall impact may be reduced when bee hosts are attacked by several competing parasites. For example, *Bombylius major* possibly parasitises *Nomada* cuckoo bees (Drake 2013a). It is also likely that there is competition between parasites of the same species. What happens underground is still largely unknown, but some evidence can be derived from the pupa cases (exuviae) of *Bombylius* species. At a large colony of Ashy Mining Bees *Andrena cineraria* in Devon, exuviae were present on the ground around the nesting area from *Bombylius discolor* as well as from *Bombylius major*. Further studies provided an approximate estimate of the rate of parasitism of *Andrena cineraria* larvae by *Bombylius major* as 8 per cent (Drake 2013b).

Conopid flies

Rather sinister-looking insects, known as fat-headed flies, may be spotted sitting quietly on flowering shrubs or fruit tree blossom in early spring. Their large heads have white parchment-like faces, and the end of the abdomen is curled under their body. They differ from shadow flies and bee-flies in being parasites of adult insects rather than of their larvae. Conopids in the sub-family Myopinae attack solitary bees, with at least nine conopid species involved in Britain. Their trick is to intercept a bee as it forages on flowers, landing on its abdomen for a split second and inserting its ovipositor under the edge of an abdominal segment to insert an egg. The resulting larva ultimately causes the death of the bee and the emergence of a new fly.

The conopid larva develops quickly inside the bee's body, going through three instars (stages) as it feeds on the living host's tissues. The presence of the mature parasite causes the bee to dig into the ground, probably at its nest site, where the conopid larva pupates and the bee dies. The adult conopid emerges through the dead host's exoskeleton, usually in the spring of the following year. Conopid eggs have a hooked extension which anchors the egg into the fat store inside the abdomen of the host. The third instar has a tapered end, allowing the larva to feed on the contents of the thorax via the host's petiole, or 'waist' (Smith 1969).

Myopa often attack *Andrena* species, but knowledge of bee hosts is as yet very sketchy. The various species of *Myopa* look rather similar, all being a rusty-brown colour, often with marks on their wings, a colouring which gives them good camouflage amongst vegetation. Ideally, a conopid of known species should be watched as it attacks a particular bee to lay its egg, but it is very difficult to

FIG 237. (a) The conopid fly *Myopa buccata* mating. (b) *Myopa hirsuta*.

identify both conopid and bee in the field. Up to five species of conopids have been observed together on a single flowering hawthorn bush, attended by six species of *Andrena*. Frequent split-second attacks on *Andrena* by female *Myopa* were observed, as well as numerous instances of male conopids intercepting females (M. R. Welch pers. comm.). It is sometimes possible to associate conopid species with host species by catching emerging conopids at nesting aggregations, since parasitised bees often die in their nest burrows. Newly emerged *Myopa hirsuta* were found in late March at a mixed aggregation of *Andrena thoracica* and *Andrena clarkella* on heathland in Norfolk, suggesting that one or both of these species is a host. The conopid was also seen on nearby willows, sidling towards pollen-collecting *Andrena* bees by making short flying hops along the slender branches. Conopids attack adult bees away from their nest sites, where the bee has no protection; bees have limited scope for defending themselves against attack, other than quick evasion. The level of parasitism by *Myopa* species can be high: Paxton *et al.* (1996) captured and dissected a sample of the Chocolate Mining Bee *Andrena scotica* from a Swedish nest site and found that between 3 per cent and 40 per cent of females contained conopid eggs or larvae of (mostly) *Myopa buccata*.

Thecophora atra is a smaller member of the Myopinae which specialises on *Lasioglossum* species as hosts. It waits around nesting aggregations as well as flowers, flying up to make mid-air assaults on these small bees. A similar species, *Thecophora occidentis*, was the subject of a study at a mixed aggregation of bees in

TABLE 14. Recorded associations between conopids and solitary bee hosts in Britain.

Common name	Species	Myopa buccata	Myopa fasciata	Myopa hirsuta	Myopa pellucida	Myopa testacea	Myopa vicaria	Physocephala rufipes	Thecophora atra	Zodion cinereum
Big-headed Mining Bee	*Andrena bucephala*	✓								
Ashy Mining Bee	*Andrena cineraria*		✓		✓	✓				
Clarke's Mining Bee	*Andrena clarkella*			✓						
Oak Mining Bee	*Andrena ferox*	✓								
Yellow-legged Mining Bee	*Andrena flavipes*									
Tawny Mining Bee	*Andrena fulva*									
Heather Mining Bee	*Andrena fuscipes*		✓							
Orange-tailed Mining Bee	*Andrena haemorrhoa*									
Buff-tailed Mining Bee	*Andrena humilis*									
Large Meadow Mining Bee	*Andrena labialis*									
Bilberry Mining Bee	*Andrena lapponica*	✓								
Small Scabious Mining Bee	*Andrena marginata*							✓		
Grey-patched Mining Bee	*Andrena nitida*									
Black Mining Bee	*Andrena pilipes*	✓								
Small Sallow Mining Bee	*Andrena praecox*						✓			
Cliff Mining Bee	*Andrena thoracica*				✓					
Orange-legged Furrow Bee	*Halictus rubicundus*									✓
Broad-faced Furrow Bee	*Lasioglossum laticeps*								✓	
Green Furrow Bee	*Lasioglossum morio*								✓	
Shaggy Furrow Bee	*Lasioglossum villosulum*								✓	

Data from Baldock & Early (2015), Falk & Lewington (2015), Else & Early (2018), Drake (2013) and the authors.

Ontario, by Knerer and Atwood (1967). The fly parasitised six furrow bee species at the study site, with hundreds of adult conopids present in the nesting area. It was found that the impact of the parasitoid on the bee population depended on the social behaviour of the host concerned. *Lasioglossum cinctipes* is a eusocial species. Ten per cent of spring emerging gynes had conopid larvae in their abdomen. In the early stages of nest construction, parasitised gynes descended into their nest, where they died. Adult conopids emerged from their bodies in June, just as new worker bees were appearing from healthy nests. Many of the worker bees were attacked on their first foraging trips, with about 70 per cent being parasitised by July. However, the presence of an internal parasite did not prevent the workers from continuing to provision nest cells, and there was no difference between the productivity of parasitised and unparasitised nests. New gynes emerging from these nests quickly went into hibernation, giving little opportunity for the conopids to attack them. However, a non-social species *Lasioglossum forbesii* was scarce and highly parasitised. Being a solitary species, each female provisioned its own cells, requiring her to spend a lot of time on flowers where she was vulnerable to conopid attack.

These observations suggest that eusocial behaviour in furrow bees can reduce the impact of conopid parasites by diverting mortality from gynes to sterile workers. Conopid parasites such as *Thecophora* can in turn manipulate their hosts' behaviour. The larva first feeds on non-essential abdominal tissues, allowing the host to remain alive and to continue providing nutrients: at this stage, parasitised workers hardly differ in behaviour from healthy ones. Before the host is killed, *Thecophora occidentis* larvae induce a change in the behaviour of the host, causing it to descend into its nest burrow, which is a safer and more favourable environment for the *Thecophora* pupa. In the Ontario study, the workers of another eusocial bee, *Halictus ligatus*, removed dead parasitised workers from inside the nest, placing them outside where they were often removed and eaten by ants, further reducing the impact on the next generation of gynes.

Cacoxenus fly, Drosophilidae

Cacoxenus indagator is a small (3 mm long) grey fly with dark red eyes often

FIG 238. *Thecophora atra* (Myopinae) at a *Lasioglossum* nest site. Its length is about 5 mm.

FIG 239. *Thecophora atra* watching a White-zoned Furrow Bee *Lasioglossum leucozonium* female approaching a poppy.

seen around bee hotels. It is closely related to the fruit fly *Drosophila melanogaster*, famed for its use in research on heredity. *Cacoxenus* differs from *Drosophila* in that its larvae feed on the pollen and nectar provisions in bees' nests rather than on rotting fruit. The *Cacoxenus* fly waits and watches until it can enter a nest cell while the cell is being provisioned by a bee. At a bee hotel under observation, the fly entered a nest of an *Osmia bicornis* while the host was inside, depositing pollen. The fly walked in with her ovipositor extended and seemed to be able to place eggs on the pollen with no response from the bee. Many eggs can be deposited in each nest cell before the bee has a chance to seal it off with a mud partition. The cleptoparasitic fly larvae develop side by side with the bee larva and share the food-store, but the bee larva's growth is greatly reduced and in many cases it dies. If the bee survives to emerge from its nest cell, the mature fly larva is able to follow the escape route created by the bee. But if left alone, an adult fly, which has weak mouthparts, is faced with one or more strong mud partitions to pass through. Observations of *Osmia bicornis* in a bee hotel with a perspex window revealed that the front wall of each brood cell is partially constructed before food is stored and an egg is laid, reducing the time that it takes to seal the vulnerable brood cell (see Chapter 3). The female bee also fills in cracks along the sides of each cell where the perspex meets the cavity. This is done by pulling a ball of wet mud backwards towards the nest entrance (perhaps aided by the horns on her head), spreading the ball with her mandibles as she moves along. Such cracks are often filled before the cells are constructed.

FIG 240. *Drosophila indagator* at a Red Mason Bee *Osmia bicornis* nest in a bee hotel.

BELOW: FIG 241. Four cells of the Red Mason Bee *Osmia bicornis* in a bee hotel. The three cells on the right contain spaghetti-like *Cacoxenus* droppings. The second cell from the right contains a small surviving bee cocoon amongst them, whereas in the other two infected cells, the bee larva has starved and disappeared. The left-hand cell is still healthy.

FIG 242. Partitions of an old nest of a Red Mason Bee *Osmia bicornis* showing the concave outer sides of the mud wall of each brood cell. Newly emerged bees and flies must break through the mud walls from left to right to emerge.

The shape of the end walls of a brood cell – flat and rough on the exit side but smooth and concave on the other – gives each emerging bee guidance about which direction to go in order to get out of the nest. *Cacoxenus* flies are apparently capable of using the same cues. Strohm (2011) used artificial nests, mimicking the natural shape of the mud partitions, to investigate how a newly emerged fly could escape. When newly emerged flies were introduced into the artificial cells, they first investigated the wall at each end. Out of 20 flies observed, 19 made more attempts to penetrate the rough and flat (correct) end than the smooth and concave (incorrect) end, and overall, there were four times as many escape attempts at the flat end than at the concave end. To break through the mud, adult flies used the inflatable fluid-filled sac on their head (the ptilinum) to crack away mud particles near the top edge of the cell partition. Once a hole was made, the fly wriggled through, using worm-like movements of its still-soft body. When there was more than one fly in a cell, they worked together in breaking down the wall. Nearly all flies escaped successfully from natural nests, sometimes passing through a series of cell partitions to do so. Eggs were laid about equally in all cells, including the first in the line, farthest from the entrance. Some mature larvae penetrated cell partitions before they pupated (their jaws are stronger than those of adults) and accumulated in the empty space (vestibule) at the front of the nest, where pupation occurred.

The timing of emergence of *Osmia bicornis* and *Cacoxenus* from a bee hotel showed that both bees and flies are protandrous (males emerge before females). Male bees and flies emerged more or less in step, whereas most female flies emerged later than female bees, allowing the flies to coincide with nest-building by the host, which starts a few days after female bees emerge. *Cacoxenus* also parasitises other members of the Megachilidae such as leafcutter bees and appears to have a significant impact on the survival of bee larvae in bee hotels.

FIG 243. An inflatable sac, the ptilinum, on the front of a *Cacoxenus indagator* head, is used to open up cracks in the mud wall of the nest partitions. It could be seen inflating and deflating in this newly emerged fly.

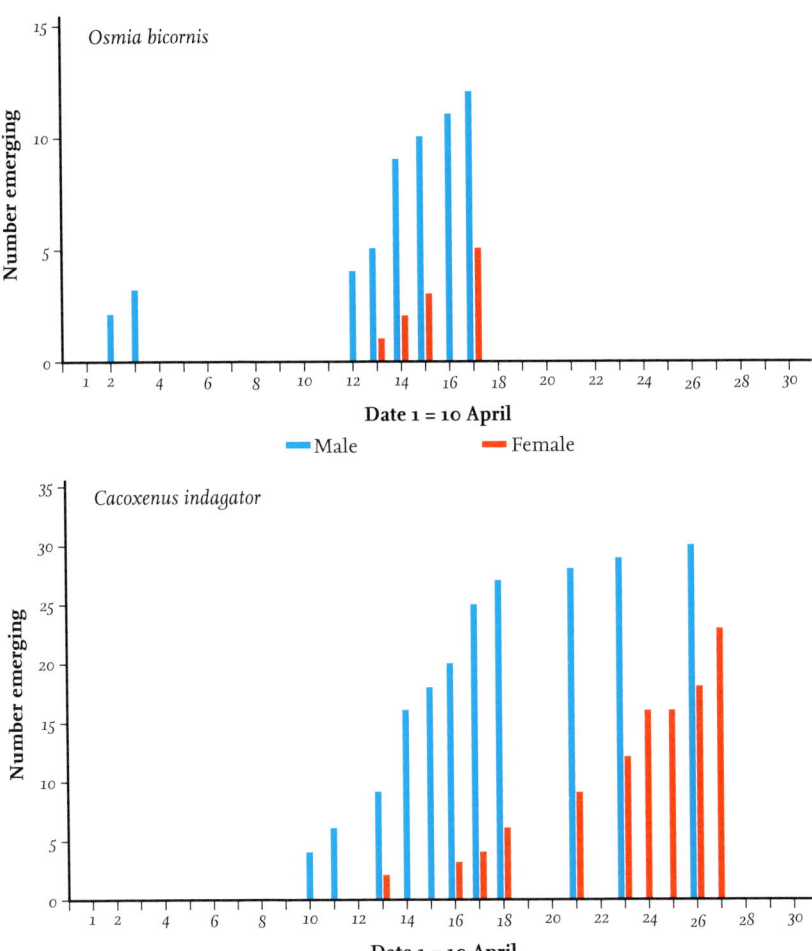

FIG 244. The emergence of the Red Mason Bee *Osmia bicornis* and *Cacoxenus indagator* from Red Mason Bee nests in a bee hotel. Columns show the cumulative number that had emerged. Blue = male, red = female.

TWISTED WING PARASITES (STREPSIPTERA)

Stylops

Female *Andrena* bees are sometimes observed crawling lethargically on flowers, seeking nectar but not collecting any pollen. On close inspection, one or more parasites can be seen, lodged between the segments of the abdomen. The

FIG 245. Buffish Mining Bee *Andrena nigroaenea* female with a *Stylops* female protruding from her abdomen. The bee was observed repeatedly turning the tip of her abdomen under her body, a behaviour which is likely to assist in the distribution of *Stylops* larvae onto the petals.

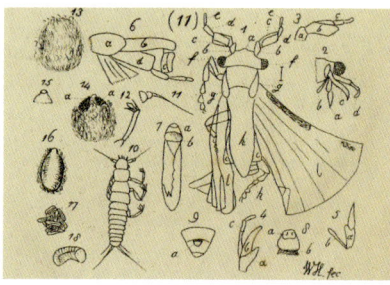

FIG 246. William Kirby's drawings of *Stylops kirbii* from *Monographia Apum Angliae*. The drawings include the adult male (with wings), a first instar larva and a female *Stylops* protruding from a bee's abdomen.

parasites belong to the genus *Stylops*, and infected bees are described as being stylopised. *Stylops* was first described by the Reverend William Kirby, the author of the first book on English bees, *Monographia Apum Angliae*, mentioned in the Introduction. He had found the parasite on the body of a Buffish Mining Bee *Andrena nigroaenea*, one of the many bee species he named. The curious appearance and behaviour of *Stylops* did not fit any existing insect group, and *Stylops* was placed in an order of its own, the Strepsiptera (twisted wings). When Kirby became president of the Entomological Society of London (now the Royal Entomological Society), *Stylops kirbii* was chosen as the society's emblem.

The female *Andrena nigroaenea* shown in Figure 245 has a poorly developed scopa on its hind tibiae, typical of stylopised female Andrenas. This change is induced by the parasite itself: stylopised female bees become more like males, whereas stylopised males become more like females. The outcome of these changes is that both bee sexes show little or no interest in reproductive behaviour and usually remain on flowers.

When mature, male and female *Stylops* partially protrude between the host's abdomen segments. Males form a puparium (a pupa case similar to that of a fly) inside the host bee and, on emerging, fly off to seek a female *Stylops* on a different bee, surviving for just a few hours. Female *Stylops* do not pupate but retain a soft white body with just their fused head and thorax visible. They have reduced mandibles and simple eyes, no gut and an egg-filled, bag-like abdomen which fills most of the abdomen of the bee. The stationary female *Stylops* secretes a

FIG 247. (a) *Stylops ater* male and (b) male mating on its specific host, the Grey-backed Mining Bee *Andrena vaga*. (c) Detail of mating. (All © John Smit)

pheromone which attracts the winged male, which alights on the bee's body and mates via the protruding head of the female *Stylops* (Tolash *et al.* 2012).

After mating, the eggs hatch inside the female *Stylops* and her abdomen becomes filled with tiny, mobile planidium larvae. There is an opening near her head where both mating and the release of larvae occur. As the host bee continues to forage, the female *Stylops* deposits hundreds of minute mobile planidium larvae onto flower heads. The larvae crawl onto other flower visitors and hitch a ride. If they end up in a solitary bee's nest, the larvae enter a developing bee larva, with two or three sometimes parasitising the same individual.

Like all internal parasites, *Stylops* larvae must somehow avoid the immune system of the host, though this has not been investigated in bees. In a similar *Stylops* which parasitises grasshoppers, the planidium larva penetrates the cuticle of a grasshopper nymph and becomes enfolded in a bag formed from the inner one-cell-thick cuticle lining (the epidermis). Once inside the bag, it sheds its skin to become a legless parasitoid, absorbing food through the bag. The thin bag is thought to allow food to enter but to protect the parasitoid from the immune system of its host (Kathirithamby *et al.* 2003).

It has been questioned whether the *Stylops* parasitoid does in fact manipulate the behaviour and morphology of its host or whether the more sluggish behaviour and reduced genital development of the host bee are simply the result of debilitation caused by the parasite's presence. There are parallels in other parasite-host relationships, a classic example being the effect of the parasitic barnacle *Sacculina carcini* on crabs. Parasitised crabs of both sexes resemble sterile females and are physiologically manipulated into nursing the parasite's eggs and larvae as if they were their own. Similar changes are brought about in bee hosts by *Stylops* larvae, reducing the size and performance of the reproductive organs of their host. The outcome is reduced or absent reproductive behaviour and secondary sexual characteristics. Males may lose the white facial patch, present in some *Andrena* species, as for example in *Andrena chrysosceles*, while, conversely, some females develop a white face marking which they normally lack. Such changes can lead to confusion with bee identification. Stylopised male Andrenas tend to be larger than healthy ones (more like females), and stylopised female bees emerge earlier than other females, coinciding with the emergence of male bees of their species (Straka *et al.* 2011). A further clue that lethargic behaviour is not just caused by a lack of energy comes from the finding that Andrenas with a female *Stylops* suddenly become frantic as *Stylops* larvae emerge, rushing from flower to flower and distributing the larvae over a wide area (Linsley & McSwain 1957, cited Beani 2006).

Reduced sexual development allows more nutrients to be available to the parasitoid larva since less energy is diverted to flying, egg formation, pollen gathering, nest building and mate seeking. Female bees also remain in the open, on flowers, rather than inside a nest, making female *Stylops* available for mating

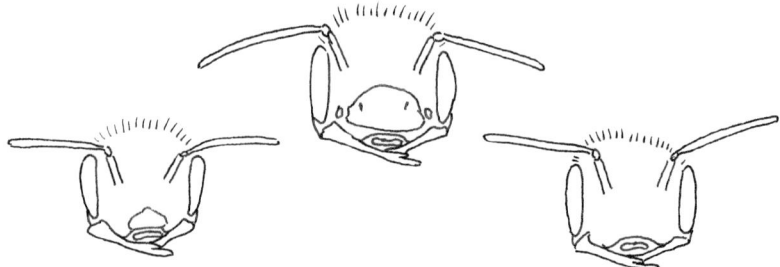

FIG 248. Hawthorn Mining Bee *Andrena chrysosceles* males. The lower two bees are stylopised, showing a reduced or absent pale face patch. The upper bee is not stylopised and shows the bee's normal appearance.

and able to disperse their larvae. It is likely that *Stylops* larvae secrete chemical signals which diffuse through the epidermis bag into the haemolymph of the host larva, changing the sex organs and behaviour of the host. In the European Paper Wasp *Polistes dominula*, a strepsipteran parasite similar to *Stylops* was found to change gene expression in the brains of its host, altering the behaviour of workers into that of queens, which played no part in nest building, making them better suited to the parasite. This has been called 'adaptive host manipulation' (Beani 2003).

Very little is known about the diversity and host specificity of *Stylops* species. This is partly because it is very difficult to match up males with particular females unless they are collected while mating. Many species were first named on the basis of their bee host, but because there was so much uncertainty, European species have been treated as all being one: *Stylops melittae* (synonymous with *Stylops kirbii*). A preliminary new classification based on DNA barcoding now suggests that there are 32 Western Palearctic species (Straka *et al.* 2015). The name *Stylops melittae* survives as the species shared by three spring-emerging bees: *Andrena nigroaenea*, *Andrena thoracica* and *Andrena flavipes*. The ability of Strepsiptera to evade the immune system of its host suggests that less host specificity will occur than in other internal parasitoids, because a species-specific chemical disguise is not needed (Kathirithamby *et al.* 2003).

Bees in the family Halictidae are attacked by Stylopids which are assigned to a related genus, *Halictoxenus*. The parasite is much smaller than *Stylops* and can be found protruding in a similar way from the abdomen of *Halictus* and *Lasioglossum* bees. In a sample of 183 *Lasioglossum parvulum* from a Cambridgeshire fen collected in March, 37 (20 per cent) were parasitised. This implies a considerable impact on the population. However, this is likely to be an overestimate because the presence of the parasite may promote earlier emergence of the bees and this was an early spring sample, captured in a water trap. Five species of *Halictoxenus* have been described (Perkins 1918). Host-parasite associations noted by Else and Edwards (2018) are *Halictoxenus arnoldi*/*Lasioglossum xanthopus*, *Halictoxenus tumulorum*/*Halictus tumulorum* and *Halictoxenus spencei*/*Lasioglossum calceatum*. This last relationship has also been recorded in the Scillies (Beavis 2022).

BEETLES (COLEOPTERA)

Oil beetles, Meloidae

The larvae of these large, rather clumsy, flightless beetles are cleptoparasites of solitary bees, mostly those in the families Andrenidae, Anthophoridae and Melittidae. Oil beetle larvae feed on the food provision in the nests of solitary

FIG 249. The Black Oil Beetle *Meloe proscarabaeus* eating the pollen of a Lesser Celandine *Ficaria verna*.

FIG 250. Triungulin larva of the Black Oil Beetle attached to a Common Furrow Bee *Lasioglossum calceatum* female.

bees as cleptoparasites. Five British species in the genus *Meloe* are still extant, some of them rare and all of them local in their distribution. Their presence at a site is usually an indication that a healthy population of mining bees is nearby. Bees that nest in large, loose colonies are likely to be most suitable (Ramsay 2002, Falk 2019; J. Walters' website http://johnwalters.co.uk/research/oil-beetles.php).

One of the more widespread species in Britain is the Black Oil Beetle *Meloe proscarabaeus* which is up to 30 mm long. In spring, they are found ambling across the ground, regularly stopping to feed on leaves or pollen, being quite selective in the plants they use. Females deposit eggs into sandy ground, which hatch into orange triungulin larvae, the name referring to the three claws on their legs. The larvae climb up plant stems, seeking flowers where bees might visit. The larvae are phoretic, meaning that they are transported to their host by another species, in this case an insect visiting a flower – the strategy used also by the much smaller Strepsiptera larvae described above. If a triungulin larva attaches itself to a female bee, it is likely to be transported back to her nest. Here, the triungulin will shed its skin, becoming a legless cleptoparasite feeding on the food-store in the brood cell of the bee. Sometimes many triungulins become attached to a single vector. Once inside the nest, the larva may devour the contents of several bee brood cells before pupating and emerging as an adult the following spring.

The chances of being carried to a suitable nest are very low, as the larva may not find a suitable insect to transport it. Ten thousand or more eggs can be laid in a beetle's lifetime, compensating for the high-risk strategy, but this high output requires a large female body. Despite the adults' restricted mobility, *Meloe* species are present on some isolated islands, such as the Hebridean and Canary Islands, by

FIG 251. (a) A Black Oil Beetle egg-laying. (b) Some of the resulting triungulins gathering on a cultivated lily. (© J. Thomas)

virtue of the phoretic stage. The large, slow-moving adult oil beetle would appear to be an easy target for predators, but the beetle has a very potent defence. It can secrete fluid containing the powerful poison, cantharidin, from its leg joints, which can cause blistering to the human skin or reportedly even kill horses, if oil beetles are inadvertently incorporated into fodder. Male oil beetles produce larger amounts of cantharidin than females and donate the poison during copulation, thereby apparently providing the eggs and perhaps the triungulins with protection

FIG 252. When disturbed, drops of liquid are released from a male oil beetle's leg joints, giving rise to the name oil beetle. (© J. Walters)

against predation and fungal attack. Female oil beetles assess the quality of a potential mate by checking the level of cantharidin in a gland on the male's head.

A curious corollary to this is that cantharidin is sometimes stolen by other insects for their own uses. Male cardinal beetles bite male oil beetles to obtain it and use it as a nuptial gift in the same way as an oil beetle (Wright 2017). Midges also collect it for their own protection. Since ancient times, humans have used cantharidin as an aphrodisiac, often with disastrous results. It is used as a traditional medicine in parts of Asia and also has potential for cancer treatments (Rauh *et al.* 2007).

Triungulins of *Meloe proscarabaeus* sometimes attach themselves in large numbers to unsuitable hosts, such as eumenid wasps (*Ancistrocerus* species). The wasps have narrow nest cavities stocked with caterpillars. Even if the oil beetle larvae were able to eat caterpillars, the nest is much too narrow for a developing oil beetle, so the larvae seem unlikely to succeed. Other oil beetles have a more sophisticated strategy. Triungulins of *Meloe franciscanus* in North America form clusters on stems, resembling the shape and size of the female of the solitary bee *Habropoda pallida*, and release a substance that mimics the female bee's sex attractant. Male *Habropoda pallida* attempt to mate with the triungulin cluster and become covered in them. The triungulins even reach out towards the male bee as it approaches. When males mate with a real female, triungulins are transferred and then transported to the female's nest. This rather bizarre arrangement must greatly increase the chances of triungulins reaching the nest of a 'correct' host species (Havernik & Saul-Gershenz 2000). Further research has demonstrated that the triungulins secrete a cocktail of hydrocarbons that exactly match a subset of those secreted by the female bee (Saul-Gershenz 2006).

Similar events are suspected in the blister beetle *Stenoria analis*, a cleptoparasite of the Ivy Bee *Colletes hederae* in Europe. In a field investigation, clusters of *Stenoria analis* triungulins, found in a meadow in northern France, were experimentally translocated to a nearby nesting aggregation of *Colletes hederae*. Male *Colletes hederae* immediately approached the larval clusters and attempted to mate, collecting triungulins on their body in the process. A mimic female pheromone was again thought to be involved (Vereecken & Mahe 2007). Within the British Isles, *Stenoria analis* is known only from the Channel Islands. It remains to be seen whether it will become established farther north and parasitise Britain's expanding population of *Colletes hederae*. The rare (or overlooked) Flame-shouldered Blister Beetle *Sitaris muralis* parasitises *Anthophora plumipes* nests in old walls in southern English counties. Triungulin larvae possibly cling to emerging male *Anthophora plumipes* and are then transferred to females during mating (Brock & Roberts 2017, Brock 2021).

TABLE 15. British representatives of the family Meloidae.

MELOIDAE: OIL BEETLES AND BLISTER BEETLES

Scientific name	Common name	Status in the British Isles	Adults active	Triungulins active	Habitat	Possible hosts
OIL BEETLES						
Meloe autumnalis	Autumnal Oil Beetle	Essex: last recorded 1952	Autumn	Spring		Unknown
Meloe brevicollis	Short-necked Oil Beetle	Very rare	March–April		Devon and Ayrshire coast	Colletes similis and Colletes floralis
Meloe cicatricosus	Scarred Oil Beetle	Southern counties: last recorded 1906				Anthophora retusa
Meloe decorus	Polished Oil Beetle	Only one British record			Gloucestershire 1870	
Meloe mediterraneus	Mediterranean Oil Beetle	Previously confused with Meloe rugosus: very rare	September–April		Grassy cliff tops: nocturnal	Not known
Meloe proscarabaeus	Black Oil Beetle	Widespread but mostly coastal	March–June	June–July	Grasslands, sea walls, heathland, coastal lagoon margins	Andrena, Melitta, Panurgus, Lasioglossum
Meloe rugosus	Rugged Oil Beetle	Rare: recent records in Dorset and Devon	October–February	Spring	Limestone areas: nocturnal	Possibly Andrena haemorrhoa, Anthophora plumipes

continued overleaf

TABLE 15. *continued*

OIL BEETLES

Scientific name	Common name	Status in the British Isles	Adults active	Triungulins active	Habitat	Possible hosts
Meloe variegatus	Rainbow Oil Beetle	Kent and Isle of Wight: last recorded 1882	Spring		Coastal grasslands	
Meloe violaceus	Violet Oil Beetle	Widespread	Spring	Spring (from previous year's eggs)	Sandy areas, including open woodland, moorland and coastal sites, mostly western Britain; also high altitude in Scotland	*Andrena* species, *Anthophora plumipes*

BLISTER BEETLES

Scientific name	Common name	Status in the British Isles	Adults active	Triungulins active	Habitat	Possible hosts
Sitaris muralis	Orange-shouldered blister beetle	Scarce: southern counties	Autumn	Spring	Old walls, often suburban	*Anthophora plumipes*
Stenoria analis	Ivy bee beetle	Channel Islands	Late summer	Autumn	Grasslands, meadows	*Colletes hederae*
Lytta vesicatoria	Spanish fly	Occasional introductions: formerly native				Ground-nesting bees

Data from Ramsay (2002), Falk (2019) and Brock & Roberts (2017).

PARASITIC WASPS: HYMENOPTERA

Sapygidae

There are two British members of this family of long, narrow, aculeate wasps, whose larvae are cleptoparasites of bees. Both wasp species are present in England as far north as Yorkshire and in Wales, reflecting the main distribution of their hosts, which are megachilid bee species. The *Sapyga* wasp lays an egg inside a bee brood cell with the aid of its pointed ovipositor, which is not exclusively a sting. The egg hatches into a jawed larva which begins by eating the bee's egg. Later instars have smaller mandibles and feed on the pollen store. *Sapyga quinquepunctata* usually parasitises *Osmia* species, especially *Osmia aurulenta*, but not apparently *Osmia bicolor* (Falk & Lewington 2015, Walters 2022; see also Chapter 3). *Sapyga quinquepunctata* is often present around bee hotels occupied by the Red Mason Bee *Osmia bicornis* and the Blue Mason Bee *Osmia caerulescens*. However, *Monosapyga clavicornis* usually targets the Large Scissor Bee *Chelostoma florisomne*. The relationship between *Monosapyga clavicornis* and *Chelostoma florisomne* was the subject of an investigation of a large nesting aggregation in the thatched roof of a field station in Denmark (Münster-Swendsen & Calabuig 2000). It is one of the few studies in which the effects of a bee parasite and the defences of its host have been quantified. A female

Chelostoma florisomne was able to remove the wasp's eggs if it discovered them in an open cell being stocked with pollen. But the wasp could also lay eggs in brood cells which had been recently sealed, by inserting the pointed tip of its abdomen through the soft nectar-and-mud partition. Bees could not remove such eggs, but they were nevertheless able to reduce the rate of parasitism by constructing empty brood cells in front of completed cells. Wasps often laid eggs into these empty cells but their newly hatched larvae became trapped and died of starvation. It was estimated that the empty cells reduced the overall rate of parasitism from 26.2 per cent to 9.5 per cent of cells.

FIG 253. *Sapyga quinquepunctata* female investigating the stem nest of a Red Mason Bee *Osmia bicornis*.

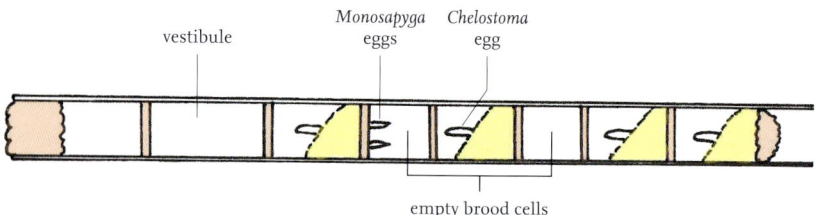

FIG 254. Diagram of a completed nest of the Large Scissor Bee *Chelostoma florisomne* made in a reed (entrance on the left, capped with mud). The rate of parasitism by *Monosapyga* wasps was 4.3 per cent in cells with an anterior empty cell but 17.5 per cent in those without one (redrawn from Münster-Swendsen & Calabuig 2000).

The activity pattern of the wasp was closely adapted to that of its host. During the morning, bees brought in about 12 pollen loads, which was enough to provision one cell, but the wasp confined its egg-laying to the afternoon, when the nest was almost complete, reducing the chances of the bee removing its eggs before the nest was sealed. After oviposition, the bee made several rapid trips to collect material to seal the nest and often then quickly made a second partition to form an empty cell. Wasps were thought to add a chemical marker to potential target nests, returning to them as soon as the host departed, then backing in to oviposit. The wasp produced small eggs in large numbers, compensating for the high losses due to the bee's defences, which were calculated to be 77 per cent of the wasp's eggs or larvae. Wasp eggs hatched in 2–3 days compared with 5 days for a bee egg. Good mobility, a well-protected head and strong mandibles allowed the wasp larva to destroy the host egg by sucking it dry over a period of 4–5 days, afterwards feeding on the pollen and nectar provision. When two wasp eggs were laid in one cell, one larva destroyed the other.

Chelostoma florisomne specialises on buttercup (*Ranunculus*) pollen, which has toxic properties. It would seem that the bee and *Monosapyga clavicornis* are both able to tolerate the toxins and this may explain the specificity of the wasp. *Osmia* species in North America which specialise on daisy family (Asteraceae) pollen, which also has some toxic properties, are not parasitised by *Sapyga* wasps. This is thought to be because the larvae cannot metabolise the toxins in the pollen. The benefits of reduced parasitism may outweigh the disadvantages of the poor nutritional quality of Asteraceae pollen (Spear *et al.* 2016). Two British *Osmia* species, the Orange-vented Mason Bee *Osmia leaiana* and the Spined Mason Bee *Osmia spinulosa*, are both Asteraceae specialists. The first is parasitised by *Sapyga quinquepunctata* but the second is apparently not. This anomaly may relate to *Osmia leaiana* using more blue Asteraceae species which are perhaps less toxic than yellow Asteraceae.

Stem-nesting bees in the subfamily Megachilidae generally leave a space (vestibule) at the nest entrance, behind the final seal, which may help to deter the entry of parasites, including cuckoo bees. In a nest of a Large-headed Resin Bee *Heriades truncorum*, in a bee nest box, the 11 brood cells averaged 1 cm in length, with the vestibule being slightly longer. The brood cell partitions in this species are constructed from resin and were less than 1 mm thick, while the final seal was a 5 mm thick resin plug impregnated with tiny pebbles (see Chapter 3). *Osmia* species which nest in snail shells are also potential targets of *Sapyga quinquepunctata* (Else & Edwards 2015). In one instance, just the innermost of three brood cells of an *Osmia aurulenta* nest was found to be parasitised by *Sapyga quinquepunctata* (Walters 2022). Red-tailed Mason Bee *Osmia bicolor* nests often have just one brood cell, well protected by pieces of dry clay pellets and other debris, extending nearly 2 cm around the interior of the shell (NWO pers. obs.). No parasitic associations are known for this bee (Else & Edwards 2015, Westrich 2018), perhaps illustrating the effectiveness of this barrier.

FIG 255. Opened Red-tailed Mason Bee *Osmia bicolor* nest showing a barrier of debris protecting the brood cell(s).

Gasteruptiidae

Wasps in this family are cleptoparasites of stem-nesting bees. The females possess a long ovipositor used to place an egg next to the egg of the host bee inside a nest cell. The parasite larva is thought to consume the egg and then to feed on the food-store (Malyshev 1964, cited Yildirim *et al.* 2004). The ovipositor is able to drill through the host's nest partitions and can be bent in different directions while seeking its target. It has sensilla at its tip, enabling it to detect its surroundings (Quicke & Fitton 1995). The most widespread species associated with solitary bees are *Gasteruption jaculator* and *Gasteruption assectator* (the Javelin Wasp and the Small Javelin Wasp), which parasitise cavity-nesting bees, including members of the family Megachilidae. Both species have also been reared from nests of the Reed Yellow-face Bee *Hylaeus pectoralis* made in galls in Common Reed *Phragmites australis* (Else 1995).

FIG 256. Javelin Wasp *Gasteruption jaculator* females at nest entrances of the Large-headed Resin Bee *Heriades truncorum*. The white-tipped sheath of the ovipositor remains outside the nest. (© V. Bartlett)

The body shape and movements of gasteruptiid wasps are very distinctive. The hind tibiae are swollen and club-shaped and contain a sensory organ, able to detect sound vibrations via its attachment to flexible parts of the tibia exoskeleton. As the wasp approaches a nesting site, it flies repeatedly up and down in front of the vertical nest surface, with its hind legs dangling. As it gets closer, it sways the hind legs from side to side, perhaps scanning for vibrations from host larvae and/or the adult host, seeking to find the former and avoid the latter (Miko *et al.* 2019). When a suitable nest is detected, the javelin wasp unsheathes the ovipositor and slides into the nest entrance 'tail first' to oviposit. The host bee may put up a defence. An online video by G. Pilkington shows a *Hylaeus* female seizing the ovipostor of a female *Gasteruption jaculator* in its mandibles and pushing her out of the nest cavity.

Mutillidae

Mutillid wasps are sometimes called velvet ants. Females are wingless with orange markings and some do look very like ants, but they are in fact aculeate wasps. There are many species worldwide but only three established in Britain: *Mutilla europaea*, *Smicromyrme rufipes* and *Myrmosa atra*. The first of these is quite large and parasitises several species of bumblebees (Owens 2020), whereas the other two attack smaller bees, often those in the genera *Halictus* and *Lasioglossum*.

Myrmosa atra is the most widespread and common species. Females are about 4 mm long, though quite variable in size, while males are about 7–8 mm. The males are black and have wings. When mating, males of some mutillids carry the female in flight, potentially taking her to new habitats such as sandy bare ground, where a variety of aculeate wasps and bees is likely to be nesting.

FIG 257. *Myrmosa atra* female.

Female *Myrmosa atra* run around rapidly where bees are nesting, frequently changing direction. They sense the ground with their antennae, stopping to examine holes in the sand and hollows under stones. They are thought to detect active nests by chemical means. When a suitable nest is found, the wasp disappears underground, perhaps for 30 minutes or more (NWO pers. obs.). Sometimes the wasp is challenged and evicted by a host bee, likely to be a guard bee if the species is eusocial. The wasp has long mandibles and is protected by its very thick exoskeleton, and, once inside a nest, it is usually ignored. It seeks out nest cells containing a prepupa or pupa, removing the tip of sealed brood cells to examine them. If suitable, an egg is laid and the cell is resealed. The mutillid larva feeds on the host bee larva as an ectoparasitoid, while attached to its surface, eventually pupating in the brood cell of the host. Multiple cells may be targeted in one nest (Brothers *et al.* 2000). Host defences are limited: sometimes workers block the burrow with soil from below as the mutillid attempts to dig down. It is possible that eusocial species have better defences against mutillids than solitary species, but evidence is limited.

Smicromyrme rufipes is a scarcer mutillid species confined to southern counties and has a variety of wasp and bee hosts. Halictid hosts include *Sphecodes* species (Richards 1980), which are cleptoparasites of other halictids, making *Smicromyrme rufipes* a hyperparasite in this instance.

Chrysididae: jewel wasps

These small parasitic wasps have bright iridescent metallic colours, generated by multiple very thin layers on the cuticle surface. The layered structure changes the colour of reflected light from red to green as the angle of incidence increases (Kroiss *et al.* 2009). The benefit of these bright colours is so far unexplained and seems unsuitable for a parasitoid, which needs to avoid detection by its host. Chrysidids comprise a very diverse group, with 490 species in Europe (Paukunnen *et al.* 2015). All are cleptoparasites or parasitoids of solitary wasps, solitary bees or sawflies and most are specific to a particular host family or genus. Those that parasitise bees include the genus *Chrysura*, which has two British representatives.

Chrysura radians occurs in England and Wales and is a parasitoid of *Osmia leaiana* (Early 2014) and possibly of other *Osmia* species. In mainland Europe, it parasitises the Viper's-bugloss Mason Bee *Hoplitis adunca*, a species which has recently established itself in the London area. *Chrysura hirsuta* parasitises three scarce *Osmia* species, namely the Mountain Mason Bee *Osmia inermis*, the Wall Mason Bee *Osmia parietina* and the Pinewood Mason Bee *Osmia uncinata*. The behaviour of *Chrysura* has been described in a North American species. Host-nest detection often involves exploring wooden surfaces with the antennae (Fig. 258). Like other Chrysididae, *Chrysura* has telescopic abdominal segments which can extend into the host nest. An egg is inserted into a bee's brood cell during the provisioning stage of the host and hatches a bit later than the bee larva. The first *Chrysura* instar is mobile and attaches itself to the host larva by its sharp mandibles, and as a parasitoid, it takes in some fluids in the host's growing stages. When the bee larva is fully grown and has spun a cocoon, the wasp larva moults to become a second instar and consumes the whole bee larva before spinning its own soft cocoon inside the tougher cocoon of the bee (Krombein 1967).

The tiny, all-green chrysidid *Trichrysis cyanea* is also a parasite of a range of solitary bees, including *Hylaeus pectoralis*, *Heriades truncorum* and *Chelostoma florisomne*.

Little is known about the specific defences of host bees against Chrysididae. The honeybee-predating wasp, the Bee Wolf *Philanthus triangulum*, attacks its

FIG 258. *Chrysura radians* female on a bee hotel.

chrysidid parasite *Hedychrum rutilans* if it is detected close to the wasp's nest burrow. The chrysidid attempts to hide in vegetation and waits for the wasp to depart, but once inside the nest, the Bee Wolf fails to notice it because the jewel wasp mimics some of the chemical components of the host's cuticle. It is likely that other chrysidids use similar chemical mimicry. Chrysidids have a very strong cuticle and are able to curl into a protective ball when threatened.

Chalcidoidea

Chalcids are tiny Hymenoptera averaging about 3 mm in length. They have no sting and are included in the taxonomic group Parasitica, alongside ichneumons, gasteruptiids, gall wasps and many others. Chalcids are extremely diverse, perhaps representing 10 per cent of all insect species. Females generally possess a long ovipositor used for placing their eggs into the nest or body of their host, which is usually the larva of another insect. Those which attack plants can be damaging to agriculture and forestry, though others have been successfully used in biological control of insect pests. Known chalcid parasites of British bee genera include *Aximopsis nodularis, Calosota vernalis, Melittobia acasta, Monodontomerus aeneus, Monodontomerus obscurus, Pteromalus apum, Pteromalus venustus* and *Torymus armatus* (Else & Edwards 2018, Gonzales et al. 2004, Noyes 2018). The main hosts are stem-nesting species of *Megachile* and *Osmia* bees.

Melittobia acasta (Eulophidae) uses British bee hosts in the genera *Apis, Bombus, Megachile* and *Osmia*, as well as aculeate wasps, flies, butterflies and moths, and can also be a hyperparasite of *Monodontomerus obscurus*. *Melittobia acasta* females use their ovipositor to puncture the skin of a mature bee larva and then feed on the exuding fluid. Further punctures are made to inject a toxin which paralyses the larva. One or more eggs are laid on the surface of the host, and the resulting legless larva develops as an ectoparasite (Balfour-Browne 1922). The adult is attracted to the nests of *Osmia* species by chemical cues from cocoons or frass, the signal probably being a particular ratio of two chemicals, namely acetic acid and acetaldehyde (Glasser & Farzan 2016, Filella et al. 2011).

The chalcid *Monodontomerus obscurus* is able to oviposit into bees' brood cells by inserting its long thin ovipositor through soft plant material and into the cocoon wall. It can have a large impact on managed populations of *Osmia* bees used for pollination of apple crops. It has been introduced inadvertently into North America, where it targets the non-native *Megachile rotundata*, used for the pollination of alfalfa crops (Kunić & Stanisavljević 2006). Some bees appear to use chemical deterrents against parasites such as chalcids. Wool Carder Bees *Anthidium manicatum* gather secretions from trichomes (small hairs) from plants such as *Pelargonium*, using specialised setae on the outer faces of their basitarsi.

FIG 259. (a) Wool Carder Bee *Anthidium manicatum* female gathering secretions from *Pelargonium* flowers. (b) The secretions appear as red spots amongst the wool of its nest. (© T. Eltz)

The secretions are transferred to the 'wool' in their brood cells and have been shown to reduce the rate of parasitism by *Melittobia acasta*, probably by masking the smell of the nest (Müller *et al*. 1996, Eltz *et al*. 2015).

Leafcutter bees (*Megachile* species) use pieces of leaf in nest construction, and it is possible that these leaves too contain parasite-repelling substances. In recent reference works (Else & Edwards 2018, Westrich 2018), over 50 plant species are listed as being used by British leafcutters. The most common is rose (*Rosa* species), used by all seven British leafcutter species. Rose was also the plant most commonly used by three North American *Megachile* species in an analysis using DNA bar-coding of nest materials (McIvor 2016). Roses do not produce any significant toxins, though they do secrete antimicrobial substances (see below). However, when insects damage leaves, many plants show a wounding response which results in a cascade of metabolic changes, making the leaves more resistant to attack. Leafcutter bees regularly return to the same plant, very often to the same leaf or leaf clusters to cut out nest material, making it likely that the leaf pieces contain deterrent compounds (Ferry *et al*. 2004). Many *Osmia* and *Hoplitis* species use leaf mastic (chewed leaves) for nest partitions and these also may be selected for their chemical as well as physical properties.

MITES (ACARINA)

Mites are Arachnids, the class which also contains ticks, spiders, harvestmen, scorpions and others. Mites can be free-living on decaying organic matter, but many are parasites of animals, plants or fungi. A wide diversity of mite species is known to be associated with bees, though just a few are bee specialists. The most widespread European bee mite is *Chaetodactylus osmiae*. The mite can often be seen on adult bees, but reproduction occurs in the bee's nest. Eggs are laid in a brood cell, hatching into larvae with three pairs of legs, which feed on the pollen and nectar provision. Adults appear after two nymph stages each with four pairs of legs. The population can increase rapidly, with up to 10 life cycles in a season, if there is sufficient food and moisture in the nest.

When food runs out, nymphs transform into a resistant stage called a hypopus (meaning 'hidden feet') which does not feed and lacks mouthparts. Two types of hypopi occur: a short-legged cyst-like form which remains dormant in the nest for months or years until a suitable bee returns, and a form with hooked legs which attaches itself to the hair of emerging bees. The latter are the mites that sometimes cover large portions of a bee's body surface. The transported mites do not feed on the adult bee but continue to do damage when a new nest is established, where they moult back into a feeding stage. These moult further to become males and females which mate and continue the cycle.

Mites can build up in numbers during warm, damp summers. Where *Osmia* populations are managed in artificial nests to pollinate fruit crops, the cocoons are removed manually from their nesting tubes and sprayed with disinfectant to keep mite numbers down (Kunić & Stanisavljević 2006). Healthy adult bees are able to

FIG 260. Mites on the propodeum of a Red Mason Bee *Osmia bicornis* male.

FIG 261. Immature phoretic mites on the declivity of tergite 1 of a Four-spotted Furrow Bee *Lasioglossum quadrinotatum* female. (© M. D. Welch)

groom mites from the accessible parts of their body. Less accessible areas such as the propodeum sometimes have smooth, shiny, hairless surfaces which perhaps limit the ability of parasites to cling on. Rather surprisingly, however, some bees and wasps carry mites in specialised mite pouches called acarinaria. In *Ctenocolletes* bees in Australia, the acarinaria are on the abdominal tergites and harbour mites which are in their non-feeding hypopus stage. A few mites are thought to dismount from the acarinaria into each brood cell, where they reproduce. Although the mite larvae feed on pollen, later mite stages feed on faeces, perhaps providing a trade-off to the bee by reducing mould growth in the nest cells. This therefore appears to be a mutualistic relationship (Houston 1987, 2018).

Acarinaria also occur in some *Xylocopa* and *Ceratina* species. Some are situated beneath the wing bases, with others at the anterior of the first tergite or within the genitalia: mites can be transmitted between the sexes during mating. Some seemingly harmful mites in acarinaria may be predatory on other mites which feed on nest provisions, or perhaps provide a food supply for beneficial mites, but the phenomenon has yet to be fully explained (Klimov 2007, O'Connor 1993).

The notorious *Varroa* mite does not affect solitary bees. It invades uncapped honeybee cells and feeds on the body fluids of the larva. The life cycle is timed to coincide with the emergence of the bee, to which the dispersal (phoretic) stage of the mite attaches itself as an external parasite. In addition to weakening honeybees directly, the mite is a vector of several bee viruses and it is possible that solitary bee mites are also vectors of disease.

Conservation of bee parasites

The survival of bee parasites depends on the presence of a healthy host population: when bee populations decline, their parasites may be threatened with extinction, along with their host. This has been termed co-extinction and is most likely to occur when parasites are highly host-specific (Stork 2010). Some bee parasites have been described as niche specialists rather than host specialists; for example, shadow flies, which parasitise a variety of ground-nesting bees (Knerer & Atwood 1967), though even here a degree of specificity comes from the fly's phenology, since parasites must coordinate their emergence with that of their hosts. The hosts of cleptoparasites can be further restricted by the pollen type collected by the host, which may be suitable only for those parasite larvae which are adapted to using it. Parasites which place their eggs or larvae directly into a host nest are likely to be more host-specific than those which simply scatter them on the ground or release them onto flowers. Internal parasites also tend to be very host-specific since they need to be resistant to their host's immune system. Three formerly widespread British oil beetle species appear to be extinct, though one has recently been re-discovered, and some scarce bee-flies are currently extending their ranges.

Bee hotels are widely marketed as a means of bee conservation in gardens and public places. A common experience is to see very rapid colonisation and success, especially of mason bees such as *Osmia bicornis*, followed by a decline towards much lower levels. This is likely to be partly a result of high rates of parasitism, since artificial nest sites are thought to be much easier for parasites to find than natural nest sites (MacIvor & Packer 2015). Dense, mixed-species aggregations also offer an opportunity for generalist parasitoids to switch from one bee species to another (MacIvor & Salehi 2014). However, the cost or benefit of nesting in aggregations in parasite defence differs between hymenopteran species, and no consistent pattern has been found (Rosenheim 1990). Concern is often expressed for the welfare of bees in bee hotels rather than for the welfare of their parasites, but this might not always be the most appropriate perspective. Some rare *Stelis* species of parasitic bees appear to be benefiting from bee hotels (see Chapter 8). This is an area of much-needed research.

Parasites and their hosts continually co-evolve their means of attack and defence: both need to 'run in order to stay in the same place' (the red queen hypothesis). Hamilton *et al.* (1990) argue that parasites have been the main driving force in the evolution and maintenance of sexual reproduction, which provides the continually changing gene combinations necessary for host adaptation to new parasitic strategies. Bees and their parasites offer an ideal model for the study of this dynamic relationship.

PREDATORS OF BEES

Predation of adult bees can be considerable, despite their ability to sting. Bees are especially vulnerable when emerging from their burrows, when mating or when in a torpid state. During warm-up by thermogenesis, any vibration of the wings or body is likely to make bees more conspicuous to predators. Bee larvae and pupae are also the victims of predation in some circumstances. Significant solitary bee predators in Britain include birds, spiders and the solitary wasp *Cerceris rybyensis*, with some bees also taken by dragonflies, social wasps and small mammals.

Birds

Birds recorded as predators of adult solitary bees include the Great Tit, Nuthatch, Redstart, Robin, Spotted Flycatcher, Wheatear and Whinchat. In a study of Stonechats (Greig-Smith & Quicke 1983), bee and wasp remains were found in 18 per cent of the droppings of fledglings, and in Red Backed Shrikes, 36 per cent of prey items impaled on thorns were bees, though the bee species were not recorded (Cramp *et al.*). In Wharfedale, Yorkshire, Mallard ducks are (somewhat surprisingly) one of the main predators of adult Heather Colletes *Colletes succinctus*. S. Saxton (pers. comm.) notes that 'whole [Mallard] families walk up and down the riverside nesting aggregations and hoover up the bees by the hundred'. Male bees swarming low over the ground were the main target. Male *Colletes succinctus* can also attract insectivorous birds such as Pied Wagtails. On the Scottish islands of Colonsay and Oronsay, Choughs have been seen digging out the larvae of *Colletes succinctus* in midwinter, having remembered the location of the nesting aggregations from earlier in the year (Clarke & Clarke 1995, cited Jardine *et al.* 2019), and in Sussex, Bearded Tits have been observed opening the nests of Reed Yellow-face Bees *Hylaeus pectoralis* in winter (P. Martin, cited Else & Edwards 2018). A further instance of bird predation where the bee species is known concerns female Ashy Mining Bees *Andrena cineraria* being taken by Blackbirds around the bees' nest burrows in a Salisbury garden. The Blackbirds considerably depleted the numbers of the bee (S.P.M. Roberts, cited Else & Edwards 2018). At a site in Norfolk, Rooks were observed taking both male and female *Anthophora plumipes* as they came in to land (M. Taylor pers. comm.). In February of a subsequent year, birds – presumed to be Rooks – had chipped away a large area of the same nesting aggregation down to the level of the brood cells (5 cm) and cleared them out (NWO pers. obs.).

Predation of the immature stages of stem-nesting solitary bees is common. *Osmia bicornis* larvae are often the victims at bee hotels where tits (*Parus* species) readily attack the nesting tubes and pull out larvae. In one instance, a Great Spotted

Woodpecker 'hacked its way through the mud caps of the *Osmia* tubes' (A. Hopkins pers. comm.). Social wasps (*Vespula* species) have also been observed preying on adult *Osmia bicornis* as they emerged from a bee hotel (C. Boyd pers. comm.).

Spiders

Spiders are major predators of bees, though records of identified solitary bees predated by known spider species are somewhat scarce. On walls in Oxford, spiders took 'a very heavy toll' of *Anthophora plumipes* females as they prospected for nest sites. The spiders concerned included *Amaurobius ferox*, *Amaurobius fenestralis* and *Eratigena* species (Stone 1990). Similarly, in Devon, an *Anthophora plumipes* male was observed being captured by a *Segestria florentina* female (Fig. 262) as the bee returned to a roosting cavity in a cob wall (Walters 2016).

The Noble False Widow Spider *Steatoda nobilis* (Theridiidae) also conceals itself around the nests of stem-nesting bees within its scaffold web (Fig. 263). This spider has extended its range from southern English counties to occupy much of England and Wales and parts of Scotland since the year 2000. Cuckoo bees such as *Coelioxys* species can also be captured around stem nests. At a coastal site in Norfolk, a *Megachile leachella* female was caught in a web made by a related spider, *Steatoda albomaculata*, while carrying a leaf segment to its nest burrow in the sand (V. Bartlett pers. comm.).

The Wasp Spider *Argiope bruennichi* (Araneidae) is a large and strikingly marked species found mostly in southern counties but extending its range northwards. It builds a large, strong, orb web amongst rank grassland. Grasshoppers are among its main prey items, but at East Tilbury Silt Lagoons,

FIG 262. Green-fanged Tube Web Spider *Seqestria florentina* female taking a Hairy-footed Flower Bee *Anthophora plumipes* male. (© J. Walters)

FIG 263. The Noble False Widow Spider *Steatoda nobilis* with Red Mason Bee *Osmia bicornis* prey. (© P. Creed)

the spider was found to capture large numbers of patrolling male Sea Aster Bees *Colletes halophilus*. Wasp spiders were common at this site, and up to three *Colletes halophilus* males were present in each web (S. Saxton 2009 & pers. comm.).

Crab spiders (Thomisidae) are well-known ambush predators, typically waiting on flowers to pounce on visiting insects. They do not build a web and simply grab the prey with their elongated front two pairs of legs. They often match the colour of the petals and, in the case of female *Misumena vatia*, can slowly alter their body colour between white and yellow, over a period of several days (Roberts 1996, Bartlett 2021). It has been generally assumed that colour-matching of the spider and flower improves the spider's chances of catching prey, by making the spider less conspicuous. Brechbühl *et al.* (2009) tested this idea in a field experiment by placing live crab spiders on flowers of three different colours (yellow, white and blue). Each of the three colours received a yellow crab spider, a white crab spider or no spider. Over 8,000 insect visits to the flowers were recorded on video. Solitary bees were found to be the most frequent visitors, making up almost half of the records. Solitary bees strongly avoided flowers with a spider, regardless of colour-matching, though some solitary bee genera reacted more strongly than others. *Lasioglossum* species avoided flowers with a spider the most and were also the most likely to be caught when they did land on a spider-attended flower. By contrast, *Halictus* species did not significantly avoid spiders but were rarely caught. Bumblebees and honeybees did not avoid flowers with a spider on them, whereas solitary bees sometimes reacted to spiders before landing, probably using visual cues. When they did land, they spent less time on the flowers than on spider-free flowers. Taking all flower visitors into

FIG 264. (a) *Misumena vatia* female on Wild Carrot *Daucus carota* on 29 July and (b) the same spider on the same plant on 2 August. (© V. Bartlett)

FIG 265. *Misumena vatia* with prey (*Colletes* species) on Oxeye Daisy *Leucanthemum vulgare*. (© V. Bartlett)

account (bees, hoverflies, beetles and others), Brechbühl's study found that colour-matched crab spiders did not have more encounters with insect visitors than non-matching spiders and did not catch more prey. Field studies suggest, however, that given the choice, yellow crab spiders do preferentially hunt on yellow flowers, while white crab spiders usually hunt on white flowers. This suggests that colour-matching could reduce the chance of the spiders themselves being seen by birds and other predators.

Females of *Thomisus onustus* are even more adaptable than *Misumena vatia*, being able to change their body colour between yellow, white and mauve. In this case, female *Thomisus onustus* are apparently able to disguise themselves to both insects and birds, despite the very differing visual systems of these two types of predators (Théry & Casas 2002). Their main prey items are hoverflies and Hymenoptera but also include other arthropods (Huseynov 2007). *Thomisus onustus* can deprive flowers of pollinators, but they can sometimes be helpful to plants when flowers are attacked by flower-eating insects (florivores), such as caterpillars. When florivores were numerous, a species of brassica (family Brassicaceae) was found to increase its emission of a volatile substance (β-ocimene), which attracted *Thomisus onustus*, which in turn captured the damaging caterpillars. Bees were attracted by the same substance, leading to spiders and bees being lured to the same flowers (Knauer *et al.* 2018). Over 70 per cent of plant families are known to emit β-ocimene as a pollinator attractant, and there are about 2,000 species of crab spiders worldwide. The authors conclude that crab spiders have a significant influence on the evolution of floral signals.

Interactions between bees, spiders and plants are clearly highly complex, and the visual systems of spiders, their prey and their predators must all be taken into consideration (Kevan *et al.* 2001, Heiling *et al.* 2005). A reduction in pollinator visits caused by spiders can potentially have a dramatic effect on flower pollination. In one investigation, models of crab spiders placed on flowers of *Rubus rolifolius* (a type of bramble) lowered seed set by over 40 per cent (Gonçalves-Souza *et al.* 2008). In a parallel Argentinian study (Gavini *et al.* 2018), models of crab spiders deterred Hymenoptera from landing on about 80 per cent of *Alstroemeria aurea* (yellow, zygomorphic) flowers, reducing seed set by 25 per cent. A literature review (Romero *et al.* 2011) showed that the presence of predators of all kinds around flowers reduced pollinator visits by an average of 36 per cent and the time spent on flowers by 50 per cent. Across a range of studies, live crab spiders deterred predators more than did models.

There have been few reports of predation on solitary bees by known spider species, and the overall effect of spiders on bee populations is hard to judge. A recent study of predation by spiders in Norfolk (Bartlett 2021) suggested that

TABLE 16. Bee predation by spiders observed at sites in Norfolk. Predation by the first two spider species was on flowers and by the remaining three in webs.

Spider species	Bee species
Misumena vatia	*Andrena bicolor, Apis mellifera, Bombus hortorum, Bombus terrestris, Colletes daviesanus, Halictus tumulorum, Sphecodes* sp.
Xysticus cristatus	*Bombus terrestris, Colletes daviesanus*
Entoplognatha ovata	*Anthophora quadrimaculata, Apis mellifera, Hylaeus pictipes, Lasioglossum punctatissimum*
Agelena labyrinthica	*Apis mellifera, Bombus* sp.
Zygiella x-notata	*Hyleus signatus*

Data from Bartlett (2021).

spiders can catch bees ranging in size from *Bombus* and *Anthophora* species to very tiny bees such as *Hylaeus pictipes*.

Larger bees are more likely to be able to struggle free from webs than smaller bees, though bumblebees do often get caught, especially when chilled by cool weather (Bartlett *loc. cit.*). *Anthophora plumipes* are sometimes able to force their way through spiders' webs when they emerge from their nest. Some spiders might even be indirectly beneficial to bees by catching bee parasites, such as the Flame-shouldered Blister Beetle *Sitaris muralis*, whose larvae are parasitoids of *Anthophora plumipes* larvae (P. Brock pers. comm.).

The Ornate Bee Fox (*Cerceris rybyensis*)

Cerceris rybyensis is one of a pair of British wasp species that prey entirely on bees. The other is the Bee Wolf *Philanthus triangulum*, which preys only on honeybees and has not been observed taking solitary bees. Aculeate wasps such as these paralyse their prey by stinging, and then place it in a burrow as food for their larvae. *Cerceris rybyensis* is active throughout July and August when many solitary bees are nesting. Its distribution is confined to England south of the Humber and parts of Wales.

Cerceris rybyensis has received relatively little study, despite its potential impact on solitary bee populations. In the summer of 2020, the opportunity arose to observe an aggregation of *Cerceris rybyensis* nests in a brick-weave garden path in Cambridgeshire. The identity of prey collected and the daily pattern of predation were recorded (Owens & Gomez de la Cuesta 2021). On fine days, wasps departed from their burrows to hunt between 9:30 and 11:30 am, and nest provisioning

FIG 266. *Cerceris rybyensis* with an Orange-legged Furrow Bee *Halictus rubicundus* female prey.

continued until 4:00–5:00 pm. They then retired to their burrows for the night, sealing the entrance behind them. On departure the following morning, wasps made classic orientation flights with ever-widening circles before heading off to collect prey. On returning from a hunting trip, wasps usually flew directly into their nest entrance carrying their prey (always a single solitary bee) beneath the body, held by the wasp's legs. If the nest entrance had become blocked, the prey was deposited while the obstruction was cleared, then retrieved with the mandibles and pulled backwards into the nest.

The predation rate was fairly constant from mid-morning until late afternoon, when the frequency of arrivals declined. A minimum estimate was that 6.7 prey items were taken on average by each wasp per day. This indicated that about 3,400 bees could potentially be taken into the aggregation of 18 *Cerceris rybyensis* nests over a four-week season, assuming ideal conditions. Estimates based on samples of cuticle debris around *Cerceris* brood cells excavated for inspection later in the year suggested that the actual rate of predation was about one-third of these projected numbers. Similar observations were made at a cluster of nest burrows on Kelling Heath, north Norfolk, in July 2021, where prey arrivals to a group of 32 nests were observed, leading to an estimated average of 4.9 bees taken by each wasp per day, assuming four hours of full activity as in the first study. About 200 *Cerceris rybyensis* nest burrows were present at the Norfolk observation site along about 350 m of track, so an estimated 27,440 bees could be taken over a four-week season, in ideal conditions. Further *Cerceris rybyensis* were nesting on other parts of the heath, so the impact of the wasp on the local bee population appeared to be considerable, though adverse weather, wasp mortality and other factors would reduce the predation rate.

Thirty-nine solitary bee species have been noted as prey items of *Cerceris rybyensis* in Britain. Most prey items (65 per cent) are halictids (two *Halictus*, 20 *Lasioglossum* and four *Sphecodes* species), but four members of the genera *Andrena*, one of *Colletes*,

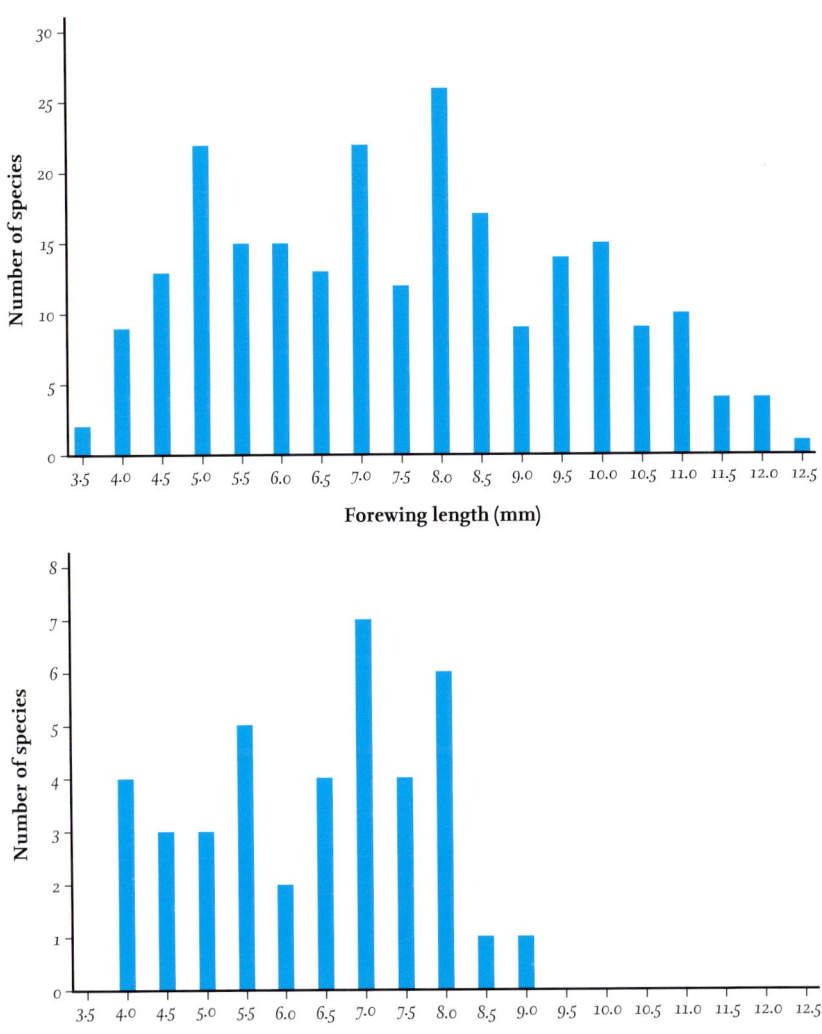

FIG 267. Frequency distribution of wing length of (a) all British bees excluding *Apis* and *Bombus* and of (b) *Cerceris rybyensis* prey. The wing length of each species was taken to be the upper range of female forewing length indicated in Falk and Lewington (2015).

one of *Melitta* and one of *Panurgus* (in the Channel Islands) are also taken as prey. No prey items other than bees have been recorded. Of 157 prey items where the sex was noted, 130 (82 per cent) were females. The size of prey seems to be critical. Based on species' forewing lengths (Falk and Lewington 2015) as a proxy for body size, the upper size range of prey is about 9 mm wing length, though size differences in *Cerceris rybyensis* females probably influence the upper limit (Fig. 267 b). Large prey items are sometimes dragged to the nest burrow, at least in the final stages, rather than being carried in flight. It would seem that there is a vacant niche for a wasp predator of larger solitary bees in Britain (Owens & Gomez de la Cuesta 2021).

It is not clear how *Cerceris rybyensis* females assess the suitability of their prey, but it is likely that they do so as they line up potential prey in flight. The Common Furrow Bee *Lasioglossum calceatum*, the Orange-legged Furrow Bee *Halictus rubicundus* and the Bronze Furrow Bee *Halictus tumulorum* have been recorded as prey at significantly more sites than other prey. Descriptions of *Cerceris rybyensis* attacking prey are few. Ian Cheeseborough (pers. comm.) reports a predation event in Shropshire, as follows: 'She flew low over flowers around a pond looking for prey, and hovered over, then dropped down to take an *Andrena bicolor*.' This matches descriptions by Marchal (1887), who observed an aggregation of *Cerceris rybyensis* amongst nests of '*Halyctus*'. During a long period of watching, he observed a small number of predation events. He describes the wasp tracking an incoming female bee carrying pollen, then, as she circled to locate her nest, it pounced on her from behind, bringing her to the ground. The bee was stung 2–3 times under the thorax while the pair rolled together in the sand. Stinging was followed by chewing of the neck region with the mandibles (malaxation). From observations of captured wasps with prey, Marchal noticed that stings were placed at the junction of the neck and head, especially in larger prey, and also at the junction of the prothorax and mesothorax, points which he judged to be halfway between the positions of ventral nerve ganglia, allowing the venom to spread to these critical points. Malaxation presumably destroys communication between the brain and the nerve cord, preventing further movement when the venom wears off, but keeping the prey alive.

Photographs taken at a cliff-top aggregation in north Norfolk show *Cerceris rybyensis* stinging its prey in the thorax while holding it down with its legs and mandibles. It then immediately moved to the head end and grasped the bee by the neck from in front, stinging it again and chewing the neck. It then orientated the prey below the body before taking flight (Fig. 268). In some instances, the wasp pounces on mating pairs of solitary bees (Early 2008).

Marchal noted that the egg is laid on the ventral side of the prey's thorax (presumably with just one prey item receiving an egg in each brood cell) (Fig. 269).

PARASITES AND PREDATORS · 359

FIG 268. *Cerceris rybyensis* stings then grasps the neck of its Common Furrow Bee *Lasioglossum calceatum* prey before re-orienting the bee for transport.

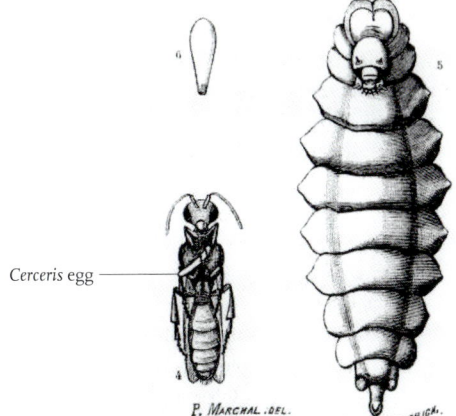

Cerceris egg

FIG 269. *Cerceris rybyensis* cocoon (upper left), *Halictus* (*Lasioglossum*) *albipes* prey with an egg (long and white) under the thorax (lower left) and an exposed *C. rybyensis* prepupa (right), illustrated by Marchal (1887).

Other solitary bee predators may include Green Tiger Beetles *Cicindela campestris*, which were observed making predation attempts on *Megachile leachella* males as they landed to bask, though none was successful. Dragonflies are also potential predators. Small mammals such as mice and shrews sometimes take solitary bee larvae as prey. For example, in the Cambridgeshire fens, Harvest Mice break open *Lipara* galls in Common Reed *Phragmites australis* stems, which frequently harbour the nests of Reed Yellow-face Bees *Hylaeus pectoralis*. (T. Reader, cited Else & Edwards 2018). Hedgehogs and badgers are known to prey upon bumblebee nests (Yalden 1976, Roberts *et al.* 2020) and are likely also to dig out solitary bee larvae.

The destiny of pollen is clearly highly varied. A small proportion reaches the stigmas of conspecific flowering plants, leading to the production of seeds, but a large quantity is transformed into flower visitors such as bees and their parasites or predators. Most significant of these for pollen-collecting solitary bees are the cuckoo bees, and these are described in the following chapter.

CHAPTER 8

Cuckoo Bees

'I have found the searching for the history of each structure or instinct an excellent aid to observation.'
—Charles Darwin, in a letter to Jean-Henri Fabre (1880)

As we saw in Chapter 7, there are many parasites waiting to exploit the hard-won food-stores in a solitary bee's nest. In that chapter, we did not mention the bees which are themselves parasitic. In Britain, about 73 bee species are cuckoos of pollen-collecting bees, comprising over a quarter of our bee fauna. Excluding cuckoo bumblebees, they come from six bee genera. All behave exclusively as cuckoos, entering other bees' brood cells and substituting their egg for that of the host. The host's egg or larva is killed, either by the adult cuckoo bee or by the cuckoo bee's larva. The cuckoo larva then feeds on the food-stores of the host bee, eventually emerging as an adult from within the host's nest. Cuckoo bees share many similarities with other bees and are believed to have evolved independently several times from pollen-collecting bees, with each cuckoo line subsequently diversifying into many species.

Cuckoo bees offer an opportunity to study the complex interactions between a cuckoo bee and its host, involving strategies and counter-strategies. How do cuckoo bees recognise the 'right' host, locate its nest and also time their entry to avoid being detected by the host – or find some way of overcoming host resistance? How do the hosts themselves limit parasitism by cuckoo bees? There are obvious parallels with the non-bee parasites such as shadow flies discussed in the previous chapter, as well as some contrasts.

TABLE 17. Cuckoo bee genera in the British Isles.

Genus	Common name	Family	Number of species
Bombus	Cuckoo bumblebees	Apidae	6
Epeolus	Variegated cuckoo bees	Apidae	2
Melecta	Mourning bees	Apidae	2
Nomada	Nomad bees	Apidae	35
Sphecodes	Blood bees	Halictidae	17
Coelioxys	Sharp-tail bees	Megachilidae	6
Stelis	Dark bees	Megachilidae	5

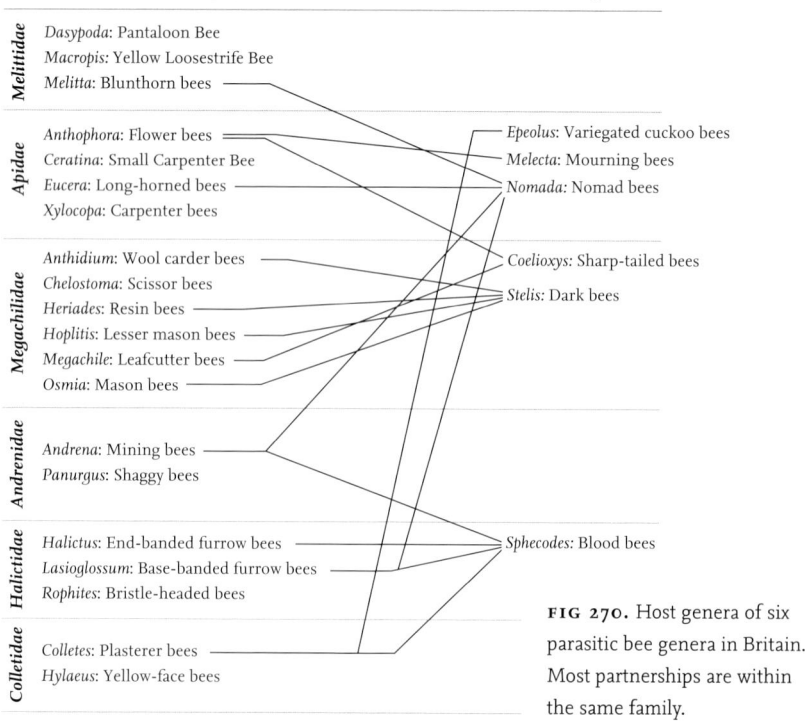

FIG 270. Host genera of six parasitic bee genera in Britain. Most partnerships are within the same family.

First we will look into the behaviour and physical adaptations of the members of each of the six cuckoo bee genera. A fully cleptoparasitic way of life involves highly adapted bees attacking other species within their own family and sometimes beyond. Each of the six solitary cuckoo bee genera present in Britain has its own box of tricks, with a range of species specialising in particular host(s). Figure 270 summarises the host genera of solitary bees parasitised by members of each cuckoo bee genus.

EPEOLUS: VARIEGATED CUCKOO BEES

These small, patterned bees are cleptoparasites of the genus *Colletes* (plasterer bees). Their bodies have spots and bars which perhaps serve to disguise them while they wait motionless on bare ground, watching their hosts' nests. Away from nesting areas, *Epeolus* can often be found taking nectar from flowers, where they can be quite conspicuous. Males and females are very similar, though males have green eyes and females brown, a feature also seen in some nomad bees whose sub-family, Nomadinae, they share.

There are 17 members of the genus *Epeolus* recognised in Europe (Bogusch & Hadrava 2018) but just two species recorded in Britain, the Red-thighed Epeolus *Epeolus cruciger* and the Black-thighed Epeolus *Epeolus variegatus*, although it is likely that each includes different genetic forms or even separate species, specialising on different hosts. Both *Epeolus* species have one generation a year, as do their hosts (they are univoltine).

FIG 271. Red thighed Epeolus *Epeolus cruciger* female at the nest entrance of a *Colletes* species.

TABLE 18. Hosts of *Epeolus* species recorded in Britain.

	Colletes cunicularius	Colletes daviesanus	Colletes floralis	Colletes fodiens	Colletes halophilus	Colletes hederae	Colletes marginatus	Colletes similis	Colletes succinctus
Epeolus cruciger				?		✓	✓		✓
Epeolus variegatus		✓		✓	✓	✓		✓	

Epeolus cruciger is strongly associated with the Heather Colletes *Colletes succinctus*, often occurring in large numbers at their nesting aggregations, which are at their peak in late summer. A cliff nesting site of *Colletes succinctus* in Norfolk is shared with the Hairy-saddled Colletes *Colletes fodiens* and the Ivy Bee *Colletes hederae*. Records from the county database show that the emergence of *Epeolus cruciger* coincides with that of *Colletes succinctus*, while the emergence of *Colletes fodiens* is earlier and coincides with that of *Epeolus variegatus*. The form of *Epeolus cruciger* associated with the Margined Colletes *Colletes marginatus* is smaller and appears several weeks earlier than the *Colletes succinctus* form (Falk & Lewington 2015). It is the subject of ongoing research and could be elevated to a separate species (Else & Edwards 2018). *Colletes marginatus* is found on coastal dunes and also inland in the East Anglian Breckland.

Colletes hederae is known to be parasitised by *Epeolus cruciger* in mainland Europe, in regions south of the Alps, but not on the near continent. A possible explanation is that this population of *Epeolus cruciger* persisted in surviving populations of *Colletes hederae* in southern Europe during the Ice Age but failed to follow the subsequent northerly range expansion of *Colletes hederae* into north-western Europe as the ice receded (Kuhlmann et al. 2007). *Colletes hederae* was first recorded in southern England in 2001 (Cross 2002), by which time it had already established some large colonies. The population had advanced as far as Norfolk by 2013 and reached Yorkshire soon afterwards (see Chapter 9). There is now strong evidence that *Epeolus cruciger* is using *Colletes hederae* as a host in Essex, where it has been reared from Ivy Bee nests (TB), and there is some indication that it is also parasitising Ivy Bee nests in Breckland (NWO). It is therefore likely that the *Colletes succinctus* form of *Epeolus cruciger* has transferred to *Colletes hederae* at some sites. Populations of *Colletes hederae* and *Colletes succinctus* overlap in emergence time and often share nesting sites, making this transition likely,

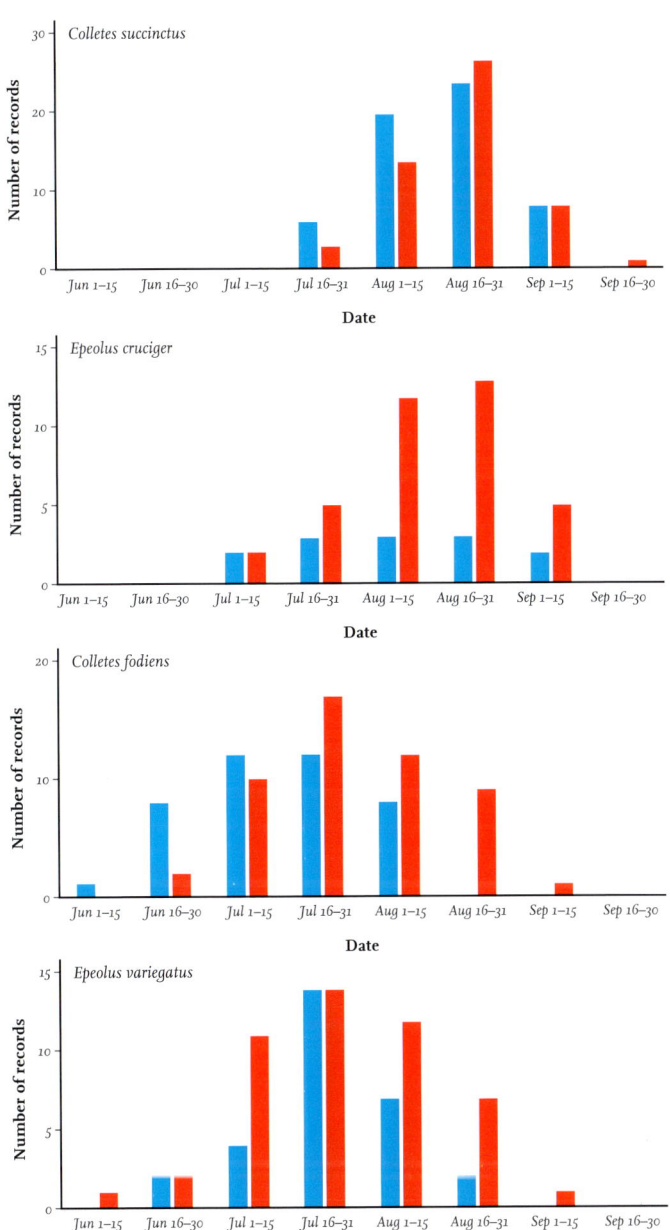

FIG 272. Phenology of two species of *Colletes* and *Epeolus*. Blue = male, red = female.

though the main emergence of *Colletes hederae* is later. Across Europe as a whole, *Epeolus cruciger* has become rare as a result of the loss of the heathland habitats needed by *Colletes succinctus* (Bogusch & Havrada 2018). In mainland Europe, *Epeolus fallax* is a parasite of *Colletes hederae*, but this species of *Epeolus* has not yet been recorded in Britain. Likewise, the closely related *Epeoloides coecutiens* (a cleptoparasite of the Yellow Loosestrife Bee *Macropis europaea*) and *Ammobates punctatus* (a cleptoparasite of the Green-eyed Flower Bee *Anthophora bimaculata*) are so far absent from our bee fauna, though present in nearby parts of mainland Europe. These cleptoparasites seem to have poor dispersal ability, perhaps partly related to their inability to be active in cool conditions (see Chapter 9).

The form of *Epeolus variegatus* that attacks the Sea Aster Bee *Colletes halophilus* peaks in September. It is quite scarce with this host in Norfolk but very numerous in Essex (Saxton 2009 & pers. comm.). It may be a different genetic form from the population associated with *Colletes fodiens*.

As with all mining bees and their parasites, the egg-laying behaviour of *Epeolus* is difficult to observe. A North American species *Epeolus compactus* was studied in California by Torchio and Burdick (1988), who removed sandstone blocks from a nesting aggregation of its host and opened them up for observation in a laboratory. The host is *Colletes kincaidii*, which makes horizontal linear nests in sandstone cliffs or walls in a similar way to Davies' Colletes *Colletes daviesanus* in Britain. As we have seen, *Colletes* nests are lined with a cellophane-like material, secreted by the female's Dufour's gland, and the pollen-nectar food provision they contain is semi-liquid.

Each *Epeolus* egg was inserted through the sloping polyester nest partition, with the flattened tip of the egg flush with the outer surface. The *Epeolus* used a

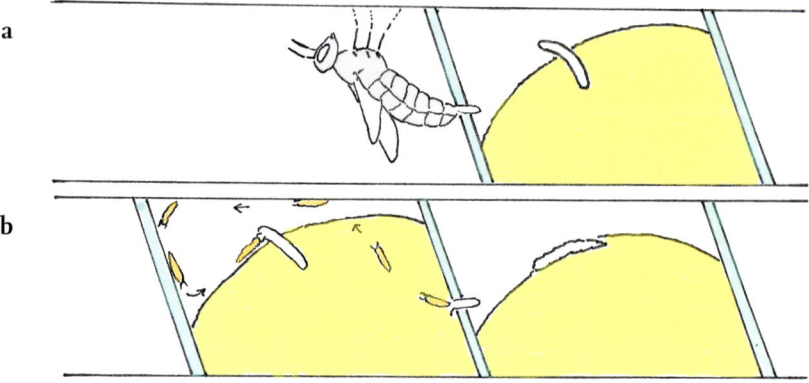

FIG 273. (a) Egg deposition by *Epeolus* and (b) hatching of the egg and destruction of the host egg (redrawn from Torchio & Burdick 1988).

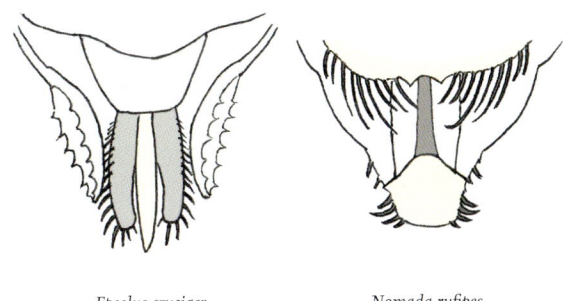

FIG 274. Terminal segments of the Red-thighed Epeolus *Epeolus cruciger* (from above) and the Black-horned Nomad Bee *Nomada rufipes* (underside), showing the hooked structures used in penetrating the host's brood cell wall, for egg-laying.

Epeolus cruciger   *Nomada rufipes*

specialised structure on the last segment (sixth sternite) of the underside of the abdomen (similar to that in *Epeolus cruciger*; see Figure 274) to cut a U-shaped hole for egg insertion, then joined the cut edge to the perimeter of the egg, using an abdominal secretion. Once deposited, only the tip of the small egg was in contact with the brood cell being stocked, where it was unlikely to be detected by the host. By the time the *Epeolus* egg hatched, the brood cell had been stocked and sealed by the host. The *Epeolus* larva then moved through the semi-liquid food provisions of the newly sealed cell to the pollen surface and destroyed the host egg, using its long, sickle-shaped mandibles. The parasite larva then proceeded to feed on the provisions, later developing into a prepupa, pupa and adult but not making a cocoon.

MELECTA: MOURNING BEES

Members of the genus *Melecta* are known as mourning bees owing to their dark colours. They are cleptoparasites of bees in the genus *Anthophora*, the flower bees. There are just two *Melecta* species known from the British Isles: the Common Mourning Bee *Melecta albifrons*, which generally targets the Hairy-footed Flower Bee *Anthophora plumipes*, and the Square-spotted Mourning Bee *Melecta luctuosa*, which targets the rare Potter Flower Bee *Anthophora retusa* but is possibly extinct in Britain. *Melecta albifrons* is, however, fairly common in southern English counties, becoming less so towards the north and with a few records in Wales. Not surprisingly, its distribution closely matches that of its host, *Anthophora plumipes*.

Anthophora plumipes generally make their nest burrows in soft walls and cliffs, though they sometimes nest in level ground. At large aggregations of the host, *Melecta albifrons* can be numerous. For example, on a warm April day, over 50 were present at a nesting aggregation on a stump of eroded cliff on a shingle beach in Norfolk. Female *Melecta* were basking, grooming or crawling slowly on the surface of the aggregation. Some were flying short distances around the densely packed nesting area, crawling into and inspecting nest entrances. Eight different pairs were seen mating during one hour's observation. It seemed that the warm weather had triggered a mass emergence of both sexes of *Melecta*. Female *Anthophora plumipes* were regularly arriving with pollen loads, which were generally taken straight into the open nest entrance. At this stage, *Melecta* females did not attempt to follow them, and the potential victims showed no response to clusters of *Melecta* close to their nest burrows.

A few days later, however, female *Anthophora plumipes* had started to display antagonism towards *Melecta albifrons* females. Sometimes they simply landed on a stationary *Melecta* and displaced it, or pursued it briefly in flight, but several instances were observed of an *Anthophora plumipes* female dragging a *Melecta* out of a nest burrow (Fig. 276). In one case, the host gripped both front legs of the cuckoo in its mandibles. There was a short period of grappling after they emerged from the nest entrance, but the *Melecta* was very passive, and photographs showed that it did not have its sting everted. Both participants seemed unharmed. However, one female *Melecta* was found walking about on the ground below the nest

FIG 275. Site of nest aggregation of the Hairy-footed Flower Bee *Anthophora plumipes*.

FIG 276. Hairy-footed Flower Bee *Anthophora plumipes* female dragging a Common Mourning Bee *Melecta albifrons* female out of the host's nest burrow.

entrances lacking its head. A female *Anthophora* was also seen to fly off from the nesting cliff with a *Melecta* female dangling beneath her, which was soon dropped. The onset of this defensive behaviour probably coincided with *Melecta* females commencing egg-laying as their ovaries matured, a few days after mating.

Similar behaviour is described by Shuckard (1886): '[*Anthophora plumipes*] is subject to the parasitism of the genus *Melecta*, whose incursions are very repugnant to them, and which they exhibit in very fierce pugnacity, for if they catch the intruder in her invasion they will draw her forth and deliver battle with great fury. I have seen both the combatants rolling in the dust, the combat and escape made perhaps easier to the *Melecta* by the load the *Anthophora* was bearing home.' Other reports come from Semichon (1904), Thorp (1969, cited Bohart 1970), and Baldock (2008). In addition to physical interactions, female *Anthophora plumipes* make loud staccato sounds inside their nest burrows from time to time, sometimes in a chorus, this perhaps being a signal that the burrow is occupied, though how the sound or vibration is detected is unclear.

Anthophora plumipes females appeared to treat *Melecta albifrons* females with respect, choosing the tip of a leg or wing to drag them from their nest and otherwise keeping their distance, though they often ended up rolling onto the ground together. The pattern of spots of *Melecta* may serve to warn the hosts of their sting. In this context, it is interesting that some *Melecta* females are largely black, closely resembling host females. Perhaps these black forms have an advantage later in the nesting cycle when the host becomes more antagonistic towards the parasite.

Melecta males and females both leave the nesting aggregation from time to time, seeking nectar from flowers such as Red Dead-nettle *Lamium purpureum*, which leaves a tell-tale orange trace of pollen on the head. Mating often occurs at the nest site and is a complex performance. The male strokes the antennae of the female from base to tip about 10 times while fanning his wings. He then leans

FIG 277. Common Mourning Bee *Melecta albifrons* females, (a) light and (b) dark forms.

back and makes thrusting movements accompanied by a rhythmic crackling buzz, seemingly made by scissoring movements of the partially folded wings. These two routines alternate, each lasting for 5–10 seconds, and mating can continue for 10 minutes or more. It is not known whether this ritual transfers a disguising pheromone (chemical signal) as reported for some nomad bee species (see below). It is perhaps more likely that female *Melecta* acquire the scent of the host simply through their development inside a host cell or by entering nest burrows. Both female and male *Melecta* retreat inside nest burrows during the night and in poor weather, though males sometimes sleep in clusters on plant stems some distance from the nesting area (Benton 2017).

Far more difficult to observe is the behaviour of *Melecta albifrons* inside the host's nest burrow. Semichon (1904) describes the behaviour of a *Melecta albifrons* female in France as she dug her way into a cell of *Anthophora fulvitarsis* (a non-British species). The material she excavated to gain entry was in small, damp lumps, suggesting that the cell had just been

FIG 278. Common Mourning Bees *Melecta albifrons* mating. The female's antennae are being stroked by the male's antennae as he whirrs his wings.

FIG 279. A Hairy-footed Flower Bee *Anthophora plumipes* completed brood cell in two halves. The hole at the lower end is the final exit point of the female bee when sealing the brood cell. The pollen store is at the top end, with an egg placed on the surface. The cell is lined with a pearly secretion from the Dufour's gland.

sealed. The *Melecta* came out after five minutes, and investigations showed that the egg had been placed on the roof of the cell at the opposite end from the pollen mass, in which the host egg was situated. The oviposition of the *Melecta* egg was not seen. Studies of other species of *Melecta* indicate that the female makes a small hole in the mud cap of a completed nest cell and then inserts her abdomen to lay one or two eggs on the inner surface of the cap or on part of the nearby wall (Thorp 1969, Bohart 1970). According to Rosen and Ding (2012), the hole is made with the tip of the abdomen rather than with the mandibles, and the egg is attached by its posterior end (the end which comes out first when it is laid). After laying, the adult *Melecta* replaces the earth in the cap.

The egg-laying behaviour and larval development of *Melecta albifrons* in *Anthophora plumipes* nests does not seem to have been described. The opportunity was taken to open and inspect a small number of nest cells from the Norfolk site in early May, and one was found with a newly hatched *Melecta* larva within. As expected, the larva was at the outer end of the cell, some distance from the food-store, which occupied about one-third of the opposite (inner) end. The long, curved host egg was half-buried in the end of the soft (rather pungent) pollen mass, the egg being much larger than the parasite larva. The mourning bee larva was making slight peristaltic movements and occasionally opening and closing its pointed mandibles. Close to the larva was the folded shed skin (chorion) of the egg from which it had hatched. The broken end of the cell with the larva in it was moved close to the host egg in order to observe the larva's response. The larva gradually

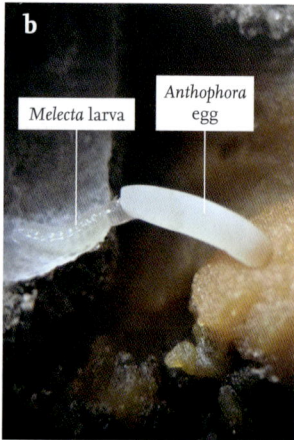

FIG 280. (a) Common Mourning Bee *Melecta albifrons* larva and eggshell (chorion) in the broken end of a Hairy-footed Flower Bee *Anthophora plumipes* brood cell. The cavity in the centre is the final point sealed by the host female and appears also to be the site of entry of the female cuckoo bee's abdomen when she penetrated the brood cell. (It is possible that the larva is a second instar, having shed its skin while still inside the eggshell.) (b) *Melecta albifrons* larva consuming the host egg. (c) Five hours later, the *Melecta* larva begins to consume pollen with its head immersed in the food-store. Only the shell of the host egg remains.

approached and reached out to the egg and punctured it with its mandibles. It then fed on the contents of the egg, which became flaccid. Five hours later, only the skin of the egg remained, and the larva had started consuming pollen.

Anthophora plumipes females secrete a whitish oily sealant from their Dufour's gland over the inner surface of each brood cell. Ring-shaped marks of the secretion suggest that she narrows the final hole in a spiral manner, somehow managing to cover all of the inside surface before retreating. In the related Australian bee *Amegilla dawsoni*, the central dimple indicates where the bee finally withdrew its tongue after spreading the lining (Houston 2018), and it seems likely that *Anthophora plumipes* females behave similarly. The entry point of the *Melecta* shows no break in the sealant around the dimple, and it appears that she uses this weak point for her own entry, while the mud is still wet, often leaving no visible damage on the inside. The smooth shape of the entry hole with no sharp breaks is consistent with egg-laying occurring while the cap is still soft, as indicated by Semichon.

The rate of parasitism is difficult to judge from counts of adult bees since there is much coming and going from the nest site. Samples of brood cells were examined in December 2020 and January 2021. These yielded a total of 78 bees, seven of which were mourning bees, indicating that about 9 per cent of nest cells were parasitised, based on this small sample. The sex ratio of the *Anthophora plumipes* sample was 36 males to 35 females. (All bees of host and parasite were adults within their brood cells at this stage and were retained for release the following season.) The brood cells of *Melecta albifrons* had a thin, soft, brown, slightly uneven skin, coating the pearly lining deposited by the host bee. The brown layer was presumably secreted by the prepupa of the mourning bee.

Melecta albifrons has the ability to find host nests quickly, as evidenced by its arrival at artificial mud nesting bricks shortly after the first year of emergence of *Anthophora plumipes* (Walters 2021; NWO pers. obs.), perhaps attracted by scent.

NOMADA: NOMAD BEES

Bees in the genus *Nomada* are, rather confusingly, sometimes called wasp-bees. Many have black and yellow bands or red and black colours which resemble wasps rather than bees. They are also known as nomad bees because of the wandering behaviour of females as they search for suitable nests to parasitise. Like most other cuckoo bees, they have very little hair. There are around 724 species of nomad bees worldwide (Ascher & Pickering 2020). *Nomada* species (family Apidae) are unusual in not being closely related to their hosts, which are usually *Andrena* species (family Andrenidae; Michener 2007). Of the 35 species of nomad bees in the British Isles, 32 have *Andrena* hosts, one targets the Long-horned Bee *Eucera longicornis*, one targets several furrow bee (*Lasioglossum*) species, and one targets several blunt-horn bee (*Melitta*) species. Of the 32 *Nomada* species which parasitise *Andrena*, 21 parasitise just one *Andrena* host, so far as is known. *Nomada* species which parasitise more than one *Andrena* usually target a group of species which belong to the same sub-genus. (A sub-genus is a cluster of closely related species within the same genus.) Being closely related, multiple hosts of the same nomad are likely to share common features, such as similar pheromones, which are recognised by the female nomad. Fabricius' Nomad Bee *Nomada fabriciana* is an exception, with hosts in four different sub-genera. The Flavous Nomad Bee *Nomada flava* (Fig. 281 d) and the Black-horned Nomad Bee *Nomada rufipes* both have hosts in two sub-genera.

TABLE 19. Hosts of the 35 *Nomada* species recorded in the British Isles. *Andrena* hosts are grouped by sub-genus. With few exceptions, most species are specific to one *Andrena* sub-genus.

Host	Sub-genus/genus	N. panzeri	N. leucophthalma	N. fusca	N. signata	N. glabella	N. ferruginata	N. armata	N. integra	N. facilis	N. rufipes	N. fabriciana	N. marshamella	N. hirtipes	N. flava
A. apicata	Andrena s.str.	x	x												
A. clarkella	Andrena s.str.	x	x												
A. fucata	Andrena s.str.	x		x											
A. fulva	Andrena s.str.	x			x										
A. helvola	Andrena s.str.	x													
A. lapponica	Andrena s.str.					x									
A. praecox	Andrena s.str.						x								
A. synadelpha	Andrena s.str.	x													
A. varians	Andrena s.str.	x													
A. hattorfiana	Charitandrena							x							
A. humilis	Chlorandrena								x						
A. fulvago	Chrysandrena								x	x					
A. denticulata	Cnemiandrena										x				
A. fuscipes	Cnemiandrena										x				
A. nigriceps	Cnemiandrena										x				
A. similllima	Cnemiandrena										x				
A. bicolor	Euandrena											x			
A. bucephala	Hoplandrena													x	
A. ferox	Hoplandrena												x		x
A. rosae	Hoplandrena												x		
A. scotica	Hoplandrena												x		
A. trimmerana	Hoplandrena												x		x
A. argentata	Leucandrena														
A. barbilabris	Leucandrena														
A. marginata	Margandrena														
A. cineraria	Melandrena														
A. nigroaenea	Melandrena												x		x
A. nitida	Melandrena														x
A. thoracica	Melandrena														
A. vaga	Melandrena														
A. falsifica	Micrandrena														
A. minutula	Micrandrena														
A. proxima	Micrandrena														
A. semilaevis	Micrandrena														
A. subopaca	Micrandrena														
A. chrysosceles	Notandrena												x		
A. nitidiuscula	Notandrena										x				
A. coitana	Oreomelissa														
A. bimaculata	Plastandrena														
A. pilipes	Plastandrena														
A. tibialis	Plastandrena														
A. nigrospina	Plastandrena														
A. labiata	Poikilandrena														
A. tarsata	Poikilandrena														
A. angustior	Ptilandrena												x		
A. dorsata	Simandrena														
A. wilkella	Taeniandrena														
A. haemorrhoa	Trachandrena														
A. flavipes	Zonandrena														
A. gravida	Zonandrena														
E. longicornis	Eucera														
L. morio	Lasioglossum														
L. nitidiusculum	Lasioglossum														
L. parvulum	Lasioglossum														
L. villosulum	Lasioglossum														
M. haemorrhoidalis	Melitta														
M. leporina	Melitta														
M. tricincta	Melitta														
		1	2	3	4	5	6	7	8	9	10	11	12	13	14

CUCKOO BEES · 375

	N. baccata	N. alboguttata	N. argentata	N. lathburiana	N. goodeniana	N. flavoguttata	N. conjungens	N. errans	N. obtusifrons	N. fulvicornis	N. subcornuta	N. guttulata	N. roberjeotiana	N. zonata	N. striata	N. ruficornis	N. fucata	N. bifasciata	N. sexfasciata	N. sheppardana	N. flavopicta
	x																				
		x																			
			x																		
				x																	
					x																
					x	x															
					x	x															
				x																	
						x	x														
							x														
						x	x														
							x														
								x													
									x	x	x										
											x										
												x									
													x								
														x							
															x						
																x					
																	x				
																		x			
																				x	
																				x	
																				x	
																					x
																					x
																					x
	15	16	17	18	19	20	21	22	23	24	25	26	27	28	29	30	32	31	33	34	35

Most nomads are univoltine (they have one generation per year), but those with bivoltine hosts are generally also bivoltine (having two generations per year). For example, the Painted Nomad Bee *Nomada fucata* (Fig. 281 a) has generations timed to target both first and second generations of the Yellow-legged Mining Bee *Andrena flavipes*. *Nomada fabriciana* has both univoltine and bivoltine populations according to the host it is targeting, *Andrena bicolor* being bivoltine and the other three hosts being univoltine. It is possible that these different populations represent cryptic races or species, differing in emergence

FIG 281. Some examples of female nomad bees: (a) the Painted Nomad *Nomada fucata*, (b) the Variable Nomad *Nomada zonata*, (c) the Bear-clawed Nomad *Nomada baccata*, (d) the Flavous Nomad *Nomada flava* and (e) the Armed Nomad *Nomada armata*.

times. There are differences in body size between *Nomada fabriciana* populations specialising on different hosts, though these could be a consequence of the size of their food ration. It is also possible that nomads targeting more than one species *within* a sub-genus are in reality different races or species. The Bilberry Nomad Bee *Nomada glabella* has recently been confirmed as a distinct species (rather than a form of *Nomada panzeri*) on the basis of DNA analysis (Falk *et al*. 2022), and other clarifications are likely to follow.

Looking at the picture from the host's side, several *Andrena* species are targeted by two different *Nomada* species, and the Buffish Mining Bee *Andrena nigroaenea* is parasitised by three. Little is known about how competing nomad species interact when they occur at the same nesting aggregations. In some cases, the nomads sharing a host have different geographical ranges.

The colourful body patterns of female nomads probably represent a warning pattern shared with other stinging insects, such as wasps. If they have similar warning colours, predators, such as birds, will learn to avoid them more quickly (Müllerian mimicry). Male nomads do not sting and probably gain protection by looking like females (Batesian mimicry). Nomads do not seem to mimic the colours of their hosts, though they sometimes belong to the same Müllerian mimicry ring. For example, the Short-spined Nomad Bee *Nomada guttulata* and its host the Red-girdled Mining Bee *Andrena labiata* are both black with a red abdomen. Despite their bright colours, nomads can be well camouflaged on some backgrounds, at least to human eyes.

The emergence times of a nomad must coordinate with those of its host. Most non-parasitic bees show protandry; that is, males emerge earlier than females. Figure 282 shows casual records submitted for Norfolk of the Orange-tailed Mining Bee *Andrena haemorrhoa* and its specific cuckoo, the Fork-jawed Nomad Bee *Nomada ruficornis*. Male *Andrena haemorrhoa* appear first from mid-March, then, in the first half of April, male and female nomads appear together, in step with the female host. On this evidence, the host appears to be protandrous, but there is little or no protandry in the nomad. The records also suggest that male and female numbers were about equal in both the host and the nomad, these being 122 male:125 female in *Andrena haemorrhoa* and 31 male:37 female in *Nomada ruficornis*, host numbers averaging about five times cuckoo numbers over the recording period. A better way of quantifying cuckoo and host numbers is to use emergence traps whereby all the bees emerging from several nests can be identified and counted. As we saw in Chapter 7, Paxton and Pohl (1999) collected all insects emerging from nests at a Tawny Mining Bee *Andrena fulva* nesting aggregation in south Wales over six seasons. The rate of parasitism by Panzer's Nomad Bee *Nomada panzeri* averaged 18 per cent; in other words,

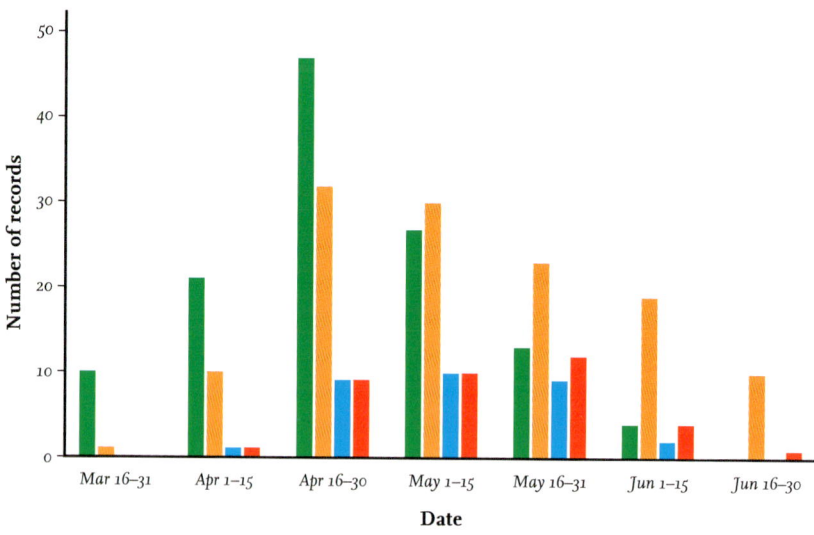

FIG 282. Records of the Orange-tailed Mining Bee *Andrena haemorrhoa* and its cuckoo, *Nomada ruficornis*. Green = *Andrena haemorrhoa* male, orange = *Andrena haemorrhoa* female, blue = *Nomada ruficornis* male, red = *Nomada ruficornis* female (records from Norfolk, 2001–19).

18 per cent of Tawny Mining Bee offspring were replaced by the nomad. This is quite close to the estimate of 22 per cent based on the Norfolk records.

The association between nomad bees and their hosts often extends to the flowers they visit. Nomads do not collect pollen but they do take nectar and this often comes from the same source as the host's pollen and nectar. For example, the Black-horned Nomad Bee *Nomada rufipes* and its host, the Heather Mining Bee *Andrena fuscipes*, both visit Ling *Calluna vulgaris*. The Silver-sided Nomad Bee *Nomada argentata* and the Armed Nomad Bee *Nomada armata* (Fig. 281 e) both visit scabious flowers (*Knautia arvensis* and *Scabiosa columbaria*), as do their respective hosts, the Small Scabious Mining Bee *Andrena marginata* and the Large Scabious Mining Bee *Andrena hattorfiana*. As noted in Chapter 5, the Short-spined Nomad Bee *Nomada guttulata* (Fig. 283 a) takes nectar from Germander Speedwell *Veronica chamaedrys*, which is also a favoured plant of its only host, the Red-girdled Mining Bee *Andrena labiata* (Fig. 283 b). To some extent, this flower-matching is a result of the shared habitat and phenology of nomad and host. Visiting the same flowers may have advantages, such as the male or female nomad finding a mate or a female nomad encountering a female host, which she might follow back to a nest. Such resource-based host-finding is likely to be associated with hosts which do not nest in aggregations and whose nest sites are correspondingly more

FIG 283. (a) Short-spined Nomad Bee *Nomada guttulata* female and (b) its host, Red-girdled Mining Bee *Andrena labiata* female, visiting Germander Speedwell *Veronica chamaedrys*.

difficult to find, especially hosts which are oligolectic (see Chapter 3). A further benefit may come from adopting the scent of the host's pollen, which could provide some disguise while in a host's nest burrow.

A further illustration of mutual adaptation comes from Clarke's Mining Bee *Andrena clarkella* and its cuckoo, the Early Nomad Bee *Nomada leucophthalma*. *Andrena clarkella* females close their nest entrance behind them after entering with a pollen load. On emerging again, a female may take as long as 25 minutes to cover and disguise the nest entrance before she leaves to collect more pollen. Observations at a site in Norfolk illustrate how the nomad copes with these obstructions. A nest burrow with the *Andrena* inside was inspected by a female *Nomada leucophthalma*. The cuckoo bee explored the dome of spoil around the nest for about one minute, seemingly becoming aware of the presence of the bee underground. She then perched on a sprig of heather 5 cm away, looking towards the nest but remaining completely still. Five minutes later, the host emerged and spent about one minute scraping loose soil over her exit hole. She then crawled about 30 cm away from the nest before taking flight (this routine was seen on two occasions, possibly being a diversionary tactic used by the bee). The nomad remained stationary for a further four minutes, then moved directly to the nest. She dug her way in at the exact point of the host's departure, taking three minutes to submerge her whole body. She remained inside, presumably egg-laying, for 20 minutes. On coming to the surface again, she spent about 30 seconds systematically pulling soil over the exit hole she had made, before flying away. Almost two hours after her own departure, the host returned with a new pollen load and entered her nest, apparently none the wiser.

On a later occasion at the same nest, the female *Andrena clarkella* appeared to have a decoy nest entrance adjacent to the true one, behaviour also reported in the Grey-backed Mining Bee *Andrena vaga* (Rezkova *et al.* 2012, Owens & Benton 2022). During a period of 25 minutes spent covering her nest before departure,

FIG 284. (a) An Early Nomad Bee *Nomada leucophthalma* female waits and watches as the head of a Clarke's Mining Bee *Andrena clarkella* female appears from a nest after depositing pollen. (b) When the host had departed and covered the nest entrance, the nomad re-excavated the burrow and entered. (c) After egg-laying, she re-covered the nest entrance.

FIG 285. Early Nomad Bee *Nomada leucophthalma* female digging into the nest of a Clarke's Mining Bee *Andrena clarkella*, with an apparent decoy nest entrance close by, excavated by the departing host. The real entrance is directly under the bee.

the female *Andrena clarkella* re-excavated the decoy hole, using some of it to level the soil around the true entrance. A female *Nomada leucophthalma* took 16 minutes, involving several tries, before finding her way in to the true nest. During this time, she approached the decoy entrance twice but did not attempt to enter. Also during this period, a second female *Nomada leucophthalma* landed on the nest spoil, and there was a slight altercation. The newly arrived female took off and returned three times but gave up without any attempt at digging.

Nomada leucophthalma appears to be specialised to *Andrena clarkella* by virtue of its emergence early in the year and its ability to find and enter the covered nests of its host. *Nomada leucophthalma* is also a parasite of the Large Sallow Mining Bee *Andrena apicata*, another willow (*Salix*) specialist. In addition, *Nomada leucophthalma* has been observed entering a nest of Early Colletes *Colletes cunicularius*, another early-spring bee specialising in willow pollen but belonging to a different genus (M. D. Welch pers. comm., Else & Edwards 2018). The host choice of *Nomada leucophthalma* could therefore relate to timing of emergence or to pollen type (or to both).

Being a parasite of a particular host requires the cuckoo to emerge at the right time of year and to be able to find and identify the 'correct' host. As mentioned, nomads might follow pollen-collecting hosts from flowers back to their nest site, or in other cases they may simply remain near the nest aggregation from which they emerged, where their hosts will be easily encountered. Each *Andrena* species secretes its own characteristic blend of chemicals onto the surface of its cuticle. A female nomad needs to 'know' which chemical blend to respond to. This knowledge could derive from the nest the nomad emerged from, or it might be innate (genetically programmed).

The antennae of bees are highly sensitive to chemical stimuli and also possess glands which secrete chemical signals (Romani *et al.* 2003). Nomad females can be observed closely approaching pollen-laden hosts at nest sites, placing their outstretched antennae close to the host's head and fanning their wings. This fanning behaviour is also typical of many nomad females as they slowly approach the entrance to a nest burrow. A possible interpretation is that the nomad is fanning her own scent towards the host, or the likely presence of the host, as a means of disguise. Alternatively, she may be sampling the scents that emanate from a potential host, its pollen load or the host nest.

Another intriguing finding is that some female nomads apparently receive host-specific cues from the male while mating. Tengö and Bergström (1977b) discovered that there is a chemical match between the Dufour's gland secretions of a female *Andrena* host and secretions made by a male nomad. This was evident in four different *Nomada-Andrena* pairs whose secretions were analysed by gas

FIG 286. Nomad females investigating pollen-laden hosts with their antennae: (a) Lathbury's Nomad Bee *Nomada lathburiana* fanning her wings with the antennae erect, with host Grey-backed Mining Bee *Andrena vaga*. (b) The Orange-horned Nomad Bee *Nomada fulvicornis* closely examining a pollen-laden Grey-gastered Mining Bee *Andrena tibialis* female. (Second photo © V. Bartlett)

chromatography. During mating, the male nomads were thought to spread secretions over the female nomad's body. The authors propose that the main function of this anointment is to disguise the female nomad, preventing attack by the host *Andrena* when the nomad approaches or enters her nest. Tengö and Bergström state: 'This relationship [*Nomada* parasitising *Andrena*] suggests a long-established coexistence of *Andrena* and *Nomada*, with plenty of time available for *Nomada* to evolve a signal system adapted to that of *Andrena*.' It is not clear why the chemical is produced by male nomads rather than the females themselves, though it was thought that it might help to ensure that a sufficient amount is spread on the female's body surface to fully disguise her when entering an *Andrena* nest. Female *Andrena* hosts were found to rarely show any aggression towards *Nomada* at nest sites. These authors also found similarities between male Blunthorn Nomad Bee *Nomada flavopicta* secretions and those of two of its *Melitta* species hosts (Tengö & Bergström 1976).

Schindler *et al.* (2018) observed mating in Little Nomad Bees *Nomada flavogutta*, Painted Nomad Bees *Nomada fucata* and Lathbury's Nomad Bees *Nomada lathburiana* (Fig. 287), either in the field or in a laboratory. In all three species, secretions were directly transferred from the male's to the female's antennae, which were stroked or entwined during mating. This behaviour had not previously been described, probably because mating in *Nomada* has rarely been observed. Scanning electron microscopy showed that swellings on the surface of the males' antennae (tyloids) are glandular, secreting pheromones in the form of a liquid or paste.

Pheromone release by male bees during mating is quite common. The pheromone sometimes behaves as an antiaphrodisiac, deterring other males

FIG 287. Lathbury's Nomad Bees *Nomada lathburiana* mating. The male has entwined one of his antennae around that of the female and is transferring a pheromone in the form of a paste. (© M. Schindler)

from approaching the female to mate (see Chapter 2). Matching between a male nomad's secretions and the specific chemical blend of the female host's cuticle must serve a different function, such as helping to disguise the female nomad during egg-laying, as proposed by Schindler *et al*. Another possibility is that the female 'learns' the identity of the host from the male's secretions. This could explain the head-to-head approach and wing-fanning by female nomads during which she could compare the potential host's chemical cues with those placed on her antennae by the male. There seem to be differences between *Nomada* species, since not all males mimic the female *Andrena* host's secretions.

When a nest of the target host has been found, the female nomad enters to deposit her egg at a time when the cell is being provisioned with food but when the host is not present to defend it. Experiments to investigate the importance of different stimuli in this behaviour were carried out with *Andrena alleghaniensis* and its cleptoparasite *Nomada pseudops* at a nest aggregation at Cornell University campus (Cane 1983). Artificial nest burrows 7 cm deep were made amongst the real nests, using a suitably sized nail. Some nail holes were left empty as controls, while others contained pollen, chilled bees or fresh bees plunged into the hole and then removed. It was found that the presence of an *Andrena alleghaniensis* female with pollen caused the most nest entries by the nomads, but pollen alone elicited more entries than just the bee (without pollen), presumably because pollen indicated a partially stocked nest without its occupant, but the presence of the occupant alone could result in physical conflict. A chilled nomad in the artificial burrow was also found to be attractive, perhaps because the presence of another nomad indicated a suitable cell to parasitise. Empty artificial burrows were inspected but not entered.

Attempts were made (NWO) to replicate some aspects of Cane's experiments. Twelve artificial nest entrances in a cluster were made using a pencil pushed into small tumuli of earth, amongst a spring-nesting aggregation of Cliff Mining Bees *Andrena thoracica* in Norfolk. This *Andrena* species does not cover its nest entrance on leaving to collect pollen. Gooden's Nomad Bee *Nomada goodeniana* females readily hovered over the experimental nest burrows and sometimes landed to inspect them, with typical wing-fanning behaviour (Fig. 288). During one hour, there were 23 instances of hovering and nine of landing beside a hole, though none entered. This evidence suggests that *Nomada goodeniana* females are initially attracted to nest burrows by sight but do not enter unless they detect host-specific odours or odours of a particular pollen type.

When real or artificial nests were approached, the female *Nomada goodeniana* typically fanned her wings, with her antennae raised and pointing ahead. She then closed her wings, moved forwards and 'looked into' the nest entrance, waving each antenna up and down alternately. It is presumably at this stage that a decision is made about whether to enter. On coming out of a real nest and at frequent intervals while searching, females groomed their bodies, especially the antennae, which are clearly of prime importance. Female *Nomada goodeniana* tended to move around a wide area of the *Andrena thoracica* nest aggregation in groups of three or four and sometimes approached the same nest entrance together, one even landing on top of another female nomad. It is possible that females gain information from each other about potential nests to parasitise, as suggested by Cane's experiments.

Nomad bee females parasitise nest cells while they are still open and being stocked with pollen. Eggs are placed in a small pocket excavated into the host

FIG 288. Gooden's Nomad Bee *Nomada goodeniana* female fanning her wings as she inspects a hole made with a pencil.

brood cell wall, usually placed at an angle to the surface (Radchenko 1981). The female curls her abdomen round to make a triangular incision in the wall, using cutting edges at the tip of the abdomen (see Fig. 274), and after laying the egg she seals the depression with a membranous covering. She may lay more than one egg and sometimes as many as four, perhaps as an insurance against eggs being destroyed by the host. The egg hatches in about a week, before the host egg, and the mobile first instar larva destroys any other *Nomada* eggs, using its pointed mandibles, then eats the host egg before turning to the food-store gathered by the host bee. No cocoon is made. The nomad probably overwinters as an adult in the host nest (Linsley & MacSwain 1955).

SPHECODES: BLOOD BEES

The name *Sphecodes* derives from the Greek *Sphex* (wasp) and *-odes* (like). Their red and black colouring is similar to that of ground-nesting wasps such as spider-hunters (Pompilidae), which have very powerful stings. By sharing this warning colour pattern with other stinging insects, *Sphecodes* probably gain added protection against predators, as noted above in *Nomada* species. The red colour derives from the pigment of the bee's cuticle rather than from blood, which is colourless in insects. Seventeen species of *Sphecodes* occur in mainland Britain, and the group has an almost worldwide distribution. Superficially, *Sphecodes* look very similar to each other and they are difficult to identify in the field, usually requiring microscopic examination for a reliable determination. This means that their behaviour and the identity of their hosts are not very well known, although many *Sphecodes* are quite common. *Sphecodes* are members of the family Halictidae and all of them are cleptoparasites, mostly specialising on other Halictidae, especially those in the genera *Lasioglossum* and *Halictus*, though there are a few British *Sphecodes* that target the genus *Andrena* and one that parasitises *Colletes*. Like most cuckoo bees, *Sphecodes* do not possess the long body hairs and other structures needed to collect pollen.

In order to understand the *modus operandi* of *Sphecodes*, we need to look at the life cycle of their hosts, since many *Lasioglossum* and *Halictus* (furrow bee) species are eusocial, with a queen and sterile workers living together as a social unit in a nest. As described in Chapter 4, only the mated gynes of these two genera survive through the winter and their first batch of spring eggs develops into workers. Further broods of workers may be produced, and all cooperate to rear a sexual brood later in the season, consisting of new gynes and males. The Sharp-collared Furrow Bee *Lasioglossum malachurum* is a eusocial species which has been well studied (Else & Edwards 2018, Habermannová *et al.* 2013, Sick *et al.* 1993, Strohm &

Bordon-Hauser 2003; see also Chapter 3). The nest burrows of this bee are made in level ground, often in large aggregations, each nest appearing as a mound of excavated earth with an open entrance in the centre. The Square-headed Blood Bee *Sphecodes monilicornis* is one of their main cuckoo species.

In the spring, *Sphecodes monilicornis* females can be seen flying and crawling around the nest burrows: like *Lasioglossum*, they have overwintered as mated adults. The cuckoo bee enters a nest burrow when the host queen is collecting pollen, destroys her eggs and larvae and lays her own eggs in any suitably stocked nest cells. In nests where the host workers have already emerged, *Sphecodes* females face a greater challenge. The queen *Lasioglossum malachurum* now remains in the nest, laying eggs, while the workers forage for pollen and nectar or act as guards at the nest entrance. The female *Sphecodes monilicornis*, which is larger than the host queen and much larger than the workers, is capable of using her mandibles to pull a guard bee from the nest entrance. She then enters the nest and attacks and kills the occupants by stinging them. Observations of nests in the laboratory (Sick *et al.* 1993) showed that she widens the openings of each stocked nest cell in turn, destroys the host egg and then replaces it with an egg of her own. She then closes the nest cell with earth, which she presses into place using her mandibles and the pygidium at the tip of her abdomen. She may remain in the nest overnight and on emerging covers over the main entrance before she departs. One female was able to lay up to five eggs in 24 hours. The emerging *Sphecodes* progeny comprise both females and males and these will mate before the female hibernates through the winter. There is evidence that *Lasioglossum malachurum* nests containing more workers receive more attacks from *Sphecodes* than those with fewer workers, and this may be one of the selection pressures limiting the size of the first brood to just 4–6 workers (Strohm & Bordon-Hauser 2003).

FIG 289. Reticulate Blood Bees *Sphecodes reticulatus* mating. The antennae are not directly involved.

Unlike some *Nomada* species, the *Sphecodes* females that have been studied do not show any chemical mimicry of their host and their method of attack depends on physical force rather than deception (Sick *et al. loc. cit.*). Likewise, mating in *Sphecodes* does not appear to involve the complex intertwining of antennae seen in *Nomada* and *Melecta*. The absence of chemical adaptation to a particular host probably allows *Sphecodes* females to attack a wider range of species.

A North American study suggested that *Sphecodes* could induce a host to become 'akinetic' (motionless), perhaps by means of a chemical attack, with the host taking minutes or hours to recover mobility. *Sphecodes* females were also observed stinging guard bees and pushing them out of the nest burrow (Knerer & Atwood 1967).

Sphecodes monilicornis and other species which target eusocial hosts seem able to modify their behaviour appropriately when attacking first or second broods, only the latter being protected by guard bees. This is one aspect of *Sphecodes* behaviour which allows them to target a wide host range. The size of a *Sphecodes* female is also significant. For example, Little Sickle-jawed Blood Bee *Sphecodes longulus* females are about 4 mm long and target very small species such as Least Furrow Bees *Lasioglossum minutissimum* and Green Furrow Bees *Lasioglossum morio*. They search a relatively small area where these species nest in loose aggregations on banks of soft earth (NWO pers. obs.).

Sphecodes females must be able to enter the host nest, and the food provision in each cell must be sufficient to support the growth of its larva. The cuckoo bee needs to be strong enough to overcome the resistance of the host and any workers present. All *Sphecodes* females have powerful mandibles and stings. The mandibles of *Sphecodes monilicornis* are particularly well developed, and the square-shaped head houses strong mandibular muscles. In addition to defence, the mandibles are used to clear a path into host-nest burrows and their pointed tips often become worn down with use. When they meet resistance from guard bees, *Sphecodes* females are able to fold their front legs into a cavity behind the head (Fig. 290), making it difficult for the hosts' mandibles to seize them. Burrows of *Lasioglossum malachurum* are often made in gravelly ground or amongst beach shingle, which perhaps helps to reduce the likelihood of attacks by *Sphecodes*.

Timing is also critical. The Spined Blood Bee *Sphecodes spinulosus* is unusual in that both sexes overwinter as immatures, such that their emergence coincides with the peak of activity of their particular host, the Orange-footed Furrow Bee *Lasioglossum xanthopus*, which is a summer-emerging species. Habitat, too, can lead to specialisation; for example, the Furry-bellied Blood Bee *Sphecodes hyalinatus* is strongly associated with southern calcareous grasslands frequented by one of its main hosts, the Chalk Furrow Bee *Lasioglossum fulvicorne*. By contrast, the Reticulate Blood Bee *Sphecodes reticulatus* is usually found on heathland, where it targets two *Andrena* heather specialists as well as the Grey-tailed Furrow Bee *Lasioglossum prasinum*.

Table 20 shows hosts of *Sphecodes* species in mainland Britain for which there is at least reasonably strong evidence. Current knowledge is very incomplete,

TABLE 20. Recorded hosts of *Sphecodes* species in Britain.

	Andrena argentata	*Andrena barbilabris*	*Andrena dorsata*	*Andrena labialis*	*Andrena flavipes*	*Colletes cunicularius*	*Halictus eurygnathus*	*Halictus rubicundus*	*Halictus tumulorum*	*Lasioglossum albipes*	*Lasioglossum brevicorne*	*Lasioglossum calceatum*	*Lasioglossum fratellum*	*Lasioglossum fulvicorne*	*Lasioglossum laevigatum*	*Lasioglossum laticeps*	*Lasioglossum lativentre*	*Lasioglossum leucopus*	*Lasioglossum leucozonium*	*Lasioglossum malachurum*	*Lasioglossum minutissimum*	*Lasioglossum morio*	*Lasioglossum nitidiusculum*	*Lasioglossum parvulum*	*Lasioglossum pauperatum*	*Lasioglossum prasinum*	*Lasioglossum pauxillum*	*Lasioglossum punctatissimum*	*Lasioglossum puncticolle*	*Lasioglossum quadrinotatum*	*Lasioglossum rufitarse*	*Lasioglossum semilucens*	*Lasioglossum sexnotatum*	*Lasioglossum sexstrigatum*	*Lasioglossum smeathmanellum*	*Lasioglossum villosulum*	*Lasioglossum xanthopus*	*Lasioglossum zonulum*
Sphecodes albilabris						✓																																
Sphecodes crassus								✓																														
Sphecodes ephippius												✓																										
Sphecodes ferruginatus													✓	✓																								
Sphecodes geoffrellus																✓	✓		✓																			
Sphecodes gibbus													✓	✓																✓								
Sphecodes hyalinatus																						✓	✓	✓							✓					✓		

CUCKOO BEES · 389

Sphecodes longulus
Sphecodes miniatus
Sphecodes monilicornis
Sphecodes niger
Sphecodes pellucidus
Sphecodes puncticeps
Sphecodes reticulatus
Sphecodes rubicundus
Sphecodes scabricollis
Sphecodes spinulosus

Data from Else & Edwards (2018) and C. Kirby-Lambert pers. comm.

FIG 290. Sandpit Blood Bee *Sphecodes pellucidus* female, showing the cavity behind the head into which the front legs can be withdrawn.

cavity for leg

but it is clear that there are several generalist species which have three or more hosts and a smaller number which appear to be specialists with just one host. Evolutionary trees established by comparing the DNA of a range of *Sphecodes* species suggest that there have been multiple switches across evolutionary time between generalists and specialists, with no hard-and-fast distinctions between these two states (Habermannová *et al. loc. cit.*). As mentioned, this flexibility probably arises from *Sphecodes*' use of host intimidation, a strategy that works on any host. Many *Sphecodes* have a primary host coupled with the ability to change if the host becomes scarce or their season is over.

In two generalist *Sphecodes* species, *Sphecodes monilicornis* and *Sphecodes ephippius*, individual bees within the same population can show different host preferences. This became apparent during field studies at mixed aggregations of halictid hosts in eastern Europe. Individual bees of these two *Sphecodes* were tracked and found to stick with the same host when visiting successive nests, passing by the nests of other host species. This was likened to 'flower constancy' whereby an oligolectic bee forages from just one flower type even when a wide choice is available: individual *Sphecodes* females were showing 'host constancy'. It is not known if these individual host choices are dependent on the nest the *Sphecodes* female emerged from by a form of imprinting, or whether there are different genetic subgroups as found in cuckoo birds, where it is well established that there are different host-specific races, each targeting a different bird species, often with egg colour patterns closely mimicking those of the host (Davies 2015, Habermannová *et al. loc. cit.*, Bogusch *et al.* 2006).

COELIOXYS: SHARP-TAIL BEES

On a hot July day in 2019, scores of Silvery Leafcutter Bee *Megachile leachella* females were observed returning with pollen to their nest burrows at a large nesting aggregation in sand dunes on the north Norfolk coast. Several small *Coelioxys* females were entering nests and sometimes emerging with a dusting of pollen. These cleptoparasitic bees were clearly targeting *Megachile leachella* nests as a place to lay their eggs. The only *Coelioxys* species previously known to be associated with this small leafcutter bee is a small form of the Square-jawed Sharp-tail Bee *Coelioxys mandibularis*, recorded from small areas of Kent and south Wales. Could this be present in Norfolk and not previously noticed? It turned out that the Norfolk *Coelioxys* were not *Coelioxys mandibularis* but a small form of the relatively common Shiny-vented Sharp-tail Bee *Coelioxys inermis*, a species with no previous records of an association with *Megachile leachella*. The small *Coelioxys inermis* closely matched the size of *Coelioxys mandibularis* parasitising *Megachile leachella* in Wales (Owens 2020a). It seems that both *Coelioxys mandibularis* and *Coelioxys inermis* have large and small forms targeting different hosts, the small form of the two cuckoos both parasitising *Megachile leachella*, though in different locations.

Across the British Isles, some *Coelioxys* species are capable of parasitising four or five different host species, sometimes from two genera – *Anthophora* and *Megachile* – and most host species have more than one *Coelioxys* species parasitising them. These observations pose questions about how a particular host is selected. Do hosts leave chemical traces which *Coelioxys* females recognise when seeking a nest to parasitise? Are size differences within a *Coelioxys* species, such as those just described, merely the effects of differences in the amount of food supplied by small and large hosts, or are there genetic differences between host-specific populations? Do female *Coelioxys* always target the species from whose nest they emerged, by some kind of environmental or genetic imprinting? Host-parasite associations will also be influenced by a host's ability to defend itself against attack.

Chemical cues are likely to be important in detecting a host nest, especially the smell of pollen and nectar within a nest burrow (Vinson *et al.* 2011). At a garden in Wales, a *Coelioxys inermis* female was observed to visit several nests in a bee hotel repeatedly and had 'an unerring ability to home in on holes containing *Megachile centuncularis*' (Becker 2018). Bee hotels can provide confirmation about hosts of sharp-tail bees if individual nest tubes are monitored (Owens *loc. cit.*, C. Boyce 2022 & pers. comm.).

Male *Coelioxys* sometimes patrol flower patches where females may come to seek nectar, though males are also seen at host nest sites. Males have spines on

TABLE 21. Known or suspected associations between *Coelioxys* species and their hosts in the British Isles.

	Anthophora bimaculata	Anthophora furcata	Anthophora quadrimaculata	Megachile centuncularis	Megachile circumcincta	Megachile leachella	Megachile ligniseca	Megachile maritima	Megachile versicolor	Megachile willughbiella
Coelioxys conoidea								✓		
Coelioxys elongata	✓				✓		✓	✓		✓
Coelioxys inermis				✓			✓	✓	✓	
Coelioxys mandibularis						✓	✓			
Coelioxys quadridentata		✓	✓							✓
Coelioxys rufescens	✓	✓			✓	✓	✓			✓

the tip of the abdomen which are possibly used in mating or territorial defence, but this has not been confirmed (see *Anthidium manicatum*, Chapter 2). Once mated, females begin looking for host nests. In June, a female Shiny-vented Sharp-tail Bee *Coelioxys inermis* was observed (NWO) searching systematically along the walls of a house, pausing when she detected irregularities in bricks and tiles. She scanned the bricks while flying 3–5 cm from the surface, slowing to make more systematic zigzag searching movements at any potential nest entrance of a *Megachile* host. By contrast, a bee of the same species at the same place in July spent most of her time taking nectar from flowers or waiting motionless near active host nests. She remained still for minutes on end, close to a bee hotel where several leafcutters were nesting, sometimes grooming or flying a short distance to a new vantage point. From time to time, she flew to a nest entrance, paused for a few seconds and sometimes entered briefly. It appeared that she was waiting for a nest cell to be at the ideal stage before laying an egg. The bee periodically flew off over the garden wall, perhaps visiting other nest sites in the locality.

The name *Coelioxys* refers to the pointed tail of the female. The point comprises the sixth tergite and fifth sternite at the tip of the abdomen and is believed to be used to puncture the leaf cap of a nest cell, in order to deposit an egg. The *Coelioxys* egg hatches quickly, and the second or third instar larva

FIG 291. Shiny-vented Sharp-tail Bee *Coelioxys inermis* female waiting near a host nest.

destroys the young host egg or larva before feeding on the food provisions. The fully grown *Coelioxys* larva spins a tough cocoon in which it remains dormant as a prepupa until the following spring. The prepupa then undergoes metamorphosis to form a pupa and finally an adult. The adult emerges by biting a circular opening in the cocoon, timed to fit with the emergence of its host. The tough cocoon is thought to protect the *Coelioxys* from being chewed by the hosts emerging from the cells behind it. The cell illustrated below was nearest the nest entrance of a Wood-carving Leafcutter Bee *Megachile ligniseca* and was followed by the emergence of a male leafcutter which pushed the whole structure out of the nest tube as he emerged.

FIG 292. Vacated cocoon of a Shiny-vented Sharp-tail Bee *Coelioxys inermis* male, exit hole on the left, within the remains of a leaf which had formed part of a Wood-carving Leafcutter Bee *Megachile ligniseca* brood cell inside a hollow stem.

Leafcutter bees' nests can be parasitised by sharp-tail bees from the early stages, with several brood cells being targeted in the same nest. In one nest observed, *Coelioxys inermis* developed in three out of the ten brood cells present, in positions 2 (the second to be completed), 7 and 10, the outer two being males and number 2 possibly a female (Owens 2020). A North American study (Rosen & Kamel 2008) found that a host cell with a female (fertilised) *Megachile* egg generally received a female *Coelioxys* egg, and a cell with a male *Megachile* egg was matched by a male *Coelioxys* egg. Thus, in both host and parasite, females developed in the inner cells and males in the outer cells of the linear nest.

The behaviour of *Megachile* females seems to be responsive to the threat of *Coelioxys* to some extent. In one observed sequence in NWO's garden, construction of a nest cell by *Megachile ligniseca* took about four minutes, involving about 10 leaf-collecting visits. The bee then departed and returned again to her nest four times within a minute, carrying no materials. Thereafter, she departed for 25 minutes, returning with a pollen load. While the host was foraging for pollen, a *Coelioxys* female had taken up a position close to the nest. The *Megachile* female was inside with her pollen load for 10 minutes, then flew out and in again twice within a few seconds, before departing to collect another piece of leaf, presumably to seal the stocked cell. It appeared that the leafcutter was making frequent checks on her nest just at the time when it was most vulnerable to *Coelioxys*. The *Coelioxys* did not appear to lay any eggs in this instance, though she partially entered the nest several times while the bee was absent and also flew to other nest entrances. The cuckoo was seen to be ejected from the nest by a returning host on one occasion. *Coelioxys* females are most often seen around host nests in fine weather. Unlike pollen-collecting bees, cleptoparasitic bees do not generate much heat through physical activity and rely mainly on basking. A final peculiarity is that both sexes of *Coelioxys* have hairy eyes, a feature possibly associated with protection from dust and pollen, rather like eyelashes (Amador *et al.* 2015).

STELIS: DARK BEES

The genus *Stelis* belongs with *Coelioxys* in the family Megachilidae, but unlike *Coelioxys*, the hosts of *Stelis* do not extend to other bee families. *Stelis* species are grey and inconspicuous, though some have bands or spots on the abdomen. The whole cuticle is covered in deep, dense punctures. There are five species known in the British Isles, one of them very recently recognised (Edwards *et al.* 2019). Some of them are difficult to find and seemingly scarce.

TABLE 22. Hosts of *Stelis* species found in the British Isles.

Species	Common name	Host
Stelis breviuscula	Little Dark Bee	*Heriades truncorum*
Stelis odontopyga	Smooth-saddled Dark Bee	*Osmia spinulosa*
Stelis ornatula	Spotted Dark Bee	*Hoplitis claviventris*
Stelis phaeoptera	Plain Dark Bee	*Osmia leaiana, Osmia bicornis*
Stelis punctulatissima	Banded Dark Bee	*Anthidium manicatum*

The Spotted Dark Bee *Stelis ornatula* provides an illustration of the typical behaviour of the genus. The host is the Welted Lesser Mason Bee *Hoplitis claviventris*, a scarce species found largely in southern counties of England and on the south coast of Wales. In June 2016, some *Hoplitis claviventris* nests were discovered on Devil's Dyke in Cambridgeshire, attended by the cleptoparasite (NWO pers. obs.).

FIG 293. (a) The Welted Lesser Mason Bee *Hoplitis claviventris* and (b) its cuckoo, the Spotted Dark Bee *Stelis ornatula*.

This was a lucky find, as *Stelis ornatula* is rarely seen. The host nests were in a cluster in the top of a sawn-off post amongst tall grass near the Lizard Orchids, which were the object of the visit. Several of the *Hoplitis* nests were already sealed with mastic (plant pulp with secretions added) of a bright brown colour. Other nest entrances were open and probably still in the stage of being excavated, since there were wood chippings around them. Several female *Hoplitis* were resting on the surface or on plant stems nearby, with some entering and leaving nests. Also present were three *Stelis ornatula* females. The host and cleptoparasite are very similar in shape, size and colour, and the white spots on the abdomen of the *Stelis* match white bands on the host, though they are more prominent.

A resemblance to the host is a common feature of the genus *Stelis*, perhaps attributable to a common evolutionary origin, or possibly to convergence by the *Stelis* as a means of avoiding detection. The *Stelis ornatula* females were waiting on the surface of the post and were ignored by the hosts, even when very close

FIG 294. (a) Three brood cells of Kirby's Lesser Mason Bee *Hoplitis leucomelana* parasitised by the Spotted Dark Bee *Stelis ornatula*, showing stages of development, with the youngest cell on the left (redrawn from Höppner 1904). (b) Larva of a Spotted Dark Bee feeding from pollen and encircling the remains of a Kirby's Lesser Mason Bee host larva. (© P. Westrich)

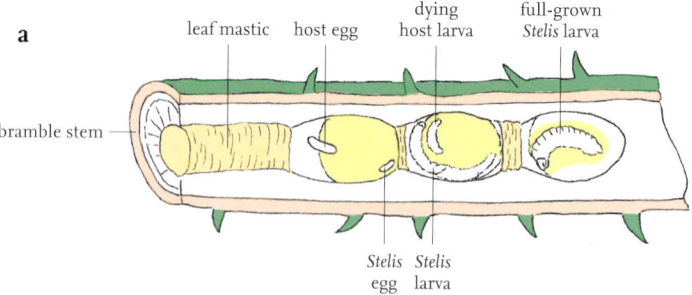

to the entrance of a nest burrow. It was not possible to see events inside the nest burrows. However, in the early 20th century, Hans Höppner, a German school teacher, recorded his careful observations of *Stelis ornatula* in the nests of Kirby's Lesser Mason Bee *Hoplitis leucomelana*, made in bramble stems (Höppner 1904). (*Hoplitis leucomelana* has only one British record, by William Kirby himself.) Höppner was able to open the stems and observe cells at different stages of development with a magnifying glass. He found that the small *Stelis* egg was laid at the base of the pollen provision before that of the host. The host's egg was laid with one end inserted into the completed ball of pollen, and the cell was then sealed with leaf mastic. A passage through the pollen indicated that the *Stelis* egg had hatched first, and the larva had worked its way through the food to the (smaller) *Hoplitis* larva and then attacked it. It inserted its mandibles into the fourth segment of the host larva and began to suck out the body contents. The host larva continued to move for quite a long time but was unable to retaliate. More recently, the larva of *Stelis ornatula* has been photographed *in situ* in Germany by Paul Westrich.

The behaviour of *Stelis ornatula* follows the pattern of nearly all members of its genus in placing its eggs into a nest while it is still being provisioned by the host (Michener 2007). *Stelis nasuta*, a small species found in mainland Europe, seems to be an exception. Fabre (2014) noted that this cuckoo bee uses its mandibles to bite into the thick mud seal of its much larger host, *Megachile pyrenaica*. *Stelis nasuta* lays several eggs at a time and then re-seals the entrance she has made. In Fabre's Provence, the repair was often made with red-coloured mud which showed thereafter as a red spot, indicating that the cell had been parasitised. More recent studies show that the number of eggs in one host cell can range from 2–12, with a corresponding variation in the size of the adults (Westrich 2018).

The Little Dark Bee *Stelis breviuscula* is an exception to the generalisation that *Stelis* species are scarce. It is a cleptoparasite of the Large-headed Resin Bee *Heriades truncorum*, a small, very active bee which nests in beetle holes and hollow stems and seals its nest with resin and pebbles. It specialises in Asteraceae flowers and these are frequently attended by its cuckoo. *Stelis breviuscula* is very similar to *Heriades truncorum* in both appearance and size. It was first recorded in the British Isles in the 1980s in West Sussex at a time when the host was rare (Else 1997, Else & Edwards 2018). Since that time, the range of *Heriades truncorum* has expanded northwards as far as Oxfordshire, Essex, Cambridgeshire and Norfolk. *Stelis breviuscula* can appear at bee hotels in gardens, where it mingles with the host, often with no apparent antagonism (TB). The cuckoo generally appears within a year or two of the arrival of its host and was seen as far north as Norwich in 2022.

FIG 295. (a) The Large-headed Resin Bee *Heriades truncorum* and (b) its cuckoo, the Little Dark Bee *Stelis breviuscula*.

FIG 296. Females of (a) the Plain Dark Bee *Stelis phaeoptera* and (b) the Banded Dark Bee *Stelis punctulatissima*. (© V. Bartlett & © M. Fogden)

The Smooth-saddled Dark Bee *Stelis odontopyga* is a more recent arrival. The first records for Britain came in 2018: a male from Kent and a female from Oxfordshire, both in June. A third specimen from the same site in Kent in 2017 was found in a museum collection (Edwards *et al*. 2019). The authors conclude that the wide separation of these first records indicates a small but widespread population of *Stelis odontopyga* which had gone unnoticed. Its host in mainland Europe is the Spined Mason Bee *Osmia spinulosa*, which nests in empty snail shells in calcareous grasslands and is another Asteraceae specialist. In Britain, *Osmia spinulosa* occurs largely in southern and eastern English counties and on the south Wales coast.

The Plain Dark Bee *Stelis phaeoptera* has also been recorded much more frequently in recent years as a result of its appearance at bee hotels, the established host in this case being the Orange-vented Mason Bee *Osmia leaiana*. However, in June 2020, a *Stelis phaeoptera* was observed emerging from a cocoon of *Osmia bicornis* at a bee hotel in a Montgomeryshire garden: a new host record for this *Stelis* (C. Boyes pers. comm.). The somewhat larger Banded Dark Bee *Stelis punctulatissima* targets the Wool Carder Bee *Anthidium manicatum* and is rarely recorded, though the impression of scarcity of *Stelis* species is likely to be partly an artefact of their dark colouring and stealthy behaviour.

CUCKOO NUMBERS

The presence of cuckoo bees is usually a good sign in that it shows that the host population is in a relatively healthy state (Sheffield *et al*. 2013). Some cuckoo bees in the British Isles are apparently very scarce and some are in danger of extinction (Table 23). It is necessary to be cautious because, as mentioned, cuckoo bees can be very difficult to spot. They may be apparent at host nests only for a short period of the breeding cycle for limited periods of the day, appearing when the weather is suitably warm. They are also well camouflaged and remain motionless for long periods. Moreover, much of their time is spent inside nests rather than outside. Populations of cuckoo bees may wax and wane: host bees and their cuckoos are in an evolutionary arms race in which their means of attack and defence are continually modified. Each cuckoo bee species is likely to be in competition with other cuckoo bees sharing the same host or with non-bee cleptoparasites (see Chapter 7). Cuckoo bees often occupy only some parts of the range of their host(s), and some potential hosts appear to have no cuckoo bees associated with them at all, perhaps because the host has acquired a (temporary) cuckoo avoidance strategy.

TABLE 23. Some scarce cuckoo bees in the British Isles.

Cuckoo species	Host	Comments
Coelioxys mandibularis	Megachile leachella	Dunes in Lancashire, Wales, Kent; host much more widespread
Coelioxys quadridentata	Anthophora and Megachile spp.	Rare; Megachile circumcincta greatly declined
Melecta luctuosa	Anthophora retusa	Probably extinct; host greatly declined
Nomada alboguttata	Andrena barbilabris	Kent and Sussex; first recorded Kent 2016; recent colonist
Nomada argentata	Andrena marginata	Rare; Dorset, Salisbury Plain, Surrey, Breckland; host declined
Nomada armata	Andrena hattorfiana	Rare; Salisbury Plain, Oxfordshire, Buckinghamshire; host declined
Nomada bifasciata	Andrena gravida	First recorded Kent 2018; perhaps a recent colonist
Nomada errans	Andrena nitidiuscula	Known from one site in Dorset; last recorded 1980s
Nomada facilis	Andrena fulvago	Recently recognised in the British Isles from seven sites in southern England
Nomada ferruginata	Andrena praecox	Very scarce until a resurgence from the 1980s
Nomada guttulata	Andrena labiata	Scattered records in southern England
Nomada obtusifrons	Andrena coitana	Declined; scattered records across the British Isles
Nomada roberjeotiana	Andrena tarsata	Declined; recent records largely from Cornwall
Nomada sexfasciata	Eucera longicornis	Declined; recent records only from south Devon
Nomada signata	Andrena fulva	Scarce; possible competition from Nomada panzeri
Nomada subcornuta	Andrena nigrospina	Rare; host declined
Nomada zonata	Andrena dorsata	First British record 2010; rapidly increasing
Sphecodes spinulosus	Lasioglossum xanthopus	Mostly southern England; probable decline
Stelis odontopyga	Osmia spinulosa	Scattered records in southern counties
Stelis phaeoptera	Osmia leaiana	Rare but recently more frequent at bee hotels

It is estimated that about 15 per cent of the world's 20,000 described bee species are cuckoo bees. If every pollen-collecting bee species had one bee cleptoparasite, this would indicate that each cuckoo bee has an average of about seven different hosts. Some hosts probably have no cuckoos at all, and others have more than one, so this calculation is very approximate. As we have seen, some cuckoo bee genera are more flexible than others in their host choices. Moreover, a cuckoo bee's hosts tend to differ between the various parts of its geographical range, partly because of variations in the population density of potential hosts and partly because of locally adapted behaviour. In addition, there is a trend towards there being a greater percentage of cuckoo bee species in high latitudes (Wcislo 1987). The cuckoo bees of the British Isles, including cuckoo bumblebees, comprise about 23 per cent of bee species; well above the world average. One suggested reason for this north–south divide is the shorter and more synchronised breeding season at high latitudes, leading to greater competition for resources; lack of resources would favour bees taking over nest provisions rather than seeking their own (Michener 2007). However, Michener offers the alternative explanation that the trend could simply be a result of the abundance of *Andrena* species in northern latitudes, each with their attendant *Nomada* cuckoo.

In the course of evolution on a world scale, cuckoo bees have arisen at least 30 times from pollen-collecting species. This is a minimum figure since we do not know how many groups of cleptoparasitic bees may have evolved and become extinct before the present time. In the British bee fauna, only the family Colletidae does not include any cuckoo bee species and there are very few from this family worldwide, with just a small number of *Hylaeus* cuckoos occurring in Hawaii and none from the genus *Colletes* itself. The lack of cuckoos in this genus has yet to be explained; perhaps *Colletes* are not well preadapted to a cuckoo lifestyle. Having strong mandibles is one prerequisite to becoming a cuckoo bee. Megachilidae lack piercing mouthparts since their mandibles are adapted to cutting or chewing leaves. They are therefore poor at digging and this confines their nests to existing cavities or soft substrates. In turn, cuckoo bees from this family in the genera *Stelis* and *Coelioxys* largely attack other Megachilidae and have not extended their host range to mining bees.

Table 24 summarises the characteristics of the six genera of cuckoo bees described in this chapter. As we have seen, cuckoo bees show a suite of common characteristics. Some of these common features are structures which pollen-collecting bees lack. The loss of the pollen-collecting hairs or scopae is perhaps the most obvious cuckoo characteristic and the easiest to explain in terms of the lack of pollen-collecting behaviour by cuckoo bees. It is analogous to the loss of functional eyes in animals that live in totally dark places, such as caves and the deep sea. The loss of unused organs saves the energy needed to produce them and economises on the genetic mechanisms that control their development, so they tend to disappear through natural selection (Michener 2007). Brief explanations for other common characteristics are given in the table. As we have seen, common characteristics across cuckoo bee genera are believed to have arisen because bees have become similarly adapted to a cleptoparasitic way of life over many millions of years, by the process known as convergent evolution. But differences between the six groups nevertheless occur because each genus has its

TABLE 24. Some characteristics of cuckoo bee genera.

	Melecta	Coelioxys	Nomada	Epeolus	Stelis	Sphecodes	Explanation
Structure							
No scopa	✓	✓	✓	✓	✓	✓	Does not collect pollen
Lack of body hair		✓	✓	✓	✓	✓	Exposed cuticle acts as warning coloration
Small Dufour's gland		✓	✓	✓	✓	✓	Does not line nest cells with secretions
Strong, punctured exoskeleton	✓	✓	✓	✓	✓	✓	Resists attacks from host
Pointed mandibles in adult	✓	✓	✓	✓	✓	✓	Assists with defence and entry to nests
Pointed/sharp tip of abdomen	✓	✓	✓	✓	?		Penetrates nest cells to lay eggs
Basitibial plate poorly developed	✓	✓	✓	✓	✓	✓	Assists with excavation of a nest burrow
Protective coloration	✓	✓	✓	✓	✓	✓	Protection or warning to host and/or predators
Eggs much smaller than those of host	✓	✓	✓	✓	✓		Small eggs easily hidden from host
Mobile larva with strong/large jaws	✓	✓	✓	✓	✓		Destroys or eats egg/larva of host or other cuckoo bee
Larva spins strong cocoon	✓	✓			✓		Prevents emerging host bees chewing through the cuckoo larva/pupa before it emerges

own way of being a cuckoo, inherited from the earliest types of its kind, which retained their special features as they diversified to parasitise a wide range of hosts. For example, *Sphecodes* species kill their hosts rather than using subterfuge. Presumably the first *Sphecodes* behaved in this way and their descendants continue to behave similarly. There are also differences within a genus, with each species adapting to its particular host(s).

Much of the behaviour of cuckoo bees is still poorly understood, but we can find out more simply by watching and waiting.

Behaviour	*Melecta*	*Coelioxys*	*Nomada*	*Epeolus*	*Stelis*	*Sphecodes*	Explanation
Attracted to odour of host bee and/or the pollen or oil it collects	✓	✓	✓	✓	✓	✓	Helps to find host or host nest
Mimics host smell	?	?	✓	?	?		Works as a disguise
Usually ignored or tolerated by host		?	✓	✓	?		A result of chemical disguise, warning colours or camouflage of cuckoo bee
Physical conflict sometimes occurs	✓	✓	✓	✓	✓	✓	Host recognises threat to its nest and retaliates
Waits motionless at nest sites	✓	✓	✓	?	✓		Reduces chances of alerting host and predation
Opens or widens nest cells	✓	✓				✓	Occurs in species which parasitise closed cells
Lays a large number of eggs which hatch quickly	✓	✓	✓	✓	✓	✓	Cuckoo larva able to destroy/eat host egg/larva
Less protandrous than host	✓	✓	✓	✓	✓		Timing of emergence of female cuckoo bee coincides with pollen-collecting phase of host

continued overleaf

TABLE 24. *continued*

	Melecta	Coelioxys	Nomada	Epeolus	Stelis	Sphecodes	Explanation
General							
Mainly targets members of the same bee family	✓	✓			✓	✓	A consequence of the evolution of cuckoo bees from pollen-collecting ancestors
Host-specific to a varying extent	✓	✓	✓	✓	✓	✓	Specialist and generalist behaviour both have advantages and disadvantages
Cuckoo requires higher temperatures than host	✓	✓	✓	✓	✓	✓	Cuckoo bees less active and have less hair

CHAPTER 9

Time, Space and Temperature

'The rapid development, as far as we can judge, of all the higher plants within recent geological times is an abominable mystery.'
—*Charles Darwin, in a letter to Joseph Hooker (1879)*

Time, space and temperature are ever-present themes in ecological studies and are especially relevant to solitary bees. In this chapter, we look first at the origins of bees in geological time, seeking possible candidates as ancestors of bees in the Jurassic and Cretaceous periods, when flowering plants first appeared on Earth. This will lead to issues more related to space, examining the diversity and distribution of our own bee fauna in the context of bee populations and diversity wordwide. Temperature is the next focus, reviewing evidence about the climatic tolerance of bees and the extent to which they are able to control the temperature of their body. We then look at how solitary bee life cycles and behaviour are synchronised with seasonal changes and examine how bees survive unfavourable conditions. Lastly, putting these ideas together, we come to the very topical issue of populations and range movements in response to climate and other factors.

THE ORIGINS OF BEES

Where do bees come from? Finding an answer will inevitably shed light on the origins of flowering plants, since the lives of bees and flowers are so intricately entwined. Firstly, we need to consider the lifestyle of bees and look for similarities

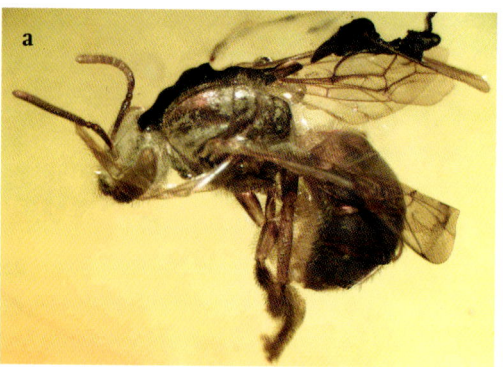

FIG 297. Examples of fossil bees: (a) *Ctenoplectrella viridiceps* in Eocene Baltic amber and (b) *Oligochlora eikworti* in Early Miocene amber, Dominican Republic. (© M. Engel)

amongst possible bee ancestors. It is well known that adult bees take all their nutrition from plants, almost entirely in the form of nectar and pollen, which they also supply to their larvae. The closest living relatives of bees are the stinging wasps. Consequently, bees and wasps are members of the same taxonomic group, the Aculeata (aculeates), in the insect order Hymenoptera – along with ants. The key characteristic of aculeates, implied by the name, is that females have the ability to sting. The earliest fossils of bees and wasps come from the Cretaceous period, which dates from 145.5–65.5 Ma (*mega-annum*, or millions of years ago), represented in Britain by strata of chalk, greensand and clay. Good wasp fossils have been found in the Wealden clay of south-east England (Baldock 2010, Radnitsyn *et al.* 1998). Fossils of wasps pre-date fossils of bees, which is consistent with bees having evolved from wasps. Aculeate wasps, however, are nearly all carnivores and use paralysed invertebrate prey to feed their larvae. It appears that the first bee-like insect was a type of wasp which transferred from hunting prey to collecting pollen.

Recent molecular studies confirm that bees living today are closely related to wasps in the families Crabronidae and Sphecidae, which, as we have seen, are classified with bees in the superfamily Apoidea. Further comparisons of genetic material from the Apoidea indicate that bees should be placed within the Crabronidae; in other words, wasps in the family Crabronidae are more closely related to bees than they are to members of other wasp families (Brady *et al.* 2009). At the same time, some misfits have been removed from the original family Crabronidae. Following this trail, molecular evidence shows that, within the revised Crabronidae, the subtribe Ammoplanina, in the tribe Pemphredonini, is the group most closely related to modern bees (Sann *et al.* 2018). There are about 130 species of Ammoplanina worldwide, with some in mainland Europe,

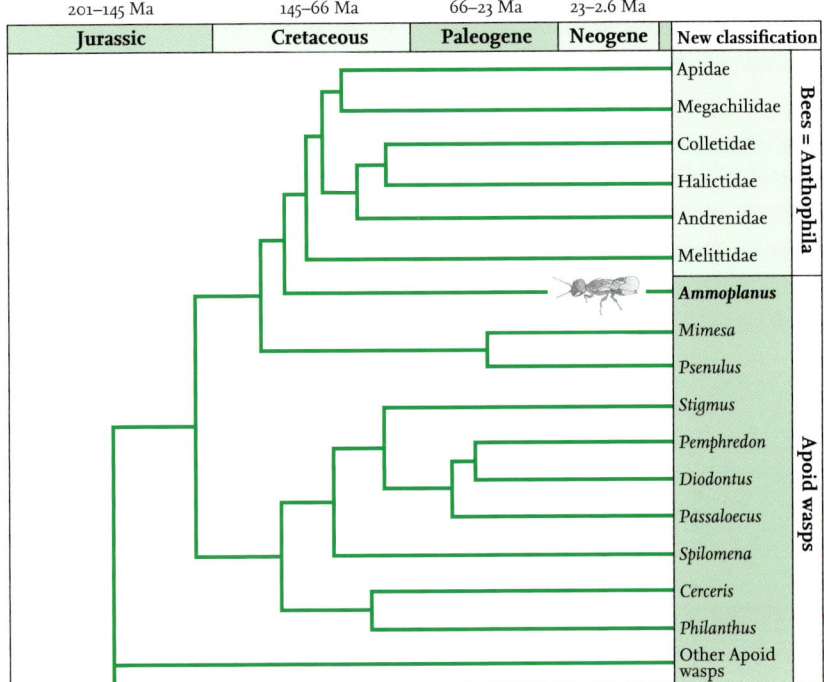

FIG 298. Evolutionary tree of modern bees and wasps (Based on Sann *et al.* 2018; the small non-British bee family Stenotritidae is omitted). (Ma = million years ago)

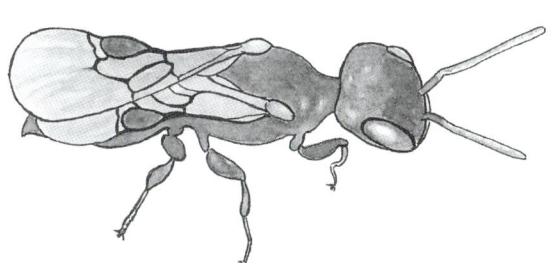

FIG 299. *Ammoplanus* species, the type of aculeate wasp most closely related to modern bees and possibly similar to bee ancestors. *Ammoplanus* is about 2 mm long and preys on thrips. The illustration is based on photographs of European species.

though none in Britain. Many belong to the genus *Ammoplanus*. All are very small – about 2 mm in length – and prey on tiny insects such as thrips ('Thysanoptera'). Thrips mostly go unnoticed but include 'thunderflies', the annoying minute black insects which rise from wheat fields on humid summer days and land on our skin. Significantly for the current story, some thrips feed on pollen (Kirk 1996).

If we follow the tree back in time, we can see that bees (Anthophila) and *Ammoplanus* shared a common ancestor in the early Cretaceous period. Evolutionary trees such as this can be generated by comparing the DNA of modern species, but this does not provide reliable dates for each of the tree's branches. It is necessary to put dates on the nodes (branching points) in order to discover when particular bee or wasp families first evolved. Dating the nodes requires the discovery of fossils which are of known age and identifiable as representatives of our modern bee families. Fortunately, there is extensive fossil evidence of Apoidea from the Cretaceous, including fossils of bee and wasp families recognisable as those present today. It seems that all modern bee families (and some now extinct) had evolved by the end of the Cretaceous, diversifying from an ancestral group present in the early Cretaceous, 120–112 Ma (Brady *et al. loc. cit.*). Therefore, the conclusion can provisionally be drawn that the first bees evolved early in the Cretaceous period from wasps, which were similar to modern *Ammoplanus*. Significantly, the Cretaceous is the time when flowering plants diversified in a dramatic way.

HUNTING FOR THE PROTOBEE

Changing a wasp into a bee requires a fairly major reconstruction. What should we look for in a 'protobee', a wasp on its way to becoming a bee? First, we might look for the most 'primitive' bees present today. For a time, the Colletidae were considered to be the ancestral group which gave rise to other bee families, partly on the basis of their lobed tongue, which resembles some wasp tongues. However, DNA analysis now places Melittidae as the ancestral group and Colletidae as a derived group. A proposed model for an ancestral bee similar to *Hylaeus* (Colletidae) – which ingested nectar and pollen and nested in a hollow stem – has therefore now been discounted. Instead, we should probably be looking for evidence of a more Melitta-like protobee with pollen-collecting hairs, provisioning larvae with semi-solid food and nesting in a burrow. A protobee might also have wide basitarsi. This part of the lower leg is narrow in wasps but widened in bees and is used to collect and manipulate pollen (Radchenko & Pesenko 1996). Recently, a fossil from Burmese (Myanmar) amber (fossilised tree resin) was described, dating from the Cretaceous (Poinar & Danforth 2006, Danforth & Poinar 2011). This insect, named *Melittosphex burmensis*, appeared to be a transitional form between wasps and bees. Critically, it apparently has a body covering of branched hairs, clearly visible in the amber (a feature unique to bees). It also has a dorso-ventrally flattened body, similar to a modern bee, but the basitarsi are narrow and wasp-like. It is about 3.5 mm long and shares some features with the Pemphredonini, the wasp group which includes *Ammoplanus*.

Melittosphex burmensis seems a good candidate for a protobee. A scenario has been proposed whereby a thrips-hunting wasp, rather like *Ammoplanus*, might transfer from hunting prey to collecting pollen and become a protobee, similar to *Melittosphex*. Thrips which have been predated by a wasp and stocked in a brood cell are very likely to be contaminated with pollen, allowing the opportunity for the wasp larvae to feed on pollen as well as the thrips' bodies. Since pollen is a rich source of protein, wasps carrying extra pollen back to brood cells would provide more protein for their offspring, potentially leading to selection for this behaviour and ultimately a transition to using pollen alone (Sann *et al. loc. cit.*). It must be added, however, that this particular fossil specimen (*Melittosphex burmensis*) has been interpreted by others as a chrysidid wasp (jewel wasp) rather than representing a stem group to bees, and it has been suggested that defects in the amber were misinterpreted as branched hairs. The earliest definitive fossil bee is *Cretotrigona* from late Cretaceous New Jersey amber (M. Engel pers. comm.). All known fossil bees have been catalogued by Michez *et al.* (2011).

Controversies over fossils do not invalidate the evolutionary scheme just described. William Kirk, a thrips specialist, comments: 'thrips were probably abundant in cycads and other gymnosperms before the evolution of the angiosperms, so perhaps wasps hunted thrips there first and then started hunting them in flowers as well, when the angiosperms [flowering plants] radiated'. Gymnosperms include modern conifers as well as cycads and dominated the world's vegetation for about 200 million years, before flowering plants appeared. The pollen of conifers is generally dispersed by the wind. Their reproductive organs do not bear petals and their seeds are not enclosed in fruits, unlike those of flowering plants. However, cycads rely on insects for pollination, and some Australian species are totally dependent on particular species of thrips. Thrips appeared in the Permian period (240 Ma) alongside cycads and may be amongst the oldest pollinators of plants (Terry 2001, Terry *et al.* 2007, Brookes *et al.* 2015). The knowledge that thrips take pollen from gymnosperms raises the possibility that bees, in a form similar to the proposed protobee, were ready and waiting as the flowering plants emerged and potentially contributed to their evolution from the start.

Charles Darwin feared that the apparently rapid evolution of the flowering plants would undermine his theory of evolution by natural selection, which he saw as a very gradual process. The problem has not been entirely resolved, though modern ideas include the proposal that 'genome downsizing' allowed flowering plants to have more compact, better organised cells, resulting in more rapid growth than their competitors (Simonin & Roddy 2018). A recent review of the fossil record alongside molecular evidence points to the origin of some flowering plant families as early as the Jurassic, with a rapid diversification following in the Cretaceous (Silvestro et al. 2021). It remains generally accepted that the abundance, diversity and beauty of flowering plants can be attributed in large part to their long associations with bees and other insects.

BEE DIVERSITY

Modern bee families apparently survived the mass extinction that ended the Cretaceous period and which saw the demise of the dinosaurs. We are part of the rich taxonomic diversity which has since arisen (yet dangerously unconcerned about our dependence on its preservation). Fortunately, there is now a wide recognition of the importance of bees as pollinators, as well as a growing appreciation of bees as a source of inspiration and scientific interest. As mentioned, there are currently around 20,000 named bee species worldwide, with an unknown number yet to be described. We urgently need a better understanding of bee biology and the factors contributing to their diversity.

A wide range of organisms, including most insects, show the highest diversity near the equator, in tropical forests, with diversity steadily declining towards the poles; there is a latitudinal gradient (Rosenweig 1995). However, bee diversity shows an unusual pattern, with relatively few species near the equator and more in temperate zones. The highest abundance and number of bee species ('hotspots') occur in seasonally dry (xeric), warm temperate areas (Michener 1979, 2007). The most diverse regions for bees have a Mediterranean climate, characterised by cool, wet winters and warm, dry summers. Apart from the Mediterranean basin itself, major bee hotspots include the Chaparral of California, the Fynbos of South Africa, the Kwongan of south-western Australia and the Matorral of Chile. These regions typically have a rich diversity of shrubs and herbaceous plants but lack extensive closed canopy forests. These hotspots have long been recognised, yet the overall world pattern of bee diversity has remained unclear, owing to regional variations in levels of knowledge: some regions may appear to be richer than others simply because they have received more attention. A recent study (Orr et al. 2021) involving the analysis and validation of nearly 6 million bee records from across the globe provides a more detailed world picture of bee biodiversity (Fig. 300).

The map confirms the existence of a strongly bimodal latitudinal gradient of bee diversity across the globe, with peaks in mid-latitudes of both the northern

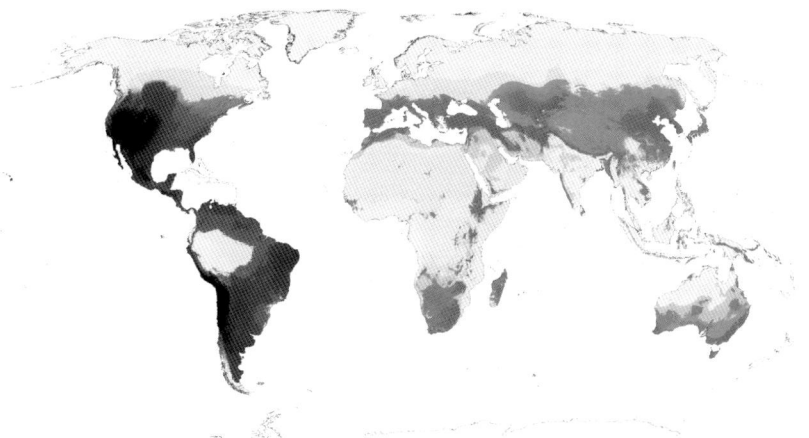

FIG 300. Estimated relative bee species richness worldwide (darker is more) (Orr *et al.* 2021). (© A. Hughes)

and southern hemispheres but a relative poverty of species in tropical zones. The study showed that 'ideal' bee habitats have plenty of sun, helping plant growth and bee temperature regulation, but also have enough moisture to allow plant and bee survival. Mediterranean-type climates as well as semi-desert environments were associated with the highest bee diversity, whereas trees and forests, even in tropical areas, did not generally support so many bee species. The warm, dry habitats 'preferred' by bees make sense in terms of collecting pollen, since pollen is gathered on body hairs and this becomes difficult in damp conditions. Any moisture causes the branched hairs of bees to cling together, reducing their ability to trap pollen grains. Electrostatic attraction also plays a large part in pollen collection but cannot operate in humid conditions (Clarke *et al.* 2017). Dry, bright weather with high rates of evaporation is ideal for foraging. The larval stages of bees also benefit from warm, xeric conditions, allowing bees to remain dry and avoid fungal attack within their brood cells.

Spring in the Mediterranean region brings a burst of flowers from February to April, when many bees are actively nesting. From May or June, daytime temperatures often reach 30 °C or more, and many herbaceous plants wither and go to seed. By this time, bees' eggs and larvae are developing within plant stems or underground, largely free from any damage by moisture during the hot, dry summer months. Further rain in autumn brings a renewal of flowering, and some bees emerge at this time and are specialists of annual plants that appear late in the season. Spring-nesting bees are now likely to have reached the prepupa stage, which is resistant to fungal attack. By contrast, Britain is cold and damp, restricting bee activity to the small proportion of warm, dry days when pollen can be collected and demanding careful selection of microhabitats for nesting. Features of nest design help to protect their occupants from moisture and fungi (see Chapter 3). It would seem self-evident that bees require habitats with plenty of bee-attracting flowers. It might, however, be more accurate to stand the argument on its head: bee-pollinated plants will tend to evolve largely in habitats which are good for bees (Rosenweig 1995). In the tropics, the highly social stingless bees (Meliponini) are abundant and diverse and may outcompete other bee genera in the tree canopy. They are tropical specialists, with a true pollen basket (corbicula) – otherwise seen only in honeybees and bumblebees – and their nests are made from water-resistant tree resin (Roubik 1989).

BEE DIVERSITY IN BRITAIN

Measuring diversity is not straighforward. We might start by counting the number of bee species in a garden, a nature reserve or a county. Could we then express bee diversity in terms of bee species per km^2? No – if we double the area under consideration, the number of species does not double: it will increase by much less. As we increase the area sampled, the increments get smaller and smaller. Figure 301 (a) provides an illustration. Points on the graph (left to right) refer to (1) a small garden (NWO), (2) the parish of Weybourne, in which the garden is situated, (3) the county of Norfolk, (4) Norfolk, Suffolk and Essex combined, (5) the east of England: Norfolk, Suffolk, Essex, Cambridgeshire, Bedfordshire and Hertfordshire (Jackson 2019), (6) England and (7) Britain: England, Wales and Scotland. These are nested areas; the smallest area fits inside the next biggest one and so on up the scale (Rosenweig *loc. cit.*). It is clear from the shape of the graph that we will get very different estimates of species per unit area depending on the size of the study area chosen. Small areas such as a garden will give much higher estimates of diversity than larger areas such as a county. Comparisons are more meaningful when the areas being studied fall on the more horizontal parts of the graph; in this example, from about the area of Norfolk (around 5,000 km^2 – point 3) upwards. Beyond this size of sampling area, the graph is more or less flat, so comparisons between the diversity of different counties, for example, do have some validity. (In the world map, described above, areas of less than 5,000 km^2 were excluded). In an ideal review of biodiversity, comparisons can be made using standardised surveys covering fixed areas, such as squares of 10 km (hectads), but this has not been attempted on a significant scale for bees. If the data on the x-axis are plotted on a logarithmic scale, we get an approximately straight line (Fig. 301 b).

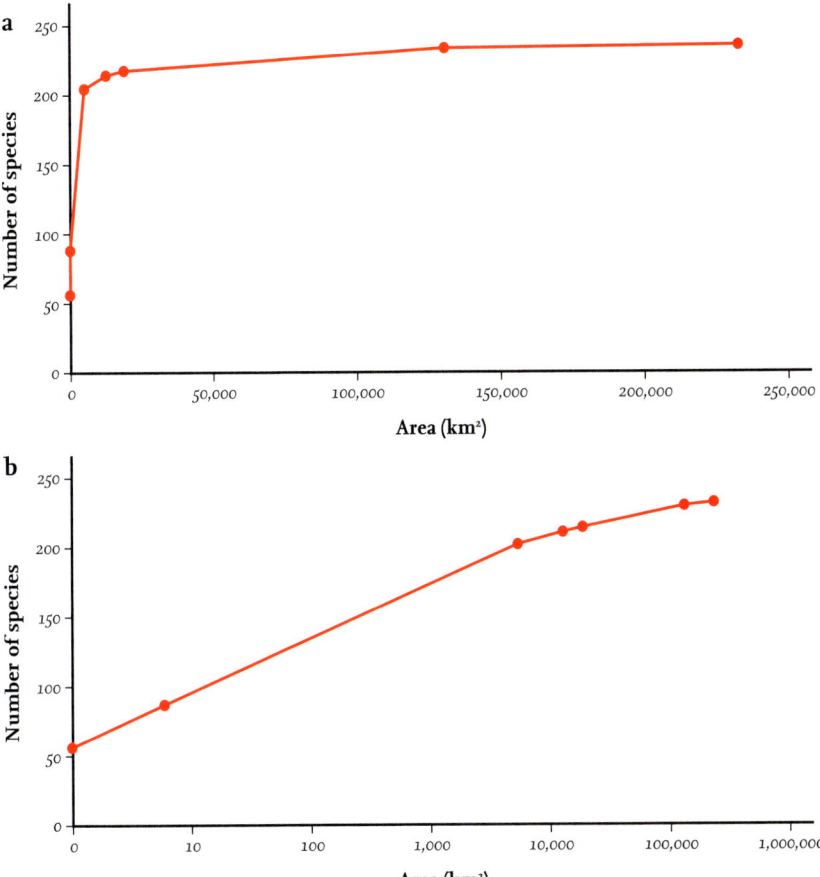

FIG 301. (a) Solitary bee species per km² in relation to area sampled (see text for explanation). (b) The same data with area plotted on a logarithmic scale.

Britain lies well to the north of Europe's main bee hotspots. Our most bee-rich habitats are those with quickly warming and well-draining soils, such as heathland, sand dunes and chalk grasslands. Much of Britain remains relatively well supplied with hedgerows and copses and these provide a wealth of flowering plants, such as Blackthorn *Prunus spinosa* and hawthorn (*Crataegus* species), much favoured by spring-emerging solitary bees. It is notable that the Brecklands of Suffolk and Norfolk, which are founded on wind-blown glacial sands, serve as a northern outpost for several bee species in Britain, for example, the Grey-tailed Furrow Bee *Lasioglossum prasinum*, the Ashy Furrow Bee *Lasioglossum sexnotatum* and, until their recent range expansion, the Green-eyed Flower Bee *Anthophora bimaculata*, the Red-tailed Mason Bee *Osmia bicolor* and the Red Bartsia Blunthorn Bee *Melitta tricincta*. Breckland also provides an inland habitat for a number of species otherwise found largely on the coast, notably the Margined Colletes *Colletes marginatus*, the Silvery Leafcutter Bee *Megachile leachella* and the Coast Leafcutter Bee *Megachile maritima*. Coastal temperatures are less extreme, and the distributions of several species show a coastal bias. In addition to those just mentioned, these include the Cliff Mining Bee *Andrena thoracica* and the Gold-fringed Mason Bee *Osmia aurulenta*. Cities behave as heat islands, and there are a few species which seem to favour cities over the countryside, notably the Four-banded Flower Bee *Anthophora quadrimaculata* (Bartlett & Bartlett 2018) and its cuckoo, the Grooved Sharp-tail Bee *Coelioxys quadridentata*. The European Orchard Bee *Osmia cornuta* first established itself in London's suburbs.

British solitary bees are clustered within favourable habitats and microhabitats, but there are also broader regional patterns, including our own local latitudinal gradient of diversity from south to north. We can quantify this by comparing the ten 100 km horizontal bands of the national grid of the Ordnance Survey map, which cross England, Wales and Scotland. The orientation of the more or less 'horizontal' national grid lines approximates closely to latitude. Figure 302 shows the number of solitary bee species present in each east–west 100 km band, based on the species maps in Else and Edwards (2018). The graph includes 234 species of solitary bees established in Britain. Bumblebees *Bombus* species and the Honeybee *Apis mellifera* are excluded. Band 1 crosses the south

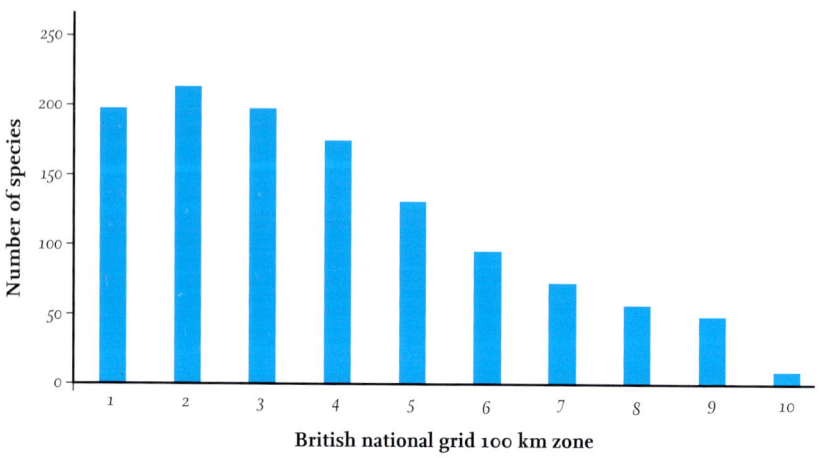

FIG 302. The number of solitary bee species recorded in each 100 km latitude zone of Britain.

coast of England (parts of Cornwall, Devon, Dorset and Hampshire), and band 10 crosses the north of Scotland (the Outer Hebrides, Sutherland and Caithness). The 100 km bands cover very different areas of land, but the consequent errors have only a small effect because the land area within every band is large enough to fall within the more horizontal part of the species/area graph. (The small area covering the south coast included in band 1 does depress the scores slightly.)

There is clearly a very marked south to north trend of declining solitary bee diversity with increasing latitude in Britain, with southern counties and south Wales being the most diverse regions for solitary bees. The three most southerly bands average over 200 solitary bee species from 26 genera, whereas the most northerly three bands average just 40 solitary bee species from four genera.

The ratio of the number of genera to the number of species remains roughly constant from south to north, suggesting that the loss of genera with latitude is simply down to chance. The percentage of cuckoo bee species remains between 20 and 30 per cent from south to north with some suggestion of a northward decline (Fig. 303). No cuckoo species has been recorded in the most northerly band (parts of Sutherland and Caithness). Only two solitary bee species, the Small Flecked Mining Bee *Andrena coitana* and the Tormentil Mining Bee *Andrena tarsata* have been recorded in Orkney, with none in Shetland. The Outer Hebrides host six solitary bee species: the Tormentil Mining Bee, the Northern Mining Bee *Andrena ruficrus*, Clarke's Mining Bee *Andrena clarkella*, the Northern Colletes *Colletes floralis*, the Heather Colletes *Colletes succinctus* and the Bloomed Furrow Bee *Lasioglossum albipes* (B. Neill pers. comm.).

Comparisons show that the latitudinal gradient of the species-rich genera *Andrena*, *Nomada*, *Lasioglossum* and *Sphecodes* keep remarkably in step from the south to the north of Britain, suggesting that the decline of diversity with latitude is related largely to climate and is not affected by differences in the behaviour and physiology of each genus or between hosts and cuckoo bees. Ground-nesting and cavity-nesting bees show similar latitudinal gradients. These observations are consistent with the much less extreme relationship between bumblebee (*Bombus* species) diversity and latitude; bumblebees are able to generate and conserve heat in their bodies very effectively (Prŷs-Jones & Corbet 2011). The main effects of latitude seem to be the limits it imposes on bee activity, foraging and food availability, and climate may also influence the survival of immature stages. In general, there is a trend towards a higher proportion of diptera species and fewer bees at higher latitudes (Olsen 2005).

In northern latitudes, the supply of food is slower and there is a shorter season (but greater day-length) in which to find it. At the northern edge of its range, a solitary bee will only just manage to maintain its numbers and will be

TABLE 25. Number of solitary bee species of each genus recorded in each 100 km national grid band of Britain (England, Wales and Scotland) and total number of solitary bee species present in the island of Ireland, Scilly, Orkney and Shetland.

Sub-family	Genus	Species in Britain *	\multicolumn{10}{c	}{Species in each 100 km grid band south to north}	Island of Ireland	Scilly	Orkney	Shetland
			1	2	3	4	5	6	7	8	9	10				
Andrenidae	Andrena	62	55	59	56	48	37	29	22	21	19	5	24	13	2	
Andrenidae	Panurgus	2	2	2	2	2	1									
Apidae	Anthophora	5	5	5	5	4	2	2	2	1			1			
Apidae	Ceratina	1		1	1											
Apidae	Epeolus	2	2	2	2	2	2	1	1							
Apidae	Eucera	2	2	2	1	1	1									
Apidae	Melecta	2	2	2	2	1	1									
Apidae	Nomada	32	28	31	28	25	16	16	11	9	9		11	3		
Apidae	Xylocopa	1	1	1	1		1									
Colletidae	Colletes	9	8	8	8	8	8	6	4	3	3	1	4	3		
Colletidae	Hylaeus	12	11	11	10	9	6	4	5	1	1		4	1		
Halictidae	Dufourea	2		2												
Halictidae	Halictus	6	5	5	3	3	2	2	2	1	1		2	2		
Halictidae	Lasioglossum	31	29	29	28	25	20	13	12	9	7	3	10	10		
Halictidae	Sphecodes	21	16	16	17	14	10	7	6	3	3		7	5		
Megachilidae	Anthidium	1	1	1	1	1	1	1	1				1	1		
Megachilidae	Chelostoma	2	2	2	2	2	2	1								
Megachilidae	Coelioxys	6	5	6	6	6	4	1	1	1	1		2			
Megachilidae	Heriades	2	1	2	1	1										
Megachilidae	Hoplitis	3	1	2	1	1	1	1								
Megachilidae	Megachile	7	7	7	7	7	7	6	4	3	3		5	2		

Sub-family	Genus	Species in Britain*	Species in each 100 km grid band south to north										Species			
			1	2	3	4	5	6	7	8	9	10	Island of Ireland	Scilly	Orkney	Shetland
Megachilidae	Osmia	13	8	10	10	9	6	5	3	5	3		2			
Megachilidae	Stelis	4	4	4	3	3	2	1								
Melittidae	Dasypoda	1	1	1	1	1	1									
Melittidae	Macropis	1	1	1	1	1										
Melittidae	Melitta	4	3	4	3	3	2	1						1		
	Total species	234	200	216	200	177	133	97	74	58	50	10	72	42	2	0
	Total genera		24	26	25	24	20	17	13	11	10	4	11	11	1	0

Data from Else & Edwards (2018) and Beavis (2013, 2022). * Data exclude species not mapped.

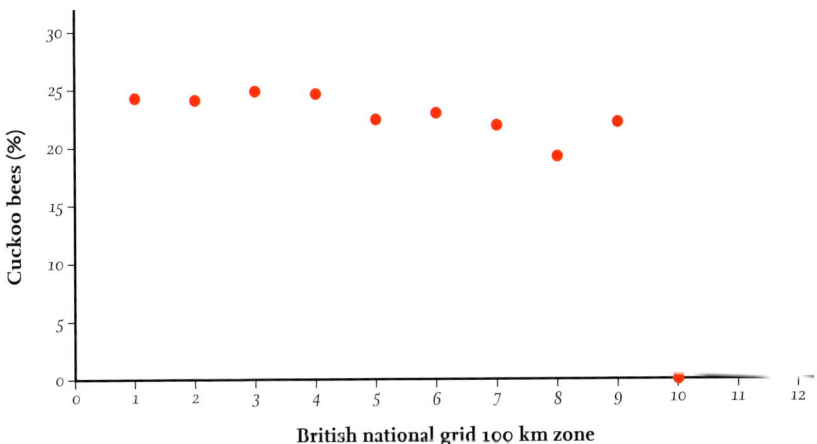

FIG 303. Percentage of cuckoo bee species in each national grid zone.

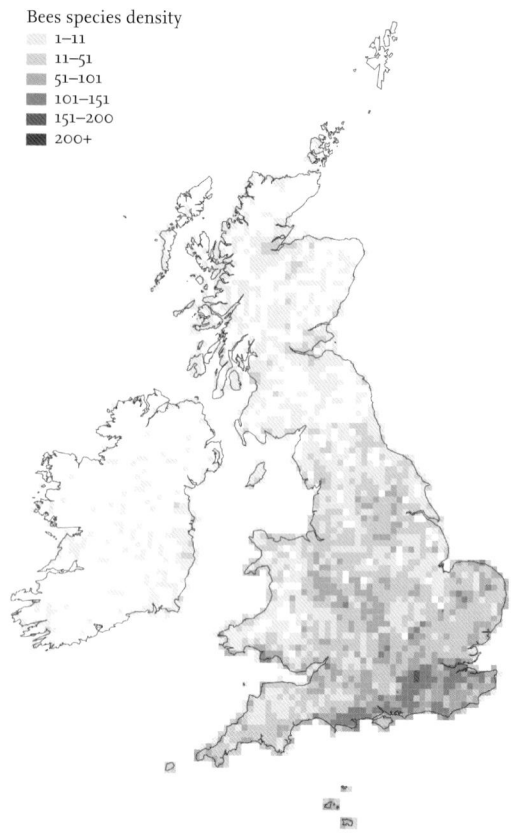

Bees species density
- 1–11
- 11–51
- 51–101
- 101–151
- 151–200
- 200+

FIG 304. The diversity of solitary bees in Britain and Ireland on a 10 km² scale, based on 630,870 records submitted to the Bees, Wasps and Ants Recording Society (BWARS) and the Irish National Biodiversity Data Centre (some Irish records are not included in the BWARS data). Darker shading indicates a higher number of species recorded per 10 km². White squares have no records, in some cases through a lack of any solitary bee recording. (© R. Edwards, BWARS)

confined to the most favourable habitats. Some solitary bee species have a strong northerly bias in their distribution, notably *Colletes floralis*, *Andrena ruficrus*, *Lasioglossum rufitarse*, *Osmia inermis*, *Osmia parietina* and *Osmia uncinata*. All have a boreo-alpine distribution extending across Europe and beyond (Else & Edwards 2018). Bees which are active at lower temperatures tend to have more effective thermogenesis (Stone 1990), but there are likely to be additional ecological factors involved, perhaps relating to pollen sources and competition.

BRITAIN AS AN ISLAND

As the Ice Age glaciers finally departed between 6,000 and 8,000 yBP, what we now know as Britain became separated from mainland Europe by water. This separation places a major limitation on the diversity of our flora and fauna. The most recent glaciation, known as the Devensian, peaked at about 21,000 yBP. During this last glacial maximum, glaciers extended to the level of what is now Swansea, Wolverhampton and Lincoln (Stringer 2006). The area south of the ice sheet was dominated by tundra vegetation, similar to that of modern Spitzbergen (Clarke et al. 2012, Barringer et al. 1991). Isolated temperate refugia may have remained, however (reviewed by Berry 2009), allowing the slim possibility that some bees survived in southern Britain and Ireland. From about 16,000 yBP, the climate suddenly warmed, allowing large mammals and Mesolithic (Middle Stone Age) hunters to return to Britain. Subsequently, there was a further sudden cooling from 13,000 yBP, seemingly preventing any human occupation for about the next 2,000 years. Evidence from pollen deposits indicates that birch and pine started to return from about 10,000 yBP, followed by a variety of warmth-dependent deciduous trees. From 9,000 to 7,500 yBP, the mean summer temperature in Britain is believed to have been more than $2\,°C$ higher than it is today, presumably allowing considerable overland colonisation by bees. Around 8,000–6,000 yBP, the low-lying land connecting Britain to mainland Europe, known as Doggerland, was gradually inundated by rising sea levels. The timing of this transition coincided with the arrival of Neolithic (New Stone Age) farmers to western Europe, bringing livestock and crops. We may imagine that the subsequent opening up of forests, bringing clearings, fields and sunny tracks, created a paradise for bees, which partially survives today.

The exact timing of the Mesolithic–Neolithic transition in Britain has been uncertain but has significance for the composition of our bee fauna. Some important evidence came in 1999, from the somewhat surprising source of a lobster ejecting stone tools from its burrow, observed by divers beneath the Solent (Isle of Wight). The lobster's excavations had penetrated the original land surface, now inundated and covered in marine sediments. Excavations revealed a Mesolithic settlement, dated at about 8,000 yBP. Among the more recent archaeological layers, there is evidence of wheat – a clear sign of the influence of farming. Marine cores around Britain suggested that there may still have been a land connection at this time and the possibility of trading with Neolithic farmers on the 'mainland', rather than wheat being grown on the Mesolithic (lobster) site. This and other evidence suggest that bees had a good opportunity to disperse by land into Britain, during very favourable climatic conditions, between 10,000 and 8,000 yBP and perhaps for longer. Despite this, many bee species which are now abundant just across the water did not make it to our shores, or if they did, they have since died out. Table 26 shows the number of bee species and genera in Britain and in some nearby European countries.

TABLE 26. Numbers of bee species and genera recorded in Britain and nearby countries (the French total includes Corsica). (Source: *Discover Life* website)

	Total species	Genera	Solitary bees	Bumblebees	Honeybee	Area/km^2
Ireland (island of)	106	13	84	20	1	84,421
Norway	211	29	175	35	1	385,207
Britain (England, Wales, Scotland)	272	28	244	27	1	229,040
Denmark	291	30	261	29	1	42,933
Netherlands	364	33	334	29	1	41,543
Belgium	405	40	373	31	1	30,689
France	952	58	906	45	1	551,695

It is immediately clear that countries along the seaboard of western Europe, at similar latitudes to Britain, have a much richer bee fauna than Britain. Denmark, the Netherlands and Belgium all host more bee species and genera than Britain, despite having much smaller land areas and a similar or smaller diversity of habitats. France has a very rich bee fauna, as a consequence of its large size and Mediterranean influence. Norway is larger than Britain but has fewer bee species, as a result of its higher latitude. Norway's bumblebee fauna is relatively rich, again demonstrating the importance of the ability to raise body temperature by thermogenesis. Ireland's bee diversity is well below Britain's, with fewer than half the number of species and genera. From about 16,000 yBP, the Irish and British ice sheets separated, preventing direct colonisation of Ireland from Britain, though some authorities propose the persistence of a temporary land bridge.

Bees have good dispersal abilities and so it requires a major geographical barrier to prevent populations from mixing. The English Channel clearly provides such a barrier, as evidenced by the smaller diversity of our bee fauna compared with mainland Europe. Island bee diversity shows a species-area curve similar to that described above: smaller islands have more species per unit area than larger ones. More distant islands receive fewer bee colonists but are more likely to develop endemic species (unique to the island) (Baldock *et al.* 2020). Despite their proximity, there can be genetic differences between British and mainland European solitary bee populations. As discussed below, the British population of Early Colletes *Colletes cunicularius* is classifed as a British subspecies *Colletes cunicularius celticus* (O'Toole 1974). Other distinctive forms of British solitary bees will probably be recognised.

BEES LIKE IT HOT – CONTROL OF BODY TEMPERATURE

Here, we move in from the broad perspective to look at how individual bees respond to changes in temperature. There is no better place to do this than a garden, where there can be plenty of opportunities to keep watch over a long period of time in varying weather conditions.

Basking

The minimum thoracic temperature of a bee for flight is about 30 °C, a temperature well above the typical air temperature occurring in Britain. During active flight, the body temperature of some bees can be similar to that of the human body (37 °C) (Stone 1990). A bee's body needs to be maintained at a high temperature as it goes about its daily routine. By watching the behaviour of individual bees, we can gain some insights into the strategies they use to control body temperature hour by hour. Males of the Blue Mason Bee *Osmia caerulescens* are 5–6 mm long and frequently bask in sunlit spots, usually choosing to land on dry vegetation or stones. In a Norfolk garden (NWO), males tend to cluster in loose groups, close to the place where they emerged from nests in a bee hotel. Each male returns repeatedly to a small number of preferred basking places, but when the sun goes in, the bees quickly disappear into holes and crevices. In full sun, the males patrol patches of flowers, using their long tongue to take nectar. When a female is sighted, they may pounce on her in an attempt to mate, though they are often rebuffed. By patrolling almost constantly, males have a good chance of encountering virgin females. The less time they waste in their short adult lives, the better. Basking would seem to be time wasted, but males were observed basking even when the shade temperature reached 25 °C.

To investigate further, individual males were watched to see how long they spent basking in the sun before they flew off again on patrol, noting the shade air temperature on each occasion. All observations were made on days with cloudless skies over a period of five weeks. Fortunately, May 2020 provided excellent weather. The results show a clear tendency for males to bask for longer periods when the air temperature was lower, suggesting that the behaviour was related to warming up rather than merely resting or grooming. This was supported by

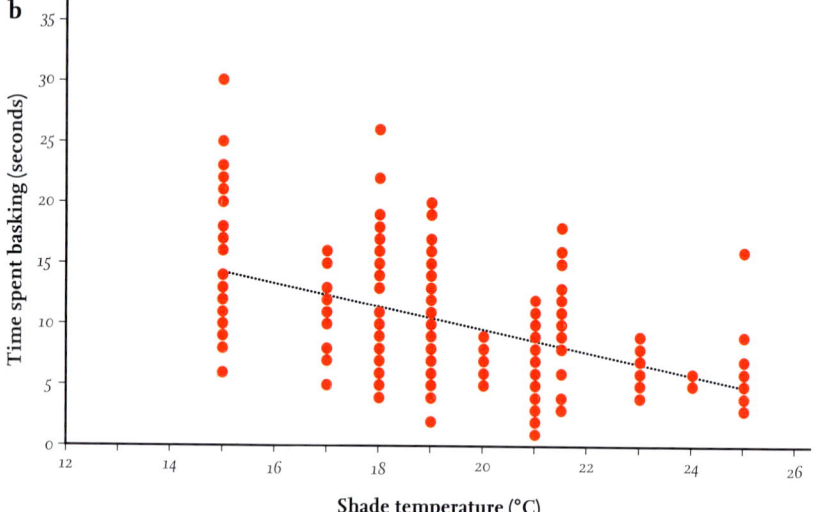

FIG 305. (a) Blue Mason Bee *Osmia caerulescens* male on one of his habitual basking places, a flint stone. (b) Mean duration of basking sessions by Blue Mason Bee males at different shade temperatures (240 observations): some points are superimposed.

the fact that individual bees compensated for the movement of the sun when selecting their basking locations on particular stones at different times of the day.

As a bee becomes hotter, its movements become faster, and this may be one benefit of basking. Each male bee is potentially in competition with other males and must be quick enough to pounce on a female before she moves or another male precedes him. Males had brief altercations with other males over basking spots, sometimes becoming briefly locked together. They occasionally scent-marked vegetation as they skimmed over it, presumably attracting females to the general area where the males were clustered. They seemed to be faithful to their patch:

one individual, marked with a spot of paint on 18 May, was regularly seen basking within the same 8 m stretch of a flower bed, often at particular spots, until 22 June. Newly emerged female *Osmia caerulescens* also basked frequently, this perhaps helping to speed up digestion and metabolism, including the maturation of their eggs. There may also be an oxygen deficit acquired during flight, which requires periods of inactivity to restore, with bees recovering more quickly at higher temperatures. Bees often pump their abdomen when they alight.

Once female mason bees have started nest construction, speed is again of the essence. Those which are the quickest and most active will produce the most offspring, since more brood cells will be completed in the limited time that weather and daylight allow. Rapid nest completion also reduces opportunities for parasites and predators. Female Fork-tailed Flower Bees *Anthophora furcata* often collect nectar from catmint (*Nepeta* species). As the air temperature rises, they are able to visit more flowers in a given time (Fig. 306; NWO). The catmint observed was about 3 m from the bees' nest burrows in a tree stump.

Pollen loads were taken back to the nests periodically, but the *Anthophora furcata* females were not observed basking. A temperature increase from 15–25 °C resulted in an approximate doubling of the rate of flower-visiting. Despite the multitude of variables affecting bee behaviour, this conforms surprisingly closely with the typical performance of all biochemical reactions in relation to changes in temperature – a doubling for a 10 °C rise. Even so, the relationship between speed of movement and

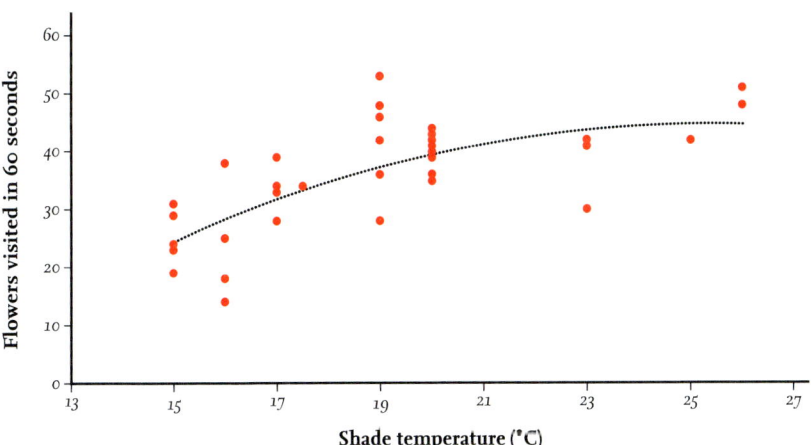

FIG 306. The number of garden catmint flowers visited per minute by Fork-tailed Flower Bee *Anthophora furcata* females at different shade temperatures. The bees were taking mostly nectar (38 observations).

temperature in ectothermic (cold-blooded) animals such as bees is not generally a simple one (Clarke 2017). Very high temperatures may be damaging, and there is a suggestion that the graph starts to flatten out at higher temperatures.

Thermogenesis

Basking in the sun can raise a bee's body temperature very effectively, but in cloudy conditions or in the shade, some bees can continue to warm up by generating heat in their flight muscles, a process known as thermogenesis. The flight muscles of a bee are situated in its thorax. Heat can be released from the muscles by the oxidation of stored sugars and/or by 'shivering' the flight muscles. Heat is then dispersed around the rest of the bee by the circulation of body fluids.

Larger bees are generally able to warm up their bodies more quickly than smaller ones, because larger objects (including bees) have a lower surface area:volume ratio and so lose heat more slowly from their body to the surroundings. Smaller bees can generate more heat per gram of flight muscle than larger bees, but this does not fully compensate for the greater rate of cooling (Stone & Willmer 1989). We can expect small bees to be more dependent on basking than large bees, since small bees are less able to warm up by thermogenesis but can warm up more quickly than larger bees by basking, as a result of their higher surface area:volume ratio. Large bees, such as *Anthophora plumipes*, do nevertheless bask, especially on sunny mornings, probably because this avoids the energy costs of thermogenesis. Most bees have a dense covering of hair which helps to insulate their bodies by trapping warm air, though, unlike most mammals and birds, they allow their body temperature to fall when they are inactive. An exposed torpid bee is in danger of predation, so warming up needs to be rapid, unless the bee is hidden. As a last resort, torpid bees can make an emergency escape from predators by dropping down into vegetation when disturbed. A rapid warm-up also saves time which could be spent foraging or seeking a mate.

During flight, an *Anthophora plumipes* loses heat to the air by convection, but this is compensated for by the heat generated by muscle movement, so the bee can often remain in flight without stopping to warm up its body. It is possible to obtain a measurement of the body temperature of a flying bee by catching it, then rapidly inserting a very fine thermocouple into its thorax, a procedure known as 'grab and stab'. The bee can be released with little harm (Unwin and Corbet 1991). Using this method, Stone (1990) found that the thorax temperature of an *Anthophora plumipes* in flight could reach up to 39 °C. In controlled laboratory studies, Stone made measurements of the body temperature of tethered flying bees while a thermocouple remained attached to their thorax. Once airborne, the bees' body temperatures soon became stable. It was found that the stable flight temperature of an *Anthophora plumipes* was higher when the air temperature around it was higher. This suggests that in very warm weather, these bees can overheat. A somewhat smaller species of bee, the Early Colletes *Colletes cunicularius*, had a lower stable flight temperature than *Anthophora plumipes* when both were tested at the same ambient temperature. In general, larger bee species can more easily keep warm during flight but are more at risk of overheating.

Osmia caerulescens is much smaller than these last two species (wing length around 8.5 mm in females and 6 mm in males) and apparently loses heat during flight more quickly than it is generated, even in hot weather, so sooner or later the bees need to land again to bask in the sun. *Osmia caerulescens* and *Anthophora furcata* require an air temperature of about 15 °C to become active, whereas Stone found that *Anthophora plumipes* could take flight when the air temperature was as low as 9 °C. Female *Anthophora plumipes* can be active at air temperatures as low as 6 °C if there is some sun (NWO pers. obs.).

Body size can vary even within a species, and this can have a considerable effect on behaviour. Stone showed that larger *Anthophora plumipes* females were able to fly at lower air temperatures than smaller ones but that smaller individuals were more tolerant of very high air temperatures. The mass of pollen and nectar loads carried by females was greater when the temperature was higher, so smaller females can be favoured by being able to continue foraging in very hot weather. On the other hand, larger females were parasitised less by their cuckoo bee, *Melecta albifrons* (see Chapter 8). Laboratory studies showed that *Melecta albifrons* were less effective at thermogenesis than their *Anthophora plumipes* hosts. Large *Anthophora plumipes* had an advantage over smaller individuals in being able to construct and seal more nest cells during periods when temperatures were too low for *Melecta albifrons* to be active.

TABLE 27. Summary of the effects of body size on temperature regulation by bees.

	Large bees	*Small bees*
Surface area/volume ratio	Smaller	Larger
Rate of heat loss	Slower	Faster
Temperature control by thermogenesis	More effective	Less effective
Temperature control by basking	Less effective	More effective
Minimum air temperature for activity	Lower	Higher
Danger of overheating	Greater	Smaller

The wing length of bees can be taken as a proxy for body mass and has the advantage of being much easier to measure. The overall distribution of wing length across all British solitary bee species approximates to a normal distribution

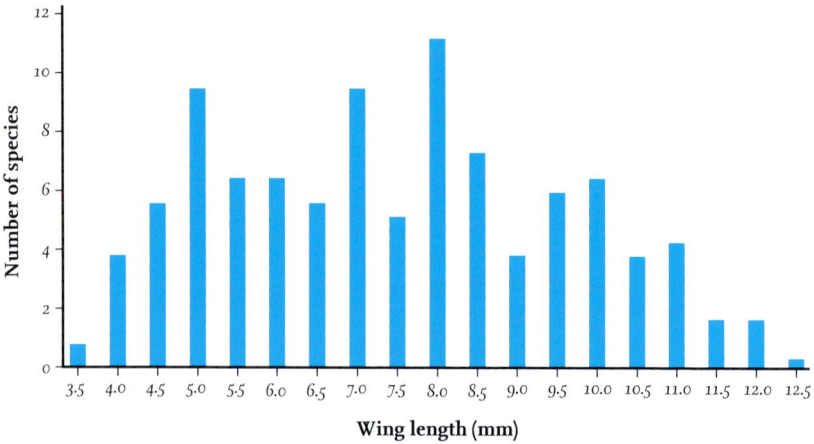

FIG 307. Frequency distribution of British solitary bee wing lengths. Wing lengths taken to be the upper range of the female (Falk & Lewington 2015). (Total = 233 species.)

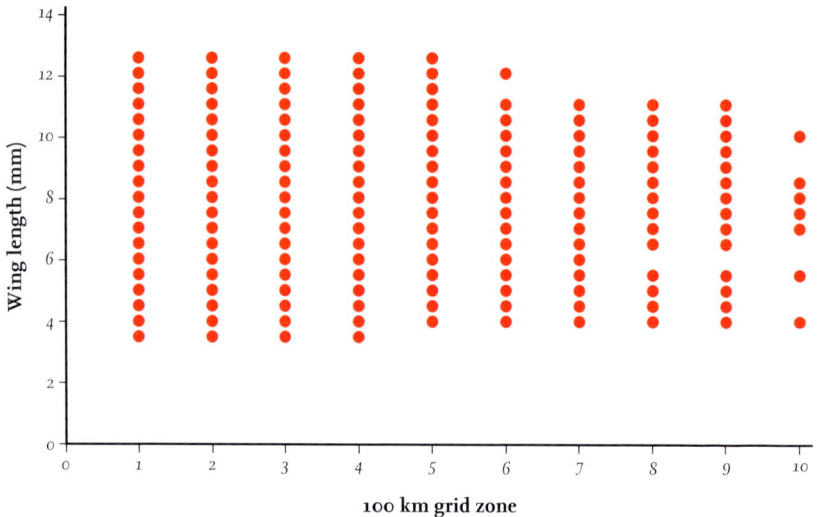

FIG 308. Wing length of bees within national grid 100 km bands from south (band 1) to north (band 10). Total = 233 species: some points are superimposed. Mean of the maximum wing length = 7.5 mm. (*Xylocopa* [22 mm] is omitted.)

(bell shape), with most bees towards the centre of the range and fewer at each extreme – a common pattern in nature. The average wing length remains virtually the same from the south to the north of Britain, though there is a suggestion that the size range is narrower in the north (Figs 307 & 308). This makes sense in that very small or very large body sizes could require the most favourable conditions. It may be no coincidence that the size of the two solitary bee species occurring in Orkney, *Andrena tarsata* and *Andrena coitana*, are both in the middle of the range. There may be latitudinal trends in size within particular bee species, but this does not seem to have been investigated. Overall, being a large or small bee species has advantages and disadvantages in relation to temperature regulation and foraging behaviour, and these operate similarly at all latitudes. Optimum size may also be related to other attributes, such as fecundity or dispersal ability. Also, as we have seen in *Anthophora plumipes*, cleptoparasites can exert selection pressures on the body size of females. Climate change will inevitably alter the way these various factors interact, in ways that are difficult to predict.

LONGEVITY – HOW LONG DO ADULT BEES LIVE?

By marking newly emerged bees we can gain some idea of their typical longevity. We might expect there to be differences between species and between males and females, while weather conditions, predation and disease are all also likely to play a part. Male *Osmia caerulescens* can live for a month or more, based on marked bees observed in a garden, as described above. Females were not marked but seemed to have an even longer lifespan, being present from mid-May to mid-July, though these were likely to have included several females with overlapping flight times and possibly a second generation.

An attempt was also made to discover the lifespan of male Cliff Mining Bees *Andrena thoracica* at a nesting aggregation on heathland. The aggregation consisted of two clusters of nests about 8 m apart, each covering about 3 m^2 of fairly level ground, with about 100 nest burrows in each. On 24 March, 20 recently emerged males from each cluster were netted and marked with a red or a yellow spot of paint, then released at their point of capture. Over the following eight days, 30 marked males were observed patrolling (including repeat observations of some individuals), all being within the cluster from which they had originated: the males seemed to be faithful to the small area of their home patch. There was no nectar source within the nest aggregations, and males apparently fed rarely or not at all. Males marked on 24 March were last seen on 3 April, suggesting a lifespan of around 10–14 days. Males of the second generation were active from

FIG 309. (a) Positions of adjacent nest aggregations of the Cliff Mining Bee *Andrena thoracica* on heathland. (b) Marked Cliff Mining Bee male at a nest burrow. This bee returned repeatedly to the same place, sometimes entering the burrow.

about 15 June until 15 July. On 24 June, 15 males were marked with a spot of paint. There was just one re-sighting, on 10 July, within 1 m of the marking place – a lifespan of a minimum of 16 days.

Female *Andrena thoracica* were present from from 22 March until 24 April, with the second generation flying from 24 June until 25 July, their timing being about 10 days behind that of males. Their survival was not tested but they seemed to live longer than males.

Anthophora plumipes has just one generation per year. A nesting aggregation on a shingle beach in north Norfolk is situated in an eroded stump of cliff and easily accessible for study (see Chapter 8). In 2020, the first males were seen from late February, with the main emergence happening in mid-March. The last, faded, survivors were observed on 9 May. Females emerged from 20 March and were seen until 28 May. On sunny mornings in late March, scores of males were on lookout, each at the entrance of an emergence hole from a nest burrow, to which they had retreated for the night. As the day warmed up, they took flight to pursue females, with perhaps 80 individuals circling around the stump of cliff at a frantic pace. Fifty were netted and painted with a red spot on the thorax, to investigate their longevity. There were just six re-sightings, all at the nest aggregation, the last being 23 days after marking. From this evidence, adult male post-emergence lifespan was about 3–4 weeks. However, Stone (1990) found that many adult males were active for six weeks and some for 9–10 weeks, while adult females lived for 5–7 weeks after emergence.

Straka *et al.* (2014) carried out field studies of the effects of foraging activity and weather on the lifespans of female *Anthophora plumipes* and Greybacked Mining Bees *Andrena vaga*.

FIG 310. Hairy-footed Flower Bee *Anthophora plumipes* male emerging from a nest burrow, after a wait of about seven months in adult form within a brood cell. The large eyes assist in rapid pursuit of females in flight.

Females were individually marked with paint the first time they were observed at the nest aggregation. Average female lifespan was estimated to be 14 days for *Anthophora plumipes* and 15 days for *Andrena vaga*, but with maximum recorded lifespans of 41 days and 38 days, respectively. Unsurprisingly, both species were found to emerge earlier in years with warmer spring weather. However, bees lived longer in cooler years, probably because their activity was less intense, since each day of foraging added to wear and tear. Moderately warm and sunny but moist conditions during active days was correlated with a longer lifespan (through less wear and tear), though very high temperatures reduced activity in both species and, in *Andrena vaga*, reduced lifespan. In the latter species, it was thought that bees could overheat while re-excavating their covered nests in hot weather.

Studies of honeybees also indicate that wear and tear have an effect on mortality, with most workers living only 3–4 weeks. When they first emerge, worker honeybees remain in the hive, tending the brood, and at this stage mortality is very low, but mortality surges when they leave the hive to forage. Since foraging workers inevitably wear out physically quite quickly and die, there is little or no selection against physiological ageing (the decline of organ function). Queen honeybees can, however, live for five years or more, and some halictid gynes (queens) survive a second winter to breed again. This contrast between the survival of inside and outside bees offers one explanation for the evolution of eusocial behaviour: females which stay in the nest (queens) experience less wear and tear, live longer and can lay more eggs (Rueppell *et al.* 2008).

Many species of mining bees (*Andrena*), flower bees (*Anthophora*) and mason bees (*Osmia*) which emerge in early spring have spent the winter in adult form inside a brood cell, sometimes within a cocoon. Overwintering as an adult seems to be riskier compared with overwintering as a prepupa, since adults are more at the mercy of winter conditions. The European Orchard Bee *Osmia cornuta* is one such early-spring species which overwinters as an adult in its cocoon and has been the subject of detailed experiments (Bosch & Kemp 2004). Cocoons experimentally chilled from early winter (October) were compared with those kept under milder conditions until November. The following spring, adults emerging from the October-chilled cocoons survived longer than those subjected to a warm October, probably because the latter had used up more of their fat reserves. Under very warm winter conditions, emerging bees were found to be unable to chew their way out of their cocoon or, if they emerged, were unable to fly.

Bees can display symptoms of 'old age' in their physical state and also in their behaviour. Hair colour often fades in just a few days, and this may be followed by hair loss through abrasion against the walls of the nest burrow, especially in female mining bees. Damage to pollen-collecting hairs will inevitably increase the time taken to stock each nest cell. Wings also become tattered, and mandibles are worn down. Mental deterioration also occurs. In a classic Russian paper on 'The Nesting Habits of Solitary Bees', Malyshev (1935) noted that 'nests built in the course of old age are distinguished mainly by the lack of completeness or regularity'. The dimensions of the nest cells were erratic or stocked with too much or too little provisions, or nests were built but no egg laid. New behaviours could arise which were not previously evident, such as attacking and destroying other nests. Examples of disorganised behaviour can occur in nesting aggregations of *Anthophora plumipes* and Sea Aster Bees *Colletes halophilus*. Females of these two species were seen wandering from nest to nest close to the end of the nesting cycle, with no apparent 'purpose' (NWO pers. obs.), and some were perhaps behaving as intraspecific cleptoparasites. However, Cerna *et al.* (2013) found that *Anthophora plumipes* females often re-used abandoned nests; nest usurpation and conflict were rare.

Ageing can have a significant impact on the rate of nest provisioning in *Osmia bicornis*. Seidelmann (2006) calculated the efficiency of provisioning by recording the total number of hours taken for each of 718 nests to be completed from start to finish, allowing for hours of darkness and cold periods (below 18 °C). Then, in the following winter, the mature cocoons from each nest were weighed. The measurements gave an efficiency score for each nest in units of mass of cocoons produced per hour of nest construction. The efficiency of provisioning was found to decline dramatically with increasing age of the bee. A further finding was that the rate of parasitism increased markedly in nest cells produced by older bees, because these were left open for longer during construction as a result of slower provisioning. It was suggested that older bees, by switching to producing male brood cells (which require less food and are completed more quickly), reduced the impact of parasites, such as the fly *Cacoxenus indagator* (see Chapter 7).

NEST SITES, ROOSTING AND SHELTER

In Britain, nesting aggregations of mining bees are generally situated on bare ground or in short grass, root plates or cliffs, where there is little or no shade, and they often face east or south-east towards the morning sun. In dense

FIG 311. Hairy-footed Flower Bee *Anthophora plumipes* nesting aggregation at a site which receives the first morning sun.

aggregations, the activities of the bees themselves can help to keep the ground free of vegetation. Some early-spring species, such as Clarke's Mining Bee *Andrena clarkella*, nest in deciduous woodland beneath the tree canopy, but before shading leaves have developed. The temperature within nesting burrows can be surprisingly stable. At a nesting aggregation of *Colletes halophilus* in Essex, soil temperatures at 10 cm depth were 18–22 °C overnight and remained in that range when the surface temperature reached 44 °C during the day (Saxton 2009).

Cavity-nesting female bees usually retreat inside their nest burrows in poor weather and at night and males sometimes also do so. The burrows are generally horizontal and relatively well protected from the rain. Some mining bees cover their nest entrances from within when heavy rain starts during the day. This is the case in *Andrena thoracica*, even though they do not generally close their nests at other times. Buff-tailed Mining Bee *Andrena humilis* females also leave their nest entrances uncovered during daylight, except in bad weather, but they do seal the entrance at nightfall, after retreating inside. Female mining bees sometimes continue to excavate their nests during wet weather and overnight when temperatures allow. This can be seen dramatically at nesting aggregations of halictids such as the Sharp-collared Furrow Bee *Lasioglossum malachurum*; ground that is flat and featureless after rain can be transformed by hundreds of tumuli by the following morning. In *Anthophora* species, bees sometimes roost together spaced along a stem, holding on by their mandibles, with their bodies suspended in the air. This occurs in *Nomada* females, too. Male blunthorn bees (*Melitta* species) have a different roosting strategy, forming dense clusters on vegetation. Sainfoin Blunthorn Bees *Melitta dimidiata* form groups of males amongst flowers and stems, usually of their food plant, Common Sainfoin *Onobrychys viciifolia*. On the Salisbury Plain, M. Hodgkiss (cited Else & Edwards 2018) watched a cluster of males on a warm, overcast June morning. At 7:45 am, the bees began vibrating, and from 8:15 am they departed, by which time the temperature had reached 16 °C. Clover Blunthorn Bee *Melitta leporina* males (Fig. 312) show similar roosting behaviour. In both these *Melitta* species, males have been found to return to exactly the same spot on successive nights. The vibrating behaviour described in Sainfoin Blunthorn Bee males suggests that they shiver their flight muscles as a means of thermogenesis. Perhaps clustering together speeds up their rate of warming by mutual insulation. Similar clusters of males occur in Common Furrow Bees *Lasioglossum calceatum*. In this case it has been suggested that a male cluster might work as a lek – a gathering of males to which females are attracted for mating (O'Toole 1999). In Australia, males of *Liptotriches australica* (Halictidae) form clusters numbering in the thousands, while males of *Mellitidia tomentifera* (Melittinae) roost in regimented formation on the surface of a leaf, all facing

FIG 312. About 30 Clover Blunthorn Bee *Melitta leporina* males roosting on catmint (*Nepeta* species) in a garden in June. Males returned to the same flower spike each evening, arriving from some distance in ones and twos, for at least three successive nights (G. Else pers. comm.). (© G. Else)

towards the leaf base (Houston 2018). It seems that clustering in males may serve multiple functions.

Male Large Shaggy Bees *Panurgus banksianus* are renowned for remaining enclosed in flower petals overnight. Most of their daylight hours are spent patrolling yellow Asteraceae flowers where females are likely to be foraging, but if wet weather interrupts their activities, they can become trapped in a torpid state when the flowers close. At a Norfolk heathland on 7 July 2020, after sunny intervals during the morning, persistent rain started at 12:30 pm. Males had been patrolling a line of about 60 Common Catsear *Hypochaeris radicata* flowers beside a footpath. As the rain intensified, 30 of the males remained on their flower as it closed and became enveloped by the petals (ray florets). Rain continued for the following two days. On the third day, 10 July, at least 10 males were still inside their flower, having spent two full days and three nights within the petals, partly exposed to the rain. As the sun returned on the third day, the males dried out and resumed patrolling.

It is clear that adult bees are able to survive for several days without feeding. A short period of inactivity caused by poor weather is not necessarily too disadvantageous, since inactive bees are likely to age more slowly, as described above. Furthermore, nectar and pollen supplies accumulate in the absence of insect visits, making foraging more profitable on their return, perhaps even leading to greater overall nesting success than constant fine weather.

FIG 313. A torpid Large Shaggy Bee *Panurgus banksianus* male cradled in a Common Catsear *Hypochaeris radicata* flower during rain.

Bee activity is reduced in hot, dry weather, presumably because of dehydration of the bees and also because there is less nectar production by flowers in dry conditions. This is very obvious in Mediterranean habitats, where carpets of flowers can be almost devoid of bees during hot weather (NWO pers. obs.). In dry conditions, bees might seek more dilute nectar, and in very dry weather, some bees, for example the Four-banded Flower Bee *Anthophora quadrimaculata*, imbibe water. The existence of microhabitats where bees can find shelter during rain, warmth during cool weather, cool patches in hot weather and damper patches in drought are vital in enabling bees to survive. Uniform, level patches of ground sown with wild flower seed mixes, though useful, are not an adequate alternative to mature meadows and pastures with anthills, scrub and damp hollows. Old commons and brownfield sites are of very great value (see Chapter 10).

Cuckoo bees can get 'caught out' by bad weather. A large nesting aggregation of the Heather Colletes *Colletes succinctus* on a cliff face in Norfolk is attended by their cuckoo bee, the Red-thighed Epeolus *Epeolus cruciger*. One fine August afternoon, clouds came over the site and the weather suddenly cooled to 17 °C. The *Epeolus* soon became torpid, and large numbers simply rolled down to the base of the cliff and remained amongst fallen nest debris, with their abdomen pumping. Cuckoo bees spend much of their time watching and waiting and seem to be highly dependent on high ambient air temperatures and basking. Stone and Willmer (1989) report that two cuckoo bee species, *Melecta albifrons* and a *Coelioxys* species, had slower warm-up rates by thermogenesis than did pollen-collecting species of similar body mass. Field experience indicates that cleptoparasitic bees are more dependent on very warm weather than pollen-collecting species.

DIAPAUSE

All bees go through the developmental stages of egg–larva–pupa–adult. Breeding needs to be synchronised with the timing of spring and the availability of pollen, while during the winter months, bees must find a way to survive while in a torpid state. Bee life cycles therefore have a stop-start nature, and the timing of each phase can be affected by the weather. When a bee's egg and its food-store have been installed and concealed in a brood cell, there is some urgency for larval development to be as rapid as possible, but less urgency to go through metamorphosis to become an adult. The chances of a bee larva hidden in its brood cell being predated are relatively small compared with, for example, a butterfly caterpillar or a grasshopper nymph feeding in the open. Yet the bee larva is still at risk from any parasites that may have found their way in. The more quickly the bee larva feeds, the less time there is for parasites to discover it and the less food that any attendant parasite will have time to consume. The bee's egg hatches in just a few days and the larva develops rapidly, shedding its skin three (possibly four) times as it grows. Once all the food has been consumed, the bee larva stops growing and releases faeces for the first time, now being known as a prepupa (see Chapter 1). In many bee species, the prepupa remains in a dormant phase called diapause, which can last through several weeks of the summer or throughout a whole winter. During winter diapause, antifreeze substances are produced and the metabolic rate slows down. There is very limited physical activity, though a prepupa is able to wriggle quite a bit. After diapause, the prepupa undergoes metamorphosis to become a pupa and finally a winged adult. Nearly all *Halictus* and *Lasioglossum* species, as well as most other spring-flying bee species, pass the winter as fully formed adults inside their brood cells and are ready to emerge as soon as temperatures are suitable.

VOLTINISM

Some bees manage two generations in one year and are termed bivoltine, while others complete just one generation and are known as univoltine. Within these two main themes, each species has its own characteristic pattern, suited to its nesting requirements and food resources. The familiar Red Mason Bee *Osmia bicornis* which so readily adopts our bee hotels is univoltine. Emergence is timed to coincide with the arrival of spring flowers. *Osmia bicornis* is polylectic, foraging opportunistically from a wide range of plant species with a deep corolla. Oligolectic (specialist) bee species are more strictly attuned to the flowering

TABLE 28. Oligolectic bees and their pollen sources. The plant family is listed, but in some cases, pollen is taken from just one or a small number of plant species within the family.

	Araliaceae	Apiaceae	Asteraceae	Brassicaceae	Campanulaceae	Cucurbitaceae	Dipsacaceae	Ericaceae	Fabaceae	Fagaceae	Liliaceae	Orobanchaceae	Primulaceae	Ranunculaceae	Resedaceae	Rosaceae	Salicaceae	Oligolectic	Polylectic	Total	% oligolectic
Andrena	2	3	2		1	2	3	2	1							1	4	21	47	68	31
Anthidium																		0	1	1	0
Anthophora																		0	5	5	0
Ceratina																		0	1	1	0
Chelostoma				1									1					2	0	2	100
Colletes	1	4						1								1		7	2	9	78
Dasypoda		1																1	0	1	100
Dufourea		1				1												2	0	2	100
Eucera									2									2	0	2	100
Halictus																		0	6	6	0
Heriades			2															2	0	2	100
Hoplitis																		0	2	2	0
Hylaeus									1						1			2	10	12	17
Lasioglossum			1													1		1	32	33	3
Macropis													1					1	0	1	100
Megachile																		0	7	7	0
Melitta				1					2		1							4	0	4	100
Osmia			3						3									6	7	13	46
Panurgus			2															2	0	2	100
Rophites																		0	1	1	0
Xylocopa																		0	1	1	0
Total	1	2	17	2	2	1	3	4	9	1	1	1	1	1	1	1	5	53	122	175	

Data from Else & Edwards (2018)

times of their particular pollen sources and depend on the presence of their specific food source at the time when they emerge from diapause. For example, Heather Mining Bees *Andrena fuscipes* time their emergence to the flowering of Heather *Calluna vulgaris* in late summer. Bivoltine bee species such as the Cliff Mining Bee are nearly always polylectic, using a range of pollen sources which differ between the two generations. Of the 175 species of pollen-collecting solitary bees, 53 species (30 per cent) are oligolectic and all of these are thought to be univoltine, with the exception of the Spined Mason Bee *Osmia spinulosa* (Walters 2022). Amongst the remaining 122 polylectic bee species, 21 (17 per cent) are known to be bivoltine, 18 of these being *Andrena* species, the others being one *Hylaeus* and two *Osmia* (Table 28). In some species, it is not known whether one or two generations occur within one season (Early 2011). Overall, oligolectic solitary bees are rarely bivoltine, but polylectic species are often bivoltine.

In recent years, there have been frequent records of spring *Andrena* species appearing again in late autumn in southern and midland English counties. Many of the sightings have involved the Chocolate Mining Bee *Andrena scotica*. Females of this bee were seen on Ivy flowers at several sites in Warwickshire in autumn 2018, some carrying pollen loads, and at Wytham Woods in Oxfordshire, Chocolate Mining Bees, Ashy Mining Bees *Andrena cineraria*, Grey-patched Mining Bees *Andrena nitida* and Hawthorn Mining Bees *Andrena chrysosceles* were all active in mid-October 2017 (S. Falk pers. comm.). These are typically spring-emerging univoltine species which overwinter as an adult inside a brood cell. It seems they had emerged precociously in the fine autumn weather, with some showing signs of nesting behaviour. Bivoltine species have also been involved. In Kent, both male and female Yellow-legged Mining Bees *Andrena flavipes* were observed visiting flowers as late as 4 December (I. Beavis pers. comm.). In this case, it appeared that a third generation of this bivoltine species was being attempted. Other autumn-appearing *Andrena* species in Hampshire have included the typically bivoltine Short-fringed Mining Bee *Andrena dorsata* and Gwynne's Mining Bee *Andrena bicolor* (G. Else pers. comm.).

DELAYED EMERGENCE (PARSIVOLTINISM)

Nesting bees are likely to perform poorly if prolonged wet or cold weather occurs during their nesting season. Mountain Mason Bees *Osmia inermis* and Orange-vented Mason Bees *Osmia leaiana* are known to sometimes emerge from their cocoons after two years rather than after one year. Such strategies are thought to reduce the chances of complete failure caused by bad weather or by heavy parasitism in any particular year. This phenomenon, of a proportion of bees in a population having delayed emergence, has been termed parsivoltinism (Torchio & Tepedino 1982). In a bee hotel in Wales, all 26 of the brood cells of *Osmia leaiana* produced in 2017 did not emerge until 2019, and from a cohort of 63 cocoons in 2019, the majority emerged in 2021. At the same site, three *Osmia bicornis* females also emerged after two years (C. Boyes 2020 & pers. comm.). Similarly, female *Colletes cunicularius* are able to overwinter either as an adult or as a prepupa. In the latter case, the overwintering prepupa becomes an adult within its brood cell in the following summer but does not emerge until the year after that (Malyshev 1927). A similar pattern occurs in Large Scissor Bees *Chelostoma florisomne*. A variation on this theme occurs in Bull-headed Furrow Bees *Lasioglossum zonulum* in which females, having already nested, go into diapause for a second winter and breed again in the following year. Likewise, Little Carpenter Bee *Ceratina cyanea* females can live up to 18 months and is the only British solitary bee that can be found as an adult at any time of the year. At the end of the summer, both sexes enter the old nest burrows, where they overwinter (Baldock 2007, 2008). In the eusocial Orange-legged Furrow Bee *Halictus rubicundus*, some larger females emerging from the first generation do not become workers but instead mate with spring-appearing males, then go into immediate diapause which lasts until the following year (Yanega 1988). A different pattern occurs in some Smooth-faced Furrow Bee *Lasioglossum fratellum* populations, in which females survive to breed in a second season (Field 1966). In both cases, the behaviour, like parsivoltinism, could be evolved mechanisms by which bees reduce the chances of complete failure due to bad weather, disease or parasites.

PROTANDRY

In many bee species, males tend to emerge before females, a biological phenomenon known as protandry. It is common to see male bees swarming at nesting aggregations for a week or two before most of the females appear on the scene. Later in the breeding cycle, females outnumber males, which tend to die first. It is, however, not easy to quantify the extent of protandry. Counting the number of males and females present, day by day, can give a misleading picture. Peaks of abundance do not necessarily coincide with peaks of emergence, since adult bees can live for three or four weeks, and their numbers accumulate until deaths start to exceed emergence. A clearer picture can be obtained by rearing offspring from bee hotels and noting the bees that emerge each day. In one cohort in NWO's garden, the first male *Osmia bicornis* emerged from a bee hotel on 11 April and the first female on 22 April.

In many solitary bee nests, female brood cells are completed before male brood cells. This generally occurs in stem-nesting bees and prevents females being destroyed as emerging males push past them. It seems paradoxical that male eggs are usually laid last, yet males emerge before females. However, it is not the time of egg-laying that determines the time taken to reach adulthood. As we have seen, for much of its life cycle a bee is in diapause, and it is the physiological control of this dormant stage (as a prepupa or newly formed adult) which determines the time of emergence. Emergence times of bees can also be influenced by internal parasites, particularly by Strepsiptera, which advance female emergence times to match those of males (see Chapter 7). Another parasite, the bacterium *Wolbachia*, present in a large number of solitary bee species (Gerth *et al.* 2011), can divert males of some invertebrates into functioning females (Hughes *et al.* 2012). This is advantageous to the bacterium since it can be passed on in the cytoplasm of eggs but not in sperm. These parasites could distort patterns of female emergence and might sometimes account for female-biased sex ratios, though this is yet to be established for bees.

Protandry is not universal amongst bees and seems to be reduced or absent in some cuckoo bees, including *Epeolus* and *Nomada* species. Males of the Oak Mining Bee *Andrena ferox* also emerge at about the same time as females (Else 2012). In this last species, mating may occur within the underground communal nesting area as the two sexes emerge together. In shaggy bees (*Panurgus* species), too, male and female flight periods are quite similar.

CLIMATE AND RANGE

Perhaps our best-known solitary bee is *Osmia bicornis*, which has already featured several times in this chapter. It is one of 12 species of the genus *Osmia* breeding in Britain (Table 29). Mason bees have a wide range of nesting styles and pollen sources and make an interesting group to compare in relation to their distribution and behaviour. In addition to the common and familiar species such as the Blue Mason Bee described above, there are several species with a much more restricted distribution. An attempt was made by the authors to find some of the scarcer species and to consider reasons for their distribution and phenology in relation to their nesting habits and habitat requirements.

For one of our first challenges, the authors met in a Kentish wood in May, seeking the rare Fringe-horned Mason Bee *Osmia pilicornis*, but were rather disheartened to find ourselves sheltering under a tree in a hailstorm. A little later the sun came out and, despite an air temperature of just 5 °C, the bees appeared.

TABLE 29. *Osmia* species in Britain and Ireland.

Osmia species		Nest	Nest material	Distribution	Flight period	Habitat	Pollen source	Aculeate parasites
Gold-fringed Mason Bee	*Osmia aurulenta*	Snail shell	Leaf mastic	Southern; coastal	Apr–Aug	Dunes, etc.	Polylectic	*Sapyga quinquepunctata*
Red-tailed Mason Bee	*Osmia bicolor*	Snail shell	Leaf mastic	Southern; favours chalk; not Ireland	Mar–July	Calcareous grassland	Polylectic	
Red Mason Bee	*Osmia bicornis*	Hollow stem, other cavities	Mud	Widespread	Apr–June	Varied	Polylectic	*Sapyga quinquepunctata*
Blue Mason Bee	*Osmia caerulescens*	Hollow stem	Leaf mastic	Widespread with southern bias; not Ireland	Apr–July	Varied	Polylectic	*Sapyga quinquepunctata*
European Orchard Bee	*Osmia cornuta*	Hollow stem	Mud	London area and Midlands	March–June	Varied, including gardens	Polylectic	
Mountain Mason Bee	*Osmia inermis*	Under rock	Leaf mastic	Central Highlands, Blair Atholl, Braemar; possibly extinct	Late May–July	260–430 m calcareous montane grassland	Mostly Common Bird's-foot-trefoil	*Chrysura hirsuta*
Orange-vented Mason Bee	*Osmia leaiana*	Hollow stem, other cavities	Leaf mastic	Southern; not Ireland	May–Aug	Varied	Oligolectic Asteraceae	*Sapyga quinquepunctata*

Wall Mason Bee	Osmia parietina	Holes in walls, limestone pavement	Plant mastic	Wales, Silverdale, Dumfries	May–July	Limestone pavement	Mostly Common Bird's-foot-trefoil	Chrysura hirsuta
Fringe-horned Mason Bee	Osmia pilicornis	Holes in wood	Leaf mastic	Sussex, Kent	Apr–June	Woodland clearings	Mostly Bugle and Ground Ivy	
Spined Mason Bee	Osmia spinulosa	Snail shell	Leaf mastic	South-eastern areas; spreading northwards	May–Sept	Dry calcareous grasslands	Oligolectic Asteraceae	
Pinewood Mason Bee	Osmia uncinata	In wood; e.g. old beetle holes	Leaf mastic	Moray Firth, East Highlands, Speyside	May–July	Open pine woodland – Abernethy	Mostly Common Bird's-foot-trefoil	Chrysura hirsuta
Cliff Mason Bee	Osmia xanthomelana	Mud pots	Mud	Isle of Wight, Lleyn Peninsula	Apr–July	Hot coastal sites, landslips	Oligolectic Fabaceae	

First seen were males, which emerged from their hiding places amongst moss on the woodland floor, then basked in the sunshine, before taking wing to patrol the area. When the sun went in, a male was seen to dig down head-first into the moss once again, then turn around with just his head showing to await the next sunny spell. The bees at this site were perhaps vulnerable to the abundant wood ants (*Formica* species) and were hiding in the moss as a means of protection.

During sunny intervals, female *Osmia pilicornis* were able to collect nectar and pollen even at these low temperatures. In view of these observations, it is curious that *Osmia pilicornis* is largely confined to southern English counties. However, it is likely that the bee's nesting requirement for dry dead wood in sunny forest clearings limits their distribution (Earwaker 2014).

By contrast, three other mason bees have northerly distributions in the British Isles. The Wall Mason Bee *Osmia parietina* is found in Wales and northern England and makes its nest in crevices in stones. We found some typical habitat amongst limestone pavement in Lancashire, with bees moving rapidly between Common Bird's-foot-trefoil *Lotus corniculatus* flowers or basking on debris. As mentioned in Chapter 3, the even rarer Mountain Mason Bee *Osmia inermis* occurs on open, flower-rich calcareous pastures in the Cairngorms region of Scotland and builds its nests using leaf mastic on the underside of stones. These nest sites seem to be chosen in the absence of other dry places to nest in open country. Despite re-visiting the area where the species was last recorded, in calcareous moorland at around 500 m, the bee was not found. Its world distribution extends around

male *Osmia pilicornis*

FIG 314. Fringe-horned Mason Bee *Osmia pilicornis* male cleaning his antennae after retreating into moss when clouds came over.

FIG 315. Habitat of the Wall Mason Bee *Osmia parietina*.

FIG 316. Wall Mason Bee *Osmia parietina* female basking.

the fringes of the Arctic Circle and also at high altitudes farther south (Else & Edwards 2018).

The closely related Pinewood Mason Bee *Osmia uncinata* has a similar geographical distribution to *Osmia inermis* but is found in clearings in pine forests and uses holes in dead wood for its nests. All three species favour the very widespread plant Common Bird's-foot-trefoil as a source of pollen and nectar. As with *Osmia pilicornis*, their distribution may be related as much to nesting requirements and pollen sources as to climate. We must also allow the possibility that we are seeing the remnants of a once more widespread distribution of some rare species.

Amongst our more common *Osmia* species, the role of climatic factors is easier to detect and some have been the focus of detailed research. In recent years, the European Orchard Bee *Osmia cornuta* has appeared in Britain for the first time and is now established as far afield as Suffolk and the Welsh borders (A. Knowles pers. comm.). It is similar to *Osmia bicornis* and has a more southerly but overlapping distribution.

In Serbia, colonies of *Osmia bicornis* and *Osmia cornuta* are widely used for the pollination of pear and apple crops. Both bee species synthesise antifreeze compounds, such as glycerol, as winter approaches. These cryoprotectants allow their body fluids to fall to very low temperatures without ice formation. Experimental studies of commercially managed cocoons of each species showed slight differences in their supercooling points (the temperature at which their body fluids froze). Samples of cocoons were tested at 15-day intervals throughout three winters, using sensitive thermocouples. As the cocoons were gradually

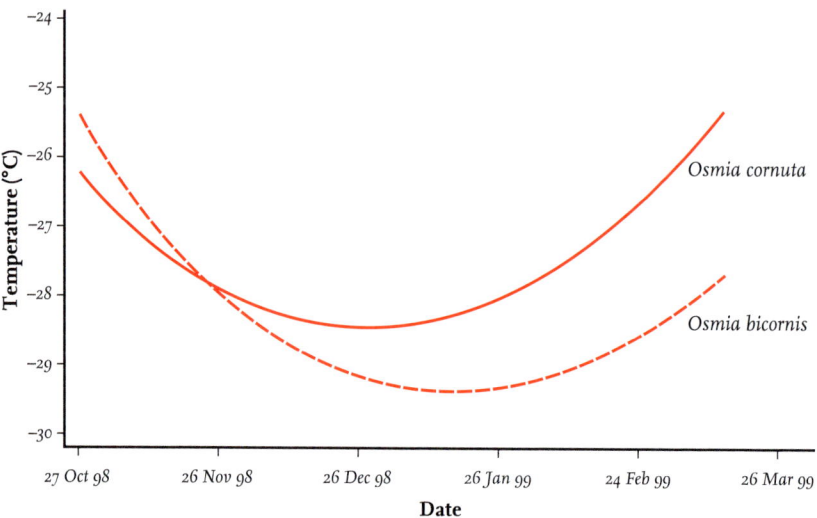

FIG 317. Changes in winter supercooling points (°C) of *Osmia cornuta* and *Osmia bicornis* cocoons (redrawn from Krunić & Stanisavljević 2006).

cooled, the supercooling point could be detected by a slight pulse of heat (heat of crystallisation) as the body fluids of the bee froze. In both species, the supercooling point decreased to a minimum in midwinter before gradually rising again to normal levels. *Osmia bicornis* had a slightly lower supercooling point than *Osmia cornuta* (a minimum of –31 °C compared with –30 °C) and their supercooling point rose more slowly in the spring. The difference between the two species in the rate of breakdown of cryoprotectants in the spring explained the earlier and more rapid spring emergence of *Osmia cornuta* compared with *Osmia bicornis* around Belgrade, where the studies were carried out. As temperatures rose above 15 °C, *Osmia cornuta* normally appeared within a few days, whereas *Osmia bicornis* appeared only when such temperatures persisted for a longer period of time (Krunić & Stanisavljević 2006).

The rapid response of *Osmia cornuta* to increasing spring temperatures means that populations farther north are likely to be caught out if a cold snap follows a few warmer days. Red Mason Bees are a bit more 'patient' and emerge only when warmer weather is firmly established. These differences confine the distribution of *Osmia cornuta* to more southerly regions, but there it gets a head start on *Osmia bicornis*. In a French study (Tasei 1973), nests of both species were kept at a constant temperature of 21 °C, and the average time taken for each stage of the life cycle was measured. In both species, there was a period of summer diapause, but this lasted much longer in *Osmia cornuta* than in *Osmia bicornis*, making the life cycle of *Osmia cornuta* longer overall by about five weeks. This difference ensures that, on average, each species arrives at the adult stage a short time before the onset of cold winter weather in the regions they inhabit. If they reach adulthood too early, they can metabolise too much of their fat reserves while waiting for the temperature to fall (Bosch *et al.* 2010). There is nevertheless a need for an interlude for physiological adjustment before winter; bees need to acclimatise to changing temperatures gradually. They also need time to synthesise antifreeze substances before their metabolic rate falls too far, so it is a fine balance.

Osmia bicornis and *Osmia cornuta* are both polylectic, enabling them to thrive more or less wherever the climate suits them. Neither is greatly restricted by nesting requirements since all they require is a hollow cavity and some mud. *Osmia caerulescens* is also common and widespread for similar reasons, needing only a hollow stem and leaves for mastic. *Osmia leaiana* is a stem nester, too, but is more of a pollen specialist, being oligolectic on Asteraceae. Three further species, *Osmia aurulenta*, *Osmia bicolor* and *Osmia spinulosa*, have more particular needs in that they nest in empty snail shells and are largely confined to calcareous habitats (Walters 2022). Lastly, the rare *Osmia xanthomelana* is described as thermophilous (warmth-loving) and is found largely on south-facing coastal sites with eroded clay cliffs and a good source of Fabaceae pollen. These comparisons of a group of closely related species show that predicting the effects of climate change is very difficult, since so many factors determine each species' distribution. Only amongst common and widespread species can we expect to see a clear link between climate and range and detect any effects of climate warming.

LOSSES AND GAINS

Solitary bees are considered to be keystone species, vital to the survival of many other species of animals and plants sharing their habitat. To what extent are we losing these vital pollinators of crops and native plants? Is climate change influencing the distribution and numbers of British bees? The publication of the *Handbook of the Bees of the British Isles* (Else & Edwards 2018) provides a recent update of changes to our bee populations. Based on this work, Table 30 lists bee species which are reported to have increased, decreased or become extinct in the British Isles over recent decades.

The overall picture is that 24 bee species show evidence of a noticeable increase in numbers and/or range over recent decades, while 29 species have noticeably declined. Amongst the declining species are some which have recently

TABLE 30. British bee species (excluding bumblebees) which are noted as changing their status over the past 50 years. Brackets indicate species which were recorded but doubtfully established in Britain.

Increased	Decreased	Extinct
Andrena cineraria	Andrena coitana	Andrena tridentata
Andrena dorsata	Andrena hattorfiana*	Dufourea halictula*
Andrena flavipes	Andrena marginata*	Dufourea minuta
Andrena florea*	Andrena nana	Eucera nigrescens*
Andrena gravida	Andrena nigrospina	Halictus maculatus
Andrena proxima*	Andrena niveata*	**Melecta luctuosa**
Anthophora quadrimaculata	Andrena rosae	**Nomada errans**
Ceratina cyanea	Andrena similis	(Andrena floricola)
Colletes cunicularius*	Andrena tarsata*	(Andrena lepida)
Heriades truncorum*	Andrena varians	(Andrena nanula)
Lasioglossum malachurum	Anthophora retusa	(Coelioxys afra)
Lasioglossum pauxillum	**Coelioxys quadridentata**	(Halictus subauratus)
Melitta tricincta*	Eucera longicornis	(Hoplitis leucomelana)
Nomada conjungens	Hylaeus pictipes	(Hylaeus punctulatissimus)
Nomada ferruginata	Lasioglossum nitidiusculum	(Lasioglossum laeve)
Nomada flavopicta	Lasioglossum sexnotatum	(Megachile ericetorum)
Nomada fucata	Lasioglossum smeathmanellum	(Megachile lapponica)
Nomada hirtipes	Megachile circumcincta	(Rophites quinquespinosus)
Nomada lathburiana	Nomada argentata	
Osmia bicolor	Nomada armata	
Osmia leaiana*	**Nomada obtusifrons**	
Sphecodes niger	**Nomada roberjeotiana**	
Stelis breviuscula	Nomada sexfasciata	
Stelis phaeoptera	Nomada signata	

Increased	Decreased	Extinct
	Osmia inermis*	
	Osmia pilicornis	
	Osmia xanthomelana	
	Sphecodes ferruginatus	
	Sphecodes gibbus	

Bold = cuckoo bee species. Data from Else & Edwards (2018)
* = oligolectic.

increased once again, including *Hylaeus pictipes* and *Lasioglossum sexnotatum*. Seven species appear to have become extinct, with all the extinctions except for *Halictus maculatus* happening during the 20th century. None of the lost species was ever abundant or widely distributed, so far as is known, except possibly *Andrena tridentata*, which may have been locally common. There are a further 11 species (shown in brackets) with a very small number of past records, representing vagrants, misidentifications or an unclear origin. The increasing and the decreasing groups include a similar range of bee genera (11 and 10, respectively), while the extinct group comprises six genera. The figures therefore suggest a steady turnover of a wide range of species rather than an overall decline. Three recent county bee faunas give a more in-depth view at a local scale. Baldock (2008) reviews changes to the status of bees in Surrey between 1902 and 2008. The numbers of 112 species (50 per cent) were judged to be unchanged, 63 (30 per cent) had increased, while 18 (10 per cent) had declined. In Kent, Allen (2009) reports that 332 species of bees and aculeate wasps (combined) were recorded in the county up to 1900. In the 20th century, 381 species were recorded during the 1980s but fewer in other decades. In the first decade of the 21st century, the score was 372 species, with several more species recorded since. In the much more agricultural county of Norfolk, Owens (2017) found that 23 bee species from 10 genera had apparently been lost from the county before the year 2000, while 17 species, also from 10 genera, had been gained post-2000. Four of the 23 'lost' species have since been rediscovered in the county.

NEW ARRIVALS

On the positive side, 16 solitary bee species have colonised or re-colonised the British Isles from mainland Europe over recent years. These may be signals of a response to climate change and are worth looking at in some detail in order to seek further clues about their causes. As we have seen, one of the species concerned is *Osmia cornuta*. At a bee hotel in Gloucestershire, two *Osmia cornuta* nests had been completed and sealed by 11 April, before any emergence from the previous year's Red Mason Bee nests (Mabbett 2020, 2021). These observations nicely fit the known difference in phenology and response to temperature between the two species described above. It seems that our warmer springs are allowing *Osmia cornuta* to expand its range and perhaps compete with *Osmia bicornis* for nest sites.

Bee hotels provide significant extra nesting habitat and also increase the chances of any newly arriving stem-nesting bees being spotted by bee-watchers. The Viper's Bugloss Mason Bee *Hoplitis adunca* was first detected at a bee hotel at Greenwich Peninsula Ecology Park, London, in 2016 and this species, too, is slowly spreading (Notton et al. 2016). The Little Dark Bee *Stelis breviuscula* also visits bee hotels. Once a rarity, this cuckoo bee is steadily tracking the spread of its host, the Large-headed Resin Bee *Heriades truncorum*, which has expanded its range this century from south-eastern counties as far as Shropshire, Cambridgeshire and Norfolk. Its close relative, the Small-headed Resin Bee *Heriades rubicola*, was recorded for the first time in Dorset in 2006, with a further record in London in 2016 (Cross & Notton 2016). Cavity-nesting bees can potentially be imported in plants or timber from abroad, and such may be the origin of this and some other newly recorded bees, such the Jersey Mason Bee *Osmia niveata*.

The arrival of the large and beautiful Violet Carpenter Bee *Xylocopa violacea* could hardly be missed. There have been sporadic records over a wide area of Britain over many years, with some evidence of successful breeding, notably three females observed collecting pollen from lavender in a Northamptonshire garden in 2010 (Baldock 2010). Recent sightings include those in Cambridge, where a female was seen gathering pollen between June and late September 2019 and a fresh male appeared in February 2020 (Tomkins 2020), and there was also evidence of a small population in Marden, Kent, in 2021 (I. Beavis pers. comm.).

A species which has apparently established itself by natural means is the Smooth-saddled Dark Bee *Stelis odontopyga*. Human transport is less likely in this case than in the other *Osmia* species mentioned, since its host, *Osmia spinulosa*, as we have seen, makes its nest in empty snail shells, though accidental transport amongst imported plants is still conceivable.

TABLE 31. Recently established or re-established solitary bee species in Britain.

Andrena polita	Records from Halling, Kent, early 20th century; last record 1934; recorded again in vicinity 2020	Hazlehurst (2021)
Andrena vaga	Records from Kent, 1940s; recolonisation Dungeness, Kent, 2008; subsequent spread to Hampshire and Essex	Else (2014a & b), Hazlehurst (2014), Else & Edwards (2018), R. Tidman pers. comm.
Colletes cunicularius	Colonisation of counties in central, southern and eastern England since 1997	Else & Edwards (2018)
Colletes hederae	Isle of Purbeck, Dorset, prior to 2001; subsequent rapid spread across England and Wales	Cross (2002)
Halictus eurygnathus	Present in Kent pre-1950; further records from Sussex and Kent since 2003	Falk (2011)
Heriades rubicola	Female carrying pollen, Dorset coast, 2006; female at Greenwich, London, 2016	Cross & Notton (2017)
Hoplitis adunca	Male and female at Greenwich, London, at drilled holes in wood; further records nearby and at Wimbledon, 2019	Notton *et al.* (2016)
Lasioglossum sexstrigatum	Sand pit, Merstham, Surrey, 2008; subsequent records from Surrey, Kent, Sussex and Suffolk	Hawkins (2011), Baldock (2008)
Nomada alboguttata	Hoo Peninsula, Kent, 2016; Rye Harbour, Sussex, 2017	Kirby-Lambert (2016)
Nomada bifasciata	Sandwich Bay and Isle of Sheppey, Kent, 2018	Falk & Earwaker (2018)
Nomada zonata	First recorded Kent, 2010; expansion as far as Norfolk and Oxfordshire by 2019	Hazlehurst (2019), Goddard (2016) cited Else & Edwards (2018), Kirby-Lambert (2016)
Osmia cornuta	Petersham, Surrey 2012; Blackheath, London, 2014; further London records from 2017; nesting, 2018; nesting Cheltenham, 2020	J. Power pers. comm., D. Gedge (reported in Else & Edwards 2018), Mabbett (2021)
Osmia niveata	One record from Kent, 1850; two further records at bee hotels in gardens in Leicestershire, 2004, 2008	BWARS website
Sphecodes albilabris	Two males at Layham, Suffolk, 2020	C. Kirby-Lambert pers. comm. (2020)

continued overleaf

TABLE 31. *continued*

Stelis breviuscula	Sussex, 1984; scarce until about 2010 then expansion in south-east counties and Essex	Else (1997), Baldock (2008), Else & Edwards (2018)
Stelis odontopyga	Pegwell Bay, Kent, 2017, 2018; Oxfordshire, 2018; Sandwich Bay, Kent, 2019	Edwards *et al.* (2019), Edwards (2019)
Xylocopa violacea	Nesting attempt Sussex, 1996; evidence of successful nesting Kent, Leicester and Cardigan, 2006; Northants, 2010; Cambridge, 2020	Roberts & Peat (BWARS website), Baldock (2010), Edwards (2015), Tomkins (2020)

Ground-nesting bees are perhaps less likely to be transported accidentally. The first Ivy Bees *Colletes hederae* were recorded near Swanage on the Dorset coast in 2001, already as an established aggregation (Cross 2002). The bee's initial appearance along the south coast is consistent with it being a natural migrant from France. The British population has exploded, typically appearing along the coast before spreading inland. Ivy Bees were recorded for the first time in Scotland and Ireland in autumn 2021. The species' expansion in Britain is part of an ongoing northern and eastern spread across Europe. The Ivy Bee is our only truly autumn bee, peaking with the Ivy blossom in October (Falk & Lewington 2015). Nesting continues well beyond the autumn equinox (21 September). Thereafter, northward-dispersing bees encounter not only cooler weather but also shorter days. These factors and the arrival of parasites may eventually place

FIG 318. Maidstone Mining Bee *Andrena polita* female. (© G. Hazlehurst)

FIG 319. Giant Blood Bee *Sphecodes albilabris* male. Males and females of this cuckoo bee emerge in the summer in the absence of their host, *Colletes cunicularius*, which is a spring species. The female cuckoo bees hibernate and emerge to parasitise their host in the following spring. (© C. Kirby-Lambert)

constraints on future population expansion. A trend to becoming more generalist (less specialised) is a characteristic of range expansion in some insects, such as butterflies (Thomas *et al.* 2001), and there is some evidence that Ivy Bees are widening their choice of pollen (TB; Westrich 2018).

The Grey-backed Mining Bee *Andrena vaga* was well established before there was general awareness of its presence: a large nesting aggregation known since 2008 at the RSPB reserve at Dungeness in Kent was recognised as this species in 2014. Numbers have steadily grown, and as of 2021, the bee had spread as far as Norfolk. Before these events, there were just a few isolated records, all from Kent, in the 1940s, suggesting that their recent establishment on the south coast arose by re-colonisation. The Fringed Furrow Bee *Lasioglossum sexstrigatum* is another

FIG 320. Ivy Bee *Colletes hederae* British distribution based on 10 km squares: (a) up to 2004, (b) up to 2014 and (c) up to 2020. (© R. Edwards, BWARS)

very recent addition to the British list. It was discovered in 2008 in a sand pit managed by the Surrey Wildlife Trust (Hawkins 2011). Subsequently, it was realised that three females collected in 2006, at Blackheath Common in Surrey, were the same species (Baldock 2008). There has since been a small number of further records from Surrey, Kent, Sussex and Suffolk. Another halictid, the Downland Furrow Bee *Halictus eurygnathus*, has re-appeared after a long absence. There were pre-1950 records from Kent and Sussex, then no further sightings until it was discovered in Sussex in 2003 (Falk 2011). It is possible that the species was present, but overlooked, during the long period of apparent absence.

The Early Colletes *Colletes cunicularius* has recently extended its British range dramatically into several central, southern and eastern counties of England, having previously been largely confined to the coasts of north-west England and Wales. It is believed that these recent range changes result, at least in part, from immigration from overseas rather than from expansion of the original British population, which has been described as a separate subspecies (O'Toole 1974). It is hoped that DNA studies will throw light on the origins of these new aggregations. Its specific cuckoo, the Giant Blood Bee *Sphecodes albilabris*, was found in Britain for the first time in Suffolk in August 2020 (C. Kirby-Lambert pers. comm.).

There are three further recent cases of cuckoo bees, all *Nomada* species, catching up with their hosts. The Large Bear-clawed Nomad *Nomada alboguttata* was first recorded at Hoo Peninsula in Kent in 2016 (Kirby-Lambert 2016) and was recorded at Rye Harbour in 2017. The host of this cuckoo bee is the Sandpit Mining Bee *Andrena barbilabris*, a widespread species associated with sandy habitats. The second new nomad bee, the Variable Nomad Bee *Nomada zonata*, was first recorded in 2010 (Hazlehurst 2019) and has spread rapidly, parasitising both generations of the Short-fringed Mining Bee *Andrena dorsata*, a very common polylectic species found in southern England and Wales. The third nomad bee to appear is the Dusky-horned Nomad Bee *Nomada bifasciata*, a cuckoo of the White-bellied Mining Bee *Andrena gravida*, a scarce bee confined to south-eastern counties. Single females were found at two separate coastal sites in Kent in 2018 (Falk & Earwaker 2018).

Understanding the causes of dispersal behaviour and range expansion in solitary bees is fundamental to their conservation. We need to know how far bees typically travel when dispersing between fragmented populations and the circumstances under which they undertake longer movements. The history of recent colonists is different in each case. Some have probably benefited from human assistance, while others have arrived through natural dispersal from mainland Europe. Of the 16 colonist species listed, three were first seen before the year 2000, six between 2001 and 2010 and seven between 2011 and 2020,

FIG 321. Chocolate Mining Bee *Andrena scotica* females from (a) Norfolk and (b) Ross and Cromarty.

suggesting a trend towards increasing northerly advance. In several cases, the source populations in Europe are known to be expanding, but this may not always be the case. Dispersal is a normal part of the life cycle of bees and other insects, with a variable proportion of newly emerged adults making longer flights, perhaps under the influence of weather, body size or high population density. The Scilly Isles, 45 km from the British mainland, have received several solitary bee colonists over recent decades, including *Colletes hederae* and the Ashy Mining Bee *Andrena cineraria*. The Yellow-legged Mining Bee *Andrena flavipes* colonised Scilly in 1997 and has since spread across the islands (Beavis 2013, 2022).

Regional differences in hair colour can occur in solitary bees, though these have not been studied systematically. For example, the Chocolate Mining Bee *Andrena scotica* has more dark hair in some Scottish localities. A subspecies of the Buffish Mining Bee *Andrena nigroaenea sarnia* is recognised in the Scilly Isles.

As we have seen, female solitary bees are quickly mated on emergence and retain within their body a store of sperm, which lasts for the whole of their reproductive life. A female solitary bee is therefore capable of establishing a new population on her own, far from her origins (Michener 2007). After mating, *Andrena thoracica* females spend several days establishing a nest burrow, which they defend from other females, before they start stocking nest cells. It is likely that their ovaries are maturing during this period, which seems also to be the time when some females disperse. Dispersal can be in any direction but is only detectable when bees move into previously unoccupied places. Bees may arrive in Britain from mainland Europe (and return) continually, but climate warming is making their survival and establishment in Britain more likely. Support for the idea of multiple colonisation comes from genetic studies of *Colletes hederae*, which show that our bees are as genetically diverse as their source populations in mainland Europe (Dellicour *et al.* 2013).

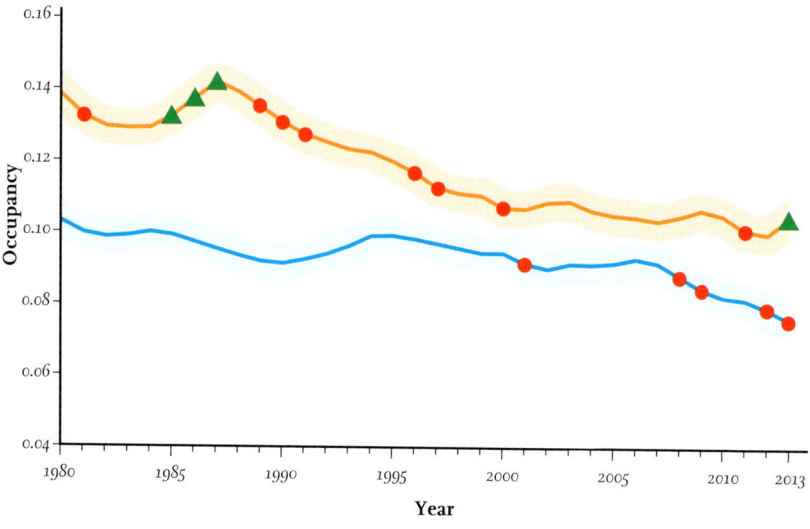

FIG 322. Changes in the occupancy of 1 km grid squares in Britain for 214 hoverfly species (orange line) and 139 bee species (blue line), between 1980 and 2013, based on national databases. Green triangles show years with notable increases; red circles show years with notable decreases.

Newly established solitary bee populations are less likely to be affected by parasites, reducing the energy and time that bees need to expend in defence and perhaps partly explaining the rapid spread of some of our new colonists. Sooner or later, parasites are likely to catch up, stabilising run-away populations. As mentioned, there is evidence that *Colletes hederae* escaped from *Epeolus cruciger* cuckoos – which were present in *Colletes hederae* refugia south of the Alps – during the postglacial range expansion of *Colletes hederae* (Kuhlmann *et al.* 2006). British *Colletes hederae* nesting aggregations, sometimes numbering in the tens of thousands, have so far seemed to be free of cuckoos, but there is recent evidence that *Epeolus cruciger* is now present in some populations in Essex and elsewhere (TB) and *Epeolus variegatus* has appeared at some *Colletes hederae* aggregations in Suffolk and Sussex (A. Knowles, M. Edwards, pers. comm.)

Solitary bee populations appear to be in a state of continuous flux and adaptation. However, against this background of gains and losses of individual bee species, there is evidence of a steady decline in bee numbers generally, which has accelerated since about 2007. The recent losses have been particularly marked amongst already scarce species and bees of upland populations (Powney *et al.* 2019) (Fig. 322). The challenge of conserving our solitary bees is discussed further in the final chapter.

CHAPTER 10

Ecology and Conservation

Enclosure like a Bonaparte let not a thing remain,
It levelled every bush and tree and levelled every hill
And hung the moles for traitors – though the brook is running still
It runs a naked brook, cold and chill.
                                    Remembrances, John Clare (1908)

INTRODUCTION

For most of us, solitary bees inhabit an unmarked space in our mental map of the natural world. If they have survived, it is by accident – no-one tried to protect them. In fact, it is a bit worse than this. The real, material spaces they currently occupy are often ones that are negatively valued by the dominant culture: mineral extraction sites, ex-industrial 'brownfield', unkempt 'wasteland', industrial spoil dumps, neglected corners, dead trees, broken-down walls, weed-filled urban locales suffering planning blight. Along with so much wildlife that has managed to survive in a human-dominated world, their survival has depended on neglect.

But is this still working? In the countryside, the industrialisation of agriculture, driven by government policy and powerful commercial pressures, has been for several decades decimating the remaining traces of inherited natural as well as humanly modified ecosystems. Ancient woodland, flower-rich meadows, lightly grazed downland, even hedgerows, orchards, copses and cottage gardens have either disappeared or become reduced to scattered remnants; the species

FIG 323. (a) Arable agriculture, north-west Essex. (b) A South Downs nature reserve with arable agriculture in the distance.

they harboured are increasingly dispersed into isolated, genetically uniform local populations, vulnerable to pesticide poisoning and local extinction. There is no longer a place for 'neglect' in the rural landscape, and even where habitats are protected, the situation is alarming. Perhaps the most striking evidence of this came from a study of 63 protected sites in Germany (Hallmann *et al.* 2017). Insects were collected in malaise traps over 27 years, from 1989 to 2016. Over that period, the total biomass of flying insects had declined by almost 81 per cent in midsummer, and rather less early and late in the year, giving an overall decline of some 75 per cent. The research method did not allow for an explanation of the decline, but as surrounding land was predominantly farmland, it seems a reasonable guess that agricultural change in the wider landscape was implicated.

FIG 324. (a) A housing development on a Dorset heath. (b) Municipal grassland management.

If agricultural intensification is the main threat in the countryside, what of our towns and cities? How do the denizens of our urban landscapes fare? As new housing estates colonise the countryside at the edges of towns, or infill neglected urban spaces, habitats are lost, to be replaced by ironic street names: Meadow View, Woodland Drive, Birch Close and the like. The combination of empowered developer interests and impoverishment of local planning authorities seems to point inexorably to the destruction or adverse management of remaining urban wildlife habitat. A long history of Victorian philanthropy and successive generations of visionary campaigners for garden settlements, urban parks, green spaces and allotments have left a valuable legacy, but it is one that is becoming increasingly difficult to defend. This is especially so in south-east England, where the commercial value of land is high. Even when urban green spaces have been preserved, the standard methods of municipal management of grassland still prevail in many towns: a mixture of 'lollipop' (often ornamental) trees and a close-mown grass monoculture.

So, the prospects for non-human life both in our towns and in the countryside look bleak – especially if we project our gaze forward to consider the advancing existential threat of climate change. The invertebrates, including solitary bees, though less noticed, are no less threatened than their more favoured furry and feathered relatives. We seem to be on the verge of an 'insect apocalypse' or, at least, a 'crisis' (Goulson 2021, Milman 2022). However, none of this is inevitable. As Goulson argues, the apocalypse can still be averted.

There are two distinct, but increasingly intertwined, sources of hope. The first of these has come to be termed 'citizen science' but is perhaps better thought of as a popular tradition of amateur natural history. As early as the 17th and 18th centuries, notable figures such as John Ray and Gilbert White were making and recording detailed observations of their local wildlife, and from the early decades of the 19th century, amateur natural history became a popular national enthusiasm, with hundreds of best-selling books and even a periodical literature. This led to the formation of many local natural history societies, where amateur naturalists organised field trips to witness, enjoy and record the denizens of woods, downs and wetlands within reach of their towns and cities. Support for these societies has ebbed and flowed in the 150 years or so since they were formed, but since the 1950s, a growing army of field naturalists has been registering alarm at detectable decline in birds, butterflies and other wildlife in their favoured rural 'hotspots'. This alarm was intensified and augmented by the beginnings of a new environmental movement, marked most famously by Rachel Carson's *Silent Spring* (1962) and her challenge to the toxic chemical onslaught being unleashed on the surface of the Earth. The amateur naturalists have not been, in the main, militant participants in the environmental movement, but their organisations eventually turned from simply enjoying their encounters with nature towards a more focused attempt to record and map the wildlife of their local area. County-wide groupings began publishing atlases of the flowering plants, butterflies, moths, dragonflies, mammals, fungi, and even bees, wasps and ants. In turn, the evidence produced by these efforts gave rise to national and international organisations devoted to raising alarm and promoting conservation measures for their favoured organisms: the Bees, Wasps and Ants Recording Society, Butterfly Conservation, the Bumblebee Conservation Trust, the British Dragonfly Society, Buglife and numerous mapping schemes complemented the already powerful advocacy for

FIG 325. (a) Amateur naturalists recording downland plants. (b) A banner used during the 2019 school strike.

birds (the British Trust for Ornithology and the Royal Society for the Protection of Birds) and the county Wildlife Trusts (formerly Naturalists' Trusts).

Taken together, these developments represent the formation of a powerful civil society coalition, capable of exercising influence at national level, but also strongly rooted in local constituencies and well placed to keep watch on threats to habitats or adverse management of wildlife-rich localities in their territory.

The second source of hope had different beginnings but has become increasingly intertwined with the first. Scientific concern about global declines in biodiversity grew in the context of a number of international initiatives, mainly coordinated through the United Nations from the 1950s onwards. As colonies overthrew or were released by the colonial powers, there was concern about what might happen to the huge reservoirs of biodiversity they held (previously, 'conservation' had been largely in the service of game-hunting interests of the colonisers). From the rich countries (global north) came a demand that conservation measures be imposed. However, this was resisted by poorer countries, who demanded assistance in their efforts at economic development (Adams 1992). The long and complex process of integrating biodiversity conservation and a certain vision of global economic justice eventually produced a conception of 'sustainable development' that was given at least partial realisation in the Earth Summit of 1992 (the UN Conference on Environment and Development, held in Rio de Janeiro). This yielded two important binding agreements: the Framework Convention on Climate Change and the Convention on Biodiversity. The latter's key ideological shift was to replace a view of biodiversity as the 'common heritage of humankind' with a recognition of national sovereignty over natural 'resources'. The convention imposed on signatory nations an obligation to build up an accurate account of their biological resources and to plan their conservation and sustainable use. At the same time, access to national biological resources on the part of outside bodies was not to be unreasonably prevented, so long as the benefits of any resulting biotechnology were equitably shared. Although cultural and aesthetic values of biodiversity were mentioned in passing, the key commitments of the convention are to recognise biodiversity as a set of economic resources and to ensure their sustainable exploitation (Grubb *et al.* 1993).

The economic argument for biodiversity conservation, especially in the context of widespread concern about an impending 'pollination crisis', provided a great stimulus for scientific research into wild pollinators and, especially, solitary bees (Ollerton 2021). This research has added immensely to our understanding of bees and their ecology and has provided evidence for field naturalists and environmentalists to use in their campaigns for local authority planning, green-space management and wider policies to be rendered more 'bee friendly'. At national government level, too, the plight of pollinators has been recognised. However, as we argue below, the advancing use of economic justifications for conservation has severe limits and may even be counter-productive, diverting attention from the most pressing ecological problems and conservation priorities.

WHAT DO BEES NEED?

Before we consider the complex and often controversial issues surrounding efforts to conserve our solitary bees, it may be worth reviewing what we've learned in the earlier chapters of this book about the conditions needed for the maintenance of flourishing populations of these insects. First, and most obviously, both adults and immature offspring are entirely dependent for their nutrition on flowering plants. Many well-meaning advocates of bee conservation act as though this were their only requirement. Bees and flowers have evolved together, in relations sometimes characterised as mutualistic, sometimes as mutually exploitative. Either way, there are complicated relations of interdependency, with bees relying on pollen, nectar and, in a few species, plant oils for food, while flowering plants mostly depend on insects for pollination, and so for sexual reproduction. Among the insect visitors to flowers, bees are generally regarded as among the most important agents (vectors) of pollen transfer.

So, an abundance of flowers in a habitat is one condition necessary to sustain a population of bees. But the evolution of the relationship between bees and flowers has introduced further complexities. Bees vary in their dietary breadth. Some of them (oligoleges) specialise in gathering pollen from a restricted range of flower types, while others are able to forage for pollen from a very wide range of flowers. Consequently, habitats capable of sustaining a diverse community of bee species can be expected to offer not just abundant pollen and nectar but also a wide diversity of flower types to complement the evolved associations between particular groups of bees and their favoured pollen hosts (see Chapters 5 and 6).

Floristic diversity is important for another reason. The timing of bee life cycles varies greatly from species to species: some fly in early spring, others later in the year, while others fly both early and late, and still others have prolonged flight periods, needing suitable flowers in bloom over several months. For the bees with more flexible relations to the flowers they visit, the availability of at least some plants in flower during their flight period is essential, but for the species with restricted pollen diets, matching the time of their flight period with that of their favoured pollen host is crucial. So, habitats capable of sustaining diverse bee communities can also be expected to have a continuous sequence of varied plants in flower from early spring through to late summer.

Apart from the demands of a seasonal climate, bees depend on suitable weather conditions to be able to continue foraging, nest-making and so on. Some species have ways of generating heat internally, and some are adapted by size or by a dense coat of hair to forage at low temperatures. Others make use of flower bells or exposed basking sites to maintain sufficient body temperature for flight. They also need shelter to withstand inclement weather and for protection overnight. Sometimes they are able to utilise the same sorts of cavities used for nesting, but flower heads and other plant structures are used by many, especially males (see Chapter 9).

The majority of bees make nests and stock brood cells with pollen and nectar for their offspring. As we have seen, these nests may be constructed in burrows in the soil or may be placed in pre-existing cavities above ground – in hollow plant stems, in the abandoned tunnels of wood-boring beetles or, in some specialised examples, in empty snail shells or in vacated plant galls. Increasingly, cavity nesters benefit from human artefacts, provided incidentally (old walls, for example; Ollerton 2021) or deliberately (bee hotels). So, habitats capable of sustaining a diverse bee community will have flat or sloping bare ground, friable soil or sand, some dead wood, hollow-stemmed herbs and shrubs, and other provisions for cavity nesters. The cuckoo bees make up a substantial minority of our bee fauna, and though they do not make their own nests, they are, of course, indirectly dependent on the ability of their more industrious relatives to find suitable nest sites. In general, they are dependent on flourishing populations of their host species (see Chapters 3 and 8).

A few species are capable of modifying existing cavities, carving out space for themselves in wood or clearing pith from hollow stems, while still others collect materials for lining their brood cells and plugging their completed nests. These materials include mud, cut leaf fragments, plant hairs, small stones and resin. Most of these are widely available in most habitats that are capable of sustaining populations of bees, but any full account of bee ecology would need to keep them in mind.

As sexual beings, solitary bees depend on opportunities for males and females to find and identify one another. As we saw in Chapter 2, male bees have many different ways of finding females, including patrolling nest sites for emerging virgin females, scouring patches of appropriate flowers, establishing territories, or scent-marking patrol routes. Most of these strategies make no extra demands on suitable habitat, though structural complexity, often including shrubs and other prominent features, seems to be important, especially for the use of scent-marked routes.

Finally, according to species, solitary bees face a range of challenges from competitors, predators, parasites and diseases. These have significant consequences in regulating the populations of affected species and can alter the composition of the bee communities at any particular site. Nest parasitism by cuckoo bees, as well as parasitism by parasitic wasps and flies, takes its toll on immature stages, while flowers are now known to be important sites for the transmission of microbial diseases as well as 'friendly' microorganisms. Although differences in pollen dietary preferences, size, flight period, temperature regulation, morphology of mouthparts and so on must produce a degree of resource partitioning among foragers, this is by no means complete. There is likely to be considerable competition among species for floral resources at any site. One aspect which has recently attracted attention is the capacity of domesticated honeybees to out-compete wild bees, especially in areas with limited floral resources and a high density of beehives (Corbet *et al.* 1995, Lindström *et al.* 2016, Norfolk *et al.* 2018).

A recent study by one of us (TB) reduced these conditions to a list of seven habitat requirements for a flourishing bee assemblage: shelter, thermoregulation (basking sites), nectar sources, pollen sources, mating habitat, nest sites and, for some species, nesting materials (Benton 2021a). At risk of offending some readers, we might have added that fewer honeybees would be a benefit to solitary bees in many habitats (Whitehouse 2021).

WHERE TO FIND BEES?

As we mentioned in Chapter 9, the diversity of bee species in Britain is very strongly linked to climate. Geographically, Britain lies well to the north of the parts of Europe with the highest bee diversity, and within Britain, there is a marked north–south gradient, with many more species recorded in the southern counties of England and a steady decline northwards. Baldock (2008) compared bee species totals (including bumblebees and the honeybee) recorded from a selection of English counties and regions, with over 200 species each from Surrey, Kent and Essex, falling to 86 in the north-east. In Chapter 9, the gradient is presented in terms of solitary bee species recorded from 100k latitudinal bands. The southernmost four (taking us from the south coast to approximately north Lincolnshire in the east and north Wales in the west) each have 200 or more recorded solitary bee species, with gradual decline to only 10 species in the northernmost band (northern Scotland and the Northern Isles). Of course, this is not a static picture. Climatic change seems to be allowing some species to extend their ranges northwards. Also, greater availability of suitable habitat in more northerly localities can compensate for climatic limitations. There is also a small number of species that have northerly or upland distributions. These are of special conservation concern and may be suffering a narrowing of their climatic 'envelope'.

Within their climatic range, bees are to be found wherever there is habitat that enables them to meet their needs. There are several natural or semi-natural habitats that are associated with flourishing bee populations, but these have been in decline in recent decades and have been to some extent replaced by other habitats, often by-products of human activity.

We begin with the natural and semi-natural habitats.

Chalk downland
The term 'downland' refers to the mainly southern English, flower-rich grasslands formed on thin alkaline soils overlying chalk deposits. They were formed over many centuries, mainly by sheep-grazing, but also in some areas grazing by rabbits has been important. The largest remaining area is the Ministry of Defence's (MoD's) Salisbury Plain, but other stretches include the Berkshire and Dorset Downs, as well as the North and South Downs. These stretch across the warmer, sunnier southern and south-eastern counties of England, offering favourable climatic conditions for bees, as well as freely draining soils, warm, south-facing slopes, and a rich botanical diversity. Soil movements on steeper slopes, footpaths, vehicle tracks, poaching by cattle and sheep, rabbit scrapings and the activities of ants provide open ground and an uneven topography for

ground-nesting bees, and the botanical diversity provides forage for a wide range of species, including pollen specialists. Notable downland species are scabious specialists the Large and Small Scabious Mining Bees *Andrena hattorfiana* and *Andrena marginata*, *Campanula* specialists the Bellflower Blunthorn Bee *Melitta haemorrhoidalis* and the Small Scissor Bee *Chelostoma campanularum*, as well as relict populations of the Downland Furrow Bee *Halictus eurygnathus* and the Potter Flower Bee *Anthophora retusa*. Baldock (2008) also includes the three snail-shell nesters, the Gold-fringed, Red-tailed and Spined Mason Bees (*Osmia aurulenta*, *Osmia bicolor* and *Osmia spinulosa*) as occurring mainly on the chalk in Surrey. Abundance of flowers in the pea family (Fabaceae) also favours species that specialise in this group of plants, notably the Clover Blunthorn Bee *Melitta leporina* and, in some places, the Long-horned Bee *Eucera longicornis*. The rare Sainfoin Blunthorn Bee *Melitta dimidiata*, which specialises in Sainfoin pollen, is almost confined to the grassland of Salisbury Plain. Baldock (*ibid.*) lists several Surrey downland sites with well over 70 recorded species, including one with 84.

Since the late 18th century, changes in agricultural economics and technology have led to a steady loss of downland to the plough (see Figure 323 b, above), but this process was greatly accelerated from the 1950s onwards, fuelled by advancements in agricultural machinery and state provision of financial incentives for agricultural intensification. Shoard (1980) gives an eloquent account of the beauty of the pristine grassland on the South Downs, and of the human pleasures associated with it, denouncing both the human and ecological loss wrought by its destruction (on a similar theme, see also Bangs 2008). Some respite has been given by such designations as Area of Outstanding Natural Beauty and Environmentally Sensitive Area and, from 2011, designation of the South Downs as a national park. Despite this, by 2011, floristically rich chalk grassland comprised only 3 per cent of the South Downs in Sussex. Combined with the encroachments of arable agriculture, but opposite in its effects, was a steep decline in grazing, most notably by rabbits in the wake of myxomatosis. Large areas of grassland have undergone succession to scrub and secondary woodland. More recent recognition of the importance of downland for biodiversity has led to remaining areas being taken into conservation management, often in an effort to save populations of butterflies such as the Silver-spotted Skipper *Hesperia comma* and Adonis Blue *Polyommatus bellargus*. In general, bees seem to have also benefited from these management practices.

An important six-year survey of the bees, wasps and flies of 15 downland sites in East Sussex by Steven Falk (2011) yielded a total of 133 species of bees, including 118 species of solitary bees. While his study focused on remaining areas of flower-rich chalk downland, his analysis showed the importance of adjacent habitats, including hedgerows, semi-improved pastures, arable margins and blocks of

FIG 326. (a) A downland nature reserve (South Downs). (b) Adonis Blue butterfly *Polyommatus bellargus* male, a priority species for downland management.

woodland. Scrub such as hazel and bramble was important for aerial-nesting species, and blackthorn and hawthorn for early forage, especially for *Andrena* species. Moderate grazing by cattle and horses, too, depending on timing, location and stock levels, might enhance total diversity of bees.

Heather heath

Heaths are commonly defined as lowland areas with acidic soils, poor in nutrients, and dominated by dwarf ericaceous shrubs, notably Ling *Calluna vulgaris*. They have been represented in literature as barren, treeless wastes,

FIG 327. (a) A New Forest heath. (b) A Suffolk heath, showing bare ground nesting habitat. (c) A vehicle rut providing nesting habitat for heathland specialist bees at Kelling Heath, Norfolk.

offering only meagre sustenance for the poor rural communities associated with them. It is now known that human activity over many centuries played an active part in both establishing and maintaining heathlands, from as early as the deforestation initiated in Neolithic times. In the 20th century and since, the lack of a significant role in rural economies has led to large-scale losses of heathland through neglect, building and tree-planting (Chadwick 1982, Webb 1986, Chatters 2021). These losses began even before the 20th century, as noted by Thomas Hardy in his preface to *Return of the Native*. The unity of the heaths was now 'disguised' by 'intrusive strips and slices brought under the plough with varying degrees of success, or planted to woodland' (Hardy [1895] 1978).

Only recently has the value of heathland as wildlife habitat become recognised and, especially, its importance for invertebrates – bees among them. Current concerns, and conflicts, focus on how to manage remaining heathland to enhance its benefits to wildlife (Cox & Bealey 2022). The best-studied heathlands include those that remain in Dorset, the New Forest, Surrey, the Suffolk Sandlings and the East Anglian Brecks. Baldock (2008) reported on the bees and wasps of more than 20 Surrey heaths, eight of which had 90 or more recorded bee species. Owens (2017) noted the importance of heathland's contribution to the Norfolk bee fauna, with 69 species recorded from Kelling Heath on the Cromer Ridge.

The main explanation for the rich bee (and wasp) fauna of heather heaths is the hot microclimate and friable, sandy soils, which are especially favourable for ground-nesting bees. Gorse and broom offer sources of pollen in spring, while from midsummer onwards, the main plants in flower are Cross-leaved Heath *Erica tetralix* and Bell Heather *Erica cinerea*, followed by Ling. However, the acidity of the soils greatly reduces the botanical diversity. A small, distinctive group of bees (including the Small Sandpit Mining Bee *Andrena argentata*, the Heather Mining Bee *Andrena fuscipes* and the Heather Colletes *Colletes succinctus*) specialise in harvesting heather pollen and have flight periods coinciding with the flowering of the heather.

These are, of course, only a small minority of the bee species that inhabit heathlands, an indication that many nesting species must forage on parts of the heath where

FIG 328. Heather Colletes *Colletes succinctus* female, a distinctive heathland specialist bee.

some vegetation succession has taken place, or in other adjacent habitats. Over several years, the late David Baldock and others made intensive studies of the aculeate insects of Farnham Heath, in Surrey (Wood & Baldock 2016, Baldock & Wood 2018, Baldock 2020). By 2019, fully 301 bee, wasp and ant species had been recorded on the site, and this total included 126 bee species, more than 60 per cent of the species list for the whole of Surrey. Commenting on the same heath, Darwin (1859, cited Baldock & Wood 2018) had noted the effect of grazing in suppressing the growth of trees on unenclosed parts of the heath. The heath had subsequently been planted with pines, and this study took place soon after the new owner, the Royal Society for the Protection of Birds (RSPB), had started to remove the trees in order to restore the open heath. As the authors point out, the abundance of aculeates was to be explained in part by the restoration of favourable nesting conditions, but also by transitional zones, bare ground, root plates, fence posts and areas of ruderal vegetation. In short, this extraordinary richness of bee and wasp species was specific to this particular site and at a possibly transitory period of time, during which the specifically heathland conditions were complemented by other habitat patches and differentiated nesting and foraging opportunities.

Lowland dry acid grassland

Recognised as a UK Biodiversity Action Plan (BAP) priority habitat from 2008, acid grassland shares several bee-friendly features with heather heath. It is characterised by its freely draining sand or gravel substrate, open texture, with bare ground, and provision of rapidly warming microclimates. A combination of low-nutrient soils and grazing pressure slows down succession to scrub and woodland, though these habitats are often present to some extent, especially at the margins.

There is suitable habitat for ground-nesting species, as in heather heath, though the small number of species associated with heather are absent. Although the flora is limited, these grasslands hold an abundance of the yellow-flowered composite plants of the daisy family, Asteraceae. These attract *Andrena* species such as the Ashy Mining Bee *Andrena cineraria*, the Grey-patched Mining Bee *Andrena nitida* and many of the furrow bees belonging to the genera *Lasioglossum* and *Halictus*.

Important areas of this habitat include moorland in Devon and Cornwall, the Pembrokeshire coast, the New Forest, the Weald of Sussex and Kent, the East Anglian Brecks and some large commons to the south and east of London. Baldock (2008) lists 12 acid grassland sites in Surrey, including Mitcham Common with 89 bee species and Kew Royal Botanic Gardens with 84 bee species, recorded in 2001–2. Follow-up surveys at Kew from 2015–18 raised the total to 107 species, with 81 species common to both surveys. Differences between the two

FIG 329. (a) Middlewick MoD Range, Essex. The site has an exceptionally rich invertebrate fauna, including at least nine Section 41 species (Species of Principal Importance), and is now allocated for housing. (b) Bare bank by the 'butts', Middlewick Range, nesting habitat for ground-nesting bees. (c) Green-eyed Flower Bee *Anthophora bimaculata* female. (d) The scale of opposition to 'development' of the site was locally unprecedented.

surveys suggested considerable fluctuations in the bee population (H. Koch pers. comm.). Similarly rich sites are to be found associated with Epping Forest, on the borders of Greater London and south-west Essex.

With little value as agricultural land, acid grassland has been subject to building development, agricultural 'improvement' or loss to scrub and woodland succession. Some sites (such as those associated with Epping Forest) are protected by Site of Special Scientific Interest (SSSI) designations, but others continue to be lost to 'development'. One such threatened site is an MoD firing range (Middlewick Range) in Colchester, Essex. Incorporated into the Local Plan when offered up by the MoD, it was proposed as a site for 1,000 houses. Its importance as an invertebrate habitat had already been recognised, and the plan was intensively resisted by a local coalition of residents, naturalists and expert ecologists. The MoD's ecological consultants offered a 'walkover' survey of the acid grassland in poor weather, together with a proposed mitigation in the shape of a promised 10 per cent Biodiversity Net Gain (BNG), to be achieved in seven years by application of acid to an arable field. Despite a devastating critique delivered by an independent consultancy and other expert submissions, and more than 1,000 objections from the public, the plan was accepted as sound at a subsequent examination. This case is among a growing number that raise important questions about the use of BNG as a restraint on destruction of biodiverse sites, and we will return to it later in the chapter (Lonsdale 2022).

Upland sites: mountain and moorland

In Britain, these habitats are principally located in the north and west, so in addition to the influence of altitude, climatic conditions differ from those that favour many bee species in southern and eastern parts of the country. The Tormentil Mining Bee *Andrena tarsata* illustrates this well. It occurs on some upland sites in Cornwall, where it has been closely studied and actively conserved by Patrick Saunders (Saunders 2020), as well as in moorland habitats in northern England, Wales and Scotland (Hargreaves 2021). This is a scarce and declining bee which could be adversely affected in its more southerly localities by climate change. Other such species include the very rare Perkin's Mining Bee *Andrena rosae*, which is coastal in Cornwall but also occurs at high altitudes in Dartmoor (Saunders 2020). The Northern Mining Bee *Andrena ruficrus* is a highly localised species of Scotland (Jardine & McGregor 2021) and northern England, reported from the South and West Pennines and near Sedgefield (B. Hargreaves & E. Lamborn pers. comm.). The Heather Mining Bee *Andrena fuscipes* inhabits moorland edge in the northern part of its range (Falk & Lewington 2015), as does the Small Flecked Mining Bee *Andrena coitana*. The Bilberry Mining Bee *Andrena*

FIG 330. (a) Bartinney Nature Reserve, Cornwall, a locality for the Tormentil Mining Bee *Andrena tarsata* and Perkin's Mining Bee *Andrena rosae*. (© P. Saunders) (b) Meall Gruaim, Perthshire, a historic site for the Mountain Mason Bee *Osmia inermis*.

lapponica occurs on moorland in its northerly localities, where it collects pollen from bilberry and other *Vaccinium* species.

Woodland

The quality of woodland habitat for bees depends on many variable features of climate, substrate and management regime. In plantations with clear-fell, or managed woodland with coppice, wide rides, glades and suitable marginal habitat, there are often bare banks, cleared ground and root plates used as nest sites and flowery areas with abundant forage. Woodlands are often rich in early-spring flowers, such as Bugle *Ajuga reptans*, Ground Ivy *Glechoma hederacea*, dead-nettles (Lamiaceae) and willows (*Salix*), with Bluebells *Hyacinthoides non-scripta*, Wood Sage *Teucrium scorodonia* and various vetches and other Fabaceae later on. A few species that make their nests in dead wood are associated with woodland, notably the rare Fringe-horned Mason Bee *Osmia pilicornis*, which uses Bugle as its main pollen host in Britain. Another very rare species, the Oak Mining Bee *Andrena ferox*, specialises in oak pollen and so is closely associated with mature

FIG 331. (a) Kent woodland, habitat of the Fringe-horned Mason Bee *Osmia pilicornis*. (b) A woodland ride in early spring, habitat of Clarke's Mining Bee *Andrena clarkella*.

oak woodland. Clarke's Mining Bee *Andrena clarkella* is a spring-flying species that collects pollen almost exclusively from willows (*Salix*) catkins and often nests among tree roots and path edges in woodland.

Wetland

The majority of bee species are associated with dry, warm habitats, but a small number are associated with wetlands. The Reed Yellow-face Bee *Hylaeus pectoralis* is one of the most interesting of these, making its nest in abandoned galls of the fly *Lipara lucens*. These are found in reedbeds, but usually in drier areas where

FIG 332. (a) Habitat of the Reed Yellow-face Bee *Hylaeus pectoralis*. (b) Wetland habitat of the Yellow Loosestrife Bee *Macropis europaea*.

there are flowers such as Creeping Thistle *Cirsium arvense*, brambles (*Rubus*) and various umbelliferous flowers (Apiaceae) (Benton 2021b). Another distinctive wetland species is the Yellow Loosestrife Bee *Macropis europaea*, which is able to waterproof its underground brood cells with plant oils provided by Yellow Loosestrife *Lysimachia vulgaris* flowers. Both males and females visit a range of other flowers for nectar.

Coastal habitats

In Britain, these are usually a mixture of natural, semi-natural and human-made habitats, often in close association with one another. In some places, cliffs provide vertical nest sites for ground-nesting bees, often with rich floristic habitat above, while in other places, there are complex dune systems with a hinterland of salt marsh, sea defences and grazing marshes. Though they are likely to be exposed to different climatic conditions, and to occasional inundation, partially vegetated coastal dunes can offer suitable habitat for a diverse bee fauna, with Shrubby Seablite *Suaeda vera* often giving a vegetation structure similar to that of the ericaceous shrubs of heather heath. An example is the extensive area of sand dune, shingle, salt marsh and sea wall at Colne Point, on the Essex coast. This site holds more than 40 species of solitary bees, as well as the BAP priority bumblebee, *Bombus muscorum*. The site is rich in mining bees (*Andrena*) and furrow bees (Halictidae), including the Squat Furrow Bee *Lasioglossum pauperatum*. Species of particular conservation interest include the Pantaloon Bee *Dasypoda hirtipes*, the Coast Leafcutter Bee *Megachile maritima*, its nest parasite, the Large Sharp-tail Bee *Coelioxys conoidea*, and the Silvery Leafcutter Bee *Megachile leachella* which here lines its nest with leaf cuttings from hawkbit. There is a vast population of the scarce and regionally important Sea Aster Bee *Colletes halophilus*, together with its nest parasite, the Black-thighed Epeolus *Epeolus variegatus*.

Coastal grazing marshes and associated sea wall habitats harbour a rich bee fauna if managed appropriately (Gardiner *et al.* 2015). In Essex, Fingringhoe Ranges (MoD) is sheep-grazed, with some mechanical mowing and rotational management of the sea wall grassland. Vehicle tracks provide valuable nesting habitat for ground-nesting bees, such as the Ridge-cheeked Furrow Bee *Lasioglossum puncticolle*, the Sharp-collared Furrow Bee *Lasioglossum malachurum* and several *Andrena* species. Abundant flowers in the pea family (Fabaceae) such as Meadow Vetchling, Bird's-foot-trefoil and Clovers (*Trifolium* species) provide forage for the Red-girdled Mining Bee *Andrena labiata*, the Large Meadow Mining Bee *Andrena labialis*, Wilke's Mining Bee *Andrena wilkella* and a strong population of the localised and declining Long-horned Bee *Eucera longicornis*. Sea walls

FIG 333. (a) Colne Point Nature Reserve, Essex. (b) Coast Leafcutter Bee *Megachile maritima* female.

extend for over 2,000 km in England and often have different soils, vegetation and microclimates on their different aspects – top, inner slope and flat 'folding' on the landward side – and this provides habitat niches for a range of coastal bee species (Gardiner, Pilcher & Wade 2015). Vehicle tracks on the foldings, and footpaths on the top, provide bare ground for mining bees, and where the management regime is appropriate, the foldings, especially, can be very flower-rich.

The vast shingle headland of Dungeness on the Kent coast provides a wide range of floral resources for bees and is well known amongst bee-watchers for the dramatic recent reappearance of the Grey-backed Mining Bee *Andrena vaga*

FIG 334. (a) Coastal grazing marsh, Essex. (b) Sea wall flood defences, Jaywick. (c) Sea wall and folding, MoD firing range, habitat of the Long-horned Bee *Eucera longicornis*. (d) Long-horned Bee male.

ECOLOGY AND CONSERVATION · 477

FIG 335. Lavernock point, south Wales, a coastal habitat for the rare Carrot Mining Bee *Andrena nitidiuscula*. (© L. Olds)

in Britain (Allen 2020; see also Chapter 3, Figs 80 a & 86). The bee had not been recorded since the 1940s before it was photographed on a sandy bank in the RSPB Dungeness reserve in 2014 by a friend of one of the authors, the latter recognising it as a species he had encountered in northern France. The sandy bank had been created on the reserve for solitary bees, and when inspected by the authors in 2018, it was inhabited by a large mixed aggregation of Grey-backed Mining Bees, Small Sallow Mining Bees *Andrena praecox*, Yellow-legged Mining Bees *Andrena flavipes*, Orange-tailed Mining Bees *Andrena haemorrhoa* and another recent coloniser of the site, Early Colletes *Colletes cunicularius*. This last species as well as *Andrena vaga* and *Andrena praecox* specialise on willow pollen which was in rich supply around the nearby pools on the reserve. The Grey-backed Mining Bee has since extended its range as far as Dorset and Norfolk.

The south Wales coast supports a very diverse range of solitary bees. Two of the most significant sites are Kenfig Dunes near Bridgend, Glamorganshire, and Pembrey Country Park a little farther west in Carmarthenshire. Kenfig has a large area of mature dunes and pools, whereas Pembrey is technically a brownfield site as it was used for explosive manufacture in the last war. Both sites are strongholds for some scarce solitary bee species, including the Gold-fringed Mason Bee *Osmia aurulenta*, which uses empty snail shells on the sand for its nest, and the Square-jawed Sharp-tail Bee *Coelioxys mandibularis* and its host, the Silvery Leafcutter Bee *Megachile leachella*, as well as the Large Sharp-tail Bee *Coelioxys conoidea* and its host, the Coast Leafcutter Bee *Megachile maritima*. Other species include the Sandpit Mining Bee *Andrena barbilabris* and the Bull-headed Furrow Bee *Lasioglossum zonulum*. Also in Glamorganshire, several populations of the rare Carrot Mining Bee *Andrena nitidiuscula* have recently been found, the first confirmed for Wales (Olds 2022). Pembrokeshire also hosts some scarce Andrenas, including the Black-headed Mining Bee *Andrena nigriceps*.

FIG 336. (a) Cliff habitat of the Potter Flower Bee *Anthophora retusa*. (b) Potter Flower Bee male.

Coastal sites in the north and north-west of Britain host several scarce and distinctive species and provide northern outposts of southern species. The Northern Colletes *Colletes floralis* is an almost exclusively coastal species of Ireland and Scotland, with a few records from Cumbria. Little *et al.* (2016) report on two coastal sites in Angus and Fife, Scotland, and Early (2013) offers an account of the aculeates of the Cumbrian coast.

CHARACTERISTICS OF SEMI-NATURAL BEE HABITATS

The above discussion of natural and semi-natural habitats that are valuable for bees yields two features, in particular, that they seem to share. The first is that they have all been subjected to long-run processes of decline, destruction and fragmentation in the face of neglect, building development or agricultural intensification. In recent decades, their value has been to some extent recognised, and the decline has slowed to some degree. Ancient woodland has been protected, and some habitats, such as flower-rich downland and heathland, have, in some places, been restored and extended. One example is Grimston Warren in west Norfolk, which was cleared of conifers by Norfolk Wildlife Trust around the year 2000 and has returned to fine *Calluna* heath. Of particular interest is the very large population of Small Sandpit Mining Bee *Andrena argentata* and its cuckoo, the Bear-clawed Nomad Bee *Nomada baccata* – each at its most northerly post in Britain. The Tormentil Mining Bee *Andrena tarsata* has appeared again after a lapse of many years. The Early Colletes *Colletes cunicularius*, a recent Norfolk colonist, the Margined Colletes *Colletes marginatus* and the scarce Green-eyed Flower Bee *Anthophora bimaculata* have all become established.

In a few places, such as Salisbury Plain and MoD ranges in north-east Essex, military occupation has incidentally offered protection from the plough and pesticides. The MoD Stanford Training Area in the Norfolk Breckland has escaped the onslaught of both pesticides and conifer planting since the 1940s. The site covers 120 square kilometres and is the focus of a Higher Level Stewardship-funded project to reinstate the intermittent cultivation with fallow breaks (brecks) once used to manage the very sandy environment. Ten percent of the Stanford Training Area (and other Breckland sites) is being cultivated by a mix of treatments and timings over a period of 10 years with the effects on fauna and flora being monitored. Evidence to date shows that 'priority species' – Breckland specialists, including solitary bees – are showing a positive response to the treatments. The rare Black-headed Mining Bee *Andrena nigriceps* has been recorded in the Brecks for the first time since the 1980s (Hawkes *et al.* 2020).

FIG 337. (a) A Breckland ride, habitat of the Red-tailed Mason Bee *Osmia bicolor*. (b) Flower-rich grassland on Salisbury Plain.

Elsewhere, such habitats usually persist as small, disconnected fragments on land that is too difficult or expensive for farm 'modernisation' or that has been secured by conservation organisations as local nature reserves.

The second common feature of natural and semi-natural sites is that even the best examples, harbouring many bee species, do not supply all the requirements of the bees on site. The great German entomologist, Paul Westrich, coined the expression 'partial habitats' to give expression to this. For bees in particular, habitat requirements do not necessarily all coexist in one place (Westrich 1996). Bees' requirements have been reduced to three by some authors: flowers, requirements for reproduction and 'supplements' (Ollerton 2021). But, as mentioned above, we have identified seven habitat requirements for at least some bees. The extent of decline and fragmentation of the natural and semi-natural habitats of bees has greatly increased the importance of the pattern of surrounding habitats for bee populations.

In short, any serious approach to bee conservation has to pay attention both to the very small scale (a broken wall, a patch of Ivy, an informal footpath, a vehicle track, and so on) and to the landscape scale: is there a mosaic of different habitat patches within flying range of the bees that contains all the resources and structural features they need? Is this mix of habitat patches on a sufficiently extensive scale to sustain viable populations of the bees, including the ones that are very specialised in their attachment to particular pollen hosts? For example, the scarce Large Scabious Mining Bee *Andrena hattorfiana* specialises in gathering pollen from Field Scabious *Knautia arvensis*. A study in southern Sweden calculated that, given an average rate of utilisation, a population of 10 nesting females would require more than 150 Scabious plants, or some 780 inflorescences (Larsson & Franzén 2007).

Where substantial areas of these primary habitats for bees have survived, usually as protected spaces managed by conservation organisations or official bodies, conservation poses questions about management methods and also habitat change in surrounding landscapes. An important study carried out on 40 sites across the island of Ireland analysed bee diversity and abundance on three spatial scales: local, landscape and regional (Murray *et al.* 2012). The sites included representative examples of five types of natural or semi-natural habitat and were all designated for their conservation value. Though bumblebees seemed less affected by landscape-scale conditions, communities of other wild bees (solitary bees, for our purposes) were strongly affected by surrounding habitats. For solitary bees, the habitat type had the greatest impact on diversity and abundance, and there was also high sensitivity to management regime. For example, in sites managed by grazing, the intensity of grazing was important; if too little, floral resources were limited by vigorous growth of grasses. Intense grazing had potentially contradictory consequences, either reducing flower diversity and abundance or, in chalk grassland, providing more nest sites through exposing bare ground. The greater sensitivity of solitary bees to features of the surrounding landscape (also shown by Steffan-Dewenter *et al.* 2002) may be related to their generally smaller size (compared to bumblebees) and shorter foraging ranges. Woodcock *et al.* (2013) also found that solitary bees (compared with honeybees and

FIG 338. Machair grassland, Inner Hebrides, managed for Corncrake *Crex crex* but with benefits for bees.

bumblebees) visited crops of Oilseed Rape *Brassica napus* more frequently where there was semi-natural habitat nearby. Their finding has some significance in relation to the use of wild bees as crop pollinators (see below).

Some conservation organisations have taken up the challenge of attempting to provide mosaics of habitat at sufficient scale through a combination of land purchase and voluntary agreements with landowners. In some places, the concept of 'living landscapes' pioneered by the county Wildlife Trusts has been successful, as has Buglife's mapping of 'B-Lines' and the RSPB's 'futurescapes' initiative, among many more local projects.

BEES IN TOWN AND COUNTRY

So far, then, the story has been one of loss and fragmentation of more traditionally recognised wildlife habitats with advancing urbanisation and conversion to intensive systems in agriculture. This leaves us with questions about how far retaining diverse populations of wild bees is compatible with the two most rapidly growing land uses: urban (Baldock *et al.* 2015, Wenzel *et al.* 2019) and industrialised agriculture.

Advancing urbanisation and industrial transformation have brought with them unintended consequences: buildings and transport networks created demand for mineral extraction and took strips of land out of agricultural production. The cycle of economic boom and bust, together with technical innovation, left in their wake derelict and abandoned land, quickly colonised by a motley array of native and alien plants (a process brilliantly depicted in Flyn 2021). Quarries, slag heaps, landfill, industrial wastelands, roadside verges and railway cuttings joined former grazing land and small remnants of semi-natural habitat in many parts of the country. Where neither public nor private investment has been forthcoming, these by-products of an otherwise nature-hostile economy have become hugely important wildlife refuges.

Perhaps the most (justly) famous of these is a large area on Canvey Island, south Essex, now a nature reserve and home to an outstanding diversity of invertebrate species, including many solitary bees. But Canvey Wick, precious as it is, owes much of its significance to its location within a much wider mosaic of habitats along both sides of the Thames, running from east London out towards the coast. By the 1990s, the area was seen as untidy, derelict wasteland, ripe for massive development, with associated housing, retail, industrial and port facilities under the broad title of the Thames Gateway.

Meanwhile, local ecologists had been reporting the uniquely important wildlife communities in the numerous interconnected sites along the Essex side of the river (Payne & Harvey 1996, Plant & Harvey 1997, Benton 2000). Harvey (2000) explained that the area had a unique combination of features favourable for invertebrates and for solitary bees in particular. South-east Essex has a distinctive climate, with hot, dry summers, high incidence of summer sunshine, and relatively warm winters. This combines with the generally south-facing aspects along the Thames, a mixture of acid and alkaline soils, including sands and gravels, London clay and chalk. There are relict areas of unimproved grassland, intermingled with areas of more recent flower-rich grassland derived from decades of colonisation of ex-industrial workings, silt deposits, and quarries. But these areas, too, are distributed among sandy banks and cliffs, patches of bare ground, and the informal habitat management achieved by bike-scrambling, occasional fires and horse-grazing. Harvey reported survey work that showed Thames Gateway sites held 96 per cent of the Essex aculeate fauna and fully 74 per cent of the national fauna. At just one site (already built on by the time of Harvey's article), there had been an 'Invertebrate Index' far greater than that of Salisbury Plain SSSI.

The climatic and geological character of the Thames Gateway landscape was the basis for a long history of human uses and impacts, including the groundwork for an oil refinery, subsequently abandoned, dumping of industrial waste, rough grazing, mineral extraction, and a range of illicit, often 'antisocial', activity, all of which contributed to the formation of an 'untidy', 'unsightly' liminal space, which at the same time provided conditions for the flourishing of a uniquely rich and diverse community of animals and plants. For the bees, there were extensive areas of flower-rich grassland, including a great diversity of both native and 'alien' wild flowers, as well as patches of scrub. Bare ground and south-facing banks and other exposures provided ample nesting habitat for mining bees, and cavity nesters were benefited by the lack of any formal management of the grassland. Overall, the absence of grass-cutting, artificial fertiliser, pesticides and herbicides was a favourable contrast to the treatment of both agricultural land and urban green space. A combination of the low-nutrient soils and informal recreation slowed natural succession, so areas of bare ground and sparsely vegetated areas remained as part of the mosaic.

As Harvey noted, by 2000 many of the most valuable habitats had already been built on or had planning permissions. In the case of what became Canvey Wick, part of the site had been purchased by the East of England Development Agency, and potentially damaging planning consents had been given. However, the survey work that had already been done and a comparison of the site with tropical rainforest, though inappropriate, helped to give publicity to the importance of the site. English Nature (under the determined leadership of Dr Chris Gibson) took up the challenge to defend the site, one of the first brownfield sites to be given such support, and an eventual compromise protected the most sensitive parts of the area as a nature reserve, owned by the Land Trust and managed by the RSPB and the invertebrate charity, Buglife.

Despite this success, and the subsequent inclusion of ecologically rich brownfield sites as priority habitats in the UK's Biodiversity Action Plan, many important habitat patches continued to fall to development. A 2011 review by Buglife (Robins & Henshall 2012) reported the loss of more than half of the important brownfield sites identified only six years previously, 'putting rare and endangered species in danger of local or national extinction'. Some large sites, such as Wat Tyler Country Park, Rainham Marshes and Benfleet Downs/Hadleigh Downs, persist, but it remains to be seen whether their rich biodiversity will be sustained as they become increasingly isolated and metapopulations of vulnerable species become fragmented.

So-called brownfield, ex-industrial sites have a special importance as refuges for wildlife, and for bees in particular, in northern Britain. Here we mention just a small number of examples. In the north-west, former quarries and landfill sites close to St Helens and Newton-le-Willows in the Sankey Valley and Mucky Mountains, to the west of Manchester, harbour many bee species, including

FIG 339. (a) Flower-rich grassland in an industrial landscape in Northwick, Canvey Island, south Essex. (b) Surprising orchids in Canvey Northwick.

the Black-headed Mining Bee *Andrena nigriceps*, the Catsear Nomad Bee *Nomada integra* and the Smooth-gastered Furrow Bee *Lasioglossum parvulum*. Farther north, close to Bolton and Blackburn, are extensively quarried and otherwise modified upland habitats in the South and West Pennines. These support localised species such as the Tormentil, Small Flecked, Bilberry and Northern Mining Bees (*Andrena tarsata, Andrena coitana, Andrena lapponica* and *Andrena ruficrus*), the Flat-ridged Nomad Bee *Nomada obtusifrons* and the Bilberry Nomad Bee *Nomada glabella*, as well as the scarce Bilberry Bumblebee *Bombus monticola*. The Lancashire Wildlife Trust's Brockholes Nature Reserve, a large former sand quarry, has a long list of bees and other aculeates (K. McCartney & B. Hargreaves pers. comm., Hargreaves & White 2021).

In the north-east, Louise Hislop (pers. comm.) makes particular mention of the Spetchells. These are a trio of long, narrow chalk spoil hills near Prudhoe in the Tyne Valley (to the west of Newcastle-upon-Tyne). The spoil forming the hills was produced by an ICI chemical factory and is capped by soil introduced from the south-west. The chalk grassland that has developed is very rich in solitary bees, with a list so far of 54 species, including what is probably the most northerly outpost of the Red-girdled Mining Bee *Andrena labiata* and a nesting population of some 250,000 Buffish Mining Bees *Andrena nigroaenea*. Ellen Lamborn (pers. comm.) has highlighted two sites in county Durham: Gravel Hole, on the outskirts of Stockton-on-Tees, and Maze Park, a local nature reserve made up of three large slag heaps. Gravel Hole is an old quarry, recently taken over by the Tees Valley Wildlife Trust. As at Canvey Island, local bike riders help to

FIG 340. Hadleigh Downs Country Park, Thames-side grassland.

FIG 341. (a) The Mucky Mountains, Newton-le-Willows, are spoil heaps formed of chemical by-products from soda production. They are a site for the Black-headed Mining Bee *Andrena nigriceps* and other scarce species. (© D. W. Owen) (b) Billinge Hill (St Helens), habitat for scarce and localised species, including the Black-headed Mining Bee and the Catsear Nomad Bee *Nomada integra*. (© B. Hargreaves)

reduce scrub encroachment. At least 11 species of *Andrena* mining bees have been recorded, as well as the Large Yellow-face Bee *Hylaeus signatus*, the Fork-tailed Flower Bee *Anthophora furcata*, the Fork-jawed Nomad Bee *Nomada ruficornis*, the Furry-bellied Blood Bee *Sphecodes hyalinatus* and many others (a total reported so far of 49 species). The *BWARS Newsletter* includes frequent updates on northern sites and species by several observers, including some of those cited above.

In south Wales, an inheritance of the former mining industry was extensive colliery spoil sites. These have been studied for more than seven years by Liam Olds for their invertebrate fauna, including bees. Between 2015 and 2021, Olds

FIG 342. (a) The Spetchells, hills formed by chalk spoil from a chemical factory in the Tyne valley, habitat of 54 species of solitary bees and showing tumuli formed by a huge aggregation of the Buffish Mining Bee *Andrena nigroaenea*. (b) Allendale valley, spoil from old lead mineshafts, with a large population of the Heather Colletes *Colletes succinctus*. (Both © L. Hislop)

recorded 100 bee species, including 83 solitary bees, from these sites. This is a list comparable with some of the best primary habitats such as acid grassland and heaths, and it includes several scarce species of mining bees, such as *Andrena apicata*, *Andrena bucephala*, *Andrena congruens*, *Andrena falsifica*, *Andrena tarsata* and their cuckoos, and the Early Colletes *Colletes cunicularius*.

Another study, this time of spoil heaps and quarries in Warwickshire, was conducted by Steven Falk (Falk 2006). Fourteen sites were studied for their assemblages of bees and wasps between 1990 and 2002, and 189 species were recorded, many of them of high conservation status. The list included some 85 species of solitary bees, and the study went on to inform the subsequent Warwickshire local Biodiversity Action Plan.

FIG 343. Habitats formed on former colliery sites in south Wales: (a) Gelli Tips, Rhondda Cynon Taf, left to regenerate naturally since the 1960s, with heath/moorland species such as the Bilberry Mining Bee *Andrena lapponica*, the Heather Mining Bee *Andrena fuscipes* and the Heather Colletes *Colletes succinctus*. (b) As above, the key nesting location for bees on this site,

including the Red-backed Mining Bee *Andrena similis*. (c) Blaenserchan Valley, near Pontypool, habitat for the scarce Long-fringed Mining Bee *Andrena congruens*. (d) Craig-Evan-Leyshon Common, near Pontypridd, habitat for numerous *Andrena* and *Lasioglossum* species and their cuckoos. (All © L. Olds)

URBAN BEES SURVIVE IF GIVEN A CHANCE

Areas like the Thames Gateway landscape would be hard to characterise as either urban or rural. A series of changes to planning law since the Town and Country Planning Act of 1947 has allowed for erosion of the boundaries between urban and rural settlements, as well as expansion of urban infrastructure into rural hinterlands and associated changes of land use around the margins of towns and cities, while urban areas continue to make significant contributions to food production. For this reason, attempts to compare urban and rural settings in their capacity to sustain rich assemblages of bees are somewhat artificial. Still, many valuable pieces of research have attempted to do this. A review of published studies comparing urban with rural pollinator diversity found equal or higher diversity in urban areas in 12 (out of 21) of those that focused on bees alone (Ollerton 2021).

Accepting, for now, the assumed urban–rural divide, the importance of urban landscapes for bees (and other invertebrates) is increasingly recognised with new sources of evidence that run counter to 'the inherited historical view of the general public, that urban environments are biological deserts' (Hall *et al.* 2017). An important international review (Wenzel *et al.* 2019) of 141 studies of pollinator responses to urbanisation revealed a very mixed set of results. Where urban habitats were compared with natural or semi-natural rural ones, they were found to be less diverse, but when compared with intensively managed agricultural habitats, they fared much better. The authors also made a significant distinction among urban landscapes: 'urban sprawl' versus densely built-up, usually central, areas. At landscape scale, urban sprawl had more green spaces with more floral resources and bee-nesting habitat. Built-up urban areas were characterised by a higher proportion (more than 50 per cent) of their land area with impervious surfaces, including buildings, streets and hard-surfaced gardens, and with correspondingly fewer floral resources and nest sites. These latter features were critical for pollinator abundance and diversity at the local scale. Urbanisation was also found to play a strong filtering role in shaping the community of flower-visiting species, depending on a range of traits, such as flight period, size, dietary breadth and nesting behaviour. Denser urban landscapes were shown to favour cavity nesters, as against ground nesters, for example. The review concluded that urban flower-visiting insects were generally sufficient to pollinate both wild and crop plants, the latter being of particular importance in countries where urban food production is crucial for the livelihoods of urban poor communities.

However, even urban centres may harbour significant bee populations. The Four-banded Flower Bee *Anthophora quadrimaculata* and its very rare cuckoo, the Grooved Sharp-tail Bee *Coelioxys quadridentata*, are unusual in being found predominantly close to the centre of cities, including London, perhaps because of the availability of old walls for nesting, alongside well-stocked gardens. This *Anthophora* is now common in Norwich, and a study in one Norwich garden found that this long-tongued species depends on flowers with a deep corolla, including catmint, Ivy-leaved Toadflax *Cymbalaria muralis*, Red Valerian *Centranthus ruber*, lavender and lobelia (Bartlett & Bartlett 2018). Some populations of solitary bees generally associated with rural habitats have survived encirclement by urban extension; for example, a population of the Large Scabious Mining Bee *Andrena hattorfiana* which continues to survive at Earlham Cemetery in Norwich.

A study of 18 grassland sites, featuring cavity-nesting bees, in Louisville, Kentucky, USA, confirmed the view that this group of bees is favoured by urbanisation (Sexton *et al.* 2021). Using bee boxes situated centrally in each of the sites, the authors studied three aspects of bee demography: fecundity, survivorship of larvae and numbers of emerging adults. Fecundity (number of eggs in each cavity) and emerging adults were positively related to the proportion of native flowers to alien species in the local environment, and survivorship was positively related to degrees of urbanisation at the landscape scale. Though the study did not investigate possible causes of this relationship, the authors considered 'enemy release' may have been at work, brood parasitism being reduced in urban habitats. A possible explanation of the negative effects of higher proportions of alien flowers is that honeybees might be favoured, outcompeting less versatile native bees.

In Britain, a study of city-centre sites in Northampton recorded 50 bee species, with the likelihood that this was an underestimate (Sirohi *et al.* 2015, Ollerton 2021). Another study, which compared a wider range of flower-visiting insects in and around 12 large towns and cities across the UK (including England, Wales and Scotland), reported broadly similar levels of abundance and diversity of species on urban sites, farmland and nature reserves. Diversity of bee species was somewhat higher in urban sites than in farmland, as might be expected, but close to that found in nature reserves (Baldock *et al.* 2015).

Generalised comparisons of urban and rural landscapes are of value, but it is also important to consider the particular socioeconomic character and historical development of each urban complex. So-called 'left-behind' post-industrial towns are likely to have abandoned quarries, spoil heaps and contaminated and derelict sites, some of which may have been colonised by a rich diversity of plants and insects, while more wealthy neighbourhoods are likely to be equipped with large gardens, extensive local parks and other green spaces.

Towns and cities with long histories and surviving archaeological features (Roman walls, ruined monasteries, large estates left by aristocratic houses and the like) often manage these within urban green spaces that can harbour valuable

FIG 344. (a) Systematic beds, Cambridge University Botanic Garden. (b) Flower-rich grassland in an urban churchyard. (c) A bank beside a shopping centre feeder road provides nesting habitat for several *Lasioglossum* and *Andrena* species, together with their cuckoos.

communities of bees and other wildlife. Other features such as cemeteries, churchyards and botanical gardens are also capable of hosting large communities of solitary bees. Seventy-four species of solitary bees have been recorded in Cambridge University Botanic Garden, for example (NWO), and 95 at Kew Royal Botanic Gardens up to 2018.

Tunbridge Wells is an ancient spa town on the Kent–Sussex border in the High Weald, with an outstandingly rich bee fauna. The following is a brief summary of the account given by local bee expert, Ian Beavis.

At the centre of the town are two large commons, which were once settled by Mesolithic people, later used as swine pastures during Saxon times, and subsequently incorporated into the Norman manor, but delineated as 'wastes' for use by commoners for grazing, harvesting gorse and quarrying for stone. Their management of the commons allowed the retention of areas of heath and acid grassland, and from the 17th century, when the town became a popular spa resort, local activists fought, successfully, to preserve the natural aspect of the commons from proposals to convert them into formal parks. These natural features, around which the town grew, became a positive asset, proudly advertised to visitors. The establishment of the Tunbridge Wells Borough Council in 1889 also brought into being the Commons Conservators, with an explicit brief to protect wild plants and animals.

As the town grew, the Borough Council ensured the creation of five new public parks, and today, these combine with the commons, two cemeteries and a woodland nature reserve to provide a network of rich habitats across the town, linking across the green belt to several significant rural woodland and heathland reserves. Each of these green spaces has associated voluntary groups, and management advice and practical help have been given by outside bodies such as the Kent High Weald Partnership and the Kent Wildlife Trust. Pre-existent habitat has been retained in the parks, and they each have areas managed for wildlife, with an emphasis on bees and other pollinators. One quite small park has such species as the Buff-tailed Mining Bee *Andrena humilis*, the White-bellied Mining Bee *Andrena gravida* and its cuckoo and, astonishingly, the rare Long-horned Bee *Eucera longicornis*, which also has a population in a larger park in the town. The cemeteries hold populations of the scarce communal bee, the Big-headed Mining Bee *Andrena bucephala*, the cuckoo bee *Nomada signata* and the very local White-bellied Mining Bee *Andrena gravida*. However, the commons support the richest diversity of bees, including many of the heathland specialists, both *Panurgus* species, many scarce or local mining bees and their cuckoos and, most recently, the spring-flying Early Colletes *Colletes cunicularius*.

The home town of one of us (TB) has an almost complete Roman wall; a ruined priory; a castle set in an extensive public park, once used by Dutch

FIG 345. Green spaces in Tunbridge Wells: (a) Calverley Grounds, habitat of the Buff-tailed Mining Bee *Andrena humilis*, the White-bellied Mining Bee *Andrena gravida* and the Long-horned Bee *Eucera longicornis*. (b) The Common, habitat of many scarce mining bee species and heathland specialists. (c) Grosvenor and Hilbert Park, habitat for the Pantaloon Bee *Dasypoda hirtipes*, the Clover Blunthorn Bee *Melitta leporina*, the Red-girdled Mining Bee *Andrena labiata* and numerous other species. (All © I. Beavis)

settlers for drying textiles, and donated to the public in the late 19th century by a local notary; an archaeological site marking the position of two civil war forts and a Bronze Age settlement, now protected as an urban nature reserve; and an extensive country park partly deriving from a former royal hunting forest. There are several thriving allotment sites, often with biodiversity-conscious users. These features and others combine with relict semi-natural habitats adjacent to the river which runs through the town, patches of grass-heath on some green spaces, together with acid grassland and ancient woodland protected by MoD management (Tansley 2019). Historically affluent suburban streets include large gardens, some of them managed in wildlife-sympathetic ways, and there are several railway lines with adjacent cuttings and embankments. In newer developments, often built at the expense of existing wildlife-rich habitats, gardens or green spaces are small or absent. Development pressures have been intense, and some of the remaining wildlife sites have been saved by active campaigns by local civil society, though many more have been lost, including the former MoD range, discussed above.

More recent awareness of the twin threats of biodiversity loss and climate change has led to some changes that run counter to the story of ecological decline in the town. A broad coalition of organisations proposed a circular walking and cycling route (the 'orbital') around the town's periphery. This is some 32 km in extent, joins up existing rights of way and offers connectivity between many of the town's green spaces. This has been enthusiastically supported by the council, but another initiative, to shift management of green spaces to enhance biodiversity, has proved more difficult. Here, the traditional urban aesthetic of close-cut lawns, with nature severely under control, has been less of a problem than was expected. The greatest obstacle has been the well-meaning but misguided pressure for mass planting of trees, irrespective of ecology and public amenity. Even so, an agreed change to an annual grass-cut at one strip of land adjacent to the Roman wall yielded reports of 25 bee species (including localised species such as the Grey-gastered Mining Bee *Andrena tibialis* and the Bryony Mining Bee *Andrena florea*) in the first year, with more species in year two (Benton 2020).

Though studies have shown the significance for insects of the relative scale of green space in urban landscapes, there is less insight into the importance of configuration: how the relative placing of different sorts of habitat across the landscape impacts on the viability of populations of different bee species. Gardens, estimated to occupy over 430,000 h in the UK (Davies *et al.* 2009), have been generally recognised as potentially of great importance as refuges for bees, with numerous reports of large numbers of species present in some. The late David Baldock recorded 121 species over 10 years in his large Surrey garden, and

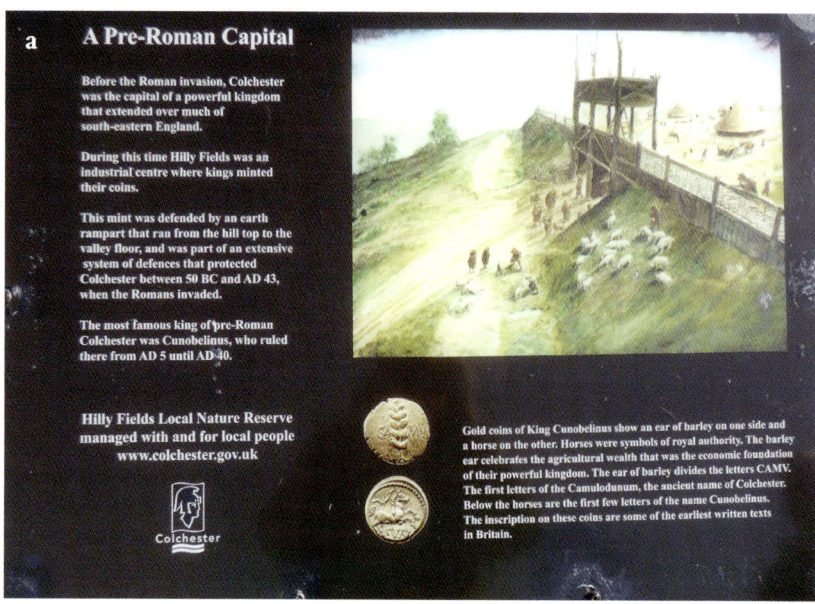

FIG 346. Surviving Colchester wildlife sites: (a) Information board for Hilly Fields, site of a Bronze Age mint and two civil war forts, now a local nature reserve of Colchester Borough Council. (b) Grass heath, Hilly Fields. (c) High Woods Country Park, part of a former royal hunting forest. (d) Cymbeline Meadows, reaching into the centre of town, formerly arable land.

others have recorded surprisingly large numbers (Archer 2013, 2014, 2016, Owens 2017, Erenler cited Ollerton 2021).

Much research and resulting advice to gardeners has been devoted to discovering which flowers to encourage for bees and other pollinators. Comba *et al.* (1999) and Corbet *et al.* (2001) report evaluations of different garden flowers from this perspective, in particular comparing native species, and closely related cultivars, as against exotic and highly modified cultivars. A more recent study by Rollings and Goulson (2019) logged insect visits to plots of 111 species of ornamental flowers over a five-year period. Flower visitors were classified as honeybees, bumblebees, solitary bees, hoverflies, Lepidoptera and 'other pollinators'. Unfortunately, the solitary bees were not identified down to species, but as a group they showed very different patterns of visitation from those of bumblebees and honeybees (top flowers for solitary bees included four Asteraceae, one Apiaceae [*Eryngium*] and a *Geranium* cultivar). Although the authors are more concerned with the overall value of different plants for pollinators, their results suggest it might be possible to 'engineer' the species mix of insect visitors to a garden by judicious encouragement of, for example, flowers favoured by solitary bees but not by honeybees. As oligolectic bees such as the Bellflower Blunthorn Bee *Melitta haemorrhoidalis* and the Bryony Mining Bee *Andrena florea* appear to be scarce in urban locales, planting campanulas, White Bryony *Bryonia dioica* and other specialist pollen hosts would be advantageous.

Rather few of the garden studies have included observations on the uses the bees make of the garden (Early 2013 is an exception). One of us (TB) was able to watch behaviour in his small town-centre garden throughout the 'lockdown year' of 2020 (Benton 2020, 2021a). A total of 53 solitary bee species was recorded in the garden. Ten of these were cuckoo species, an indication of strong local populations of their host species. The remaining 43 species were watched to see how many of their seven habitat requirements (see above) were met within the garden. Eight species were present throughout their flight period and were nesting in the garden (most using bee hotels). It seems that suitably managed gardens can meet all the habitat requirements of this group of bees, though viable populations would require extensive areas of suitably managed gardens. A second group, also of eight species, was using the garden for most of their requirements but apparently not nesting. Many of these were ground-nesting species. The third, and largest, group (17 species) were relatively frequent visitors but used the garden for a limited range of requirements – usually to take nectar or pollen when particularly favoured plants came into flower. A fourth group, comprising 10 species, were 'occasional' visitors, some staying for a day or two, others being seen only once.

Groups two and three were species that either met most of their needs in the garden but were going elsewhere for the remainder (typically for nest sites) or were making use of external resources for most of their needs but using the garden for supplementary requirements. For the garden to play either of these roles, there must be suitable habitat patches within foraging range of both groups of species. Females of the Hairy-footed Flower Bee *Anthophora plumipes* were nesting in the nearby Roman wall, as well as in cracks in neighbouring brickwork, while the nearest suitable nesting habitat for some of the ground-nesting species was the traditionally managed grassland of the nearby ruined priory. Asteraceae specialists may have been making use of a churchyard some 200 m away, and early species (notably the 10 *Andrena* species) may have been using willow (*Salix*) catkins growing by a railway line, also some 200 m away.

The fourth group are of particular interest. Although their use of the garden was very limited and ephemeral, it could nevertheless have been of great ecological significance. Some of these species were relatively habitat-specialist, with suitable habitat patches within one or two kilometres. For these species, it seems that gardens might play a significant role as 'stepping-stones', enabling some gene flow between sub-populations of a more widespread meta-population.

FIG 347. Town-centre garden, Colchester.

Others were species without known populations within 10 or more kilometres. It is likely these were dispersing from established or declining habitats in the wider region. In recent years, several species currently expanding their British ranges have arrived in this area (including the Bryony Mining Bee *Andrena florea*, the Ivy Bee *Colletes hederae* and the Large-headed Resin Bee *Heriades truncorum*). It is likely that gardens have provided 'oases' facilitating these changes of distribution. This lockdown 'snapshot' offers evidence of very low-density movements that, without such intense observation in a small locality, might be undetectable.

THREATS TO URBAN BEES (AND OTHER URBAN WILDLIFE)

Having recognised the value of urban habitats for bees, it is necessary also to recognise their vulnerability. There remains a long-standing 'urban aesthetic' that favours uniform, close-cropped turf on public green spaces. This aesthetic seems to be on the decline in some areas, in favour of attempts by some local authorities to create flower-rich meadows, but, unless these are carefully introduced, traditional expectations can quickly generate public resistance. Another threat to public green spaces is the well-meant but potentially disastrous mania for mass tree-planting without concern for pre-existing biodiversity. Abandoned, 'waste' and generally despised brownfield sites that have, in many places, become important wildlife refuges are an open goal for developers where planning guidance favours development on brownfield rather than greenfield (even where the latter may be featureless arable monoculture). While private gardens remain important as refuges for wildlife, including bees, there is a growing tendency for new-build to provide little or no garden, and in many older properties, gardens have been replaced with hard standing to serve as car ports or just to avoid the bother of gardening for people with busy lives.

A recent international review by Finoglio *et al.* (2021) offers a conceptual model for understanding the relations between drivers of urban insect decline and diversity loss and the traits of the species themselves. They identify five such drivers: impervious surfaces, habitat fragmentation, urban heat island effects, pollution and exotic plants. These drivers, including interactions between them, are likely to vary in their consequences for insects, depending on the characteristics of the latter. Among the relevant traits are body size, life history stages, phenology, dispersal abilities, degree of dietary specialism and vulnerability to predators and parasites.

Applying these ideas to bees, the spread of impervious surfaces reduces the area that might otherwise provide both forage and nesting habitat

FIG 348. How (not) to manage urban green space? (a) The motor mower. (b). Municipal mowing plus tree-planting.

for ground-nesting bees, and there are likely to be indirect hydrological consequences also, affecting ground-nesting species in particular. Bees might be expected to cope with habitat fragmentation better than other, less mobile, taxa, but here again much depends on the pattern of spatial fragmentation and the characteristics of the bee species themselves. Studies are equivocal on the effects of body size (Wenzel *et al.* 2019), but, in general, limits will be set by the foraging range of the species concerned, and by dietary breadth. If the pollen host of an oligolectic species is sparsely distributed, this must place extra demands on the bee, in terms of locating pollen sources and meeting the energetic costs of return flights with pollen loads. The review by Wenzel *et al.* (*ibid.*) indicates that generalist foragers are advantaged over specialist ones in urban landscapes, though both the Ivy Bee *Colletes hederae* and the Bryony Mining Bee *Andrena florea* flourish in the urban sites monitored by Benton (2020): *Colletes hederae* because its pollen host is abundant and *Andrena florea* despite the sparsity of its pollen host!

Relatively little is known about the effects of heightened urban temperatures, though in Britain many species are close to the northern limit of their climatic range. These species may well be advantaged by warmer urban climates, and, indeed, may be using urban habitat as stepping-stones in their range expansion. Wenzel *et al.* note a tendency for late-flying species of bees to be advantaged by the urban heat island effect. However, a study of three protected sites in Munich, which compared bee diversity in the 1990s with that in 2018, found markedly increased diversity in two and some decline in a third. The authors found no correlation between local extinction risk and traits such as body size, length of flight period, seasonality, nesting preferences and pollen specialisation. In two sites, there was a positive correlation between species survival and habitat breadth – that is, the ability of the species to utilise three or more of six general habitat types (Hofmann & Renner 2020).

The significance of pollution as a driver of bee decline is difficult to assess. However, if the use of herbicides and pesticides is included in this category, then this must have a significant role. In many areas, local authorities have been under pressure to reduce or eliminate the use of environmental poisons such as glyphosate, but pesticides are still in heavy use in urban gardens: ironically, garden centres typically display shelves of bird feeders and bug hotels opposite shelves stocked with impressive arrays of chemical poisons. Evidence of toxins in the brood stores of urban bumblebees points to significant pesticide loads, possibly even deriving in part from flea treatments of domestic pets (Goulson 2021). Finally, evidence so far is equivocal on the impact of the frequency of exotic, alien and horticulturally modified plants in urban habitats. Studies of pollinators (often used as a general term for flower-visiting insects) suggest that

FIG 349. A contested urban planning site.

abundance and diversity are positively correlated with additional floral resources, independently of their origin. However, some studies relate the predominance of generalist species of bees in urban areas to the displacement of native flower species important for species with more specialist dietary requirements (Fenoglio et al. 2021, Wenzel et al. op. cit.).

In general, it seems that urban landscapes are capable of sustaining highly diverse communities of bees, depending on the quantity, configuration and management of a range of types of green space. The important change that now seems to be under way in many towns and cities in Britain and across mainland Europe is a transition to more biodiversity-friendly management of public green spaces. This usually involves reduced mowing and, often, attempts to enhance botanical diversity. Where conversion from arable agriculture is concerned, sowing wild flower mixes may be beneficial, but in pre-existing grasslands, simple changes to the management regime have been shown to be very effective in increasing botanical and invertebrate diversity.

A citizen science study carried out in and near Dresden, Germany, compared insect biomass, abundance and diversity in nine 'butterfly meadow' sites with

FIG 350. Spear Thistles *Cirsium vulgare* on an urban street following cessation of weedkiller use.

that in nine 'intensively managed lawns' (Wintergerst *et al.* 2021). A total of 260 insect species belonging to a wide range of taxa were collected by sweep-netting, the butterfly meadow habitats yielding far higher results than the intensively mown ones across all taxa. In the case of bees, from 7–22 species were recorded on individual butterfly meadow sites, contrasted with up to two, but usually none, in the intensively managed grasslands. The intensively managed sites were subject to the standard municipal frequent mowing across the whole site, but the butterfly meadows were cut at most three times per year and the cuttings removed, leaving up to 30 per cent of the area uncut at each mowing. Although the bee species recorded were predominantly polylectic (or 'polyphagous' according to the report), oligolectic species, including the Large Scabious Mining Bee *Andrena hattorfiana* and the Hairy-saddled Colletes *Colletes fodiens*, were recorded, and cuckoo bees were well represented. Benefits of broadly similar shifts of urban grassland management have been reported elsewhere (e.g. in Saltdean, Sussex; Garbuzov *et al.* 2014).

ECOLOGY AND CONSERVATION · 507

FIG 351. Signs of change? (a) Effects of reduced grass-cutting along a town centre riverside. (b) Unimproved flower-rich grassland, saved from tree-planting by local activists. (c) An information board, explaining the purpose of reduced grass-cutting.

Owing to their relatively high cognitive and perceptual capabilities, adaptability and mobility, bees seem to be more resilient than other guilds of flower visitors in the face of potentially adverse drivers of change in urban habitats, and they respond well to favourable urban grassland management. However, the evidence suggests that the composition of urban bee communities is shaped by such traits as nesting requirements and dietary breadth, with pollen-host generalists and cavity nesters favoured over ground-nesting and oligolectic species, but much depends on the particular histories and habitat configurations of the towns and cities concerned.

BEES IN THE COUNTRYSIDE

The natural and semi-natural habitats mentioned earlier in this chapter are, or were, mainly located in the countryside, and it is important to recognise that despite large-scale loss, fragmentation and deterioration, these negative processes have occurred unevenly across different regions. In some areas, environmental designations have offered some protection, whilst in others, withdrawal of investment, technical difficulties, patterns of land ownership and other factors have favoured persistence of less-intensive forms of agriculture, abandonment, or leisure use. As in the case of the Thames Gateway, these processes have tended to obscure the division between town and country, but the greatest spur to recent research has been alarm about a supposed 'pollination crisis' resulting from major declines in flower-visiting insects, including bees. As we saw in Chapters 5 and 6, pollination is a key ecological function in which bees play a major part. Pollination by animals (in Britain, mainly by insects) is necessary for the successful reproduction of the great majority of flowering plant species, but the great increase in research effort on pollination since the early 1990s (Ollerton 2021) was provoked by a threat to the pollination of a specific group of plants: those used in human diets.

The rural focus, then, was on agricultural and horticultural systems and their dependence on what was now thought of as a service; one that could be carried out by a wide range of invertebrate taxa, including bees, which now came to be defined as 'pollinators' in terms of their contribution to the humanly crucial 'service' of crop pollination. Of course, researchers have always been insistent that the service was also to wild plant species, but the rationale adopted by funding bodies (and many conservation organisations) was the importance for human nutrition and, often, the economic value of the service performed (Losey & Vaughan 2006, Juniper 2015).

As mentioned in the introduction to this chapter, data collected by voluntary organisations was providing evidence of dramatic losses of abundance, geographical range and diversity of some insect groups, most notably butterflies and bumblebees, well before this was registered in the post-Rio world as a potential threat to pollination, and so to human food supplies (Williams 1982, 1986, 1989). By the 1990s, systematic research effort, often drawing on the survey work of amateur entomologists, was providing evidence of declines in bees and other pollinating insects (Falk 1991, Banaszak 1992, O'Toole 1993), and these declines were increasingly interpreted in terms of habitat fragmentation in agricultural landscapes.

Honeybees were widely assumed to be the most important pollinators, with bumblebees (usually the Buff-tailed Bumblebee *Bombus terrestris*) used for pollination of crops requiring buzz pollination (see Chapters 5 and 6) and back-up pollination for orchards. Threats to honeybees from the *Varroa* mite, and so-called colony collapse disorder, concentrated minds on the risks of dependence on such a narrow range of species for a crucially important 'ecosystem service'. The dangers of an actual or potential pollination crisis came to be recognised.

A productive debate about the precise nature of this crisis and the availability of relevant evidence was provoked by Ghazoul (2005). He argued that alarm about risks to pollination might be exaggerated or, at least, premature. So far as human food was concerned, most staple crops are not insect pollinated, and crops that are insect pollinated are mainly grown in smaller-scale, non-intensive, low-input systems. Meanwhile, concern about pollination of wild flowers was misplaced, as most were pollinated by generalist species, while conservation issues were concentrated on rare, usually specialist, pollinator species. Interestingly, Ghazoul still advocated caution, admitting that future environmental changes could not be predicted, and he agreed that monitoring of pollinator abundance was still necessary. This challenge produced an immediate reposte from Dewenter *et al.* (2005), in which they drew attention to studies of pollinator (especially bee) declines across Europe and the USA; the dependence of many vegetable and forage crops on pollination for seed production (overlooked by Ghazoul); the rate at which less intensive horticultural systems were being displaced by intensive, larger-scale ones; and evidence showing the importance of pollinator diversity even for generalist crop plants and wild flowers.

Where there was agreement, however, was on the need for more research, and this has been arriving ever since. A study by Beismeijer *et al.* (2006) used existing distribution atlases for bees, hoverflies and wild plants in Britain and the Netherlands to compare their status across two time periods (before 1980 and 1980 and later). The data showed reductions in bee diversity per $10\,km^2$

over the two periods in both countries, with bee communities increasingly dominated by a small number of species. Oligolectic and univoltine bees declined disproportionately (as did long-tongued bees in the Netherlands data). Over the same period, plants dependent on bee pollination also declined disproportionately: a suggestive parallel, although the data could not demonstrate a causal connection.

A subsequent review article by Potts *et al.* (2010) reported international studies confirming the general theme of parallel declines in pollinating insects, especially bees, and obligate animal-pollinated cross-breeding plants, both crop and wild species. Although it was to be expected that plants dependent on specialist pollinators might be most vulnerable to declines in diversity of pollinators, there was little evidence of this. Pollination networks commonly include asymmetrical interactions between specialist and generalist plants and their pollinators, and it seems that some species may be lost without damage to the network. The notion that some species in networks can be considered 'redundant' is open to the criticism that a distinction needs to be made between insect visitors, in general, and those which are actual pollen vectors. Even if pollination networks do have a degree of resilience in the event of loss of particular species, as core ecological functions are carried out mainly by generalist pollinators, this should not lead to complacency. As Potts *et al.* point out, environmental changes can still disrupt such networks and lead to tipping points and subsequent collapse.

Evidence of widespread declines of both wild and managed pollinating insects was already accumulating by the late 1990s (Kearns *et al.* 1998), and the potential significance of this for human food security and agricultural profitability was clearly recognised. The UN Convention on Biodiversity spawned a global initiative for the conservation of pollinators in 2000, and in the British Isles, a coalition of research councils, the Welcome Institute and the Scottish Government established an Insect Pollinator Initiative to coordinate action to conserve pollinators in Britain. A subsequent 'knowledge exchange' was organised, bringing together conservation organisations, farming and agri-business interests and research scientists to identify key knowledge needs for pollinator conservation (Dicks *et al.* 2012). Although there were differences between the more traditional conservation and newer ecosystem services approaches, an agreed set of priorities was established. These included the need to monitor floral resources in agricultural landscapes, establishing the importance of pollinator diversity (i.e. estimating the vulnerability of pollination systems to loss of particular species), understanding basic pollination biology, the impact of pesticides and the economic benefits of pollination. An early output of the initiative was Vanbergen *et al.* (2013), which provided a comprehensive analysis of the decline of insect flower visitors firmly within the ecosystem services perspective. The review emphasises the importance of insect pollination of crops for both human health and nutrition and identifies the key drivers of decline. Usefully, the study argues for interdisciplinary work (including across the natural science–social science boundary) to understand the interactions between drivers of decline (for example, interactions between pesticides, nutritional limitations and resistance to pathogens in affecting pollinator health).

Another study under the umbrella of the Insect Pollinator Initiative (Baude *et al.* 2016) offered a quantitative estimate of the extent and periodicity of decline and stabilisation of floral resources between the 1930s and 2007, noting the decline of floristic habitats such as calcareous grassland and shrub heathland in the face of the expansion of arable cultivation with few floral resources. Somewhat paradoxically, they conclude that while *total* floral resources grew from 1998, the *diversity* of these resources stabilised at low levels from 1990 onwards. Unfortunately, the study did not include floral resources in urban areas and measured only nectar provision, but their estimates that a mere four plant species provide over 50 per cent of the nectar resource and that a mere 22 provide over 90 per cent are striking.

Subsequent research has focused on the multiple drivers of pollinator decline in agricultural systems. Intensification is a generalised culprit, but there are distinguishable aspects, each of which may impact on pollinator decline. Industrialisation involves a larger scale of operation, using machinery for tillage, nutrient, herbicide and pesticide application, and harvesting which, in turn, requires a simplification of the landscape, eliminating obstacles to machinery and enabling efficiency of the various operations. Removal of habitat fragments such as copses, wetlands and hedgerows causes isolation of small relict populations of bees, hoverflies and other pollinating insects, undermining the connectivity needed to sustain meta-populations and leading to genetic uniformity and consequent loss of resilience in remaining local populations.

These impacts are likely also to be exacerbated by spray drift (and other forms of environmental diffusion of toxins) and vulnerability to both pesticides and, indirectly, further loss of wild flowers (Goulson *et al.* 2015). An interesting study by Ollerton *et al.* (2014) used data collected by BWARS to identify rates of extinction of bees and flower-visiting wasps in Britain over time. Extinctions in the latter part of the 19th century could have been related to the increasing use of nitrogenous fertiliser and associated decline of wild flower meadows, while another rise in extinction rates, from the late 1920s to the late 1950s, coincided with state-sponsored drives for agricultural productivity and food security.

Since then, extinction rates have been very low. The authors speculate that this might have to do with the more ecologically sensitive species having been already dispensed with or, more optimistically, that compensatory conservation measures, such as flower-rich field margins, might have been effective. Extinction, of course, is only one of a number of aspects of insect decline that might impact pollination. Change in spatial distribution is another, and the same BWARS data shows a very mixed picture, with some species in decline, while others are expanding their range in Britain (see Chapter 9). Again, population size and relative stability over seasons are also relevant to pollination but are very difficult to monitor (Carvell *et al.* 2015, O'Connor *et al.* 2019, Breeze *et al.* 2020).

Agricultural intensification, in addition to progressive elimination of semi-natural habitats, involves high levels of chemical inputs to increase yield per unit area of agricultural land. These inputs include artificial fertilisers, such as nitrogenous compounds, with well-known deleterious effects, as well as chemical herbicides, fungicides and insecticides. These are designed either to replace biological controls previously provided by non-crop ecosystems, or to eliminate plant species that would otherwise compete with crops. From the late 1930s onwards, a new generation of insecticides was widely promoted and applied on a massive scale. Organochlorines and organophosphates (DDT being the best-known) are neurotoxins with devastating effects on insect populations. Initially this made them popular with many farmers, but before long it was recognised that their environmental persistence and spread through food webs was transferring their toxicity to birds, mammals and humans. Rachel Carson's work (1962) was one of the first to draw public attention to this toxic assault on global terrestrial and aquatic ecosystems, but intensive retaliation from the agrochemical industry delayed the introduction of appropriate legislation and limited its scope for many years.

Since the early 1990s, a new wave of neurotoxin insecticides has been unleashed into agricultural systems. Between 1990 and 2016, pesticide applications to agricultural land increased by 70 per cent, with an average of 16 applications to each unit of cropland. The new group of insecticides, neonicotinoids are, like their predecessors, neurotoxins. Alarms were raised by losses of bee colonies coincident with the spread of these chemical agents, and they were banned in the EU for use on insect-pollinated crops in 2013 (Goulson 2021). Subsequent large-scale field research has confirmed the destructive effects of these chemicals on bee populations, and although the research has commonly focused on bumblebees and honeybees, there has been significant evidence that solitary bees are, if anything, more seriously affected (Rundlöf *et al.* 2015, Woodcock *et al.* 2016).

These discoveries have raised important concerns about the evidential basis for regulatory control of new agricultural chemicals, as both wild and managed pollinating insects are subjected to repeated doses of a range of different chemical agents, while there are no tests for the combined effects of 'cocktails'. Standard levels of permitted dosage are linked only to death rates, but it is now recognised that sublethal doses of these neurotoxins unsurprisingly cause cognitive and learning loss as well as disruption of navigational abilities in bees. These effects are obviously disastrous both for the ecosystem function of bees as pollinators and their ability to find their way between nest and forage sources (Stanley *et al.* 2016). Again, although these studies are mainly concentrated on honeybees and bumblebees, there is no reason to suppose solitary bees are any less disastrously affected, implying that risk assessments should involve these 'other bees' (i.e. bees other than honeybees and bumblebees). This has been confirmed in a study of two *Osmia* species in which larval ingestion of a neonicotinoid produced both lethal and sublethal effects, dependent on dosage (Claus *et al.* 2021).

Finally, it had been assumed that the use of these pesticides as seed dressings on wind-pollinated cereal crops would be environmentally benign. This was called into question by the discovery of neonicotinoids in wild flower pollen and nectar on arable farms using this treatment. Application of the chemicals as seed dressings had seemed to avoid the problems of drift into the wider environment associated with sprays. However, early research by the agrochemical industry itself had shown that an average of only 5 per cent of the toxins were taken up by the germinating plant. The great majority remained in the soil, and as they are both water soluble and slow to degrade, they became steadily accumulated and then dispersed to marginal land, hedgerows, verges and remaining semi-natural habitats surrounding cropped land. Here they have been taken up in the roots of wild plants and subsequently appear in the pollen loads and crop contents of bees (Botías *et al.* 2015). This evidence led to the EU issuing a ban on the outdoor use of the three main neonicotinoids, including use on wind-pollinated crops such as cereals, in 2018. Unfortunately, farmers can still apply for short-term derogations, and at the time of writing, the UK government has already employed its freedom from EU regulation to grant a derogation for the use of these chemicals on sugar beet in some circumstances. A recent analysis (Brühl *et al.* 2021) of the bodies of insects collected on conservation areas adjacent to agricultural land across Germany found residues of 47 pesticides. In 16 of 21 of the conservation areas, the study detected residues of neonicotinoids, still present in 2020.

Ellis *et al.* (2018) identify what they see as a contradiction posed by the demand for pollination services by some farming systems. They argue that for those crops

FIG 352. Agricultural spraying in action. There is a 10 m conservation(!) headland between the hedge and the tractor.

which depend on insect pollination (i.e. most fruit and vegetable crops, herbs and flowers), the requirement for pollination presents a barrier to intensification. As demand for these products grows, the pollination services provided by small-scale beekeepers and native bee species are insufficient, and attempts to compensate for this with commercially provided pollination services encounter further obstacles. The consequences for the health of commercially managed honeybees used for mobile pollination services (at their most extreme in the USA) are well known (Simone-Finstrom *et al.* 2016). Meanwhile, the loss of semi-natural habitat deprives both wild and managed bee populations of nutrition outside the flowering period of crop plants. There are also concerns about the spread of infectious diseases between managed honeybee colonies and wild bee species through sharing of forage plants.

Proposals for conserving and enhancing pollinator and, especially, bee diversity in farmed landscapes have generally emphasised the importance of providing forage sources, usually in the form of sown wild flower margins. These have featured widely, especially in the EU, among the stewardship schemes funded by state subsidies. In Britain, a National Pollinator Strategy was initiated in 2014 and was followed up by advice on creating and managing pollinator habitat (Nowakowski & Pywell 2016). This provided a useful introduction to the biology of social and solitary bees, as among the most important pollinators, and advised on the various ways of enhancing the necessary floral resources, without undermining the productivity of farming. The importance of floral resources throughout spring and summer was stressed, with the alternatives of green hay or use of commercial flower seed mixes mentioned. The former requires nearby flower-rich habitat as a source of seed and has some technical difficulties, so the discussion centres on the use of varied seed mixes in conservation strips. It is admitted that these do not effectively cover the bee 'hunger gap' in early spring, and the advice is to rely on headlands and hedgerows. The need for nesting habitat is mentioned but given little detailed attention.

Research on the use of such measures for enhancing populations of bees and other insects for pollination of crops has been limited in Britain to a rather small number of insect-pollinated crop plants. In Britain, Oilseed Rape *Brassica napus* is widely grown in intensive agricultural systems. This crop benefits from insect pollination (Bommarco *et al.* 2012), and one study has investigated the relative significance of visitation by bumblebees, honeybees and solitary bees (Woodcock *et al.* 2013). The frequency of contact with floral stigmas, and the availability of free pollen grains in the body hairs, led to estimated greater pollinating efficiency by solitary bees compared to both bumblebees and honeybees: 34 per cent probability of pollen transfer per visit for honeybees, compared to 71.3 per cent for solitary bees. However, this was offset by the greater frequency of visitation by honeybees to crops situated close to hives. One response is to move hives around the farmed landscape, to increase the efficiency of honeybee pollination. An alternative (preferable, we think) would be to increase the area of semi-natural grassland and nesting sites for solitary bees in croplands.

Internationally, the cultivation of insect-pollinated crops has expanded over recent decades (threefold since 1961; Aizen & Harder 2009). Senapathi *et al.* (2021) have reviewed 43 studies of pollinator communities visiting 21 insect-pollinated crops across six continents. The review focused on evidence for resilience of pollinator communities across seasons, the significance of changes in abundance of dominant pollinators in particular communities, and the effects of pollinator diversity on crop production. Though, as discovered in other research, a small number of dominant pollinators in any community were responsible for the great majority of the pollination, crop yields were still benefited by pollinator diversity. More diverse pollinator communities were also more stable from season to season and were buffered from large changes in abundance of dominant pollinators. Diversity was also shown to be more important with increases in cropland area. Of particular relevance to our topic, 'other bees' were reported to be the dominant taxa in pollinator communities at 42 per cent of the sites included in the studies. Similar conclusions were reached by other studies.

Devkota, dos Santos and Blochtein (2021), for example, found that yield from mustard crop in Nepal was highly dependent on insect (mainly bee) pollination and in particular benefited from pollinator diversity. Diversity was greatest close to natural vegetation, with honeybees displacing 'non-Apis' bees at greater distances. Norfolk, Eichhorn and Gilbert (2016) found that flowering ground vegetation in smallholder almond orchards in arid climates benefited wild pollinators and increased fruit set.

Bumblebees and honeybees have long been recognised as important to the pollination of fruit trees, such as apple, pear and plum. However, as early as the 1940s, V. H. Chambers (1946) was observing the foraging of several *Andrena* species on orchard trees. He pointed out that while honeybees and bumblebees mixed their pollen with nectar and packed the resulting mass into their corbiculae, *Andrena* females carried their pollen loads loose and also had numerous pollen grains attached to the rest of their body hairs as they foraged. This suggested to him that they might be important pollinators, and he accordingly collected full pollen loads of bees returning to their nests in or close to orchards in south Bedfordshire. The pollen loads of four *Andrena* species whose flight period coincided with the flowering of the trees contained at least some fruit tree pollen. In three of these species, more than half the loads contained *Prunus* pollen, and in two of them, a majority of the loads which included *Prunus* pollen were made up of more than 50 per cent *Prunus*. The Tawny Mining Bee *Andrena fulva* (*armata* in those days) therefore seemed to be a sufficiently constant forager on *Prunus* to be a significant pollinator, while Blackthorn Mining Bee *Andrena varians* showed high levels of constancy to both *Prunus* and *Pyrus* (pear).

In contrast to the above studies, much of the recent research and advice has been addressed to the decline of insects in large-scale arable monocultures. With the exception of Oilseed Rape, the crops concerned are predominantly wind-pollinated, so enhancement of bee habitats is presumably motivated by conservation objectives, including other ecological functions, rather than addressing pollination limitation specifically. Studies have demonstrated, unsurprisingly, that farms with wild flower conservation strips have more pollinating insects than ones which do not. However, one recent study conducted in the Czech Republic used a pollination network approach to compare sown field margins with nearby semi-natural habitats (Hadrava *et al.* 2022). Interestingly, the pollination networks in both cases were estimated to be functionally robust in the face of possible co-extinctions (i.e. of pollinator and its host plant). However, from the point of view of conservation, the semi-natural habitats sustained a greater diversity of species, probably because of their greater structural diversity and longevity. More directly connected to our theme,

FIG 353. Conservation strip in an agricultural landscape.

insect visitors to the conservation strips were almost entirely honeybees and bumblebees, with very little sign of solitary bees.

This finding confirms a study of the use of farmland by solitary bees in southern England (Wood, Holland & Goulson 2017). These authors found that solitary bees made relatively little use of flowers in sown conservation strips, and that farms without these enhancements did not host significantly fewer species than those that did. Solitary bees were more likely to collect pollen from flowers in the wider environment, such as along hedgerows, headlands and other uncropped spaces. These authors propose development of seed mixes that include plants favoured by solitary bees and early-flowering plants that would be useful for early species, such as many *Andrena* species, as well as providing suitable habitat in the wider farm environment.

Subsequent work by Goulson and his colleagues has taken further the search for seed mixes that might better support solitary bees. In one study (Nichols, Goulson & Holland 2019), transect walks were used to quantify bee visits to 45 wild flower species grown for commercial purposes in a farm near Bath, Somerset. A subset of 14 flower species was visited by 92.5 per cent of wild bee species, while 18 plant species covered the observed foraging of all 40 bee species. Of these, 32 were solitary bees and the rest bumblebees. Interestingly, flowers usually considered attractive to bumblebees were not much visited by bumblebees, and only two flower species found to be most attractive to bees in this study were shared with the most common commercial wild flower seed mixes. Clearly these findings point to the possibility of far more favourable seed mixes, especially for solitary bees. However, the methodology used did not allow for distinguishing between bee visits for pollen and for nectar, so pollen generalists may have been favoured over oligolectic species (though the presence of some Asteraceae and Campanulaceae in the proposed mix would have catered for at least some of these). Overall, the solitary bee species observed were predominantly the ground-nesting Andrenidae and Halictidae. This points to the importance of nesting habitat and, especially, of nest sites for cavity nesters, if solitary bees are to be conserved in farmland.

In addition to wild flower strips, provision, or retention, of hedgerow, scrub and woodland-edge habitats can be of value for potential nest sites and also offers a wider range of forage resources, especially in early spring and late summer. Jacobs *et al.* (2009) studied the effect of a range of pollinating insects on seed and fruit set for five hedgerow plants, by excluding insects from some during the flowering period and comparing the results with open-flowering plants as a control. Though Dog Rose *Rosa canina* and Bramble were largely unaffected, exclusion of insects from Blackthorn, Hawthorn and Ivy significantly reduced

FIG 354. Semi-natural habitat, farming and a power station side by side on Thames-side.

fruit set. As these plants provide winter forage for many farmland birds and small mammals, the wider ecological importance of maintaining a reliable population of insect pollinators is clear.

In general, the spread of intensive agricultural systems has been negative for solitary bee abundance and diversity. Provision of environmental enhancements to encourage pollinators, such as sown wild flower strips, can increase pollinator abundance and benefit yield for some insect-pollinated crops. However, currently common commercial seed mixes offer little for pollinator diversity, particularly for solitary bees. Recent research points to the feasibility of alternative seed mixes to support environmental initiatives that aim to achieve wider conservation goals, such as enhanced biodiversity, as distinct from provision of pollination services to crops. However, to provide the habitat requirements for species-rich communities of solitary bees would involve integration of agricultural systems with interspersed semi-natural habitat, covering up to 30 per cent of farmland (Kearns *et al.* 1998). For this to be successful, agricultural practice itself would need to shift in the direction of sustainability: i.e. reduction in the use of artificial fertilisers, pesticides and herbicides, and introducing integrated pest management and reduced tillage.

Since leaving the EU, proposals under development by the British government and the devolved regions indicate future funding for new approaches to land management, including 'sustainable farming', 'local nature recovery' and 'landscape recovery'. To a considerable extent, these proposals replicate the main schemes funded under the EU Common Agricultural Policy, though the landscape recovery projects seem to be more ambitious, even mentioning 'rewilding' as part of their remit. Unfortunately, these projects seem intended for only very limited areas (up to 3 per cent) of the total farmed land and are aimed at rather distant targets, well beyond the relevant electoral cycles. It remains to be seen how far and in what ways these initial sketches are implemented and integrated into practical changes in agricultural practice.

Pending major change in the way our farmland is managed, those of us lucky enough to have a garden may still do our bit to provide habitat for wildlife – and for bees in particular.

HELPING GARDEN BEES

As we argued above, gardens may have a significant role not only in sustaining local populations of bees but also in serving as 'stepping-stones' for species extending their range. Almost any wild bee can turn up in a garden. Lists made by ourselves and 10 fellow bee-watchers in private gardens, allotments and botanic gardens gave a total of 157 solitary bee species. The sites were mostly in southern counties but included Edinburgh Botanic Garden and a garden in the Outer Hebrides. Representatives of all major British bee genera were recorded except for *Eucera*. Oligolectic solitary bees were well represented, for example by *Macropis europaea*, *Andrena florea*, *Andrena hattorfiana*, *Melitta haemorrhoidalis* and *Hylaeus signatus*. The highest garden count was 119 species in a Suffolk garden.

Gardens typically contain many exotic plant species but they can be as rich in solitary bees as semi-natural green spaces. As we have seen (Benton 2021a), solitary bees visit gardens for one or more of a range of habitat requirements, the most obvious being flowers on which to forage. It is well known that horticulturally modified garden plants often have inaccessible pollen and nectar and, despite their dramatic forms and colours, may provide little or no food for bees. By contrast, plants in their wild form, even if exotic, are likely to be of use to bees, which usually learn quickly how to access their rewards.

It is important to provide food resources for solitary bees throughout the season, including a wide range of species and floral forms. One easy method is to mow different parts of a lawn on a 3–6-week rotation, throughout the season,

TABLE 32. Solitary bee species recorded in 13 or more of 18 garden bee lists.

Andrena bicolor	*Hylaeus hyalinatus*
Andrena dorsata	*Hylaeus pictipes*
Andrena fulva	*Hylaeus signatus*
Andrena haemorrhoa	*Lasioglossum calceatum*
Andrena nigroaenea	*Lasioglossum morio*
Andrena nitida	*Megachile centuncularis*
Andrena scotica	*Megachile ligniseca*
Anthidium manicatum	*Megachile willughbiella*
Anthophora furcata	*Nomada fabriciana*
Anthophora plumipes	*Nomada goodeniana*
Chelostoma campanularum	*Nomada marshamella*
Halictus tumulorum	*Osmia bicornis*
Hylaeus communis	*Osmia caerulescens*

allowing their component plants to flower. The best lawns can support 40 or more plant species. These will establish themselves naturally – there is no need to use seed mixes on lawns and it is much better to encourage locally growing wild flowers than alien forms (see https://www.plantlife.org.uk/uk). Wild plant species richness in a garden will improve with time, especially if wild corners are left to themselves.

In addition to food resources, nesting places for solitary bees are easy to provide, in the form of various designs of bee hotels. As noted repeatedly in this book, artificial nesting sites are ideal places for attracting solitary bees and for observing their behaviour.

The nesting tubes for bee hotels can include a range of diameters between 3–4 mm and 10–12 mm and should be at least 15 cm in length. They can be made from hollow stems, such as bamboo, or drilled into blocks of wood. The almost universal occupant, *Osmia bicornis*, requires tubes of about 7 mm in diameter, but it is well worth experimenting with different dimensions and designs. Mud-brick nesting accommodation, designed for *Anthophora plumipes* and its cuckoo, *Melecta albifrons*, are easily made from a mixture of clay soil, small pebbles and dry grass stems, stirred together with water. The mud is then placed in a suitable frame for drying. Holes can be made when the mud is still wet, using a stick or cane of 8–10 mm in diameter. The depth of the brick should be about 15 cm with the holes about 10 cm in depth (Walters 2021).

FIG 355. A bee hotel for cavity-nesting bees, placed in a public green space. (© M. D. Welch)

TABLE 33. Solitary bees observed using bee hotels.

Hollow stems	Mud brick
Anthidium manicatum	Anthophora plumipes
Anthophora furcata	Melecta albifrons
Coelioxys inermis	Megachile willughbiella
Coelioxys rufescens	
Heriades truncorum	
Hylaeus communis	
Hylaeus hyalinatus	
Hylaeus pictipes	
Hylaeus signatus	
Megachile centuncularis	
Megachile ligniseca	
Megachile willughbiella	
Megachile versicolor	
Osmia bicornis	
Osmia caerulescens	
Osmia cornuta	
Osmia leaiana	
Stelis breviuscula	
Stelis phaeoptera	

Any areas of bare ground, such as worn pathways across the garden, make excellent mining bee habitats. More ambitious options include providing a heap or bank of sandy soil or digging a dry ditch, with most of the spoil on the south-facing side. Smooth stones can be placed on the banks to attract basking bees.

However, 'tending our own gardens', as recommended by Voltaire (1759), may not be sufficient of itself to defend wildlife habitat on the scale needed, in the face of growing 'development' pressures. Gardeners are also citizens and have a legitimate part to play in addressing wider issues of land use and management.

FIG 356. Wool Carder Bee *Anthidium manicatum* male looking out from a wooden bee hotel.

BELOW: **FIG 357.** A mud brick providing solitary bee nest habitat. Roof tiles protect the structure from the rain. The brick immediately attracted several nesting Hairy-footed Flower Bees *Anthophora plumipes*.

TABLE 34. Some suggestions of plants for solitary bees, including wild and cultivated species.

Ajuga reptans	Bugle	*Lysimachia* species	Yellow Loosestrife
Aquilegia vulgaris	Columbine	*Malus* species	Apple
Bellis perennis	Daisy	*Myosotis sylvatica*	Wood Forget-me-not
Bryonia dioica	White Bryony	*Onobrychis viciifolia*	Sainfoin
Calendula arvensis	Field Marigold	*Ononis spinosa*	Spiny Restharrow
Campanula species	Bellflowers	*Potentilla* species	Cinquefoil
Centaurea species	Knapweed	*Papaver rhoeas*	Common Poppy
Chamerion angustifolium	Rosebay Willowherb	*Primula veris*	Cowslip
Cirsium species	Thistles	*Primula vulgaris*	Primrose
Cornus sanguinea	Dogwood	*Prunus* species	Plum/Bullace
Crataegus monogyna	Hawthorn	*Pulmonaria officinalis*	Lungwort
Crepis vesicaria	Beaked Hawksbeard	*Pyracantha*	Firethorn
Echium vulgare	Viper's Bugloss	*Ranunculus*	Buttercup species
Epilobium species	Willowherbs	*Reseda lutea*	Wild Mignonette
Escallonia species	Escallonia	*Reseda luteola*	Weld
Calluna vulgaris	Ling Heather	*Rosa canina*	Dog Rose
Eryngium species	Sea Holly	*Rosmarinus officinalis*	Rosemary
Geranium species	Geranium	*Rubus fruticosus*	Bramble
Glechoma hederacea	Ground-ivy	*Salix* species	Willow
Heracleum sphondylium	Hogweed	*Salvia pratensis*	Meadow Clary
Hypochaeris radicata	Common Catsear	*Scabiosa columbaria*	Small Scabious
Jacobaea senecio	Common Ragwort	*Solidago* species	Goldenrod
Knautia arvensis	Field Scabious	*Sonchus* species	Sowthistles
Lamium album	White Dead-nettle	*Symphoricarpos* species	Snowberry
Lamium purpureum	Red Dead-nettle	*Taraxacum officinale*	Dandelion
Lathyrus pratensis	Meadow Vetchling	*Thymus serpyllum*	Thyme
Leontodon hispidus	Rough Hawkbit	*Trifolium repens*	White Clover
Lobelia erinus	Lobelia	*Veronica chamaedrys*	Germander Speedwell
Lotus corniculatus	Common Bird's-foot-trefoil		

FIG 358. A dry ditch in sandy ground, used by several species of solitary bees.

BEES AND THE PLANNING SYSTEM

The Town and Country Planning Act of 1947 was a major step forwards in attaining some democratic input into decisions concerning changes in the wider living environment of residential communities. However, one weakness from the standpoint of nature conservation was the distinction between 'land use', which was subject to planning law, and 'land management', which was not. This opened the way to the subsequent spread of intensive agriculture and left such practices as the clearance of woods and hedgerows, 'improvement' of flower-rich grasslands, drainage of wetlands and so on outside the regulatory system except in formally designated sites, which were the primary responsibility of the Nature Conservancy, as it was then called. Of course, Britain was not alone in favouring the spread of intensive systems at the expense of natural and semi-natural habitats. This has occurred worldwide, regardless of local rules and regulations.

Subsequent changes to planning law and guidance have mostly weakened the powers of local planning authorities, while empowering developer and landowner interests. However, growing concern about biodiversity loss has recently been reflected in the planning framework in the shape of a requirement that a condition of planning consent for a site rich in biodiversity should be a commitment to enhance biodiversity by 10 per cent, with assurance that the improved habitat status can be secured for 30 years. This is termed Biodiversity Net Gain (BNG). Defra, the relevant government department, provides a 'metric' to enable existing biodiversity on a threatened site and proposed mitigation to be measured. Though it seems permissible for developers to invent their own metrics (see the case mentioned above, involving MoD evaluation of an acid grassland site in Colchester, Essex), the metric offered by Defra at the time of writing is subject to devastating criticisms in relation to insect, and especially pollinator, conservation. Falk (2021) argues that flowering plants designated as indicating deteriorating sites are often among the most important for insect visitors, that general specifications (e.g. on the extent of bare ground) need to be varied according to other properties of a site, that important habitats, such as spoil heaps and sandpits, are excluded, and that the requirements of many taxa for mosaics of habitat are unrecognised by the rigid compartmentalisation of the approach. In general, Falk argues that this metric stands in direct contradiction to the aims of the National Pollinator Strategy.

Though Falk does not argue against the BNG approach as such, it is open to more general questions. In the context of a global crisis of biodiversity loss, combined with escalating 'drivers' of threats to wildlife habitats, it is clear that a much stronger framework of law is required. A start might be to address the so-called 'mitigation hierarchy': avoid development, mitigate it, or compensate for habitat loss. There are major philosophical problems with applying a numerical metric of any sort to the biodiversity of a site, which would be required for a percentage increase to be measured. Number of species? Number of threatened species? Number of ecologically functional species? Number of culturally favoured species? Number of scientifically interesting species? Local valuation of the site? Ecological services provided by it? Aesthetic judgement of the landscape? All of these questions can only be answered on the basis of judgements of value applied to the evidence and qualitative evaluation across incommensurables.

In short, BNG replaces what would be a matter of open public debate and decision-making with a technical measure that hides key value decisions as if they were grounded in objective science. The political philosopher John O'Neill takes this objection further. He distinguishes between beings or places that are valued *de re*, for themselves, or *de dicto*, in this context, merely for the services they provide. He gives the example of public objections to the destruction of an 800-year-old woodland to make way for a service station. Some objections involved questioning the practical possibility of replicating the biodiversity of the woodland in the proposed compensatory offer. But others called into question the offer of compensation itself as failing to recognise the unique particularity of the woodland and its place in people's lives:

> However, the problem also lies in the fact that no new site could replicate the particular valued history of the woodland that would be lost in Smithy Wood. Even if it were possible to keep a woodland with the same features, something would be lost. It is this particular place with the history that is embodied in it that matters. The de re value cannot be compensated with the gain of woodland elsewhere.
> (O'Neill 2017)

The particularity of places and the histories (human as well as non-human) embodied in them provide a potentially powerful basis for refusing to go down the 'mitigation hierarchy': sites which are rich in biodiversity are likely to be so because of a prolonged history of incidental or deliberate management, and because of unique constellations of social, economic, climatic, geological and geographical conditions. They have a value *'de re'* as O'Neill puts it, and so the lower tiers of the mitigation hierarchy just do not apply: 'development' should be, simply, avoided. An alternative approach to planning would involve a democratically accountable provision of a network of connected and accessible habitat patches, appropriately managed, and constituting a securely protected 'green infrastructure' for each community.

WHAT IS THE POINT OF PROTECTING BEES?

The idea that wildlife had an economic value, and that this might provide arguments for its conservation, was quite absent from the ethos of the popular conservation movements until Britain's own Biodiversity Action Plan was introduced. This provided an impetus for local authorities to devise their own action plans and for mapping biodiversity, as well as for assessing the threats faced by particular species at local and national levels. The integration of natural processes into orthodox economic thinking had long preceded this moment, however, but it was not until the beginning of the 1970s that the independent discipline of environmental economics was fully established. A report published in 1970, on critical environmental problems, is widely regarded as the first to introduce the concept of 'environmental services' (SCEP 1970). Nature as a whole (or, in some versions, nature insofar as it affects human welfare) is treated as a bundle of capital resources, which we draw upon in the form of 'services' of various sorts. These services might include supply of goods (raw materials or, in the case of honeybees, honey), or they might be regulatory, including climatic stability and, of direct relevance to this book, pollination.

By the 1990s, as the significance of global loss of biodiversity came to be recognised, it was incorporated into environmental economics as an economic threat. The subsequent establishment of biodiversity action plans was justified in terms of addressing the economic costs of ecological degradation. Although cultural and spiritual benefits were included in standard accounts of ecosystem services, they could be included only in terms of an economic (monetary) valuation, along with the more obvious valuation of services with a direct economic impact. So, now environmentalists had a new weapon with which to influence governments and businesses: economic valuation of the services provided by nature and costs of over-exploitation and damage to 'natural capital'.

Pollination was soon recognised as a key ecosystem service, and by the late 1990s, often under the auspices of the UN Convention on Biodiversity, national governments were investing in academic research on plant-pollinator interactions and the prevailing threats to pollinators, notably habitat loss, climate change and pesticide poisoning. Britain's Insect Pollinators Initiative devoted £10 million for research into pollinators, funding nine projects. Still, the emphasis was on honeybees, bumblebees and butterflies, but some projects widened out to a recognition of the importance of solitary bees, hoverflies and other insects that have a role to play in pollination. Urban areas were also included, in addition to the more obviously economically significant focus on agriculture.

This focus on 'pollinators' as providers of an economically important service has justified the funding of important academic research, much of it cited in this book, which has greatly enhanced our understanding of the ecological relations of plants and their many insect visitors. This has drawn upon and also contributed to the knowledge gained by amateur field naturalists. There has emerged an intermingling of the 'citizen science' of amateur natural history, biological scientific research and environmental movement activity at local, national and even international levels. It is possible that such broad alliances may do something to offset the damaging consequences of both intensive agriculture and the failings of traditional municipal land management and the planning system.

However, there are reasons to be cautious. An article, authored by 59 leading researchers, and published in 2015, drew on evidence from 90 studies of insect-pollinated crop systems across five continents (Kleijn *et al.* 2015). The conclusions (consistently with much of the research cited in this chapter) were that crops were generally pollinated by a small number of dominant species. These species were common, represented only a tiny proportion of the regional fauna, and were readily encouraged by available environmental measures in farms. Species considered threatened were rarely present among crop-pollinator communities, and even when present, made little contribution to pollination. The authors took account of evidence suggesting that diversity of pollinator communities contributed both to yield and to stability over successive seasons but argued that, on the larger scale, this was less significant; also that intensive systems were likely to have already eliminated local source populations of threatened species and that many of these would, in any case, be pollen specialists with non-crop pollen hosts.

The clear conclusion of this study is that while agriculture does benefit from insect crop pollination, and also from environmental schemes on farms designed to enhance pollinator populations, this is an inadequate response to the requirements of pollinator conservation. More targeted schemes are required, focusing on threatened species and habitats and justified by more traditional arguments for wildlife conservation, rather than the economic value of ecological services.

We endorse that conclusion, but it is worth taking the argument a little further. First, a distinction can be made between arguing for conservation on the basis of the provision of an 'ecosystem service' and a potentially wider argument based on the diverse ecological functions of flower-visiting insects, including wild bees, independently of whether those functions are thought of as 'services' to humans or their economic benefit. Although pollination of wild plants is usually mentioned but passed over without much elaboration in advocacies of the ecosystem services approach, it is arguably of fundamental importance. If, as estimated, some 94 per cent of flowering plants are either dependent on, or benefit from, insect pollination, and if flowering plants are overwhelmingly the dominant green plants across terrestrial ecosystems, then the value of insect pollination far outweighs whatever 'service' is delivered to human crop systems. The flourishing of green plants is essential as the main basic level for all terrestrial food webs and as crucial to nutrient cycling, atmospheric composition, and climate stability. Pollination, then, is a key ecosystem function, sustaining the whole web of terrestrial life of which humans are a part – if currently a profoundly disruptive part.

This approach to using insights from ecology to justify conservation measures by attempting to calculate the economic benefits of 'ecosystem services' is highly questionable: not because these benefits do not exist, but because they are overwhelmed by the wider importance for life on Earth of ecological functions such as pollination. The analogy between industrial or financial capital, and 'natural capital', which is required for this approach, is unsustainable and amounts to what philosophers call a 'category mistake'. In practice, to make conservation depend on the favourable outcome of an economic calculation is a high-risk strategy, especially in an economic system driven by what one of its greatest intellectual advocates describes as 'creative destruction' (Schumpeter 1942).

Finally, if it is both philosophically incoherent and practically too risky to rely on an economic argument for conservation of biodiversity, what are the alternatives? Kleijn *et al.* present the alternative as a moral case: comparable to taking care of the elderly, or conserving ancient buildings or works of art. This is an argument that will find favour with many, but it may be that there is no single decisive grounding for conservation. In a diverse culture, there are numerous potential resources. People's daily engagement with nature, in the shape of a walk in the local park or spending time in the garden (especially valued during the pandemic), forms part of our identity and sense of well-being. Our literary heritage is full of representations of nature, as are the visual arts and music. What would a future generation make of these if their referents and sources of inspiration were irreplaceably lost? There are more philosophical arguments, too, in terms of, for example, an extension of rights to non-human beings, 'biophilia', or deep ecology (Benton 1998). A powerful secular case comes from one of the two founders of modern evolutionary theory:

> *It is interesting to contemplate an entangled bank, clothed with many plants of many kinds, with birds singing in the bushes, with various insects flitting about, and with worms crawling through the damp earth, and to reflect that these elaborately constructed forms, so different from each other, and dependent on each other in so complex a manner, have all been produced by laws acting around us.... There is grandeur in this view of life, with its several powers, having been originally breathed into a few forms or into one; and that, while this planet has gone cycling on according to the fixed law of gravity, from so simple a beginning endless forms most beautiful and most wonderful have been, and are being, evolved.*
>
> (Darwin [1859] 1968)

The question remains: will hoped-for cultural shifts in favour of strong action to protect biodiversity be capable of generating sufficient popular demand for the radical changes in food production, consumption, transport and wider economic organisation that are needed?

List of British Solitary Bees

ENGLISH TO SCIENTIFIC NAME

Aldworth Mining Bee	*Andrena lepida*	Broad-faced Mining Bee	*Andrena proxima*
Alfken's Mini-miner	*Andrena alfkenella*	Broad-margined Mining Bee	*Andrena synadelpha*
Armed Nomad Bee	*Nomada armata*		
Ashy Furrow Bee	*Lasioglossum sexnotatum*	Bronze Furrow Bee	*Halictus tumulorum*
		Brown-footed Leafcutter Bee	*Megachile versicolor*
Ashy Mining Bee	*Andrena cineraria*		
Backthorn Mining Bee	*Andrena varians*	Bryony Mining Bee	*Andrena florea*
Banded Dark Bee	*Stelis punctulatissima*	Buff-banded Mining Bee	*Andrena simillima*
		Buffish Mining Bee	*Andrena nigroaenea*
Banded Mud Bee	*Megachile ericetorum*	Buff-tailed Mining Bee	*Andrena humilis*
Bare-saddled Blood Bee	*Sphecodes ephippius*	Bull-headed Furrow Bee	*Lasioglossum zonulum*
Bare-saddled Colletes	*Colletes similis*		
Barham Mini-miner	*Andrena nana*	Burbage Mining Bee	*Andrena lathyri*
Bear-clawed Nomad Bee	*Nomada baccata*	Carrot Mining Bee	*Andrena nitidiuscula*
Bellflower Blunthorn Bee	*Melitta haemorrhoidalis*	Catsear Nomad Bee	*Nomada integra*
		Chalk Furrow Bee	*Lasioglossum fulvicorne*
Big-headed Mining Bee	*Andrena bucephala*		
Bilberry Mining Bee	*Andrena lapponica*	Chalk Yellow-face Bee	*Hylaeus dilatatus*
Bilberry Nomad Bee	*Nomada glabella*	Chilterns Mini-miner	*Andrena floricola*
Black Mining Bee	*Andrena pilipes*	Chocolate Mining Bee	*Andrena scotica*
Black-headed Leafcutter Bee	*Megachile circumcincta*	Clarke's Mining Bee	*Andrena clarkella*
		Cliff Furrow Bee	*Lasioglossum angusticeps*
Black-headed Mining Bee	*Andrena nigriceps*		
Black-horned Nomad Bee	*Nomada rufipes*	Cliff Mason Bee	*Osmia xanthomelana*
Black-thighed Epeolus	*Epeolus variegatus*		
Bloomed Furrow Bee	*Lasioglossum albipes*	Cliff Mining Bee	*Andrena thoracica*
Blue Mason Bee	*Osmia caerulescens*	Clover Blunthorn Bee	*Melitta leporina*
Blunthorn Nomad Bee	*Nomada flavopicta*	Coast Leafcutter Bee	*Megachile maritima*
Blunt-jawed Nomad Bee	*Nomada striata*	Common Furrow Bee	*Lasioglossum calceatum*
Small-headed Resin Bee	*Heriades rubicola*		
Broad-banded Nomad Bee	*Nomada signata*	Common Mini-miner	*Andrena minutula*
Broad-faced Furrow Bee	*Lasioglossum laticeps*	Common Mourning Bee	*Melecta albifrons*
		Common Yellow-face Bee	*Hylaeus communis*

Common name	Scientific name	Common name	Scientific name
Coppice Mining Bee	Andrena helvola	Hairy Yellow-face	Hylaeus hyalinatus
Dark Blood Bee	Sphecodes niger	Hairy-footed Flower Bee	Anthophora plumipes
Dark-winged Blood Bee	Sphecodes gibbus	Hairy-saddled Colletes	Colletes fodiens
Davies' Colletes	Colletes daviesanus	Hawksbeard Mining Bee	Andrena fulvago
Downland Furrow Bee	Halictus eurygnathus	Hawk's-beard Nomad Bee	Nomada facilis
		Hawthorn Mining Bee	Andrena chrysosceles
Dull-headed Blood Bee	Sphecodes ferruginatus	Heather Colletes	Colletes succinctus
		Heather Mining Bee	Andrena fuscipes
Dull-vented Sharp-tail Bee	Coelioxys elongata	Impunctate Mini-miner	Andrena subopaca
Dusky-horned Nomad Bee	Nomada bifasciata	Ivy Bee	Colletes hederae
Early Colletes	Colletes cunicularius	Jersey Mason Bee	Osmia niveata
Early Nomad Bee	Nomada leucophthalma	Kirby's Lesser Mason Bee	Hoplitis leucomelana
		Kirby's Mining Bee	Andrena afzeliella
Fabricius' Nomad Bee	Nomada fabriciana	Kirby's Nomad Bee	Nomada subcornuta
False Margined Blood Bee	Sphecodes miniatus	Large Bear-clawed Nomad Bee	Nomada alboguttata
Five-spined Rophites	Rophites quinquespinosus		
		Large Gorse Mining Bee	Andrena bimaculata
Flat-ridged Nomad Bee	Nomada obtusifrons	Large Meadow Mining Bee	Andrena labialis
Flavous Nomad Bee	Nomada flava	Large Sallow Mining Bee	Andrena apicata
Fork-jawed Nomad Bee	Nomada ruficornis	Large Scabious Mining Bee	Andrena hattorfiana
Fork-tailed Flower Bee	Anthophora furcata	Large Scissor Bee	Chelostoma florisomne
Four-banded Flower Bee	Anthophora quadrimaculata		
		Large Shaggy Bee	Panurgus banksianus
Four-spotted Furrow Bee	Lasioglossum quadrinotatum	Large Sharp-tail Bee	Coelioxys conoidea
		Large Yellow-face Bee	Hylaeus signatus
Fringed Furrow Bee	Lasioglossum sexstrigatum	Lathbury's Nomad Bee	Nomada lathburiana
		Least Furrow Bee	Lasioglossum minutissimum
Fringe-horned Mason Bee	Osmia pilicornis		
Fringeless Nomad Bee	Nomada conjungens	Little Carpenter Bee	Ceratina cyanea
Furry-bellied Blood Bee	Sphecodes hyalinatus	Little Dark Bee	Stelis breviuscula
Furry-claspered Furrow Bee	Lasioglossum lativentre	Little Nomad Bee	Nomada flavoguttata
Fuscous Nomad Bee	Nomada fusca		
Geoffroy's Blood Bee	Sphecodes geoffrellus	Little Sickle-jawed Blood Bee	Sphecodes longulus
Giant Blood Bee	Sphecodes albilabris		
Golden Furrow Bee	Halictus subauratus	Little Yellow-face Bee	Hylaeus pictipes
Gold-fringed Mason Bee	Osmia aurulenta	Lobe-spurred Furrow Bee	Lasioglossum pauxillum
Gooden's Nomad Bee	Nomada goodeniana		
Green Furrow Bee	Lasioglossum morio	Long-faced Furrow Bee	Lasioglossum punctatissimum
Green-eyed Flower Bee	Anthophora bimaculata		
Grey-backed Mining Bee	Andrena vaga	Long-fringed Mini-miner	Andrena niveata
Grey-banded Mining Bee	Andrena denticulata	Long-fringed Mining Bee	Andrena congruens
Grey-gastered Mining Bee	Andrena tibialis	Long-horned Bee	Eucera longicornis
Grey-patched Mining Bee	Andrena nitida	Long-horned Nomad Bee	Nomada hirtipes
Grey-tailed Furrow Bee	Lasioglossum prasinum	Maidstone Mining Bee	Andrena polita
		Margined Colletes	Colletes marginatus
Grooved Sharp-tail Bee	Coelioxys quadridentata	Marsham's Nomad Bee	Nomada marshamella
		Mountain Mason Bee	Osmia inermis
Groove-faced Mining Bee	Andrena angustior	Northern Colletes	Colletes floralis
Gwynne's Mining Bee	Andrena bicolor	Northern Mining Bee	Andrena ruficrus

Oak Mining Bee	*Andrena ferox*	Sharp-collared Furrow Bee	*Lasioglossum malachurum*
Onion Yellow-face Bee	*Hylaeus punctulatissimus*	Sheep's Bit Dufourea	*Dufourea halictula*
Orange-footed Furrow Bee	*Lasioglossum xanthopus*	Sheppard's Nomad Bee	*Nomada sheppardana*
Orange-horned Nomad Bee	*Nomada fulvicornis*	Shingle Yellow-face Bee	*Hylaeus annularis*
Orange-tailed Mining Bee	*Andrena haemorrhoa*	Shiny Dufourea	*Dufourea minuta*
		Shiny-gastered Furrow Bee	*Lasioglossum laeve*
Orange-vented Mason Bee	*Osmia leaiana*		
Orchard Mason Bee	*Osmia cornuta*	Shiny-margined Mini-miner	*Andrena semilaevis*
Painted Mining Bee	*Andrena fucata*		
Painted Nomad Bee	*Nomada fucata*	Shiny-vented Sharp-tail Bee	*Coelioxys inermis*
Pale-tailed Mining Bee	*Andrena tridentata*		
Pantaloon Bee	*Dasypoda hirtipes*	Short-fringed Mining Bee	*Andrena dorsata*
Panzer's Nomad Bee	*Nomada panzeri*	Short-horned Furrow Bee	*Lasioglossum brevicorne*
Patchwork Leafcutter Bee	*Megachile centuncularis*	Short-horned Yellow-face Bee	*Hylaeus brevicornis*
Perkin's Mining Bee	*Andrena rosae*		
Pinewood Mason Bee	*Osmia uncinata*	Short-spined Nomad Bee	*Nomada guttulata*
Plain Dark Bee	*Stelis phaeoptera*	Sickle-jawed Blood Bee	*Sphecodes puncticeps*
Plain Mini-miner	*Andrena minutuloides*	Silver-sided Nomad Bee	*Nomada argentata*
		Silvery Leafcutter Bee	*Megachile leachella*
Potter Flower Bee	*Anthophora retusa*	Six-banded Nomad Bee	*Nomada sexfasciata*
Purbeck Nomad Bee	*Nomada errans*	Small Flecked Mining Bee	*Andrena coitana*
Red Bartsia Blunthorn Bee	*Melitta tricincta*	Small Gorse Mining Bee	*Andrena ovatula*
Red Mason Bee	*Osmia bicornis*	Small Sallow Mining Bee	*Andrena praecox*
Red-backed Furrow Bee	*Lasioglossum laevigatum*	Small Sandpit Mining Bee	*Andrena argentata*
		Small Scabious Mining Bee	*Andrena marginata*
Red-backed Mining Bee	*Andrena similis*	Small Scissor Bee	*Chelostoma campanularum*
Red-girdled Mining Bee	*Andrena labiata*		
Red-horned Mini-miner	*Andrena nanula*	Small Shaggy Bee	*Panurgus calcaratus*
Red-tailed Blood Bee	*Sphecodes rubicundus*	Small Shiny Furrow Bee	*Lasioglossum semilucens*
Red-tailed Mason Bee	*Osmia bicolor*	Smeathman's Furrow Bee	*Lasioglossum smeathmanellum*
Red-thighed Epeolus	*Epeolus cruciger*		
Reed Yellow-face Bee	*Hylaeus pectoralis*	Smooth-faced Furrow Bee	*Lasioglossum fratellum*
Reticulate Blood Bee	*Sphecodes reticulatus*		
Ridge-cheeked Furrow Bee	*Lasioglossum puncticolle*	Smooth-gastered Furrow Bee	*Lasioglossum parvulum*
Large-headed Resin Bee	*Heriades truncorum*	Smooth-saddled Dark Bee	*Stelis odontopyga*
Rough-backed Blood Bee	*Sphecodes scabricollis*	Southern Bronze Furrow Bee	*Halictus confusus*
Rufescent Sharp-tail Bee	*Coelioxys rufescens*		
Rufous-footed Furrow Bee	*Lasioglossum rufitarse*	Spined Blood Bee	*Sphecodes spinulosus*
		Spined Hylaeus	*Hylaeus cornutus*
Sainfoin Blunthorn Bee	*Melitta dimidiata*	Spined Mason Bee	*Osmia spinulosa*
Sandpit Blood Bee	*Sphecodes pellucidus*	Spotted Dark Bee	*Stelis ornatula*
Sandpit Mining Bee	*Andrena barbilabris*	Square-headed Blood Bee	*Sphecodes monilicornis*
Scarce Black Mining Bee	*Andrena nigrospina*		
Sea Aster Bee	*Colletes halophilus*	Square-headed Furrow Bee	*Halictus maculatus*
Shaggy Furrow Bee	*Lasioglossum villosulum*	Square-jawed Sharp-tail Bee	*Coelioxys mandibularis*

English Name	Scientific Name	English Name	Scientific Name
Square-spotted Mourning Bee	*Melecta luctuosa*	Welted Lesser Mason Bee	*Hoplitis claviventris*
Squat Furrow Bee	*Lasioglossum pauperatum*	White-bellied Mining Bee	*Andrena gravida*
		White-footed Furrow Bee	*Lasioglossum leucopus*
Swollen-thighed Blood Bee	*Sphecodes crassus*	White-jawed Yellow-face Bee	*Hylaeus confusus*
Tawny Mining Bee	*Andrena fulva*		
Thick-margined Mini-miner	*Andrena falsifica*	White-lipped Yellow-face Bee	*Hylaeus incongruus*
Tormentil Mining Bee	*Andrena tarsata*	White-zoned Furrow Bee	*Lasioglossum leucozonium*
Tormentil Nomad Bee	*Nomada roberjeotiana*	Wilke's Mining Bee	*Andrena wilkella*
Trimmer's Mining Bee	*Andrena trimmerana*	Willowherb Leafcutter Bee	*Megachile lapponica*
Tuberculate Long-horned Bee	*Eucera nigrescens*	Willughby's Leafcutter Bee	*Megachile willughbiella*
Tufted Furrow Bee	*Lasioglossum nitidiusculum*	Wood-carving Leafcutter Bee	*Megachile ligniseca*
Turquoise Furrow Bee	*Lasioglossum cupromicans*	Wool Carder Bee	*Anthidium manicatum*
Variable Nomad Bee	*Nomada zonata*	Yellow Loosestrife Bee	*Macropis europaea*
Violet Carpenter Bee	*Xylocopa violacea*	Orange-legged Furrow Bee	*Halictus rubicundus*
Viper's Bugloss Mason Bee	*Hoplitis adunca*	Yellow-legged Mining Bee	*Andrena flavipes*
Wall Mason Bee	*Osmia parietina*	Yellow-shouldered Nomad Bee	*Nomada ferruginata*
Water-dropwort Mining Bee	*Andrena ampla*		

SCIENTIFIC TO ENGLISH NAME

Scientific Name	English Name	Scientific Name	English Name
Andrena afzeliella	Kirby's Mining Bee	*Andrena denticulata*	Grey-banded Mining Bee
Andrena alfkenella	Alfken's Mini-miner	*Andrena dorsata*	Short-fringed Mining Bee
Andrena ampla	Water-dropwort Mining Bee	*Andrena falsifica*	Thick-margined Mini-miner
Andrena angustior	Groove-faced Mining Bee	*Andrena ferox*	Oak Mining Bee
Andrena apicata	Large Sallow Mining Bee	*Andrena flavipes*	Yellow-legged Mining Bee
Andrena argentata	Small Sandpit Mining Bee	*Andrena florea*	Bryony Mining Bee
Andrena barbilabris	Sandpit Mining Bee	*Andrena floricola*	Chilterns Mini-miner
Andrena bicolor	Gwynne's Mining Bee	*Andrena fucata*	Painted Mining Bee
Andrena bimaculata	Large Gorse Mining Bee	*Andrena fulva*	Tawny Mining Bee
Andrena bucephala	Big-headed Mining Bee	*Andrena fulvago*	Hawksbeard Mining Bee
Andrena chrysosceles	Hawthorn Mining Bee	*Andrena fuscipes*	Heather Mining Bee
Andrena cineraria	Ashy Mining Bee	*Andrena gravida*	White-bellied Mining Bee
Andrena clarkella	Clarke's Mining Bee		
Andrena coitana	Small Flecked Mining Bee	*Andrena haemorrhoa*	Orange-tailed Mining Bee
Andrena congruens	Long-fringed Mining Bee	*Andrena hattorfiana*	Large Scabious Mining Bee

Andrena helvola   Coppice Mining Bee
Andrena humilis   Buff-tailed Mining Bee
Andrena labialis   Large Meadow Mining
                            Bee
Andrena labiata   Red-girdled Mining Bee
Andrena lapponica   Bilberry Mining Bee
Andrena lathyri   Burbage Mining Bee
Andrena lepida   Aldworth Mining Bee
Andrena marginata   Small Scabious Mining
                            Bee
Andrena minutula   Common Mini-miner
Andrena minutuloides   Plain Mini-miner
Andrena nana   Barham Mini-miner
Andrena nanula   Red-horned Mini-miner
Andrena nigriceps   Black-headed Mining
                            Bee
Andrena nigroaenea   Buffish Mining Bee
Andrena nigrospina   Scarce Black Mining Bee
Andrena nitida   Grey-patched Mining
                            Bee
Andrena nitidiuscula   Carrot Mining Bee
Andrena niveata   Long-fringed
                            Mini-miner
Andrena ovatula   Small Gorse Mining Bee
Andrena pilipes   Black Mining Bee
Andrena polita   Maidstone Mining Bee
Andrena praecox   Small Sallow Mining Bee
Andrena proxima   Broad-faced Mining Bee
Andrena rosae   Perkin's Mining Bee
Andrena ruficrus   Northern Mining Bee
Andrena semilaevis   Shiny-margined
                            Mini-miner
Andrena scotica   Chocolate Mining Bee
Andrena similis   Red-backed Mining Bee
Andrena simillima   Buff-banded Mining Bee
Andrena subopaca   Impunctate Mini-miner
Andrena synadelpha   Broad-margined Mining
                            Bee
Andrena tarsata   Tormentil Mining Bee
Andrena thoracica   Cliff Mining Bee
Andrena tibialis   Grey-gastered Mining
                            Bee
Andrena tridentata   Pale-tailed Mining Bee
Andrena trimmerana   Trimmer's Mining Bee
Andrena vaga   Grey-backed Mining Bee
Andrena varians   Backthorn Mining Bee
Andrena wilkella   Wilke's Mining Bee
Anthidium manicatum   Wool Carder Bee
Anthophora bimaculata   Green-eyed Flower Bee
Anthophora furcata   Fork-tailed Flower Bee
Anthophora plumipes   Hairy-footed Flower Bee

Anthophora   Four-banded Flower Bee
  quadrimaculata
Anthophora retusa   Potter Flower Bee
Ceratina cyanea   Little Carpenter Bee
Chelostoma   Small Scissor Bee
  campanularum
Chelostoma florisomne   Large Scissor Bee
Coelioxys conoidea   Large Sharp-tail Bee
Coelioxys elongata   Dull-vented Sharp-tail
                            Bee
Coelioxys inermis   Shiny-vented Sharp-tail
                            Bee
Coelioxys mandibularis   Square-jawed Sharp-tail
                            Bee
Coelioxys   Grooved Sharp-tail Bee
  quadridentata
Coelioxys rufescens   Rufescent Sharp-tail Bee
Colletes cunicularius   Early Colletes
Colletes daviesanus   Davies' Colletes
Colletes floralis   Northern Colletes
Colletes fodiens   Hairy-saddled Colletes
Colletes halophilus   Sea Aster Bee
Colletes hederae   Ivy Bee
Colletes marginatus   Margined Colletes
Colletes similis   Bare-saddled Colletes
Colletes succinctus   Heather Colletes
Dasypoda hirtipes   Pantaloon Bee
Dufourea halictula   Sheep's Bit Dufourea
Dufourea minuta   Shiny Dufourea
Epeolus cruciger   Red-thighed Epeolus
Epeolus variegatus   Black-thighed Epeolus
Eucera longicornis   Long-horned Bee
Eucera nigrescens   Tuberculate
                            Long-horned Bee
Halictus confusus   Southern Bronze
                            Furrow Bee
Halictus eurygnathus   Downland Furrow Bee
Halictus maculatus   Square-headed
                            Furrow Bee
Halictus rubicundus   Orange-legged
                            Furrow Bee
Halictus subauratus   Golden Furrow Bee
Halictus tumulorum   Bronze Furrow Bee
Heriades rubicola   Small-headed Resin Bee
Heriades truncorum   Large-headed Resin Bee
Hoplitis adunca   Viper's Bugloss Mason
                            Bee
Hoplitis claviventris   Welted Lesser Mason
                            Bee
Hoplitis leucomelana   Kirby's Lesser Mason Bee
Hylaeus annularis   Shingle Yellow-face Bee

Scientific name	Common name
Hylaeus brevicornis	Short-horned Yellow-face Bee
Hylaeus communis	Common Yellow-face Bee
Hylaeus confusus	White-jawed Yellow-face Bee
Hylaeus cornutus	Spined Hylaeus
Hylaeus dilatatus	Chalk Yellow-face Bee
Hylaeus hyalinatus	Hairy Yellow-face
Hylaeus incongruus	White-lipped Yellow-face Bee
Hylaeus pectoralis	Reed Yellow-face Bee
Hylaeus pictipes	Little Yellow-face Bee
Hylaeus punctulatissimus	Onion Yellow-face Bee
Hylaeus signatus	Large Yellow-face Bee
Lasioglossum albipes	Bloomed Furrow Bee
Lasioglossum angusticeps	Cliff Furrow Bee
Lasioglossum brevicorne	Short-horned Furrow Bee
Lasioglossum calceatum	Common Furrow Bee
Lasioglossum cupromicans	Turquoise Furrow Bee
Lasioglossum fratellum	Smooth-faced Furrow Bee
Lasioglossum fulvicorne	Chalk Furrow Bee
Lasioglossum laeve	Shiny-gastered Furrow Bee
Lasioglossum laevigatum	Red-backed Furrow Bee
Lasioglossum laticeps	Broad-faced Furrow Bee
Lasioglossum lativentre	Furry-claspered Furrow Bee
Lasioglossum leucopus	White-footed Furrow Bee
Lasioglossum leucozonium	White-zoned Furrow Bee
Lasioglossum malachurum	Sharp-collared Furrow Bee
Lasioglossum minutissimum	Least Furrow Bee
Lasioglossum morio	Green Furrow Bee
Lasioglossum nitidiusculum	Tufted Furrow Bee
Lasioglossum parvulum	Smooth-gastered Furrow Bee
Lasioglossum pauperatum	Squat Furrow Bee
Lasioglossum pauxillum	Lobe-spurred Furrow Bee
Lasioglossum prasinum	Grey-tailed Furrow Bee
Lasioglossum punctatissimum	Long-faced Furrow Bee
Lasioglossum puncticolle	Ridge-cheeked Furrow Bee
Lasioglossum quadrinotatum	Four-spotted Furrow Bee
Lasioglossum rufitarse	Rufous-footed Furrow Bee
Lasioglossum semilucens	Small Shiny Furrow Bee
Lasioglossum sexnotatum	Ashy Furrow Bee
Lasioglossum sexstrigatum	Fringed Furrow Bee
Lasioglossum smeathmanellum	Smeathman's Furrow Bee
Lasioglossum villosulum	Shaggy Furrow Bee
Lasioglossum xanthopus	Orange-footed Furrow Bee
Lasioglossum zonulum	Bull-headed Furrow Bee
Macropis europaea	Yellow Loosestrife Bee
Megachile centuncularis	Patchwork Leafcutter Bee
Megachile circumcincta	Black-headed Leafcutter Bee
Megachile ericetorum	Banded Mud Bee
Megachile lapponica	Willowherb Leafcutter Bee
Megachile leachella	Silvery Leafcutter Bee
Megachile ligniseca	Wood-carving Leafcutter Bee
Megachile maritima	Coast Leafcutter Bee
Megachile versicolor	Brown-footed Leafcutter Bee
Megachile willughbiella	Willughby's Leafcutter Bee
Melecta albifrons	Common Mourning Bee
Melecta luctuosa	Square-spotted Mourning Bee
Melitta dimidiata	Sainfoin Blunthorn Bee
Melitta haemorrhoidalis	Bellflower Blunthorn Bee
Melitta leporina	Clover Blunthorn Bee
Melitta tricincta	Red Bartsia Blunthorn Bee
Nomada alboguttata	Large Bear-clawed Nomad Bee

Nomada argentata	Silver-sided Nomad Bee	*Osmia caerulescens*	Blue Mason Bee
Nomada armata	Armed Nomad Bee	*Osmia cornuta*	Orchard Mason Bee
Nomada baccata	Bear-clawed Nomad Bee	*Osmia inermis*	Mountain Mason Bee
Nomada bifasciata	Dusky-horned Nomad Bee	*Osmia leaiana*	Orange-vented Mason Bee
Nomada conjungens	Fringeless Nomad Bee	*Osmia niveata*	Jersey Mason Bee
Nomada errans	Purbeck Nomad Bee	*Osmia parietina*	Wall Mason Bee
Nomada fabriciana	Fabricius' Nomad Bee	*Osmia pilicornis*	Fringe-horned Mason Bee
Nomada facilis	Hawk's-beard Nomad Bee	*Osmia spinulosa*	Spined Mason Bee
Nomada ferruginata	Yellow-shouldered Nomad Bee	*Osmia uncinata*	Pinewood Mason Bee
		Osmia xanthomelana	Cliff Mason Bee
Nomada flava	Flavous Nomad Bee	*Panurgus banksianus*	Large Shaggy Bee
Nomada flavoguttata	Little Nomad Bee	*Panurgus calcaratus*	Small Shaggy Bee
Nomada flavopicta	Blunthorn Nomad Bee	*Rophites quinquespinosus*	Five-spined Rophites
Nomada fucata	Painted Nomad Bee		
Nomada fulvicornis	Orange-horned Nomad Bee	*Sphecodes albilabris*	Giant Blood Bee
		Sphecodes crassus	Swollen-thighed Blood Bee
Nomada fusca	Fuscous Nomad Bee		
Nomada glabella	Bilberry Nomad Bee	*Sphecodes ephippius*	Bare-saddled Blood Bee
Nomada goodeniana	Gooden's Nomad Bee	*Sphecodes ferruginatus*	Dull-headed Blood Bee
Nomada guttulata	Short-spined Nomad Bee	*Sphecodes geoffrellus*	Geoffroy's Blood Bee
Nomada hirtipes	Long-horned Nomad Bee	*Sphecodes gibbus*	Dark-winged Blood Bee
		Sphecodes hyalinatus	Furry-bellied Blood Bee
Nomada integra	Catsear Nomad Bee	*Sphecodes longulus*	Little Sickle-jawed Blood Bee
Nomada lathburiana	Lathbury's Nomad Bee		
Nomada leucophthalma	Early Nomad Bee	*Sphecodes miniatus*	False Margined Blood Bee
Nomada marshamella	Marsham's Nomad Bee		
Nomada obtusifrons	Flat-ridged Nomad Bee	*Sphecodes monilicornis*	Square-headed Blood Bee
Nomada panzeri	Panzer's Nomad Bee		
Nomada roberjeotiana	Tormentil Nomad Bee	*Sphecodes niger*	Dark Blood Bee
Nomada ruficornis	Fork-jawed Nomad Bee	*Sphecodes pellucidus*	Sandpit Blood Bee
Nomada rufipes	Black-horned Nomad Bee	*Sphecodes puncticeps*	Sickle-jawed Blood Bee
		Sphecodes reticulatus	Reticulate Blood Bee
Nomada sexfasciata	Six-banded Nomad Bee	*Sphecodes rubicundus*	Red-tailed Blood Bee
Nomada sheppardana	Sheppard's Nomad Bee	*Sphecodes scabricollis*	Rough-backed Blood Bee
Nomada signata	Broad-banded Nomad Bee	*Sphecodes spinulosus*	Spined Blood Bee
		Stelis breviuscula	Little Dark Bee
Nomada striata	Blunt-jawed Nomad Bee	*Stelis odontopyga*	Smooth-saddled Dark Bee
Nomada subcornuta	Kirby's Nomad Bee		
Nomada zonata	Variable Nomad Bee	*Stelis ornatula*	Spotted Dark Bee
Osmia aurulenta	Gold-fringed Mason Bee	*Stelis phaeoptera*	Plain Dark Bee
Osmia bicolor	Red-tailed Mason Bee	*Stelis punctulatissima*	Banded Dark Bee
Osmia bicornis	Red Mason Bee	*Xylocopa violacea*	Violet Carpenter Bee

Glossary

For anatomical features, see also Figure 6 in the Introduction.

Abdomen The hindmost of the three sections of an insect's body. In bees, the first segment of the abdomen is fused with the rear part of the thorax, leaving a 'waist' between the first and second segments of the abdomen. For convenience in this book, we use the term 'abdomen' to refer to the part of the abdomen behind the waist. Strictly speaking, this is the **metasoma**.
Acarinarium A pouch present on the body of some bees which contains mites, possibly of mutual benefit to the bee and mite.
Actinomorphic A flower or animal with radial symmetry (circular).
Arolium A soft pad of tissue at the end of the final tarsus segment, between the claws. It assists in clinging to a surface.
Basal vein A vein on the forewing, useful for the identification of a bee's genus.
Basitarsus The first segment of the tarsus, nearest to the tibia, often larger than the other three tarsus segments. Synonymous with 'metatarsus'.
Basitibial plate A small plate at the base of the tibia which assists in bracing a bee against the shaft of a nest burrow.
Bivoltine Having two generations in a year.
Brood cell A cavity constructed by a bee in which an egg is laid on a store of pollen and nectar. The brood cell may be lined with a waterproof secretion and is separated from adjacent cells by a partition of mud, leaf mastic or other material.
Cleptoparasite A parasite which 'steals' the food of another animal. In the case of bees this is usually the food-store in a brood cell of another bee. A cuckoo bee is a cleptoparasite.
Clypeus A part of the face below the antennae, delineated by sutures and sometimes coloured cream or white.
Corbicula A true pollen basket on the hind tarsus, as seen in bumblebees and honeybees. Solitary bees do not possess a corbicula.
Corolla The petals of a flower.
Cuckoo bee A bee which does not collect pollen and lays its eggs in the brood cells of pollen-collecting bees.
Cuticle The hard outer skeleton of an insect.
Diapause A dormant phase under physiological control, often over the winter but also occurring at other times. In solitary bees it occurs most often in the prepupa stage.
Dufour's gland A gland in the abdomen which synthesises and secretes a variety of compounds. Functions of the secretions include brood cell linings, food for larvae, pheromones and identity signals.
Eusocial Social behaviour involving an egg-laying fertile female (gyne/queen) and sterile female workers which forage for food provisions and construct brood cells.
Femur The third main section of the leg.
Gyne The primary female reproductive caste in social insects. Gynes are capable of laying eggs and have the potential, as queens, to control the behaviour of non-reproductive females (workers) in a colony. In some species,

colonies may be dominated by more than one gyne.
Hermaphrodite Having male and female sex organs on the same individual, usually referring to flowering plants.
Holotype The single type specimen used to describe a species or subspecies.
Inquiline An animal which lives in the nest or burrow of another animal species, often as a parasite.
Labrum The upper lip, which covers the folded mouthparts.
Mandibles The jaws.
Mesosoma The middle section of a bee's body consisting of the thorax and the propodeum.
Metasoma The third section of the body comprising the parts of the abdomen posterior to the narrow part ('waist') of the bee.
Mining bee A bee species which excavates a nest burrow in the ground. Also used in a more confined sense to indicate the genus *Andrena* (mining bees).
Monolectic Adjective for a bee which collects pollen from just one species of plant.
Nototribic Flowers placing pollen on the upper surface of a visiting insect.
Oligolectic Adjective for a bee which collects pollen only from a few closely related species of plants.
Oligolege Noun for an oligolectic bee.
Parasite An organism that lives in or on a living organism of another species (its host) and causes it harm.
Parasitoid A parasite which lives on another organism (its host) while the host is growing, then finally kills it and is free-living as an adult.
Phoretic Dispersal on the body of another animal.
Plumose Feather-like or branched (with reference to hairs).
Polylectic Adjective for a bee which collects pollen from a wide range of plant species.
Prepupa The mature stage of a bee larva after it has defecated. A resting or hibernating stage which precedes transformation into a pupa.
Propodeum The first segment of the abdomen, fused with the rear part of the thorax in bees and other aculeates.
Pterostigma A pigmented spot on the leading edge of the forewing which helps to stabilise the wing. Its colour can be useful in identification.
Pygidium/pygidial plate A raised flat area on tergite 6, often triangular and surrounded by a ridge. Used for smoothing the lining of a brood cell and moving earth within a nest burrow. Typical of females but also present in some males.
Rima A longitudinal line bearing microscopic hairs on tergite 5 of female Halictids.
Scopa Specialised long pollen-collecting hairs of solitary bees.
Seta (pl. setae) Stiff hairs attached to the cuticle of a bee or other insect.
Sternite The ventral (lower) plates of the cuticle of the metasoma.
Sternotribic Flowers placing pollen onto the underside of a visiting insect.
Submarginal cell Compartment (cell) of the forewing posterior to the marginal cell. In solitary bees, there are either two or three submarginal cells.
Tarsus The final section of a leg, made of the basitarsus and four further segments, the final one bearing the arolium and claws.
Tergite The dorsal (upper) plates of the cuticle of the metasoma.
Thorax The middle part of the body of an insect, controlling three pairs of legs and, in bees, two pairs of wings. In bees, the first segment of the abdomen is fused with the rear part of the thorax. This is termed **propodeum**, and the whole structure is termed the **mesosoma**. However, for convenience, in this book, we generally refer to the mesosoma as the thorax, except where this is likely to cause confusion.
Tibia The fourth main section of a leg (coxa, trochanter, femur, **tibia**, tarsus).
Univoltine Having one generation in a year.
Venation The pattern of the veins of the wings.
Worker A sterile female in a colony of social insects, usually involved in rearing young and collecting food.
Zygomorphic A flower with bilateral symmetry (flattened from side to side), usually having a deep corolla or spur.

See also: https://www.bwars.com/sites/www bwars.com/files/diary_downloads/Britain%27_Bees_Chapter_4_Glossary_0.pdf

References and Bibliography

Adams, W. M. (1990). *Green Development*. Routledge, London.

Ågren, L. (1978). Flagellar sensilla of two species of Andrena (Hymenoptera: Andrenidae). *International Journal of Insect Morphology and Embryology* 7(1), 73–79.

Aigner, P. A. (2006). The evolution of specialized floral phenotypes in a fine-grained pollination environment. In N. M. Waser & J. Ollerton (eds.) *op. cit.*, 23–46.

Aizer, M. A. & Harder, L. D. (2009). The global stock of domesticated honey bees is growing slower than agricultural demand for pollination. *Current Biology* 19, 915–918.

Alcock, J. (1980). Natural selection and the mating systems of solitary bees. *American Scientist* 68(2), 146–153.

Alcock, J. (2000). The natural history of a miltogrammine fly, *Miltogramma rectangularis* (Diptera: Sarcophagidae). *Journal of the Kansas Entomological Society* 73(4), 208–219.

Alcock, J. & Houston, T. (1996). Mating systems and male size in Australian Hylaeine bees (Hymenoptera: Colletidae). *Ethology* 102(4), 591–610.

Allen, G. (2009). *Bees, Wasps and Ants of Kent*. Kent Field Club.

Allen, G. (2020). *Bees, Wasps and Ants of Kent*, 2nd edition. Kent Field Club.

Almeida, E.A.B. (2008). Colletidae nesting cell biology (Hymenoptera: Apoidea). *Apidologie* 39, 16–29.

Amador, G. L., Durand, F., Mao, W. *et al.* (2015). Insects have hairy eyes that reduce particle deposition. *European Physical Journal Special Topics* 224, 3361–3377.

Andersson, M. (1984). The evolution of eusociality. *Annual Review of Ecology and Systematics* 15, 165–190.

Archer, M. (2004). The wasps and bees (Hymenoptera: Aculeata) of York's Victorian cemetery in Watsonian Yorkshire. *Naturalist* 129, 145–153.

Archer, M. (2013). The solitary bees and wasps (Hymenoptera: Aculeata) of a suburban garden in Leicester, England, over 27 years. *Entomologist's Monthly Magazine* 149, 93–121.

Archer, M. (2014). The solitary wasps and bees (Hymenoptera: Aculeata) of urban and suburban gardens. *Entomologist's Monthly Magazine* 150, 169–179.

Armbruster, W. S. (2014). Floral specialization and angiosperm diversity: phenotypic divergence, fitness trade-offs and realized pollination accuracy. *AoB PLANTS* 6.

Ascher, J. S. & Pickering, J. (2020). Discover Life bee species guide and world checklist (Hymenoptera: Apoidea: Anthophila). https://www.discoverlife.org/mp/20q?act=x_guide_credit&guide=Apoidea_species.

Aspiazu, C., Bosch, J., Viñuela, E. *et al.* (2019). Chronic exposure to field-realistic pesticide combinations via pollen and nectar: effects on feeding and thermal performance in a solitary bee. *Scientific Reports* 9, 13770.

Ayasse, M. (1994). The meaning of male odor in the mating biology of sweat bees

(Hymenoptera: Halictidae). *Les Insectes Sociaux. Proceedings of the International Congress IUSSI* 12, 88. University of Paris Nord.

Ayasse, M. & Dutzler, G. (1998). The functions of pheromones in the mating biology of *Osmia* bees (Hymenoptera: Megachilidae). *Social Insects at the Turn of the Millennium. Proceedings of the International Congress IUSSI* 13, 42. Flinders University, Adelaide.

Ayasse, M., Engels, W., Hefetz, A., Lübke, G. & Francke, W. (1990). Ontogenetic patterns in amounts and proportions of Dufour's gland volatile secretions in virgin and nesting queens of *Lasioglossum malachurum* (Hymenoptera: Halictidae). *Zeitschrift für Naturforschung* 45c, 709–714.

Ayasse, M., Engels, W., Hefetz, A., Tengö, J. & Lübke, G. (1993). Ontogenetic patterns of volatiles identified in Dufour's gland extracts from queens and workers of the primitively eusocial halictine bee, *Lasioglossum malachurum* (Hymenoptera: Halictidae). *Insectes Sociaux* 40, 41–58.

Ayasse, M., Engels, W., Lübke, G. & Francke, W. (1999). Mating expenditures reduced via female sex pheromone modulation in the primitively eusocial halictine bee, *Lasioglossum (Evylaeus) malachurum* (Hymenoptera: Halictidae). *Behavioral Ecology and Sociobiology* 45, 95–106.

Ayasse, M., Paxton, R. J. & Tengö, J. (2001). Mating behaviour and chemical communication in the Order Hymenoptera. *Annual Review of Entomology* 46, 31–78.

Ayasse, M., Schiestl, F. P., Paulus, H. F. *et al.* (2000). Evolution of reproductive strategies in the sexually deceptive orchid *Ophrys sphegodes*: how does flower-specific variation of odor signals influence reproductive success? *Evolution* 54(6), 1995–2006.

Baker, D. B. (1998). The hymenoptera collections of W. E. Shuckard and the dispersal of his type material. *Contributions to Entomology* 48(1), 157–174.

Baker, H. G. (1948). Significance of pollen dimorphism in late-glacial *Armeria*. *Nature* 161, 770–771.

Baldock, D. W. (2007). The importance of dead wood. *Bees, Wasps and Ants Recording Society Newsletter*, Autumn, 20–23.

Baldock, D. W. (2008). *Bees of Surrey*. Surrey Wildlife Trust.

Baldock, D. W. (2010a). Another possible nesting record of the Violet Carpenter-bee *Xylocopa violacea*. *Bees, Wasps and Ants Recording Society Newsletter*, Autumn, 20.

Baldock, D. W. (2010b). *Wasps of Surrey*. Surrey Wildlife Trust.

Baldock, D. W. (2020). Farnham Heath Nature Reserve RSPB: passing the 300 species mark. *Bees, Wasps and Ants Recording Society Newsletter*, Spring, 17–21.

Baldock, D. W. & Early, J. P. (2015). *Soldierflies, their allies and Conopidae of Surrey*. Surrey Wildlife Trust.

Baldock, D. W., Livory, A. & Owens, N. W. (2020). The Bees and Wasps of the Balearic Islands (Hymenoptera: Chrysidoidea, Vespoidea, Apoidea) with a discussion of aculeate diversity in Mediterranean and Atlantic archipelagos. *Entomofauna, Zeitschrift für Entomologie* 25, 202 pp.

Baldock, D. W. & Wood, T. (2018). Farnham Heath Revisited, 2017. *Bees, Wasps and Ants Recording Society Newsletter*, Spring, 8–13.

Baldock, K.C.R. *et al.* (2015). Where is the UK's pollinator diversity? The importance of urban areas for flower-visiting insects. *Proceedings of the Royal Society B* 282, 20142849.

Balfour-Browne, F. (1922). On the life history of *Melittobia acasta*, Walker; a chalcid parasite of bees and wasps. *Parasitology* XIV, 349–371.

Banaszak, J. (1996). Ecological basis of conservation of wild bees. In A. Matheson *et al.* (eds.) *op. cit.*, 55–62.

Bang, D. (2008). *A Freedom to Roam Guide to the Brighton Downs*. Bangs: Brighton.

Barrett, S.C.H. (2010). Darwin's legacy: the forms, function and sexual diversity of flowers. *Philosophical Transactions of the Royal Society B* 365(1539), 351–368.

Barrett, S.C.H., & Shore, J. S. (2008). New insights on heterostyly: comparative biology, ecology and genetics. In V. E. Franklin-Tong (ed.), *Self-incompatibility in flowering plants – evolution, diversity, and mechanisms*, 3 32. Springer, Berlin.

Barringer, C., Bell. M., Bennett, J. *et al.* (1991). *East Anglian Studies II: History of the Environment*. Open University.

Barth, F. G. [1982] (1991). *Insects and Flowers: The Biology of a Partnership*. Princeton University Press.

Bartlett, J. (2022) Pollen: Lifting the veil. *BeeCraft*, May, 29–31.

Bartlett, J. & Bartlett, V. (2018). The Four-banded Flower Bee *Anthophora quadrimaculata* Panzer (Hymenoptera, Apidae) in Norfolk. *Transactions of the Norfolk and Norwich Naturalists' Society* 51(1), 38–44.

Bartlett, V. (2021). Some observations on spider predation with particular reference to solitary bees and other aculeates. *Transactions of the Norfolk and Norwich Naturalists' Society* 54(1), 1–3.

Batra, S.W.T. & Norden, B. B. (1996). Fatty food for their broods: how *Anthophora* bees make and provision their cells (Hymenoptera: Apoidea). *Memoirs of the Entomological Society of Washington* 17, 36–44.

Baude, M. *et al.* (2016). Historical nectar assessment reveals the fall and rise of Britain in bloom. *Nature* 530(7588), 85–88.

Beani, L. (2006). Crazy wasps: when parasites manipulate the *Polistes* phenotype. *Annales Zoologici Fennici* 43, 564–574.

Beavis, I. C. (2013). A revised list of the bees, wasps and ants of Scilly. *Isles of Scilly Bird and Natural History Review* 2012, 171–182.

Beavis, I. C. (2022). Hymenoptera and Diptera of Scilly – Update. *Isles of Scilly Bird and Natural History Review* 2021, 189–193.

Becker, R. (2018a). Interplay between *Coelioxys inermis* (Kirby, 1802) and *Megachile centuncularis* (Linnaeus, 1758) in a nest-box. *Bees, Wasps and Ants Recording Society Newsletter*, Spring, 16–18.

Becker, R. (2018b). Aculeates on an upland farm in Wales. *Bees, Wasps and Ants Recording Society Newsletter*, Autumn, 27–28.

Benitez-Vieyra, S., Hempel de Ibarra, N., Wertlen, A. M. & Cocucci, A. A. (2007). How to look like a mallow: evidence of floral mimicry between Turneraceae and Malvaceae. *Proceedings of the Royal Society B* 274(1623), 2239–2248.

Bennett, A. W. (1883). The constancy of insects in their visits to flowers. *Zoological Journal of the Linnean Society* 17(100), 175–185.

Benton, T. (1998). Rights and justice on a shared planet: more rights or new relations? *Theoretical Criminology* 2(2), 149–175.

Benton, T. (2000). *The Bumblebees of Essex*. Lopinga, Wimbish, Essex.

Benton, T. (2006). *Bumblebees*. New Naturalist 98. HarperCollins, London.

Benton, T. (2012). *Grasshoppers and Crickets*. New Naturalist 120. HarperCollins, London.

Benton, T. (2017). *Solitary Bees*. Naturalists' Handbooks 33. Pelagic, Exeter.

Benton, T. (2020). Town centre green spaces. *Nature in North-East Essex*, 51–59. Colchester Natural History Society.

Benton, T. (2021a). How do solitary bees use urban gardens? A lockdown study. *British Wildlife* 32(7), 509–516.

Benton, T. (2021b). Notes on the ecology and nesting behaviour of the Reed Yellow-face Bee *Hylaeus pectoralis* in Essex. *Essex Naturalist* 38, 281–288.

Benton, T. & Fremlin, M. (2018). *Colletes hederae* Schmidt & Westrich, 1993 (the Ivy Bee) in Colchester: some observations on phenology, development and behaviour. *Essex Naturalist* 35, 179–184.

Bergmark, L., Borg-Karlson, A.-K. & Tengö, J. (1984) Female characteristics and odour cues in mate recognition in *Dasypoda altercator* (Hymenoptera: Melittidae). *Nove acta Regiae societas scientiarum Uppsaliensis* 3, 137–143.

Berry, R. J. (2009). *Islands*. New Naturalist 109. HarperCollins, London.

Betts, C. (1986). *The Hymenopterist's Handbook*. Amateur Entomologists' Society, Hanworth.

Biesmeijer, J. C., Roberts, S.P.M., Reemer, M. *et al.* (2006). Parallel declines in pollinators and insect-pollinated plants in Britain and the Netherlands. *Science* 313(5785), 351–354.

Birkhead, T. (2018). *The Wonderful Mr Willughby*. Bloomsbury, London.

Bischoff, I. (2003). Population dynamics of the solitary digger bee *Andrena vaga* Panzer (Hymenoptera: Andrenidae) studied using mark-recapture and nest counts. *Population Ecology* 45(3), 197–204.

Bischoff, I., Eckelt, E. & Kuhlmann, M. (2004). On the biology of the Ivy-Bee *Colletes hederae* Schmidt & Westrich, 1993 (Hymenoptera: Apidae). *Bonner Zoologische Beiträge* 53(1–2), 27–36.

Bischoff, I., Feltgen, K. & Breckner, D. (2003). Foraging strategy and pollen preferences of *Andrena vaga* (Panzer) and *Colletes cunicularius* (L.) (Hymenoptera:

Apidae). *Journal of Hymenoptera Research* 12, 220–237.

Boesi, R., Polidori, C. & Andrietti, F. (2009). Searching for the right target: Oviposition and feeding behavior in *Bombylius* bee flies (Diptera: Bombyliidae). *Zoological Studies* 48(2), 141–150.

Bogusch, P., Astapenkova, A. & Heneberg, P. (2015). Larvae and nests of six aculeate Hymenoptera (Hymenoptera: Aculeata) nesting in reed galls induced by *Lipara* spp. (Diptera: Chloropidae) with a review of species recorded. *PLoS ONE* 10(6), e0130802.

Bogusch, P. & Hadrava, J. (2018). European bees of the genera *Epeolus* Latreille, 1802 and *Triepeolus* Robertson, 1901 (Hymenoptera: Apidae: Nomadinae: Epeolini): taxonomy, identification key, distribution, and ecology. *Zootaxa* 4437(1), 1–60.

Bogusch, P., Kratochvíl, L. & Straka, J. (2006). Generalist cuckoo bees (Hymenoptera: Apoidea: *Sphecodes*) are species-specialist at the individual level. *Behavioral Ecology and Sociobiology* 60, 422–499.

Bohart, G. E. (1970). The Evolution of Parasitism amongst Bees. *Faculty Honor Lectures*, Paper 18. Utah State University.

Bommarco, R., Marini, L. & Vaissière, B. (2012). Insect pollination enhances seed yield, quality and market value in oilseed rape. *Oecologia* 169(4), 1025–1032.

Boomsma, J. J. (2007). Kin selection versus sexual selection: why the ends do not meet. Current Biology 17, R673–R683.

Borg-Karlson, A.-K., Tengö, J., Valterová, I. et al. (2003). (s)-(+)-Linalool, a mate attractant pheromone component in the bee *Colletes cunicularius*. *Journal of Chemical Ecology* 29(1), 1–14.

Bosch, J. & Kemp, W. P. (2004). Effect of pre-wintering and wintering temperature regimes on weight loss, survival, and emergence time in the mason bee *Osmia cornuta* (Hymenoptera: Megachilidae). *Apidologie* 35, 469–479.

Bosch, J., Sgolastra, F. & Kemp, W. (2010). Timing of eclosion affects diapause development, fat body consumption and longevity in *Osmia lignaria*, a univoltine, adult-wintering solitary bee. *Journal of Insect Physiology* 56, 1949–1957.

Bosch, J. & Vicens, N. (2002). Body size as an estimator of production costs in a solitary bee. *Ecological Entomology* 27, 129–137.

Botías, C., David, A., Horwood, J. et al. (2015). Neonicotinoid residues in wildflowers, a potential route of chronic exposure for bees. *Environment Science and Technology* 49, 12731–12740.

Boyes, C. (2020). Two-year life cycle in *Osmia leaiana*: ISO. Amiet et al. 2004 in Mid-Wales. *Bees, Wasps and Ants Recording Society* Newsletter, Spring, 38–39.

Boyes, C. (2022) Short notes on solitary bees and wasps in a Montgomeryshire (VC47) garden during 2021. *Bees, Wasps and Ants Recording Society* Newsletter, Spring, 34.

Brady, S. G., Larkin, L. & Danforth, B. N. (2009). Bees, ants and stinging wasps (Aculeata). In S. B. Hedges & S. Kumar (eds.), *The Timetree of Life*. Oxford University Press.

Brady, S. G., Sipes, S., Pearson, A. & Danforth, B. N. (2006). Recent and simultaneous origins of eusociality in halictid bees. *Proceedings of the Royal Society* B 273(1594), 1643–1649.

Brandt, K., Dötterl, S., Francke, W., Ayasse, M. & Milet-Pinheiro, P. (2017). Flower visitors of *Campanula*: are oligoleges more sensitive to host-specific floral scents than polyleges? *Journal of Chemical Ecology* 43(1), 4–12.

Brechbühl, R., Casas, J. & Bacher, S. (2009). Ineffective crypsis in a crab spider: a prey community perspective. *Proceedings of the Royal Society* B 277 (1682), 739–746.

Breeze, T. D. et al. (2020). Pollinator monitoring more than pays for itself. *Journal of Applied Ecology* 58(1), 44–57.

Briggs, D. & Walters, S. M. (1997). *Plant Variation*, 3rd edition. Cambridge University Press.

Brock, P. D. (2014). *A Comprehensive Guide to Insects of Britain and Ireland*. Pisces, Newbury.

Brock, P. D. (2021). *Britain's Insects. A field guide to the insects of Britain and Ireland*. Wild Guides, Princeton University Press.

Brock, P. D. & Roberts, S.P.M. (2017). An update on *Sitaris muralis* (Forster) (Mcloidae) in the New Forest and significant new populations in Wiltshire. *The Coleopterist* 26(3), 143–146.

Brongniart, A. (1839). Note sur poils collecteurs des Campanules et sur le mode de fécondation des ces plantes. *Annales des Sciences Naturelles: Botanique* 2(12), 244–247.

Brookes, D. R., Hereward, J. P., Terry, L. & Walter, G. H. (2015). Evolutionary dynamics of a cycad obligate pollination mutualism – Pattern and process in extant Macrozamia cycads and their specialist thrips pollinators. *Molecular Phylogenetics and Evolution* 93, 83–93.

Brothers, D. J., Tschuch, G. & Burger, F. (2000). Associations of mutillid wasps (Hymenoptera, Mutillidae) with eusocial insects. *Insectes Sociaux* 47, 201–211.

Brühl, C. A. *et al.* (2021). Direct pesticide exposure of insects in nature conservation areas in Germany. *Scientific Reports* 11, 24144.

Buglife website: Oil beetles: https://www.buglife.org.uk/sites/default/files/Buglife%20Oil%20Beetle%20guide.pdf.

Bull, A. (2015). *Looking at Brambles*. Catton Print, Norwich.

Butler, C. G. (1965). Sex attraction in *Andrena flavipes* Panzer (Hymenoptera: Apidae), with some observations on nest-site restriction. *Proceedings of the Royal Entomological Society of London A* 40(4–6), 77–80.

Cane, J. H. (1981). Dufour's gland secretions in the cell linings of bees (Hymenoptera: Apoidea). *Journal of Chemical Ecology* 7(2), 403–410.

Cane, J. H. (1983). Olfactory evaluation of *Andrena* host nest suitability by kleptoparasitic *Nomada* bees (Hymenoptera: Apoidea). *Animal Behaviour* 31, 138–144.

Cane, J. H. (2020). A brief review of monolecty in bees and benefits of a broadened definition. *Apidologie* 52, 17–22.

Cane, J. H. & Sipes, S. (2006). Characterizing floral specialization by bees: analytical methods and a revised lexicon for oligolecty. In N. M. Waser & J. Ollerton (eds.), *op. cit.*, 99–122.

Cane, J. H. & Tengö, J. (1981). Pheromonal cues direct mate-searching behaviour of male *Colletes cunicularius* (Hymenoptera: Colletidae). *Journal of Chemical Ecology* 7(2), 427–436.

Cardinal, S., Buchmann, S. L. & Russell, A. L. (2018). The evolution of floral sonication, a pollen foraging behaviour used by bees (Anthophila). *Evolution* 72(3), 590–600.

Carrie, R.J.G., George, D. R. & Wäckers, F. L. (2012). Selection of floral resources to optimise conservation of agriculturally-functional insect groups. *Journal of Insect Conservation* 16, 635–640.

Carson, R. (1962). *Silent Spring*. Hamilton, London.

Castro, S., Loureiro, J., Ferrero, V., Silveira, P. & Navarro, L. (2013). So many visitors and so few pollinators: variation in insect frequency and effectiveness governs the reproductive success of an endemic milkwort. *Plant Ecology* 214, 1233–1245.

Celary, W. (2006). Biology of the solitary ground-nesting bee *Melitta leporina* (Panzer, 1799) (Hymenoptera: Apoidea: Melittidae). *Journal of the Kansas Entomological Society* 79(2), 136–145.

Cerna, K., Zemenova, M., Machackova, L., Kolinova, Z. & Straka, J. (2013). Neighbourhood society: nesting dynamics, usurpations and social behaviour in solitary bees. *PLoS ONE* 8(9), e73806.

Chadwick, L. (1982). *In Search of Heathland*. Dobson, London and Durham.

Chambers, V. H. (1946). An examination of the pollen loads of *Andrena*: the species that visit fruit trees. *Journal of Animal Ecology* 15(1), 9–21.

Chandler, P. J. (ed.) (2010). *A Dipterist's Handbook*, 2nd edition. Amateur Entomologist 15.

Chatters, C. (2021). *Heathland*. Bloomsbury Wildlife, London.

Chinery, M. (2005). *Complete British Insects*. HarperCollins, London.

Chittka. L. (1996). Does bee colour vision predate the evolution of flower colour? *Naturwissenschaften* 83, 136–158.

Chittka, L. (1997). Bee colour vision is optimal for coding flower colour, but flower colours are not optimal for being coded - Why? *Israel Journal of Plant Sciences* 45, 115–127.

Chittka, L., Beier, W., Hertel, H., Steinmann, E. & Menzel, R. (1992). Opponent colour coding is a universal strategy to evaluate the photoreceptor inputs in hymenoptera. *Journal of Comparative Physiology A* 170, 545–563.

Chittka, L. & Thomson, J. D. (2001). *The Cognitive Ecology of Pollination: Animal Behaviour and Floral Evolution*. Cambridge University Press.

Chittka, L., Thomson J. D. & Waser N. M. (1999). Flower constancy, insect psychology, and plant evolution. *Naturwissenschaften* 86, 361–377.

Chittka, L., Vorobyev, M., Shmida, A. & Menzel, R. (1993). Bee colour vision – the optimal system for the discrimination of flower colours with three spectral photoreceptor types? In K. Wiese, F. G. Gribakin, A. V. Popov & G. Renninger (eds.). *Sensory Systems of Arthropods*, 211–218.

Chittka, L. & Waser, N. M. (1997). Why red flowers are not invisible to bees. *Israel Journal of Plant Sciences* 45, 169–183.

Claessens, J. & Kleynen, J. (2011). *The Flower of the European Orchid. Form and Function.* Claessens & Kleynen (self-published).

Claessens, J. & Kleynen, J. (2013). The pollination of European orchids: Part 2. *Cypripedium* and *Cephalanthera*. *Journal of the Hardy Orchid Society* 10(4), 114–120.

Clare, E. L., Schiestl, F. P., Leitch, A. R. & Chittka, L. (2013). The promise of genomics in the study of plant-pollinator interactions. *Genome Biology* 14, 207–218.

Clark, C. D., Hughes, A.L.C., Greenwood, S. L., Jordan, C. & Sejrup, H. P. (2012). Pattern and timing of retreat of the last British-Irish Ice Sheet. *Quaternary Science Reviews* 44, 112–146.

Clarke, A. (2017). *Principles of Thermal Ecology: temperature, energy and life.* Oxford University Press.

Clarke, D., Morley, E. & Robert, D. (2017). The bee, the flower, and the electric field: electric ecology and aerial electroreception. *Journal of Comparative Physiology A* 203, 737–748.

Classen-Bockhoff, R. (2007). Floral construction and pollination biology in the Lamiaceae. *Annals of Botany* 100(2), 359–360.

Claus, G., Pisman, M., Spanoghe, P., Smagghe, G. & Eeraerts, M. (2021). Larval oral exposure to thiacroplid: dose-response toxicity testing in solitary bees, *Osmia* spp. (Hymenoptera: Megachilidae). *Ecotoxicology and Environmental Safety* 215, 112145.

Collautti, R. I., Ricciardi, A., Grigorovich, I. A. & MacIsaac, H. J. (2004). Is invasion success explained by the enemy release hypothesis? *Ecology Letters* 7, 721–733.

Comba, L., Corbet, S. A., Hunt, L. & Warren, B. (1999). Flowers, nectar and insect visits: evaluating British plant species for pollinator-friendly gardens. *Annals of Botany* 83, 369–383.

Conrad, T. & Ayasse, M. (2018). The differences in the vibrational signals between male *O. bicornis* from three countries in Europe. *Journal of Low Frequency Noise, Vibration and Active Control* 38(2), 871–878.

Conrad, T., Paxton, R., Assum, G. & Ayasse, M. (2018). Divergence in male sexual odor signal across populations of the Red Mason Bee, *Osmia bicornis*, in Europe. *PLoS ONE* 13(2), e0193153.

Conrad, T., Paxton, R. J., Barth, F. G., Francke, W. & Ayasse, M. (2010). Female choice in the red mason bee, *Osmia rufa* (L.) (Megachilidae). *Journal of Experimental Biology* 213, 4065–4073.

Conrad, T., Vidkjaer, N. H. & Ayasse, M. (2017). The origin of the compounds found on males' antennae of the red mason bee, *Osmia bicornis* (L.). *Chemoecology* 27, 207–216.

Corbet, S. A. (1970). Insects on Hogweed flowers: a suggestion for a student project. *Journal of Biological Education* 4, 133–143.

Corbet, S. A. (1978). Bee visits and the nectar of *Echium vulgare* L. and *Sinapis alba* L. *Ecological Entomology* 3, 25–37.

Corbet, S. A. (1995). Insects, plants and succession: advantages of long-term set-aside. *Agriculture, Ecosystems and Environment* 53, 201–217.

Corbet, S. A. (1996). Which bees do plants need? In A. Matheson *et al.* (eds.) *op. cit.*, 105–113.

Corbet, S. A. (2006). A typology of pollination systems: implications for crop management and the conservation of wild plants. In N. M. Waser & J. Ollerton (eds.), *op. cit.*, 315–340.

Corbet, S. A., Beament, J. & Eisikowitch, D. (1982). Are electrostatic forces involved in pollen transfer? *Plant, Cell & Environment* 5, 125–129.

Corbet, S. A., Bee, J., Dasmahapatra, K., Gale, S., Gorringe, E. *et al.* (2001). Native or exotic? Double or single? Evaluating plants for pollinator-friendly gardens. *Annals of Botany* 87, 219–232.

Corbet, S. A. & Huang, S. Q. (2014). Buzz pollination in eight bumblebee-pollinated *Pedicularis* species: does it involve vibration-induced triboelectric charging of pollen grains? *Annals of Botany* 114(8), 1665–1674.

Corbet, S. A., Saville, N. M., Fussell, M., Prŷs-Jones, O. E. & Unwin, D. M. (1995). The competition box; a graphical aid to forecasting pollinator performance. *Journal of Applied Ecology* 32(4), 707–719.

Cox, J. & Bealey, C. (2022). Effects of grazing on heathland: evidence of benefits from a controlled experiment. *British Wildlife* 33(6), 400–409.

Cramp, S. *et al.* (1977–94). *Handbook of the Birds of Europe, the Middle East and North Africa: Birds of the Western Palearctic*. 9 vols. Oxford University Press.

Cross, I. (2002). *Colletes hederae* (Smith & Westrich) (Hymenoptera, Apidae) new to mainland Britain with notes on its ecology in Dorset. *Entomologist's Monthly Magazine* 138, 201–203.

Cross, I. (2008). Observations on egg-laying by *Bombylius minor* L. *Larger Brachycera Recording Scheme Newsletter* 27(2). *Bulletin of the Dipterists Forum* 65.

Cross, I. & Notton, D. G. (2017). Small-headed Resin Bee *Heriades rubicola* (Hymenoptera, Megachildiae) new to Britain. *British Journal of Entomology and Natural History* 30, 1–6.

Cruden, R. W. (2000). Pollen grains: Why so many? *Plant Systematics and Evolution* 222, 143–165.

Curtis, J. (1823). *British Entomology*: Vol IV. Hymenoptera: Part II. London. https://www.biodiversitylibrary.org/item/185419 page/1/mode/1up.

Danforth, B. N. (1990). Provisioning behavior and the estimation of investment ratios in a solitary bee, *Calliopsis (Hypomacrotera) persimilis* (Cockerell) (Hymenoptera: Andrenidae). *Behavioral Ecology and Sociobiology* 27, 159–168.

Danforth, B. N. (2002). Evolution of sociality in a primitively eusocial lineage of bees. *Proceedings of the National Academy of Sciences USA* 99(1), 286–290.

Danforth, B. N., Cardinal, S., Praz, C., Almeida, D. & Michez, D. (2013). The impact of molecular data on our understanding of bee phylogeny and evolution. *Annual Review of Entomology* 58, 57–78.

Danforth, B. N., Conway, L. & Shuqing, J. (2003). Phylogeny of eusocial *Lasioglossum* reveals multiple losses of eusociality within a primitively eusocial clade of bees (Hymenoptera: Halictidae). *Systematic Biology* 52(1), 23–36.

Danforth, B. N., Minckley, R. L. & Neff, J. L. (2019). *The Solitary Bees: Biology, Evolution and Conservation*. Princeton University Press.

Darwin, C. R. (1862). *On the Various Contrivances by which British and Foreign Orchids are Fertilised by Insects*. John Murray, London.

Darwin, C. R. (1876). *The Effects of Cross and Self Fertilisation in the Vegetable Kingdom*. John Murray, London.

Darwin, C. R. (1879). Letter to Dr. J. D. Hooker. https://cudl.lib.cam.ac.uk/view/MS-DAR-00095-00485/2.

Darwin, C. R. (1884). *The different forms of flowers on plants of the same species*. John Murray, London.

Davies, N. (2015). *Cuckoo: Cheating by Nature*. Bloomsbury, London.

Davis, E. S., Murray, T. E., Fitzpatrick, U., Brown, M.J.F. & Paxton, R. J. (2010). Landscape effects on extremely fragmented populations of a rare solitary bee, *Colletes floralis*. *Molecular Ecology* 19, 4922–4935.

Dawson, R. (2018). *Anthophora furcata* nest building at Doverston Reservoir. *Bees, Wasps and Ants Recording Society Newsletter*, Autumn, 23–24.

Dellicour, S., Mardulyn, O. J., Hardy, O. J. *et al.* (2014). Inferring the mode of colonization of the rapid range expansion of a solitary bee from multilocus DNA sequence variation. *Journal of Evolutionary Biology* 27, 116–132.

Devkota, K., dos Santos, C. F. & Blochtein, B. (2021). Higher richness and abundance of flower-visiting insects close to natural vegetation provide contrasting effects on mustard yields. *Journal of Insect Conservation* 25, 1–11.

Dicks, L. V. *et al.* (2012). Identifying key knowledge needs for evidence-based conservation of wild insect pollinators: a collaborative cross-sectoral exercise. *Insect Conservation and Diversity* 6(3), 435–446.

Dicks, L. V., Corbet, A. & Pywell, R. F. (2002). Compartmentalisation in plant-insect flower visitor webs. *Journal of Animal Ecology* 71, 32–43.

Dicks, L. V., Showler, D. A. & Sutherland, W. J. (2010). *Bee Conservation: Evidence for the Effects of Conservation*. Pelagic, Exeter.

Dimond, R., Falk, S., Saunders, P. & Whitehouse, A. (2014). *South West Bees Project. Survey Report 2014*. Buglife South West, Plymouth.

Discover Life World Bee Diversity. https://www.discoverlife.org/nh/cl/counts/Apoidea_species.html

Dobson, H.E.M. (1987). Role of flower and pollen aromas in host-plant recognition by solitary bees. *Oecologia* 72, 618–623.

Dobson, H.E.M., Ayasse, M., O'Neal, K. & Jacka, J. (2012). Is flower selection influenced by chemical imprinting to larval food provisions in the generalist bee *Osmia bicornis* (Megachilidae)? *Apidologie* 43(6), 698–714.

Doorn, W. G. van & Meeteren, U. van (2003). Flower opening and closure: a review. *Journal of Experimental Botany* 54, 1801–1812.

Drake, C. M. (2013). *Bombylius discolor* Mikan and *B. major* Linnaeus (Diptera, Bombyliidae) at an *Andrena cineraria* (Linnaeus) colony (Hymenoptera, Andrenidae). *Dipterists Digest* 20, 3–13.

Duffield, R. M., Fernandes, A., Lamb, C., Wheeler, J. W. & Eickwort, G. C. (1981). Macrocyclic lactones and isopentyl esters in the Dufour's gland secretion of halictine bees (Hymenoptera: Halictidae). *Journal of Chemical Ecology* 7, 319–331.

Dukas, R. (2001). Effects of predation risk on pollinators and plants. In L. Chittka & J. D. Thomson (eds.) *op. cit.*, 214–236.

Dutzler, G. & Ayasse, M. (1996). The function of female sex pheromones in mate selection of *Osmia rufa* (Megachilidae) bees. *Proceedings of the 20th International Congress of Entomology* 376.

Early, J. (2008). Nesting behaviour by *Cerceris rybyensis* (Linnaeus, 1771). *Bees, Wasps and Ants Recording Society Newsletter*, Autumn, 16–17.

Early, J. (2011). Evidence that *Hylaeus communis* (Nylander), 1852 and *Trypoxylon clavicerum* Lepeletier & Serville, 1828 can be bivoltine in Britain. *Bees, Wasps and Ants Recording Society Newsletter*, Autumn, 27–30.

Early, J. (2013a). Aculeates of the south Cumbrian coast. *Bees, Wasps and Ants Recording Society Newsletter*, Spring, 48–53.

Early, J. (2013b) *My Side of the Fence*. Early, Reigate.

Early, J. (2014). *Chrysura radians* in a Surrey garden. *Bees, Wasps and Ants Recording Society Newsletter*, Autumn, 13–14.

Early, J. (2020). Sheep's-bit *Jasione montana* L., 1753 as a foraging resource. *Bees, Wasps and Ants Recording Society Newsletter*, Autumn, 26–35.

Earwaker, R. (2014). *Species factsheet: Fringe-horned Mason Bee*. The Royal Society for the Protection of Birds, Sandy.

Eder, J. & Rembold, H. (eds.) (1987). *Chemistry and Biology of Social Insects*. Verlag J. Peperny, Munich.

Edwards, M. (1996). Optimizing habitats for bees in the United Kingdom – a review of recent conservation action. In A. Matheson et al. (eds.) *op. cit.*, 35–45.

Edwards, M. (2013a). Bees in Britain – are they all important pollinators? *British Wildlife* 25(1), 12–15.

Edwards, M. (2013b). *An Introduction to the Wild Bees of Scotland*. Scottish Natural Heritage, Perth.

Edwards, M. (2015). *Xylocopa* on the move? *Bees, Wasps and Ants Recording Society Newsletter*, Spring, 23.

Edwards, M. (2019). *Stelis odontopyga* Noskiewicz, 1926: iso. Edwards: 2019 (Hymenoptera: Megachilinae) recognised in Britain. *Bees, Wasps & Ants Recording Society Newsletter*, Spring, 34.

Edwards, M., Hazlehurst, G. & Wright, I. (2019). *Stelis odontopyga* Noskiewicz (Hymenoptera: Megachilinae) new to Britain. *British Journal of Entomology and Natural History* 32, 1–5.

Edwards, R. (ed.) (1997). *Provisional Atlas of the Aculeate Hymenoptera of Britain and Ireland*: Part 1. Biological Records Centre/BWARS, Huntingdon.

Edwards, R. (ed.) (1998). *Provisional Atlas of Aculeate Hymenoptera of Britain and Ireland*: Part 2. Biological Records Centre/BWARS Huntingdon.

Edwards, R. & Broad, G. (eds.) (2005). *Provisional Atlas of the Aculeate Hymenoptera of Britain and Ireland*: Part 5. Biological Records Centre/BWARS, Huntingdon.

Edwards, R. & Broad, G. (eds.) (2006). *Provisional Atlas of the Aculeate Hymenoptera of Britain and Ireland*: Part 6. Biological Records Centre/BWARS, Huntingdon.

Edwards, R. & Roy, H. E. (eds.) (2009). *Provisional Atlas of the Aculeate Hymenoptera of Britain and Ireland*: Part 7. Biological Records Centre/BWARS, Wallingford.

Edwards, R. & Roy, H. E. (eds.) (2012). *Provisional Atlas of the Aculeate Hymenoptera of Britain and Ireland*: Part 8. Biological Records Centre/BWARS, Wallingford.

Edwards, R. & Telfer, M. G. (eds.) (2001). *Provisional Atlas of the Aculeate Hymenoptera of Britain and Ireland*: Part 3. Biological Records Centre/BWARS, Huntingdon.

Edwards, R. & Telfer, M. G. (eds.) (2002). *Provisional Atlas of the Aculeate Hymenoptera of Britain and Ireland*: Part 4. Biological Records Centre/BWARS, Huntingdon.

Eikwort, G. E. (1975). Gregarious nesting of the mason bee *Hoplitis anthocopoides* and the evolution of parasitism and sociality among Megachilid bees. *Evolution* 29, 142–150.

Ellis, J. S., Knight, M. E., Darvill, B. & Goulson, D. (2006). Extremely low effective population sizes, genetic structuring and reduced genetic diversity in a threatened bumblebee species, *Bombus sylvarum* (Hymenoptera: Apidae). *Molecular Ecology* 15, 4375–4386.

Ellis, R. A., Weiss, T., Suryanarayanan, S. & Beilin, K. (2020). From a free gift of nature to a precarious commodity: bees, pollination services, and industrial agriculture. *Journal of Agrarian Change* 2020, 1–23.

Ellis, W. N. & Ellis-Adams, A. C. (1993). To make a meadow it takes a clover and a bee: the entomophilous flora of N.W. Europe and its insects. *Bijdragen tot de Dierkunde* 63, 193–220.

Else, G. R. (1995a). The distribution and habits of the small carpenter bee *Ceratina cyanea* (Kirby 1802) (Hymenoptera: Apidae) in Britain. *British Journal of Entomology and Natural History* 8(1), 1–6.

Else, G. R. (1995b). The distribution and habits of the bee *Hylaeus pectoralis* Förster 1871 (Hymenoptera: Apidae) in Britain. *British Journal of Entomology and Natural History* 8(2), 43–47.

Else, G. R. (1998). The status of *Stelis breviuscula* (Nylander) (Hymenoptera: Apidae) in Britain, with a key to the British species of *Stelis*. *British Journal of Entomology and Natural History* 10(4), 214–216.

Else, G. R. (2014a). *Andrena vaga* Panzer nesting in Hampshire. *Bees, Wasps and Ants Recording Society Newsletter*, Spring, 28.

Else, G. R. (2014b). *Andrena vaga* Panzer nesting in east Kent. *Bees, Wasps and Ants Recording Society Newsletter*, Spring, 29.

Else, G. R. & Edwards, M. (1996). Observations on *Osmia inermis* (Zetterstedt) and *O. uncinata* (Gerstäcker) (Hymenoptera, Apidae) in the central Scottish Highlands. *Entomologist's Monthly Magazine* 132, 291–298.

Else, G. R. & Edwards, M. (2018). *Handbook of the Bees of the British Isles*. The Ray Society, London.

Else, G., Hopkins, A. J. & Roberts, S.P.M. (2020). Sleeping behaviour and nesting sites of *Melitta dimidiata* ISO.AMIET ET AL.: 2007 (Apidae Melittinae) on Salisbury Plain, Wiltshire. *Bees, Wasps and Ants Recording Society Newsletter*, Spring, 24–25.

Elton, C. (1927). *Animal Ecology*. Sidgwick & Jackson, London.

Eltz, T., Küttner, J., Lunau, K. & Tollrian, R. (2015). Plant secretions prevent wasp parasitism in nests of wool-carder bees, with implications for the diversification of nesting materials in Megachilidae. *Frontiers in Ecology and Evolution* 2(86), 1–7.

Erenler, H. (2013). The diversity of pollinators in the gardens of large English country houses. PhD thesis, University of Northampton.

Evans, H. E. (1966). The accessory burrows of digger wasps. *Science* 152(3721), 465–471.

Fabre, J. H. (Translated 1914). *The Mason Bees*. https://www.gutenberg.org/files/2884/2884-h/2884-h.htm

Faegri, K. & van der Pijl, L. (1979). *The Principles of Pollination Ecology*, 3rd edition. Pergamon, Oxford.

Falk, S. J. Flickr Collection, Apoidea (bees). https://www.flickr.com/photos/63075200@N07/collections/72157631518508520/.

Falk, S. J. Flickr Collection, Meloidae (oil beetles and blister beetles). https://www.flickr.com/photos/63075200@N07/collections/72157631685918812/.

Falk, S. J. (1991). *A Review of the Scarce and Threatened Bees, Wasps and Ants of Great Britain*. Nature Conservancy Council, Peterborough.

Falk, S. J. (2006). The modern bee and wasp assemblages (Hymenoptera: Aculeata) of Warwickshire's calcareous quarries and spoilheaps, and the conservation issues facing them. *British Journal of Entomology and Natural History* 19(1), 7–33.

Falk, S. J. (2011). *A survey of the bees and wasps of fifteen chalk grassland and chalk heath sites within the East Sussex South Downs*. https://www.bwars.com/content/bees-and-wasps-east-sussex-south-downs-steven-falk-2011

Falk, S. J. (2021). Comments on the biodiversity net gain metric 3.0. www.stevenfalk.co.uk.

Falk, S. J. & Earwaker, R. (2018). Dusky-horned Nomada Bee, *Nomada bifasciata*, new to Britain (Hymenoptera: Apidae). *British Journal of Entomology and Natural History* 32, 170–175.

Falk, S. J., Johannson, N. & Paxton, R. J. (2022). DNA and morphological characterisation of the Bilberry Nomad Bee *Nomada glabella sensu* Stöckhert *nec* Thomson in Britain with discussion of the remaining variation within *N. panzeri*. *British Journal of Entomology and Natural History* 35, 91–112.

Falk, S. J. & Lewington, R. (2015). *Field Guide to the Bees of Great Britain and Ireland*. British Wildlife Field Guides. Bloomsbury, London.

Fenoglio, M. S., Calviño, A., González, E., Salvo, A. & Videla, M. (2021). Urbanisation drivers and underlying mechanisms of terrestrial insect diversity loss in cities. *Ecological Entomology* 46, 757–771.

Fenster, C. B, Armbruster, W. S., Wilson, P., Dudash, M. R. & Thomson, J. D. (2004). Pollination syndromes and floral specialisation. *Annual Review of Ecology Evolution & Systematics* 35, 375–403.

Ferry, N., Edwards, M. G., Gatehouse, J. A. & Gatehouse, A.M.R. (2004). Current Opinion in Biotechnology 15, 155–161.

Field, J. (1996). Patterns of provisioning and iteroparity in a solitary halictine bee, *Lasioglossum (Evylaeus) fratellum* (Perez) with notes on *L. (E.) calceatum* (Scop.) and *L. (E.) villosulum* (K.). *Insectes Sociaux* 43(2), 167–182.

Field, J., Paxton, R. J., Soro, A. & Bridge, C. (2010). Cryptic plasticity underlies a major evolutionary transition. *Current Biology* 20, 2028–2031.

Filella, I., Bosch, J., Llusià, J., Peñuelas, A. & Peñuelas, J. (2008). Chemical cues involved in the attraction of the oligolectic bee *Hoplitis adunca* to its host plant *Echium vulgare*. *Biochemical Systematics and Ecology* 39, 498–508.

Filella, I., Jordi, B., Llusià, J., Seco, R. & Peñuelas, J. (2011). The role of frass and cocoon volatiles in host location by *Monodontomerus aeneus*, a parasitoid of megachilid solitary bees. *Environmental Entomology* 40(1), 126–131.

Fitzpatrick, U., Murray, T. E., Paxton, R. J. & Brown, M.J.E. (2007). Building on IUCN regional red lists to produce list of species of conservation priority: a model with Irish bees. *Conservation Biology* 21(5), 1324–1332.

Fitzpatrick, U., Stout, J. *et al*. (2015). *All Ireland Pollinator Plan 2015–2020*. National Biodiversity Data Centre Series 3, Waterford.

Flyn, C. (2021). *Islands of Abandonment*. HarperCollins, London.

Fogden, M. (2022). *Pollinators in Crisis. How we can all give them a helping hand*. Pisces, Newark.

Formstone, B. (2019). Marford quarry, Denbighshire vice-county 50 (SJ357650) *Bees, Wasps and Ants Recording Society Newsletter*, Autumn, 4–8.

Free, J. B. (1970/1993). *Insect Pollination of Crops*. Academic, London.

Friedel, A., Paxton, R. J. & Soro, A. (2017). Spatial patterns of relatedness within nesting aggregations of the primitively eusocial sweat bee *Lasioglossum malachurum*. *Insectes Sociaux*. DOI 10.1007/s00040-017-0559-6.

Fründ, J., Dormann, F. & Tscharntke, T. (2011). Linné's floral clock is slow without pollinators – flower closure and plant-pollinator interaction webs. *Ecology Letters* 14, 896–904.

Gallai, N., Salles, J. M., Settele, J. *et al*. (2009). Economic valuation of the vulnerability of world agriculture confronted with pollinator decline. *Ecological Economics* 68, 810–821.

Gardiner, T. (2012). How does mowing of grassland on sea wall flood defences affect insect assemblages in eastern England? In Wen-Jun Zhang (ed.), *Grassland: Types, Biodiversity and Impacts*. Nova Science, USA.

Gardiner, T., Pilcher, R. & Wade, M. (2015). *Sea Wall Biodiversity Handbook*. Environment Agency/Risk Placement Services, Cambridge.

Gareau, B. J., Huang, X. & Gareau, T. P. (2018). Social and ecological conditions of cranberry production and climate change attitudes in New England. *PLoS ONE* 13(12), e0207237.

Gathmann, A. & Tscharntke, T. (2002). Foraging ranges of solitary bees. *Journal of Animal Ecology* 71, 757–764.

Gavini, S. A., Quintero, C. & Tadey, M. (2018). Predator and floral traits change pollinator behaviour, with consequences for plant fitness. *Ecological Entomology* 43(6), 731–741.

Gegear, R. J. & Laverty, T. M. (2001). The effect of variation among floral traits on the flower constancy of pollinators. In L. Chittka & J. D. Thomson (eds.) *op. cit.*, 1–20.

Gerth, M., Geissler, A. & Bleidorn, C. (2011). *Wolbachia* infections in bees (Anthophila) and possible implications for DNA barcoding. *Systematics and Biodiversity* 9(4), 319–327.

Ghazoul, J. (2005) Buzziness as usual? Questioning the global pollination crisis. *Trends in Ecology and Evolution* 20(7), 367–373.

Gibson, C. (2004). Canvey Wick – a model of sustainable development. *Talk of the Thames*, Autumn/Winter, 8.

Giurfa, M., Eichmann, B. & Menzel, R. (1996). Symmetry perception in an insect. *Nature* 382(6590), 458–461.

Glasser, S. K. & Farzan, S. (2016). Host-associated volatiles attract parasitoids in a native solitary bee *Osmia lignaria* Say (Hymenoptera: Megachilidae). *Journal of Hymenoptera Research* 51, 249–256.

Gómez, J. M. & Zamora, R. (2006). Ecological factors that promote the evolution of generalisation in pollination systems. In N. M. Waser and J. Ollerton (eds.), *Plant Pollinator Interactions, from Specialisation to Generalisation*, 145–166. University of Chicago Press.

Gonçalves-Souza, T., Omena, P. M., Souza, J. C. & Romero, G. Q. (2008). Trait-mediated effects on flowers: artificial spiders deceive pollinators and decrease plant fitness. *Ecology* 89, 2407–2413.

Gonzales, J. N., Teran, J. B. & Matthews, J. W. (2004). Review of the biology of *Melittobia acasta* (Walker) (Hymenoptera: Eulophidae) and additions on development and sex ratio of the species. *Caribbean Journal of Science*, 40(1), 52–61.

Goodell, K. (2003). Food availability affects *Osmia pumila* (Hymenoptera: Megachilidae) foraging, reproduction, and brood parasitism. *Oecologia* 134, 318–427.

Goulson, D. (2021). *Silent Earth: Averting the Insect Apocalypse*. Jonathan Cape, London.

Goulson, D., Nichols, E., Botías, C. & Rotheray, E. L. (2015). Bee declines driven by combined stress from parasites, pesticides and lack of flowers. *Science* 347(6229), 1255957.

Greig-Smith, P. W. & Quicke, D.L.J. (1983). The diet of nestling Stonechats. *Bird Study* 30(1), 47–50.

Grubb, M., Koch, M., Thomson, K., Munson, A. & Sullivan, F. (1993). *The 'Earth Summit' Agreements: A Guide and Assessment*. Earthscan, London.

Habermannová, J., Bogusch, P. & Straka, J. (2013) Flexible host choice and common host switches in the evolution of generalist and specialist cuckoo bees (Anthophila: *Sphecodes*). *PLoS ONE* 8(5), e64537.

Hadrava, J., Talašová, A., Straka, J. *et al.* (2022). A comparison of wild bee communities in sown flower strips and semi-natural habitats: a pollination network approach. *Insect Conservation and Diversity*, 15(3), 312–324.

Haider, M., Dorn, S. & Müller, A. (2013). Intra- and interpopulational variation in the ability of solitary bee species to develop on non-host pollen: implications for host range expansion. *Functional Ecology* 27(1), 255–263.

Hall, D. M. *et al.* (2017). The city as a refuge for insect pollinators. *Conservation Biology* 31(1), 24–29.

Hallmann, C. A. *et al.* (2017). More than 75 percent decline over 27 years in total flying insect biomass in protected areas. *PLoS ONE* 12(10), e0185809.

Hamilton, W. D., Axelrod, R. & Tanese, R. (1990). Sexual reproduction as an adaptation to resist parasites (a review). *Proceedings of the National Academy of Sciences USA* 87(9), 3566–3573.

Hammer M. & Menzel, R. (1995). Learning and memory in the honeybee. *Journal of Neuroscience* 15, 1617–1630.

Hanley, M. E. & Wilkins, J. P. (2015). On the verge? Preferential use of road-facing hedgerow margins by bumblebees in agro-ecosystems. *Journal of Insect Conservation* 19(1), 67–74.

Hao, K., Tian, Z.-X., Wang, Z.-C. & Huang, S.-Q. (2020). Pollen grain size associated with pollinator feeding strategy. *Proceedings of the Royal Society B* 287(1933), 20201191.

Harder, L. D. & Wilson, W. G. (1994). Floral evolution and male reproductive success: optimal dispensing schedules for pollen dispersal for animal-pollinated plants. *Evolutionary Ecology* 8, 542–559.

Hardy, T. [1878] (1979). *The Return of the Native.* Penguin, Harmondsworth.

Hargreaves, B. & White, S. (2021). *The Aculeate Hymenoptera (Bees, Wasps and Ants) of Lancashire and North Merseyside.* Lancashire and Cheshire Fauna Society, Rishton, Lancashire.

Harmon-Threatt, A. (2020). Influence of nesting characteristics on health of wild bee communities. *Annual Review of Entomology* 65, 39–56.

Harvey, M. C. (2006). *Potential insect pollinators of Red Helleborine* Cephalanthera rubra, *at Windsor Hill, Buckinghamshire.* Unpublished report.

Harvey, P. (2000). The East Thames Corridor: a nationally important invertebrate fauna under threat. *British Wildlife* 12(2), 91–98.

Harvey, P. (2004). *Why Brownfield Sites are Important for Invertebrates.* Central Association of Bee-keepers, Teddington.

Harvey, P. & Smith, D. (2005). Colne Point: a summary of invertebrate survey in 1990–91 and 2004. *Essex Naturalist* 22, 83–89.

Hawkes, R. W., Smart, J., Brown, H. *et al.* (2020). Experimental evidence that novel land management interventions inspired by history enhance biodiversity. *Journal of Applied Ecology* 58, 905–918.

Hawkins, R. D. (2011). *Lasioglossum sexstrigatum* (Hymenoptera: Apidae, Halictinae) new to Britain. *British Journal of Entomology and Natural History* 24, 90–92.

Hazlehurst, G. (2014). *Andrena vaga* at the RSPB Reserve, Dungeness in Kent. *Bees, Wasps and Ants Recording Society Newsletter,* Autumn, 23.

Hazlehurst, G. (2019). Early records for *Nomada zonata* Panzer (Hymenoptera: Apidae) in Britain. *British Journal of Entomology and Natural History* 32, 175.

Hazlehurst, G. (2021). The rediscovery of the mining bee *Andrena polita* (Smith) (Hymenoptera, Apidae) in Britain. *British Journal of Entomology and Natural History* 34, 135–136.

Hefetz, A. (1998). Exocrine glands and their products in non-*Apis* bees: chemical, functional and evolutionary perspectives. In R. K. Vander Meer, M. D. Breed, K. E. Espelie, M. L. Winston (eds.), *Pheromone Communication in Social Insects,* 236–256. CRC Press, Boca Raton.

Heiling, A. M., Chittka, L., Cheng, K. & Herberstein, M. E. (2005). Colouration of crab spiders: substrate choice and prey attraction. *Journal of Experimental Biology* 208, 1785–1792.

Heinrich, B. (1979). *Bumblebee Economics.* Cambridge University Press.

Hennessy, G., Goulson, D. & Ratnieks, F.L.W. (2020). Population assessment and foraging ecology of nest aggregations of the rare solitary bee, *Eucera longicornis* at Gatwick Airport, and implications for their management. *Journal of Insect Conservation* 24, 947–960.

Hennessy, G., Gouldson, D. & Ratnieks, F.L.W. (2021). Population assessment and foraging ecology of the rare solitary bee *Anthophora retusa* at Seaford Head nature reserve. *Journal of Insect Conservation* 25, 49–63.

Herbert, W. (1837). *Amaryllidaceae; with a treatise on cross-bred vegetables.* James Ridgway & Sons, London.

Herrera, C. M. (1996). Floral traits and plant adaptation to insect pollinators: a devil's advocate approach. In D. G. Lloyd & S.C.H. Barrett (eds.), *Floral Biology: Studies in Floral Evolution in Animal-pollinated Plants.* Chapman & Hall, New York.

Herrera, C. M. & Medrano, M. (2017). Pollination consequences of simulated intrafloral microbial warming in an early-blooming herb. *Flora* 232, 142–149.

Hicks, D. M. *et al.* (2016). Food for pollinators: quantifying the nectar and pollen

resources of urban flower meadows. *PLoS ONE* 11(6), e0158117.
Hill, P.S.M., Wells, P. H. & Wells, H. (1997). Spontaneous flower constancy and learning in honey bees as a function of colour. *Animal Behaviour* 54, 615–627.
Hofmann, M. & Renner, S. S. (2020). Bee species decrease and increase between the 1990s and 2018 in large urban protected sites. *Journal of Insect Conservation* 24, 637–642.
Honey, H. (2020). All the fun of the fair – creating a nest box and studying the beneficiaries. *Bees, Wasps & Ants Recording Society Newsletter*, Autumn, 39–41.
Höppner, H. (1904). Zur Biologie der Rubus-Bewohner. II. *Osmia parvula* Duf. Et Perr., *Osmia leucomelana* K. und ihr Schmarotzer *Stelis ornatula* Nyl. *Allgemeine Zeitschrift für Entomologie* 9(7/8), 129–134.
Houston, T. (1987). The symbiosis of acarid mites, genus *Ctenocolletacarus* (Acarina, Acariformes), and Stenotritid Bees, Genus *Ctenocolletes* (Insecta, Hymenoptera). *Australian Journal of Zoology* 35(5), 459–468.
Houston, T. (2018). *A Guide to the Native Bees of Australia*. CSIRO, South Victoria, Australia.
Howell, G. J., Slater, A. T. & Knox, R. B. (1993). Secondary Pollen Presentation in Angiosperms and its Biological Significance. *Australian Journal of Botany* 41, 417–438.
Hughes, D. P., Brodeur, J. & Thomas, F. (2012). *Host Manipulation by Parasites*. Oxford University Press.
Huie, L. H. (1916). The habits and life history of *Hylemia grisea*, Fall., an anthomyiid fly new to the Scottish fauna. *Scottish Naturalist* 49, 13–20.
Huseynov, E. F. (2007). Natural prey of the crab spider *Thomisus onustus* (Araneae: Thomisidae), an extremely powerful predator of insects. *Journal of Natural History* 41, 37–40.
Jackson, L. (2019). East of England Bee Report: A report on the status of threatened bees in the region with recommendations for conservation action. Buglife – The Invertebrate Conservation Trust, Peterborough.
Jacobs, J. H., Clark, S. J., Denholm, I. *et al.* (2009a). Pollination biology of fruit-bearing hedgerow plants and the role of flower-visiting insects in fruit-set. *Annals of Botany* 104, 1397–1404.
Jacobs, J. H., Clark, S. J., Denhom, I. *et al.* (2009b) Pollinator effectiveness and fruit set in common ivy, *Hedera helix* (Araliaceae). *Arthropod-Plant Interactions* 4, 19–28.
Jardine, D. & McGregor, A. (2021). *Andrena ruficrus*: ISO. PERKINS:1919 in Argyll. *Bees, Wasps and Ants Recording Society Newsletter*, Autumn, 16–18.
Jardine, D., Peacock, M., Vaughan, M. & Fisher, I. (2019). Choughs on Colonsay and Oronsay 1984–2018. *British Birds* 112, 390–398.
Johnson, S. D., Peter, C. I. & Agren, J. (2004). The effects of nectar addition on pollen removal and geitonogamy in the non-rewarding orchid *Anacamptis morio*. *Proceedings of the Royal Society B* 271(1541), 803–809.
Jones, N. & Cheeseborough, I. (2014). *A Provisional Atlas of the Bees, Wasps and Ants of Shropshire*. Field Studies Council.
Jordano, P., Bascompte, J. & Olesen, J. M. (2006). The ecological consequences of complex topology and nested structure in pollination wcbs. In N. M. Waser & J. Ollerton (eds.), *op. cit.*, 173–199.
Juniper, T. (2015). *What Nature Does for Britain*. Profile, London.
Kaltenpoth, M., Göttler, W. Hernzer, G. & Strohm, E. (2005). Symbiotic bacteria protect wasp larvae from fungal infection. *Current Biology* 15, 475–479.
Kathirithamby, J., Ross, L. D. & Johnston, J. (2003). Masquerading as self? Endoparasitic Strepsiptera (Insecta) enclose themselves in host-derived epidermal bag. *Proceedings of the National Academy of Sciences USA* 100(13), 7655–7659.
Kathirithamby, J., Delgado, J. A. & Collantes, F. (2015). Class Insecta: Order Strepsiptera. http://sea-entomologia.org/IDE@/revista_62B.pdf
Kay, Q.O.N., Lack, A. J., Bamber, F. C. & Davies, C. R. (1984). Differences between sexes in floral morphology, nectar production and insect visits in a dioecious species *Silene dioica*. *New Phytologist* 98, 515–529.
Kearns, C. A., Inouye, D. W. & Waser, N. M. (1998) Endangered mutualisms: the conservation of plant-pollinator

interactions. *Annual Review of Ecology and Systematics* 29, 83–112.

Keasar, T., Bilu, Y., Motro, U. & Shmida, A. (1997). Foraging choices of bumblebees on equally rewarding artificial flowers of different colours. *Israel Journal of Plant Sciences* 45, 223–233.

Keller, B., Thomson, J. B. & Conti, E. (2014). Heterostyly promotes disassortative pollination and reduces sexual interference in Darwin's primroses: evidence from experimental studies. *Functional Ecology* 28, 1413–1425.

Kevan, P. G., Chittka, L. & Dyer, A. G. (2001). Limits to the salience of ultraviolet: lessons from colour vision in bees and birds. *Journal of Experimental Biology* 204, 2571–2580.

Kim, J.-Y. (1999). Influence of resource level on maternal investment in a leaf-cutter bee (Hymenoptera: Megachilidae). *Behavioral Ecology* 10, 552–556.

King, C., Ballantyne, G. & Willmer, P. (2013). Why visitation is a poor proxy for pollination: measuring single-visit pollen deposition, with implications for pollination networks and conservation. *Methods in Ecology and Evolution* 4, 811–818.

Kirby, Rev. W. (1802). *Monographia Apum Angliae* or an attempt to divide into their natural genera and families such species of the Linnaean genus *Apis* as have been discovered in England. Published by J. Raw. https://www.biodiversitylibrary.org/item/41177 page/40/mode/1up

Kirby-Lambert, C. (2016). *Nomada alboguttata* Herrich-Schäffer, 1839 new to the British Isles and *Nomada zonata* Panzer, 1798 first record for mainland Britain. *Bees, Wasps and Ants Recording Society Newsletter*, Autumn, 29–30.

Kirk, W.D.J. (2006). *A colour guide to pollen loads of the honey bee*. International Bee Research Association.

Kirk, W.D.J. & Hopkins, A. J. (1996). Thrips. Naturalists' Handbooks 25. Richmond, Oxford.

Kirk, W.D.J. & Howes, F. N. (2012). *Plants for Bees*. International Bee Research Association.

Kleijn, D. & Sutherland, W. J. (2003). How effective are European agri-environmental schemes in conserving and promoting biodiversity? *Journal of Applied Ecology* 40(6), 947–969.

Kleijn, D., Winfree, R., Bartomeus, I. *et al.* (2015). Delivery of crop pollination services is an insufficient argument for wild pollinator conservation. *Nature Communications* 6(1), 7414.

Klimov, P. B., Vinson, S. B. & O'Connor, B. M. (2007). Acarinaria in associations of apid bees (Hymenoptera) and chaetodactylid mites (Acari). *Invertebrate Systematics* 21, 109–136.

Knauer, A., Bakhtiari, M. & Scheistl, F. P. (2018). Crab spiders impact floral-signal evolution indirectly through removal of florivores. *Nature Communications*, 1367.

Knerer, G. & Atwood, C. E. (1967). Parasitization of halictine bees in Southern Ontario. *Proceedings of the Entomological Society of Ontario*, 97, 103–110.

Knight, T. A. (1799). An account of some experiments on the fecundation of vegetables. *Philosophical Transactions of the Royal Society of London* 89, 195–204.

Knight. T. A. (1841). Philosophical & Horticultural Papers. *Transactions of the Royal & Horticultural Societies*, with a sketch of his life. Longmans, London.

Knowles, A. (2020). The history and ecology of *Lasioglossum sexnotatum* (Kirby) (Hymenoptera: Apidae, Halictinae) in Great Britain. *British Journal of Entomology and Natural History* 33, 113–120.

Konzmann, S., Koethe, S. & Lunau, K. (2019). Pollen grain morphology is not exclusively responsible for pollen collectability in bumble bees. *Scientific Reports* 9, 4705.

Kooi, C. J. van der, Vallejo-Marin, M. & Leonhardt, S. D. (2020). Mutualisms and (A)symmetry in plant-pollinator interactions. *Current Biology* 31, R91–R99.

Krombein, K. V. (1967). *Trap-nesting Wasps and Bees: Life Histories, Nests and Associates*. Smithsonian Press.

Krunić, M. & Stanisavljević, L. (2006). *The Biology of European Orchard Bee Osmia cornuta*. Faculty of Biology, University of Belgrade.

Krunić, M., Stanisavljević, L., Pinzauti, M. & Felicioli, A. (2005). The accompanying fauna of *Osmia cornuta* and *Osmia rufa* and effective measures of protection. *Bulletin of Insectology*. 58(2), 141–152.

Kuhlmann, M., Else, G. R., Dawson, A. & Quicke, D.L.J. (2007). Molecular, biogeographical and phenological evidence for the existence of three western European sibling species in the *Colletes succinctus* group (Hymenoptera: Apidae). *Organisms, Diversity and Evolution* 7(2), 155–165.

Lack, A. (1976). Competition for pollinators and evolution in *Centaurea*. *New Phytologist* 77, 787–792.

Larkin, L. L., Neff, J. L. & Simpson, B. B. (2008). The evolution of a pollen diet: host choice and diet breadth of *Andrena* bees (Hymenoptera: Andrenidae). *Apidologie* 39, 133–145.

Larsen, O. N., Gleffe, G. & Tengö, J. (1986). Vibration and sound communication in solitary bees and wasps. *Physiological Entomology* 11(3), 287–296.

Larsson, M. & Franzén, M. (2007). Critical resource levels of pollen for the declining bee *Andrena hattorfiana* (Hymenoptera, Andrenidae). *Biological Conservation* 134(3), 405–414.

Laundré, J. W., Hernández, L. & Ripple, W. (2010). The Landscape of Fear: Ecological Implications of Being Afraid. *The Open Ecology Journal* 3, 1–7.

Legewie, H. (2014). Zum problem des tierischen Parasitismus. I.Teil: Die lebensweise der Schmarotzerbiene *Sphecodes monilicornis* K. (=*Subquadratus* SM.) (Hym. Apid.). *Zeitschrift für Morphologie und Ökologie der Tiere* 4, 430–464.

Leins, P. & Urbar, C. (2006). Secondary pollen presentation syndromes of the Asterales – a phylogenetic perspective. *Botanische Jahrbücher für Systematik, Pflanzengeschichte und Pflanzengeographie* 127, 83–103.

Leonard, A. S., Dornhaus, A. & Papaj, D. R. (2012). Why are floral signals complex? An outline of functional hypotheses. In S. Patiny (ed.), *Evolution of plant-pollinator relationships*, 261–282. Cambridge University Press.

Lind, H. (1968). Nest-provisioning cycle and daily routine of behaviour in *Dasypoda plumipes* (Hym., Apidae). *Entomologiske Medelelser* 36(4), 343–372.

Linsley, E. G. (1958). The ecology of solitary bees. *Hilgardia* 27, 543–599.

Linsley, E. G. & MacSwain, J. W. (1955). The habits of *Nomada opacella* Timberlake with notes on other species (Hymenoptera: Anthophoridae). *The Wasmann Journal of Biology* 13(2), 253–276.

Linsley, E. G. & MacSwain, J. W. (1957). Observations on the habits of *Stylops pacifica* Bohart. *University of California Publications in Entomology* 11, 395–430, cited L. Beani (2006).

Litman, J. R., Danforth, B. N., Eardley, C. D. & Praz, C. J. (2011). Why do leafcutter bees cut leaves? New insights into the early evolution of bees. *Proceedings of the Royal Society B* 278(1724), 3593–3600.

Little, B., Little, S. & Webb, J. (2016). Observations of *Lasioglossum nitidiusculum* from Angus and Fife, Scotland. *Bees, Wasps and Ants Recording Society Newsletter*, Spring, 22–25.

Lonsdale, D. (2022). Editorial. *Invertebrate Conservation News* 103, 1–3.

López-Uribe, M. M., Morreal, S. J., Santiago, C. K. & Danforth, B. N. (2015). Nest Suitability, Fine-Scale Population Structure and Male-Mediated Dispersal of a Solitary Ground Nesting Bee in an Urban Landscape. *PLoS ONE* 10(5), e0125719.

Lubbock, J. (1875). *On British Wild Flowers considered in Relation to Insects*. Macmillan & Co., London.

Lubbock, J. (1882). *Ants, Bees and Wasps*. Kegan Paul, London.

Lunau, K., Piorek, V., Krohn, O. & Pacini, E. (2015). Just spines – mechanical defence of malvaceous pollen against collection by corbiculate bees. *Apidologie* 46, 144–149.

Mabbett, T. (2020). https://www.naturetrek.co.uk/news/discovery-of-european-orchard-bee-osmia-cornuta-in-my-cheltenham-garden.

Mabbett, T. (2021). Discovery of *Osmia cornuta*: ISO Amiet et al.: 2004 in a Cheltenham garden during 2020. *Bees, Wasps and Ants Recording Society Newsletter*, Spring, 24.

Macdonald, M. (2009). Notes on *Andrena ruficrus* in Scotland. *Bees, Wasps and Ants Recording Society Newsletter*, Autumn, 24–26.

Maciel de Almeida Correia, M. (1981). Contribution à l'étude de la biologie d'*Heriades truncorum* L. (Hymenoptera; Apoidea; Megachilidae). III Aspect éthologique. *Apidologie* 12(3), 221–256.

MacIvor, J. S. & Packer, L. (2015). 'Bee hotels' as tools for native pollinator

conservation: a premature verdict? *PLoS ONE* 10(3), e0122126.
MacIvor, J. S. & Salehi, B. (2014). Bee species-specific nesting material attracts a generalist parasitoid: implications for co-occurring bees in nest box enhancements. *Environmental Entomology* 43(4), 1027–1033.
Malyshev, S. I. (1927). Lebensgeschichte des *Colletes cunicularius* L. *Zeitschrift für Morphologie und Ökologie der Tiere* 9, 390–409.
Malyshev, S. I. (1935). The nesting habits of solitary bees. A comparative study. *Revista Española de Entomologia* II(3): 201–309.
Mant, J., Brändli, C., Vereecken, N. J. *et al.* (2005). Cuticular hydrocarbons as sex pheromone of the bee *Colletes cunicularius* and the key to its mimicry by the sexually deceptive orchid, *Ophrys exultata*. *Journal of Chemical Ecology* 31(8), 1765–1787.
Marchal, P. (1887). Etude sur l'instinct du *Cerceris ornata*. *Archives de zoologie expérimentale et Générale*, 27–60.
Matheson, A., Buchmann, S. L., O'Toole, C., Westrich, P. & Williams, I. H. (1996). *The Conservation of Bees*. Academic, London.
McCall, A. & Irwin, R. E. (2006). Florivory: the intersection of pollination and herbivory. *Ecology Letters* 9, 1351–1365.
Memmott, J., Waser, N. M. & Price, M. V. (2004). Tolerance of pollination networks to species extinctions. *Proceedings of the Royal Society B* 271(1557), 2605–2611.
Menezes, C., Vollet-Neto, A., Marsaioli, A. J. *et al.* (2015). A Brazilian social bee must cultivate fungus to survive. *Current Biology* 25, 2851–2855.
Menzel, R. (2001). Behavioral and neural mechanisms of learning and memory as determinants of flower constancy. In L. Chittka & J. D. Thomson (eds.), *op. cit.*, 21–40.
Menzel, R., Gumbert, A., Kunze, J., Shmida, A. & Vorobyev, M. (1997). Pollinators' strategies in finding flowers. *Israel Journal of Plant Sciences* 45, 141–156.
Michener, C. D. (1979). Biogeography of the Bees. *Annals of the Missouri Botanical Garden* 66(3), 277–347.
Michener, C. D. (2000/2007). *The Bees of the World*, 2nd edition. Johns Hopkins University Press, Baltimore.
Michez, D. & Patiny, S. (2005). World revision of the oil-collecting bee genus *Macropis* Panzer 1809 (Hymenoptera: Apoidea:

Melittidae) with a description of a new species from Laos. *Annales de la Société Entomologique de France* 41(1), 15–28.
Michez, D., Patiny, S. & Danforth, B. N. (2009). Phylogeny of the bee family Melittidae (Hymenoptera: Anthophila) based on combined molecular and morphological data. *Systema Entomologica* 34(3), 574–597.
Michez, D., Patiny, S., Rasmont, P. Timmerman, K. & Vereecken, N. J. (2008). Phylogeny and host plant evolution in Melittidae s.l. (Hymenoptera: Apoidea). *Apidologie* 39(1), 146–162.
Michez, D., Vanderplanck, M. & Engel, M. S. (2011). Fossil Bees and their Plant Associates. In S. Patiny (ed.), *Evolution of Plant-Pollinator Relationships*, 103–164. Cambridge University Press.
Miko, I. *et al.* (2019). Fat in the leg: function of the expanded hind leg in gasteruptiid wasps (Hymenoptera: Gasteruptiidae). *Insect Systematics and Diversity* 3(1), 1–16.
Milet-Pinheiro, P., Ayasse, M., Dobson, H.E.M. *et al.* (2013). The Chemical Basis of Host-Plant Recognition in a Specialized Bee Pollinator. *Journal of Chemical Ecology* 39, 1347–1360.
Milet-Pinheiro, P., Ayasse, M. & Dötterl, S. (2015). Visual and olfactory floral cues of *Campanula* (Campanulaceae) and their significance for host recognition by an oligolectic bee pollinator. *PLoS ONE* 10(6), e0128577.
Milet-Pinheiro, P., Herz, K., Dötterl, S. & Ayasse, M. (2016). Host choice in a bivoltine bee: how sensory constraints shape innate foraging behaviours. *BMC Ecology* 16, 20.
Mill, R. R. (2012). Biodiversity recording at Royal Botanic Garden Edinburgh. *Sibbaldia: The International Journal of Botanic Garden Horticulture* 10, 149–169.
Milman, O. (2022). *The Insect Crisis*. Atlantic, London.
Minckley, R. L. & Roulston, T. H. (2006). Incidental mutualisms and pollen specialization among bees. In N. M. Waser & J. Ollerton (eds.), *op. cit.*, 69–98.
Mitra, A. (2013). Function of the Dufour's gland in solitary and social hymenoptera. *Journal of Hymenoptera Research* 35, 33–58.
Montgomery, C., Vuts, J. V., Woodcock, C. M. *et al.* (2021). Bumblebee electric charge stimulates floral volatile emissions in

Petunia integrifolia but not in *Antirhhinum majus*. *The Science of Nature* 108, 44.

Müller, A. (1995). Morphological specialisations in central European bees for the uptake of pollen from flowers with anthers hidden in narrow corolla tubes (Hymenoptera: Apoidea). *Entomologia Generalis* 20(1–2), 43–57.

Müller, A. (1996). Convergent evolution of morphological specialisations in Central European bee and honey wasp species as an adaptation to the uptake of pollen from nototribic flowers (Hymenoptera, Apoidea and Masaridae). *Biological Journal of the Linnean Society* 57(3), 235–252.

Müller, A. & Bansac, N. (2004). A specialized pollen-harvesting device in western palaearctic bees of the genus *Megachile* (Hymenoptera, Apoidea, Megachilidae). *Apidologie* 35(3), 329–337.

Müller, A. & Kuhlmann, M. (2018). Pollen hosts of western Palaearctic bees of the genus *Colletes* (Hymenoptera: Colletidae): the Asteraceae paradox. *Biological Journal of the Linnean Society* 95(4), 719–733.

Müller, A., Topfl, W. & Amiet, F. (1996). Collection of extrafloral trichome secretions for nest wool impregnation in the solitary bee *Anthidium manicatum*. *Naturwissenschaften* 83, 230–232.

Müller, A. & Weibel, U. (2020). A scientific note on an unusual hibernating stage in a late-flying European bee species. *Apidologie* 51, 436–438.

Müller, C. B. (1994). Parasitoid induced digging behaviour in bumblebee workers. *Animal Behaviour* 48, 961–966.

Müller, H. (1883). *The Fertilization of Flowers by Insects*. Translated and edited by D. W. Thompson. Macmillan & Co., London.

Münster-Swendsen, M. (1970). The nesting behaviour of the bee *Panurgus banksianus* Kirby (Hymenoptera, Andrenidae, Panurginae). *Insect Systematics and Evolution* 1(2), 93–101.

Münster-Swendsen, M. & Calabuig, I. (2000) Interaction between the solitary bee *Chelostoma florisomne* and its nest parasite *Sapyga clavicornis* – empty cells reduce the impact of parasites. *Ecological Entomology* 25, 63–70.

Murray, T. E., Fitzpatrick, U., Byrne, A. et al. (2012). Local-scale factors structure wild bee communities in protected areas. *Journal of Applied Ecology* 49, 998–1008.

Musche, M. & Conradt, L. (2001). Ecological and evolutionary processes at expanding range margins. *Nature* 411(6837), 577–581.

Muth, F., Papaj, D. R. & Leonard, A. S. (2015). Bees remember flowers for more than one reason: pollen mediates associative learning. *Animal Behaviour* 111, 93–100.

Neal, P. R., Dafni, A. & Giurfa, M. (1998). Floral symmetry and its role in plant-pollinator systems: Terminology, Distribution and Hypotheses. *Annual Review of Ecology and Systematics* 29, 345–373.

Ne'eman, G., Shavit, O., Shaltiel L. & Shmida, A. (2006). Foraging by Male and Female Solitary Bees with Implications for Pollination. *Journal of Insect Behavior* 19(3), 383–401.

Neff, J. L. (2008). Components of nest provisioning behaviour in solitary bees (Hymenoptera: Apoidea). *Apidologie* 39, 30–45.

Neff, J. L. & Simpson, B. B. (1997). Nesting and foraging behaviour of *Andrena (Callandrena) rudbeckiae* Robertson (Hymenoptera: Apoidea: Andrenidae) in Texas. *Journal of the Kansas Entomological Society* 70, 100–113.

Nichols, R. N., Goulson, D. & Holland, J. M. (2019). The best wildflowers for wild bees. *Journal of Insect Conservation* 23, 819–830.

Nicolson, S. W. (2009). Water homeostasis in bees, with the emphasis on sociality. *Journal of Experimental Biology* 212, 429–434.

Nieto, A. (2015). Celebrating 50 Years of The IUCN Red List of Threatened Species. IUCN Red List Status of European Bees. https://www.iucn.org/sites/dev/files/import/downloads/ana_nieto_presentation_20_april_2015.pdf

Norden, B., Batra, S.W.T., Fales, H. M., Hefetz, A. & Shaw, G. J. (1980). *Anthophora* bees: unusual glycerides from maternal Dufour's glands serve as larval food and cell lining. *Science* 207(4435), 1095–1097.

Norfolk, O., Eichhorn, M. P. & Gilbert, F. (2016). Flowering ground vegetation benefits wild pollinators and fruit set of almond within arid smallholder orchards. *Insect Conservation and Diversity* 9(3), 236–243.

Notton, D. G. & Dathe, H. H. (2008). William Kirby's types of *Hylaeus* Fabricius (Hymenoptera, Colletidae) in the collection of the Natural History

Museum, London. *Journal of Natural History* 42(27–28), 1861–1865.

Notton, D. G., Tang, C. Q. & Day, A. R. (2016). Viper's Bugloss Mason Bee, *Hoplitis adunca*, new to Britain (Hymenoptera, Megachilidae, Megachilinae, Osmiini). *British Journal of Entomology and Natural History* 29, 134–143.

Nowakowski, M. & Pywell, R. (2016). *Habitat Creation and Management for Pollinators*. Wildlife Farming Co. & Centre for Ecology and Hydrology.

Noyes, J. S. (2018). *Universal Chalcidoidea database*. Natural History Museum, London. http://www.nhm.ac.uk/chalcidoids.

Nuttman, C. V., Semida, F. M., Zalat, S. & Willmer, P. G. (2006). Visual cues and foraging choices: bee visits to floral colour phases in *Alkanna orientalis*. (Boraginaceae). *Biological Journal of the Linnean Society* 87, 427–435.

Nuttman, C. V. & Willmer, P. G. (2003). How does insect visitation trigger floral colour change? *Ecological Entomology* 28, 467–424.

Nuttman, C. V. & Willmer, P. G. (2008). Hoverfly visitation in relation to floral colour change (Diptera: Syrphidae). *Entomologia Generalis* 31, 33–47.

Nye, W. P. (1980). Notes on the Biology of *Halictus (Halictus) farinosus* Smith (Hymenoptera: Halictidae). USDA ARR-W-11/May 1980.

O'Connor, B. M. (1993). The mite community associated with *Xylocopa latipes* (Hymenoptera: Anthophoridae: Xylocopinae) with description of a new type of acarinarium. *International Journal of Acarology* 19, 159–166.

Olds, L. (2022). First record of *Andrena nitidiuscula*: ISO. Perkins: 1919 in Wales. *Bees, Wasps & Ants Recording Society Newsletter*, Spring, 28–29.

Ollerton, J. (2017). Pollinator diversity: distribution, ecological function, and conservation. *Annual Review of Ecology, Evolution, and Systematics* 48, 353–376.

Ollerton, J. (2021). *Pollinators and Pollination: Nature and Society*. Pelagic, Exeter.

Ollerton, J., Erenler, H., Edwards, M. & Crockett, R. (2014). Pollinator declines. Extinction of aculeate pollinators in Britain and the role of large-scale agricultural change. *Science* 346(6215), 1360–1362.

Ollerton, J., Killick, A., Lamborn, E., Watts, S. & Whiston, S. (2007). Multiple Meanings and Modes: On the Many Ways to be a Generalist Flower. *Taxon* 56(3), 717–728.

Ollerton, J., Winfree, R. & Tarrant, S. (2011). How many flowering plants are pollinated by animals? *Oikos* 120, 321–326.

Olsen, K. (2005). *The Faroe Islands*. In *Island Biology: Essays in Evolutionary Ecology*. Aarhus University, Denmark.

Orr, M. C., Hughes, A. C., Chesters, D., Pickering, J. & Zhu, C. (2021). Global Patterns and Drivers of Bee Distribution. *Current Biology*, 31, 1-8.

O'Toole, C. (1974). A new subspecies of the vernal bee, *Colletes cunicularius* (L.) (Hymenoptera: Colletidae). *Journal of Entomology Series B, Taxonomy* 42, 163–169.

O'Toole, C. (1993). Diversity of native bees and agroecosystems. In J. LaSalle & I. D. Gauld (eds.), *Hymenoptera and Biodiversity*. CAB International, Wallingford.

O'Toole, C. (2012). *Plants for Bees*. In W.D.J. Kirk & F. N. Howes *op. cit.*: ch. 4.

O'Toole, C. (2013). *Bees: A Natural History*. Firefly, New York and Ontario.

O'Toole, C. & Raw, A. (1991). *Bees of the World*. Blandford, London.

Owens, N. W. (2017). *The Bees of Norfolk*. Pisces, Newbury.

Owens, N. W. (2018). Some observations on pollen collection by the Pantaloon Bee *Dasypoda hirtipes* (Hymenoptera, Apidae) in relation to the opening and closing times of the flowers it visits. *Transactions of the Norfolk and Norwich Nature Society* 51(1), 45–55.

Owens, N. W. (2020a). Two hosts confirmed for *Coelioxys inermis*: iso. Amiet *et al.*, 2004, in Norfolk. *Bees, Wasps and Ants Recording Society Newsletter*, Autumn, 18–21.

Owens, N. W. (2020b). *The Bumblebee Book*. Pisces, Newbury.

Owens, N. W. & Benton, T. (2022). Decoy nest entrances in *Andrena Clarkella*. *Bees, Wasps and Ants Recording Society Newsletter*, Spring, 16–19.

Owens, N. W. & Gomez de la Cuesta, R. (2021). Some observations of the behaviour of *Cerceris rybyensis*. *Bees, Wasps and Ants Recording Society Newletter*, Autumn, 24–35.

Owens, N. W. & Owens, F. A. (2015). Some observations of pollen vectors in two *Epipactis* orchids. *Journal of the Hardy Orchid Society* 12(3), 92–99.

Owens, N. W. & Welch, M. D. (2017). The scarce cleptoparasitic satellite fly *Miltogramma germari* (Diptera, Sarcophagidae) in Norfolk. *Transactions of the Norfolk and Norwich Nature Society* 15(1), 9–13.

Owens, N. W. & Welch, M. D. (2020a). Observations of the cleptoparasitic fly *Miltogramma punctata* Meigen 1824 (Diptera Sarcophagidae) at nest aggregations of *Colletes halophilus*: ISO. Guichard: 1974 in Norfolk. *Bees, Wasps and Ants Recording Society Newsletter*, Spring, 26–29.

Owens, N. W. & Welch, M. D. (2020b). The pNT cleptoparasite *Leucophora sericea* Robineau-Desvoidy (Diptera Anthomyiidae) in Norfolk and an update on the aculeate hosts of British *Leucophora*. *Dipterists Digest* 27(2), 231–236.

Packer, L. (1991). The evolution of social behaviour and nest architecture in sweat bees of the subgenus *Evylaeus* (Hymenoptera: Halictidae): a phylogenetic approach. *Behavioral Ecology and Sociobiology* 29, 153–160.

Packer, L., Sampson, B., Lockerbie, C. and Jessome, V. (1989). Nest architecture and brood mortality in four species of sweat bee (Hymenoptera; Halictidae) from Cape Breton Island. *Canadian Journal of Zoology* 67, 2864–2870.

Pain, S. (2015). Sweet deceit. *New Scientist* 226(3018), 42–45.

Parker, A. J., Tran, J. L., Ison, J. L. et al. (2015). Pollen packing affects the function of pollen on corbiculate bees but not non-corbiculate bees. *Arthropod-Plant Interactions* 9(2), 197–203.

Parker, R. (2008) Observations of a mason bee, *Osmia bicolor*, concealing its nest in a snail shell (7–24 May 2008). *Suffolk Natural History* 44, 51–52.

Patiny, S., Michez, D. & Danforth, B. N. (2007). Phylogenetic relationships and host-plant evolution within the basal clade of Halictidae (Hymenoptera: Apoidea). *Cladistics* 23, 1–15.

Paukunnen, J., Bert, A., Soon, V., Ødegaard, F. & Rosa, P. (2015). An illustrated key to the cuckoo wasps (Hymenoptera, Chrysididae) of the Nordic and Baltic countries, with description of a new species. *ZooKeys* 548, 1–116.

Paxton, R. J. (2005). Male mating behaviour and mating systems in bees: an overview. *Apidologie* 36, 145–156.

Paxton, R. J., Ayasse, M., Field, J. & Soro, A. (2002). Complex sociogenetic organization and reproductive skew in a primitively eusocial sweat bee, *Lasioglossum malachurum*, as revealed by microsatellites. *Molecular Ecology* 11, 2405–2416.

Paxton, R. J., Kukuk, P. F. & Tengo, J. (1999). Effects of familiarity and nestmate number on social interactions in two communal bees, *Andrena scotica* and *Panurgus calcaratus* (Hymenoptera, Andrenidae). *Insectes Sociaux* 46, 109–118.

Paxton, R. J. & Pohl, H. (1999). The tawny mining bee, *Andrena fulva* (Müller) (Hymenoptera, Andreninae), at a South Wales field site and its associated organisms: Hymenoptera, Diptera, Nematoda and Strepsiptera. *British Journal of Entomology and Natural History* 12, 57–67.

Paxton, R. J. & Tengö, J. (1996). Intranidal mating, emergence and sex ratio in a communal bee *Andrena jacobi* Perkins 1921 (Hymenoptera: Andrenidae). *Journal of Insect Behavior* 9, 421–440.

Paxton, R. J., Tengö, J. & Hedström, L. (1966). Dipteran parasites and other associates of a communal bee, *Andrena scotica* (Hymenoptera: Apoidea), on Öland, SE Sweden. *Entomologisk Tidskrift* 117, 165–178.

Paxton, R. J., Thorén, P. A., Tengö, J., Estoup, A. & Pamilo, P. (1996). Mating structure and nestmate relatedness in a communal bee, *Andrena jacobi* (Hymenoptera: Andrenidae), using microsatellites. *Molecular Ecology* 5, 511–519.

Payne, R. G. & Harvey, P. R. (1996). The natural history of the Thames Terraces at West Tilbury. *Essex Naturalist* 13, 121–130.

Perkins, V. R. (1884). On a singular habit of *Osmia bicolor* Sch. *Entomologist's Monthly Magazine* 21, 67–68.

Perkins, R.C.L. (1918). *Halictoxenus arnoldi*, an undescribed British stylopid. *Entomologist's Monthly Magazine* 54, 107–108.

Petanidou, T., Kallimanis, A. S., Tzanopoulos, J., Sgardelis, S. P. & Pantis, J. D. (2008). Long-term observations of a pollination network: fluctuation in species and interactions, relative invariance of network structure and implications for

estimates of specialisation. *Ecology Letters* 11, 564–575.
Peters, R. S. *et al.* (2017). Evolutionary history of the Hymenoptera. *Current Biology* 27, 1013–1018.
Phipps, J. (1948). The nest of *Macropis labiata* (F.) (Hym., Apidae). *Entomologist's Monthly Magazine* 84, 56.
Pijl, L. van der & Dodson, C. H. (1966). *Orchid flowers – their Pollination and Evolution*. University of Miami Press, Coral Gables.
Pilkington, G. (2017). Do female leafcutter bees give a 'hands off' signal to males? https://nurturing-nature.co.uk.
Pilkington, G. (2018) See yellow-face bee making its cellophane-like nest. https://nurturing-nature.co.uk.
Pilkington, G. (2020). *Heriades truncorum* (Large-headed Resin Bee) dabbing pollen onto its abdomen when foraging. https://nurturing-nature.co.uk.
Pilkington, G. (2021). *Gasteruption jaculator* and the *Hylaeus* bee battle. https://www.youtube.com/watch?v=pvDX-4WEgRo.
Plant, C. W. & Harvey, P. R. (1997). *Biodiversity Action Plan. Invertebrates of the South Essex Thames Terrace Gravels – Phase 1: Characterisation of the Existing Resource*. Report no. BS/055/96 in library of English Nature, Peterborough.
Potts, S. G., Biesmeijer, J. C., Kremen, C. *et al.* (2010). Global pollinator declines: trends, impacts and drivers. *Trends in Ecology and Evolution* 25(6), 345–353.
Power, J. (2023). *The Bees of Sussex*. In prep.
Powney, G. D., Carvell, C., Edwards, M. *et al.* (2019). Widespread loss of pollinating insects in Britain. *Nature Communications* 10, 1018.
Praz, C. J., Genoud, D., Vaucher, K. *et al.* (2022). Unexpected levels of cryptic diversity in European bees of the genus *Andrena* subgenus *Taeniandrena* (Hymenoptera, Andrenidae): implications for conservation. *Journal of Hymenoptera Research* 91, 375–428.
Praz, C. J., Müller, A. & Dorn, S. (2008). Specialized bees fail to develop on non-host pollen: do plants chemically protect their pollen? *Ecology* 89(3), 795–804.
Proctor, M. & Yeo, P. (1973). *The Pollination of Flowers*. New Naturalist 54. Collins, London.

Proctor, M., Yeo, P. & Lack, A. (1996). *The Natural History of Pollination*. New Naturalist 83. Collins, London.
Prokop, P. & Fedor, P. (2015). Why do flowers close at night? Experiments with the Lesser Celandine *Ficaria verna* Huds (Ranunculaceae). *Biological Journal of the Linnean Society* 118, 698–702.
Prosi, R., Wiesbauer, H. & Müller, A. (2016). Distribution, biology and habitat of the rare European osmiine bee species *Osmia (Melanosmia) pilicornis* (Hymenoptera, Megachilidae, Osmiini). *Journal of Hymenoptera Research* 52, 1–36.
Prŷs-Jones, O. E. & Corbet, S. A. (2011). *Bumblebees* (3rd edition). Naturalists' Handbooks 6. Pelagic, Exeter.
Quicke, D.L.J. & Fitton, M. G. (1995) Ovipositor steering mechanisms in parasitic wasps of the families Gasteruptiidae and Aulacidae (Hymenoptera). *Proceedings of the Royal Society B* 261(1360), 99–103.
Radchenko, V. G. (1981). Nesting of bees of four species of the genus *Andrena* F. (Hymenoptera, Andrenidae). *Entomologicheskoe Obozrenie* 60(4), 766–774.
Radnitsyn, A. P., Jarzembowski, E. A. & Ross, A. J. (1998). Wasps (Insecta: Vespida=Hymenoptera) from the Purbeck and Wealden (Lower Cretaceous) of southern England and their biostratigraphical and palaeoenvironmental significance. *Cretaceous Research* 19(3–4), 329–391.
Raguso, R. A. (2001). Floral scent, olfaction and scent-driven foraging behaviour. In L. Chittka & J. D. Thomson (eds.) *op. cit.*, 83–105.
Ramsay, A. (2002). Identification: British Oil Beetles. *British Wildlife* 14(1), 27–30.
Rasmont, P., Ebmer, A. W., Józef Banaszak, J. & van der Zanden, G. (1995). Hymenoptera Apoidea Gallica. Liste taxonomique des abeilles de France, de Belgique, de Suisse et du Grand-Duché de Luxembourg. *Bulletin de la Société entomologique de France* 100, 1–100. Updated by Apoidea Gallica working group, 2017.
Rauh, R., Kahl, S., Boechzelt, H. *et al.* (2007). Molecular biology of cantharidin in cancer cells. *Chinese Medicine* 2, 8.
Raw, A. (1972). The biology of the solitary bee *Osmia rufa* (L.) (Megachilidae). *Transactions*

of the *Royal Entomological Society of London* 24(3), 213–229.
Raw, A. (1988). Nesting biology of the leafcutter bee *Megachile centuncularis* (Hymenoptera, Megachilidae). *Entomologist* 107(1), 52–56.
Rezkova, K., Žákova, M., Žákova, Z. & Straka, J. (2012). Analysis of nesting behaviour based on daily observation of *Andrena vaga* (Hymenoptera: Andrenidae). *Journal of Insect Behaviour* 25(1), 24–47.
Richards M. H., French, D. & Paxton, R. J. (2005). It's good to be queen: classically eusocial colony structure and low worker fitness in an obligately social sweat bee. *Molecular Ecology* 14, 4123–4133.
Richards, O. W. (1980). Handbooks for the Identification of British Insects, Vol. 6. Part 3(b): *Scolioidea, Vespoidea and Sphecoidea, Hymenoptera, Aculeata*. Royal Entomological Society of London.
Ricketts, T. H., Regetz, J., Steffan-Dewenter, I. *et al.* (2008). Landscape effects on crop pollination services: are there general patterns? *Ecology Letters* 11, 499–515.
Rivest, S. & Forrest, J.R.K. (2020). Defence compounds in pollen: why do they occur and how do they affect the ecology and evolution of bees? *New Phytologist* 225, 1053–1064.
Robbirt, K. M. (2012). Phenological responses of British orchids and their pollinators to climate change: an assessment using herbarium and museum collections. PhD thesis, University of East Anglia.
Roberts, B. R., Cox, R. & Osborne, J. L. (2020). Quantifying the relative predation pressure on bumblebee nests by the European badger (*Meles meles*) using artificial nests. *Ecology and Evolution* 10(3), 1613–1622.
Roberts, M. J. (1996). *Spiders of Britain and Europe*. HarperCollins, London.
Roberts, S.P.M. & Peat, L. *Xylocopa* in Britain. https://www.bwars.com/content/xylocopa-britain
Roberts, S.P.M., Potts, S. G., Biesmeijer, K. *et al.* (2011). Assessing continental-scale risks for generalist and specialist pollinating bee species under climate change. *Biorisk* 6, 1–18.
Robins, J. & Henshall, S. (2012). The state of brownfields in the Thames Gateway. *Essex Naturalist* 29, 77–88.

Rollings, R. & Goulson, D. (2019). Quantifying the attractiveness of garden flowers to pollinators. *Journal of Insect Conservation* 23, 803–817.
Romani, R., Isidoro, N., Riolo, P. & Bin, F. (2003). Antennal glands in male bees: structures for sexual communication by pheromones? *Apidologie* 34, 603–610.
Romero, G. Q., Antiqueira, P.A.P. & Koricheva, J. (2011). A Meta-Analysis of Predation Risk Effects on Pollinator Behaviour. *PLoS ONE* 6(6), e20689.
Rooijakkers, E. F. & Sommeijer, M. J. (2009). Gender specific brood cells in the solitary bee *Colletes halophilus* (Hymenoptera; Colletidae). *Journal of Insect Behavior* 22(6), 492–500.
Rosenheim, J. A. (1990). Density-dependent parasitism and the evolution of aggregated nesting in the solitary hymenoptera. *Annals of the Entomological Society of America* 83(3), 277–286.
Rosenweig, M. L. (1995). *Species Diversity in Space and Time*. Cambridge University Press.
Roubik, D. W. (1989). *Ecology and Natural History of Tropical Bees*. Cambridge University Press.
Roulston, T. H. & Cane, J. H. (2000). The effect of diet breadth and nesting ecology on body size variation in bees (Apiformes). *Journal of the Kansas Entomological Society* 73(3), 129–142.
Roulston, T., Cane, J. H. & Buckmann, S. L. (2000). What governs protein content of pollen: pollinator preferences, pollen-pistil interactions, or phylogeny? *Ecological Monographs* 70, 617–634.
Rozen J. G. & Ding, L. (2012). Cleptoparasitic Behaviour and Immatures of the Bee *Melecta duodecimaculata* (Apoidea, Apidae, Melectinae). Appendix: Tribal Descriptions of the Anthophorini and Melectini Based on their Mature Larvae. *American Museum Novitates* 3746, 1–24.
Rozen, J. G. & Kamel, S. M. (2008). Hospicidal Behaviour of the Parasitic Bee *Coelioxys* (*Allocoelioxys*) *coturnix*, Including Descriptions of Its Larval Instars (Hymenoptera: Megachilidae). *American Museum Novitates* 3636, 1–15.
Rueppell, O., Bachelier, C., Fondrk, M. K., Page Jr., R. E. (2007). Regulation of life history determines lifespan of worker

honeybees (*Apis mellifera* L.). *Experimental Gerontology* 42, 1020–1032.

Rumeu, B., Smith, D. J., Hawes, D. J. & Ings, T. C. (2018). Zooming into plant-flower visitor networks: an individual trait-based approach. *PeerJ* 6, e5618.

Rundlöf, M., Andersson, G., Bommarco, R. *et al.* (2015). Seed coating with a neonicotinoid insecticide negatively affects wild bees. *Nature* 521, 77–80.

Russell, A. L., Golden, R. E., Leonard, A. S. & Papaj, D. (2016). Bees learn preferences for plant species that offer only pollen as a reward. *Behavioural Ecology* 27(3), 731–740.

Sann, M., Niehuis, O., Peters, R. S. *et al.* (2018). Phylogenomic analysis of Apoidea sheds new light on the sister group of bees. *BMC Evolutionary Biology* 18(1), 71.

Saul-Gershenz, L. S. & Millar, J. G. (2006) Phoretic nest parasites use sexual deception to obtain transport to their host's nest. *Proceedings of the National Academy of Sciences USA* 103(38), 14039–14044.

Saunders, E. (1896). *The Hymenoptera Aculeata of the British Islands*. L. Reeve & Co., London.

Saunders, P. (2020). Tormentil Mining Bee on Cornwall Wildlife Trusts Bartinney Nature Reserve. http://kernowecology.co.uk/Publications/Bartinney%20report%202017.pdf

Saxton, S. (2009). *Colletes halophilus* at East Tilbury Silt Lagoons. *Bees, Wasps and Ants Recording Society Newsletter*, Autumn, 19.

Scheuchl, E. & Willner, W. (2016). *Taschenlexicon Der Wildbienen Mitteleuropas.* Verlag Quelle & Meyer, Wiebelsheim.

Schiestl, F. P. & Ayasse, M. (2000). Post-mating odor in females of the solitary bee, *Andrena nigroaenea* (Apoidea, Andrenidae), inhibits male mating behaviour. *Behavioural Ecology and Sociobiology* 48, 303–307.

Schiestl, F. P. & Ayasse, M. (2001). Post-pollination emission of a repellent compound in a sexually deceptive orchid: a new mechanism for maximising reproductive success? *Oecologia* 126, 531–534.

Schiestl, F. P., Ayasse, M., Paulus, H. F. *et al.* (2000). Sex pheromone mimicry in the early spider orchid (*Ophrys sphegodes*): patterns of hydrocarbons as the key mechanism for pollination by sexual deception. *Journal of Comparative Physiology A* 186(6), 567–574.

Schindler, M., Hofmann, M. M., Wittmann, D. & Renner, S. (2018). Courtship behaviour in the genus *Nomada* – antennal grabbing and possible transfer of male secretions. *Journal of Hymenoptera Research* 65, 47–59.

Schlindwein, C., Wittmann, D., Martins, C.F., Hamm, A., Alves de Siqueira, J., Schiffler, D., & Isabel Machado, C. (2004). Pollination of *Campanula rapunculus* L. (Campanulaceae): How much pollen flows into pollination and into reproduction of oligolectic pollinators? *Plant Systematics and Evolution* 250: 147–156.

Schumpeter, J. A. (1942). *Capitalism, Socialism and Democracy*. HarperCollins, London.

Sedivy, C., Müller, A. & Dorn, S. (2011). Closely related pollen generalist bees differ in their ability to develop on the same pollen diet: evidence for physiological adaptations to digest pollen. *Functional Ecology* 25(3), 718–725.

Sedivy, C., Praz, C. J., Müller, A., Widmer, A. & Dorn, S. (2008). Patterns of host-plant choice in bees of the genus *Chelostoma*: the constraint hypothesis of host-range evolution in bees. *Evolution* 62(10), 2487–2507.

Seidelmann, K. (2006). Open-cell parasitism shapes maternal investment patterns in the Red Mason bee *Osmia rufa*. *Behavioral Ecology* 17(5), 839–848.

Seidelmann, K. (2014). Behavioural induction of unreceptivity to mating from a post-copulatory display in the red mason bee, *Osmia bicornis*. *Behaviour* 151, 1687–1702.

Seidelmann, K., Ulbrich, K. & Mielenz, N. (2010). Conditional sex allocation in the Red Mason bee, *Osmia rufa*. *Behavioral Ecology and Sociobiology* 64, 337–347.

Semichon, L. (1904). Sur la ponte de la *Melecta armata* Panzer (Hyme.). *Bulletin de la Société Entomologique de France* 9(12), 188–189.

Senapathi, D. *et al.* (2021). Wild insect diversity increases inter-annual stability in global crop pollinator communities. *Proceedings of the Royal Society B* 288(1947), 20210212.

Severinghaus, L. L., Kurtak, B. H. & Eikwort, G. C. (1981). The reproductive behaviour of *Anthidium manicatum* (Hymenoptera: Megachilidae) and the significance of size for territorial males. *Behavioral Ecology and Sociobiology* 9, 51–58.

Sexton, A. N., Benton, S., Browning, A. C. & Emery, S. M. (2021). Reproductive patterns of solitary cavity-nesting bees responsive to both local and landscape factors. *Urban Ecosystems* 24, 1271–1280.

Sheffield, C. S., Pindar, A., Packer, L. & Kevan, P. G. (2013). The potential of cleptoparasitic bees as indicator taxa for assessing bee communities. *Apidologie* 44, 501–510.

Shoard, M. (1980). *The Theft of the Countryside*. Temple Smith, London.

Shuckard, W. E. (1866). *British bees. An introduction to the natural history and economy of the bees indigenous to the British Isles*. Lovell Reeve and Co., London. https://www.gutenberg.org/files/56201/56201-h/56201-h.htm

Sick, M., Ayasse, M., Tengö, J. et al. (1994). Host-parasite relationships in six species of *Sphecodes* bees and their halictid hosts: nest intrusion, intranidal behavior and Dufour's gland volatiles (Hymenoptera: Halictidae). *Journal of Insect Behavior* 7, 101–117.

Silvestro, D., Bacon, C. D., Ding, W. et al. (2021). Fossil data support a pre-Cretaceous origin of flowering plants. *Nature, Ecology and Evolution* 5, 449–457.

Simmons, A. D. & Thomas, C. D. (2004). Changes in dispersal during species' range expansions. *The American Naturalist*, 164(3), 378–395.

Simone-Finstrom, M., Li-Byarlay, H., Huang, M. H. et al. (2016). Migratory management and environmental conditions affect lifespan and oxidative stress in honey bees. *Scientific Reports* 6, 32023.

Simonin, K. A. & Roddy, A. B. (2018). Genome downsizing, physiological novelty, and the global dominance of flowering plants. *PLoS Biol* 16(1), e2003706.

Smit, J. T. & Smit, J. (2003). *Hylaeus signatus* (Panzer, 1798) new for the fauna of Madeira, with notes on its feeding behaviour (Hymenoptera: Apidae). *Entomofauna* 24(11), 165–167.

Smith B. H. & Ayasse, M. (1987). Kin-based male mating preferences in two species of halictine bee. *Behavioural Ecology and Sociobiology* 20(5), 313–318.

Smith, B. H. & Weller, C. (1989). Social competition among gynes in halictine bees: the influence of bee size and pheromones on behavior. *Journal of Insect Behavior* 3, 397–411.

Smith, K.G.V. (1969). Handbooks for the Identification of British Insects, Vol. X. Part 3(a): *Diptera, Conopidae*. Royal Entomological Society of London.

Smith, R. M., Gaston, K. J., Warren, P. H. & Thompson, K. (2006a). Urban domestic gardens (VII): environmental correlates of invertebrate abundance. *Biodiversity and Conservation* 15, 2515–2545.

Smith, R. M., Warren, P. H., Thompson, K. & Gaston, K. J. (2006b). Urban domestic gardens (VI): environmental correlates of invertebrate species richness. *Biodiversity and Conservation* 15, 2415–2438.

Sommeijer, M. J., Neve, J. & Jacobusse, C. (2012). The typical development cycle of the solitary bee *Colletes halophilus*. *Entomologische Berichten* 72(1–2), 52–58.

Soro, A., Ayasse, M., Zobel, M. U. & Paxton, R. J. (2009). Complex sociogenetic organization and the origin of unrelated workers in a eusocial sweat bee, *Lasioglossum malachurum*. *Insectes Sociaux* 56, 55–63.

Soro, A., Ayasse, M., Zobel, M. U. & Paxton, R. J. (2011). Kin discriminators in the eusocial sweat bee *Lasioglossum malachurum*: the reliability of cuticular and Dufuor's gland odours. *Behavioral Ecology and Sociobiology* 65, 641–653.

Soro, A., Field, J., Bridge, C., Cardinal, S. C. & Paxton, R. J. (2010). Genetic differentiation across the social transition in a polymorphic sweat bee, *Halictus rubicundus*. *Molecular Ecology* 19, 3351–3363.

Spear, D. M., Silverman, S. & Forrest, J.R.K. (2016). Asteraceae pollen provisions protect *Osmia* mason bees (Hymenoptera: Megachilidae) from brood parasitism. *The American Naturalist*. 87(6), 797–803.

Spider and Harvestman Recording Scheme. http://srs.britishspiders.org.uk.

Stanley, D. A., Russell, A. L., Morrison, S. J., Rogers, C. & Raine, N. E. (2016). Investigating the impacts of field-realistic exposure to neonicotinoid pesticide on bumblebee foraging homing ability and colony growth. *Journal of Applied Ecology* 53, 1440–1449.

Stebbins, G. L. (1970). Adaptive radiation of reproductive characteristics in angiosperms. 1. Pollination mechanisms. *Annual Review of Ecology and Systematics* 1, 307–326.

Steffan, S. A, Dharampal, P. S., Danforth, B. N. *et al.* (2019). Omnivory in bees: elevated trophic positions among all major bee families. *The American Naturalist* 194(3), 414–421.

Steffan-Dewenter, I., Klein, A.-M., Gaebele, V., Alfert, T. & Tscharntke, T. (2006). Bee diversity and plant-pollinator interactions in fragmented landscapes. In N. M. Waser & J. Ollerton (eds.) *op. cit.*, 387–487.

Steffan-Dewenter, I., Münzenberg, U., Bürger, C., Thies, C. & Tscharntke, T. (2002). Scale-dependent effects of landscape structure on three pollinator guilds. *Ecology* 83, 1421–1432.

Steffan-Dewenter, I., Potts, S. G. & Packer, L. (2005) Pollinator diversity and crop pollination services are at risk. *Trends in Ecology and Evolution* 20(12), 651–652.

Steitz, I., Kingwell, C., Paxton, R. J. & Ayasse, M. (2018). Evolution of caste-specific chemical profiles in halictid bees. *Journal of Chemical Ecology* 44, 827–837.

Steitz, I., Paxton, R. J., Schulz, S. & Ayasse, M. (2021). Chemical variation among castes, female life stages and populations of the facultative eusocial sweat bee *Halictus rubicundus* (Hymenoptera: Halictidae). *Journal of Chemical Ecology* 47, 406–419.

Step, E. (c. 1920) *Insect Artizans and their Work.* Hutchinson, London.

Step, E. (1932). *Bees, Wasps, Ants and Allied Insects of the British Isles.* Frederick Warne, London and New York.

Stone, G. N. (1990). Endothermy and Thermoregulation in Solitary Bees. PhD thesis, Oxford University.

Stone, G. N. (1993). Endothermy in the solitary bee *Anthophora plumipes* (Hymenoptera: Anthophoridae): independent measures of thermoregulatory ability, costs of warm-up and the role of body size. *Journal of Experimental Biology* 174, 299–320.

Stone, G. N. (1995). Female foraging responses to sexual harassment in the solitary bee *Anthophora plumipes. Animal Behaviour* 50, 405–412.

Stone, G. N., Loder, P.M.J. & Blackburn, T. M. (1995). Foraging and courtship behaviour in males of the solitary bee *Anthophora plumipes* (Hymenoptera: Anthophoridae): thermal physiology and the role of body size. *Ecological Entomology* 20, 169–183.

Stone, G. N. & Willmer, P. G. (1989). Warm-up rates and body temperature in bees: the importance of body size, thermal regime and phylogeny. *Journal of Experimental Biology* 147, 303–328.

Stork, N. (2010). Re-assessing current extinction rates. *Biodiversity and Conservation* 19(2), 357–371.

Straka, J., Cema, K., Machackova, L., Zemenova, M. & Keil, P. (2014). Life span in the wild: the role of activity and climate in natural populations of bees. *Functional Ecology* 28, 1235–1244.

Straka, J., Juzova, K. & Nakase, Y. (2015). Nomenclature and taxonomy of the genus Stylops (Strepsiptera): an annotated preliminary world checklist. *Acta Entomologica Musei Nationalis Pragae* 55(1), 305–332.

Straka, J., Rezkova, K., Batelka, J. & Kratochvil, L. (2011). Early nest emergence of females parasitised by Strepsiptera in protandrous bees (Hymenoptera Andrenidae). *Ethology, Ecology and Evolution* 23, 97–109.

Stringer, C. (2006). *Homo Britannicus. The Incredible Story of Human Life in Britain.* Allen Lane, Penguin, London.

Strohm, E. (2011). How can cleptoparasitic drosophilid flies emerge from the closed brood cells of the red mason bee? *Physiological Entomology* 36, 77–83.

Strohm, E. & Bordon-Hauser, A. (2003). Advantages and disadvantages of large colony size in a halictid bee: the queen's perspective. *Behavioural Ecology* 14(4), 546–553.

Strudwick, T. (2011) The bees of Norfolk: a provisional county list. *Transactions of the Norfolk and Norwich Naturalists' Society* 44(1), 34–57.

Stubbs, A. & Drake, M. (2001). *British Soldierflies and their Allies.* British Entomological and Natural History Society, Reading.

Study of Critical Environmental Problems (SCEP) (1970). *Man's Imprint on the Global Environment.* MIT, Cambridge, Mass.

Šturm, R., Rexhepi, B. *et al.* (2021). Hay-meadow vibroscape and interactions within insect vibrational community. *iScience* 24(9), 103070.

Swift, J. [1726] (1970) *Gulliver's Travels*. Penguin, Harmondsworth.

Tang, T., Wu, D. & Huang, S.-Q. (2019). Bumblebee rejection of toxic pollen facilitates pollen transfer. *Current Biology* 29, 1401–1406.

Tansley, D. (ed.) (2019). *Roman River Living Landscape: habitats and species 2009–2019*. Essex Wildlife Trust, Great Wigborough.

Tengö, J. (1979). Odour-released behaviour in *Andrena* male bees (Apoidea, Hym.). *Zoon* 7, 15–48.

Tengö, J., Ågren, L., Bauer, B., Isaksson, R., Liljefors, T. et al. (1990). *Andrena wilkella* male bees discriminate between enantiomers of cephalic secretion components. *Journal of Chemical Ecology* 16, 429–441.

Tengö, J. & Bergström, G. (1976a). Odor correspondence between *Melitta* males and females of their nest parasite *Nomada flavopicta*. *Journal of Chemical Ecology* 2, 57–65.

Tengö, J. & Bergström, G. (1976b). Comparative analysis of lemon-smelling secretions from heads of *Andrena* F. (Hymenoptera: Apoidea) bees. *Comparative Biochemistry and Physiology* 57B, 179–188.

Tengö, J. & Bergström, G. (1977a). Comparative analyses of complex secretions from heads of *Andrena* bees (Hym., Apoidea). *Comparative Biochemistry and Physiology* 57B, 197–202.

Tengö, J. & Bergström, G. (1977b). Cleptoparasitism and odor mimetism in bees: do *Nomada* males imitate the odor of *Andrena* females? *Science* 196(4294), 1117–1119.

Tengö, J., Eriksson, J., Borg-Karlson, A.-K., Smith, B. H. & Dobson, H. (1989). Mate-locating strategies and multi-modal communication in male mating behaviour of *Panurgus banksianus* and *P. calcaratus* (Apoidea: Andrenidae). *Journal of the Kansas Entomological Society* 61(4), 338–395.

Tepedino, V. J. & Torchio, P. F. (1994). Founding and usurping: equally efficient paths to nesting success in *Osmia propinqua* (Hymenoptera: Megachilidae). *Annals of the Entomological Society of America* 87(6), 946–953.

Teppner, H. & Brosch, U. (2015). Pseudo-oligolecty in *Colletes hederae* (Apidae-Colletinae, Hymenoptera). *Linzer Biologische Beiträge* 47(1), 301–306.

Terry, I. (2001). Thrips: the primeval pollinators? *Thrips and Tospoviruses: Proceedings of the 7th Annual Symposium on Thysanoptera*, 157–162.

Terry, I., Walter, G. H., Moore, C., Roemer, R. & Hull, C. (2007). Odor-mediated push-pull pollination in cycads. *Science* 318(5847), 70.

Théry, M. & Casas, J. (2002). Predator and prey views of spider camouflage. *Nature* 415(6868), 133.

Thomas, C. D., Bodsworth, E. J., Wilson, R. J. et al. (2001). Ecological and evolutionary processes at expanding range margins. *Nature* 411(6837), 577–581.

Thorp, R. W. (1969). Ecology and Behaviour of *Melecta separata callura* (Hymenoptera: Anthophoridae). *The American Midland Naturalist* 82(2), 338–345.

Tolash, T., Kehl, S. & Dötterl, S. (2012). First Sex Pheromone of the Order Strepsiptera: (3R,5R,9R)-3,5,9-Trimethyldodecanal in *Stylops melittae* Kirby, 1802. *Journal of Chemical Ecology* 38, 1493–1503.

Tomkins, S. P. (2020). A recent arrival: the violet carpenter bee. *Nature in Cambridgeshire* 62, 62–63.

Torchio, P. F. (1989). Biology, immature development and adaptive behaviour of *Stelis montana*, a cleptoparasite of *Osmia* (Hymenoptera: Megachilidae). *Annals of the Entomological Society of America* 82(5), 616–632.

Torchio, P. F. & Burdick, D. J. (1988). Comparative notes on the biology and development of *Epeolus compactus* Cresson, a cleptoparasite of *Colletes kincaidii* Cockerell (Hymenoptera: Anthophoridae, Colletidae). *Annals of the Entomogical Society of America* 81(4), 626–636.

Torchio, P. F. & Tepedino, V. J. (1982). Parsivoltinism in three species of *Osmia* bees. *Psyche: A Journal of Entomology* 89(3–4), 221–238.

Torchio, P. F., Throstle, G. E. & Burdick, D. J. (1988). The nesting biology of *Colletes kinkaidii* Cockerell (Hymenoptera: Colletidae) and the development of its immature forms. *Annals of the Entomological Society of America* 81(4), 605–625.

Tur, C., Vigalondo, B., Trøjelsgaard, K., Olesen, J. M. & Traveset, A. (2014). Downscaling pollen-transport networks to the level of individuals. *Journal of Animal Ecology* 83, 306–317.

Turner, A. (2021). Honeybees are voracious: is it time to put the brakes on the boom in beekeeping? *Guardian*, 24 July.

Twerd, L. & Banaszac-Cibicka, W. (2019) Wastelands: their attractiveness and importance for preserving the diversity of wild bees in urban areas. *Journal of Insect Conservation* 23, 573–588.

Unwin, D. M. & Corbet, S. A. (1991). *Insects, plants and microclimate*. Naturalists' Handbooks 15. Richmond, Oxford.

Vaissière, B. E. & Vinson, S. B. (1993). Pollen morphology and its effect on pollen collection by honey bees, *Apis Mellifera* L. (Hymenoptera: Apidae), with special reference to Upland Cotton, *Gossypium hirsutum* L. (Malvaceae). *Grana* 33(3), 128–138.

Vallejo-Marín, M. (2019). Buzz-pollination: studying bee vibrations on flowers. *New Phytologist* 224, 1068–1074.

Vanbergen, A. J. *et al.* (2013). Insect Pollinators Initiative. Threats to an ecosystem service: pressures on pollinators. *Frontiers in Ecology and the Environment* 11(5), 251–259.

Veits, M. *et al.* (2019) Flowers respond to pollinator sound within minutes by increasing nectar sugar concentration. *Ecology Letters* 22(9), 1483–1492.

Vereecken, N. J. & Mahe, G. (2007). Larval aggregations of the blister beetle *Stenoria analis* (Schaum) (Coleoptera: Meloidae) sexually deceive patrolling males of their host, the solitary bee *Colletes hederae* Schmidt & Westrich (Hymenoptera: Colletidae). *Annales de la Société Entomologique de France* 43(4), 493–496.

Vinson, S. B., Frankie, G. & Rao, A. (2011). Field behavior of parasitic *Coelioxys chichimeca* (Hymenoptera: Megachilidae) toward the host bee *Centris bicornuta* (Hymenoptera: Apidae). *Apidologie* 42, 117–127.

Vorobyev, M., Gumbert, A., Kunze, J., Giurfa, M. & Menzel, R. (1997). Flowers through insect eyes. *Israel Journal of Plant Sciences* 45, 93–101.

Voudo, A. D., Patch, H. M., Mortensen, D. A., Tooker, J. F. & Grozinger, C. M. (2016). Macronutrient ratios in pollen shape bumble bee (*Bombus impatiens*) foraging strategies and floral preferences. *Proceedings of the National Academy of Sciences USA* 113(28), E4035–E4042.

Voudo, A. D., Tooker, J. F., Grozinger, C. M. & Patch, H. M. (2015). Bee nutrition and floral resource restoration. *Current Opinion in Insect Science* 10, 133–141.

Voudo, A. D., Tooker, J. F., Patch, H. M. *et al.* (2020). Pollen protein: lipid micronutrient ratios may guide broad patterns of bee species floral preferences. *Insects* 11, 132.

Voulgari-Kokota, A., Ankenbrand, M. J., Grimmer, G., Steffan-Dewenter, I. & Keller, A. (2019). Linking pollen foraging of megachilid bees to their nest bacterial microbiota. *Ecology and Evolution* 9(18), 10788–10800.

Walters, J. (2016). *Segestria florentina* (Rossi) preying on hairy-footed flower bees *Anthophora plumipes* (Pallas). *Newsletter of the British Arachnological Society* 136, 7.

Walters, J. Oil beetles. http://johnwalters.co.uk/research/oil-beetles.php

Walters, J. (2021). Cob brick bee hotels and observations of the nesting behaviour of *Anthophora plumipes* (Apidae: Apinae). *Bees, Wasps and Ants Recording Society Newsletter*, Spring, 6–9.

Walters, J. (2022). Species of *Osmia* using snail shells. *Bees, Wasps and Ants Recording Society Newsletter*, Spring, 5–11.

Wardhaugh C. W. (2015). How many species of arthropods visit flowers? *Arthropod-Plant Interactions* 9, 547–565.

Waser, N. M. (1986). Flower constancy: definition, cause and measurement. *American Naturalist* 127, 593–603.

Waser, N. M. (2006). Specialisation and generalisation in plant-pollinator interactions: a historical perspective. In N. M. Waser & J. Ollerton (eds.), loc. cit., 3–18.

Waser, N. M., Chittka, L., Price, M. V., Williams, N. M. & Ollerton, J. (1996). Generalization in pollination systems, and why it matters. *Ecology* 77, 1043–1060.

Waser, N. M. & Ollerton, J. (eds.) (2006). *Plant-Pollinator Interactions: From Specialization to Generalization*. University of Chicago Press.

Webb, N. R. (1986). *Heathlands*. Collins, London.

Weiss, K., Grimakin, F. G., Popov, A. V. & Renninger, G. (eds.) (1993). *Sensory Systems of Arthropods*. Birkhäuser, Basil, Boston and Berlin.

Weiss, M. R. (1991). Floral colour changes as cues for pollinators. *Nature* 354(6350), 227–229.

Welch, M. D. & Owens, N. W. (2019). The association of the pNS satellite fly *Miltogramma germari* Meigen (Diptera, Sarcophagidae) with the pantaloon bee *Dasypoda hirtipes* Fabricius (Hymenoptera, Melittidae). *Dipterists Digest* 26, 209–218.

Welch, M. D. & Owens, N.W. (2022). Field observations of the rare cleptoparasitic satellite fly *Leucophora sericea* (Diptera: Anthomyiidae) and its host *Andrena flavipes* (Hymenoptera: Andrenidae) in Norfolk. *British Journal of Entomology and Natural History* 35, 55–61.

Wells, H. & Wells, P. H. (1983). Honey bee foraging ecology: optimal diet, minimal uncertainty or individual constancy? *Journal of Animal Ecology* 52, 829–836.

Wells, H. & Wells, P. H. (1986). Optimal diet, minimal uncertainty and individual constancy in the foraging of honeybees, *Apis mellifera*. *Journal of Animal Ecology* 55, 881–891.

Wenzel, A., Grass, I., Belavadi, V. V. & Tscharntke, T. (2019). How urbanization is driving pollinator diversity and pollination – a systematic review. *Biological Conservation* 241, 108321.

Westphal, C., Steffan-Dewenter, I. & Tscharntke, T. (2003). Mass flowering crops enhance pollinator densities at a landscape scale. *Ecology Letters* 6, 961–965.

Westrich, P. (1996). Habitat requirements of central European bees and the problems of partial habitats. In A. Matheson *et al.* (eds.), *op. cit.*, 1–16.

Westrich, P. (2008) Flexible pollen collecting behaviour in the otherwise strictly oligolectic bee *Colletes hederae* Schmidt & Westrich (Hymenoptera: Apidae). *Eucera* 1(2), 17–29.

Westrich, P. (2018). *Die Wildbienen Deutschlands*. Ulmer, Stuttgart.

Westrich, P. & Schmidt, K. (1987). Pollen analysis, an auxiliary tool to study the collecting behaviour of solitary bees. *Apidologie* 18, 199–213.

Whitehouse, A. (2021). Save the bees, but which ones? https://www.buglife.org.uk/blog.

Williams, I. H. (1996). Aspects of bee diversity and crop pollination in the European Union. In A. Matheson *et al.* (eds.), *op. cit.*, 63–80.

Williams, P. H. (1982). The distribution and decline of British bumblebees (*Bombus* Latr.) *Journal of Apicultural Research* 21(4), 236–245.

Willmer, P. (2011). *Pollination and Floral Ecology*. Princeton University Press.

Willmer, P. G., Cunnold, H. & Ballantyne, G. (2017). Insights from measuring pollen deposition: quantifying the pre-eminence of bees as flower visitors and effective pollinators. *Arthropod-Plant Interactions* 11, 411–425.

Winston, M. L. & Espelie, K. E. (eds.) (2020). *Pheromone Communication in Social Insects*. Westview, Boulder, Col.

Wintergerst, J., Kästner, T., Bartel, M., Schmidt, C. & Nuss, M. (2021). Partial mowing of urban lawns supports higher abundances and diversities of insects. *Journal of Insect Conservation* 25, 797–808.

Wise, M. J., Cummins, J. J. & De Young, C. D. (2008). Compensation for floral herbivory in *Solanum carolinense*: identifying mechanisms of tolerance. *Evolutionary Ecology* 22, 19–37.

Wittmann, D. & Blochtein, B. (1995). Why males of leafcutter bees hold the females' antennae with their front legs while mating. *Apidologie* 26(3), 181–196.

Wittmann, D., Schindler, M., Blochtein, B. & Barouz, D. (2004). Mating in bees: how males hug their mates. *Proceedings of the 8th IBRA International Conference on Tropical Bees and VI Encontro sobre Abelhas*, 374–380.

Wittwer, B., Hefetz, A., Tovit, S. *et al.* (2017). Solitary bees reduce investment in communication compared with their social relatives. *Proceedings of the National Academy of Sciences USA* 114(25), 6569–6574.

Wolf, T. J. & Schmid-Hempel, P. (1989). Extra loads and foraging life span in honeybee workers. *Journal of Animal Ecology* 58, 943–954.

Wolton, R. J. (2022). *Leucophora sponsa* (Meigen) (Diptera, Anthomyiidae) is a cleptoparasite of the bee *Lasioglossum parvulum* (Schenck) (Hymenoptera, Halictidae). *Dipterists Digest* 29(2), 225–229.

Wood, T. (2015). A note on satellite fly (Diptera: Miltogramminae) avoidance behaviour in the bee *Dasypoda hirtipes*. *Bees, Wasps and Ants Recording Society Newsletter*, Spring, 11.

Wood, T. & Baldock, D. (2016). Going for gold at Farnham Heath RSPB Reserve, Tilford, Surrey. *Bees, Wasps and Ants Recording Society Newsletter*, Autumn, 33–38.

Wood, T. J., Holland, J. M. & Goulson, D. (2017). Providing foraging resources for solitary bees on farmland: current schemes for pollinators benefit a limited suite of species. *Journal of Applied Ecology* 54(1), 323–333.

Wood, T. J. & Roberts, S.P.M. (2017). An assessment of historical and contemporary diet breadth in polylectic *Andrena* bee species. *Biological Conservation* 215, 72–80.

Wood, T. J. & Roberts, S.P.M. (2018). Constrained patterns of pollen use in Nearctic *Andrena* (Hymenoptera: Andrenidae) compared with their Palaearctic counterparts. *Biological Journal of the Linnean Society* 124, 732–746.

Woodcock, B. A., Edwards, M., Redhead, J. *et al.* (2013). Crop flower visitation by honeybees, bumblebees and solitary bees: small scale behavioural differences linked to landscape scale responses. *Agriculture, Ecosystems and Environment* 171, 1–8.

Woodcock, B. A. & Isaac, N.J.B. (2019). Widespread losses of pollinating insects in Britain. *Nature Communications* 10, 1018.

Woodcock, B. A., Isaac, N.J.B., Bullock, J. M. *et al.* (2016). Impacts of neonicotinoid use on long-term population changes in wild bees in England. *Nature Communications* 7, 12459.

Wright, R. (2017). Beetles (Wildlife Reports). *British Wildlife* 28(5), 365–367.

Yalden, D. W. (1976). The Food of the Hedgehog in England. *Acta Theriologica* 21–30, 401–424.

Yanega, D. (1988). Social plasticity and early-diapausing females in a primitively social bee. *Proceedings of the National Academy of Sciences USA* 85(12), 4374–4377.

Yanega, D. (1989). Caste determination and differential diapause within the first brood of *Halictus rubicundus* (Hymenoptera: Halictidae) in New York. *Behavioral Ecology and Sociobiology* 24, 97–107.

Yanega, D. (1992). Does mating determine caste in sweat bees? (Hymenoptera: Halictidae). *Journal of the Kansas Entomological Society* 65, 231–237.

Yanega, D. (1993). Environmental influences on male production and social structure in *Halictus rubicundus* (Hymenoptera: Halictidae). *Insectes Sociaux* 40, 169–180.

Yeates, D. K. & Greathead, D. (1997). The evolutionary pattern of host use in the Bombyliidae (Diptera): a diverse family of parasitoid flies. *Biological Journal of the Linnean Society* 60, 149–185.

Yildirim, E., Coruh, S., Kolarov, J. & Madl, M. (2004). The Gasteruption (Hymenoptera: Gasteruptiidae) of Turkey. *Linzer Biologische Beiträge* 36(2), 1349–1352.

Zobel, M. U. & Paxton, R. J. (2007). Is big the best? Queen size, usurpation and nest closure in a primitively eusocial sweat bee (*Lasioglossum malachurum*). *Behavioral Ecology and Sociobiology* 61, 435–447.

Index

SPECIES INDEX

Page numbers in **bold** refer to information contained in figure captions. Page numbers in *italics* refer to information contained in tables. Species merely listed in tables have not been indexed.

Acarina (mites) 346–348, *347*, **349**
Acer pseudoplanatus (Sycamore) 232
Actium lappa (Greater Burdock) 283
Aculeata (aculeates) 2, **3**, 406
aculeate wasps **3**, 102, 305, 339–341, 355, 406, **407**
Adonis Blue butterfly (*Polyommatus bellargus*) 464, **465**
Aglais io (Peacock Butterfly) 101
Aglais urticae (Small Tortoiseshell Butterfly) 101
Ajuga reptans (Bugle) 154, 237, **248**, 472
Alcea rosea (Hollyhock) 200
Alexanders (*Smyrnium olusatrum*) 253
Alstroemeria aurea 354
Amaurobius species (spiders) 351
Amegilla dawsoni 372
Ammobates punctatus 366
Ammophila pubescens (Heath Sand Wasp) 102
Ammoplanina 406–407
Ammoplanus 407, **407**, 408, **409**
Andrena (mining bees) **5**, 16, 47–48, 57
anatomy/characteristics 7, 13, 28, **34**, 293, 295–296
chemical signals, pheromones 64–65, 88, 90, 381–382
cleptoparasites of (*Nomada*) 48, 373, 377–381, **378**, 379, **380**
emergence timing 28, 430, 437
flower types associated 201, 226, 290, 291, 292, 293
foraging behaviour 251, 290–296
habitats (heath, grassland) 467, 468
losses/gains over recent decades 446
males patrolling female emergence sites 69–71, **70**, 427
mating behaviour 89, 90–91, **91**, 98–99
nests/nesting behaviour **6**, 16, 47, 117–123, **118**, 121, 123, 306, 432, **494**

aggregations 118, **118**, 121, 307, **428**, 431–432
architecture **114**
concealed entrances 119, 120, **121**, 122, 432
non-resource-based male patrolling **84**, 84–85
North American species 290, 291
oligolectic species 200, 269, 270, 272, 290–296
parasites
Bombylius 317, **318**, **318**, 319
conopids 322–323, **323**, 324
Leucophora 306, 307, **307**, 308, **308**, 309, 310–311
Myopa 322–323, **323**, 324
Stylops **330**, 330–332, **331**, 333
phylogeny 290
pollen and nectar collection 187, 201, **201**, **292**, 293, **294**, 516
Asteraceae pollen 220, 264, 290, 291, **291**, 292
pollen transport **255**, 257, 293
polylectic species 269, 272, 290, 291
reduced sexual development, stylopised 332, **332**
species number 33, 47, 416
subgenera 47
Andrena afzeliella (Kirby's Mining Bee) 47
Andrena agilissima (Violet-winged Mining Bee) 312
Andrena alleghaniensis 383
Andrena apicata (Large Sallow Mining Bee) 381, 489
Andrena argentata (Small Sandpit Mining Bee) 69, 71, 90–91, **91**, 113, **113**, 118, **118**, 467, 480
Andrena barbilabris (Sandpit Mining Bee) 69, 90, 118, 119, 452, 479

Andrena bicolor (Gwynne's Mining Bee) **84**, 207, 261, 287, 318, 376, 437
pollen collection, flower types 185, 192, 228, 229, 261, 287
Andrena bicornis 293
Andrena bimaculata (Large Gorse Mining Bee) **84**, 293
Andrena bucephala (Big-headed Mining Bee) 74, 104, 159, **160**, 489, 495
Andrena chrysosceles (Hawthorn Mining Bee) 332, **332**, 437
Andrena cineraria (Ashy Mining Bee) 69, **84**, **220**
habitats and distribution 208, 437, 453, 468
nests/nesting behaviour 111, 118
parasites, and predators 322, 350
pollen collection, flower types 208, 219, **220**, 437
Andrena clarkella (Clarke's Mining Bee) **70**, 71, **121**, 323
cuckoo bee (*Nomada*) behaviour 379, **380**, 381
habitats, and distribution 415, **472**, 473
nest entrance closure/decoy 121, **121**, 311, 379, **380**, 381
nests/nesting behaviour 121, 432
Andrena coitana (Small Flecked Mining Bee) 415, 427, 470, 487
Andrena congruens (Long-fringed Mining Bee) 489, **491**
Andrena dorsata (Short-fringed Mining Bee) **84**, 437, 452
Andrena falsifica (Thick-margined Mini-miner) 489
Andrena ferox (Oak Mining Bee) 48, **64**, 74, 104, 159, 439, 472
Andrena flavipes (Yellow-legged Mining Bee) 69, **70**, 84, 309, 376, 437, 453, 478
habitats, and distribution 453, 478

SPECIES INDEX · 569

nests/nesting behaviour 117, 118, 320, 478
Andrena florea (Bryony Mining Bee) 76, **76**, **294**, **295**, 497, 500, 502
 in gardens/urban spaces 497, 500, 502, 504, 520
 pollen collection, flower type 30–31, 76, 260, 263, 293, 294, **294**, **295**
Andrena fulva (Tawny Mining Bee) 10, 69, 118, **118**, **204**, 311, **312**, **318**, 377, 516
Andrena fulvago (Hawksbeard Mining Bee) 207, 220
Andrena fuscipes (Heather Mining Bee) 76, 84, **295**, 295–296, 378, 437, 467, 470, **490**
Andrena gravida (White-bellied Mining Bee) 452, 495, **496**
Andrena haemorrhoa (Orange-tailed Mining Bee) 28, 84, **85**, 120, 213, **214**, 377, **378**, 478
Andrena hattorfiana (Large Scabious Mining Bee) 47, 89, 119, **119**, 201, 378, 464, 506
 distribution and habitats 464, 493, 506, 520
 foraging behaviour 119, **119**, 292, 378
 nests/nesting behaviour 119
 pollen collection/transport 89, 200, 257, 482
Andrena humilis (Buff-tailed Mining Bee) **6**, 206, **206**, 291, 432, 495, **496**
Andrena labialis (Large Meadow Mining Bee) 474
Andrena labiata (Red-girdled Mining Bee) 191, 207, **208**, 377, 378, **379**, 474, 487, **496**
Andrena lapponica (Bilberry Mining Bee) 470–471, 487, **490**
Andrena marginata (Small Scabious Mining Bee) **48**, 120, 200, 201, **201**, 257, **292**, 292–293, 378, 464
Andrena nigriceps (Black-headed Mining Bee) **215**, 479, 480, 487, **488**
Andrena nigroaenea (Buffish Mining Bee) 84, 186, 212, 330, 377, **489**
 distribution, chalk grassland 487, **489**
 Early Spider Orchid mimicking 211, **212**
 flower fidelity, pollen specialist 186, 219, 293
 males patrolling nest sites 69, 84
 nectar collection, from Thrift 186
 scent-marking, chemical signals 84, 84, 90, 98–99, 100
 Stylops parasitic on 330, **330**
Andrena nigroaenea sarnia (Buffish Mining Bee subspecies) 453
Andrena nitida (Grey-patched Mining Bee) 18, **47**, **48**, 78, 84, **85**, 253, **254**, 437, 468
Andrena nitidiuscula (Carrot Mining Bee) 479
Andrena ovatula (Small Gorse Mining Bee) 47

Andrena pilipes (Black Mining Bee) 118, **181**
Andrena polita (Maidstone Mining Bee) 449, **450**
Andrena praecox (Small Sallow Mining Bee) 84, 478
Andrena rosae (Perkin's Mining Bee) 470, **471**
Andrena ruficrus (Northern Mining Bee) 415, 418, 470, 487
Andrena scotica (Chocolate Mining Bee) 74, **85**, 104, 159, 311–312, 319, 323, 437, **453**
Andrena similis (Red-backed Mining Bee) **491**
Andrena simillima (Buff-banded Mining Bee) **48**
Andrena subopaca (Impunctate Mini-miner) **5**, 78, 191
Andrena synadelpha (Broad-margined Mining Bee) 84
Andrena tarsata (Tormentil Mining Bee) 191, **191**, 309, 415, 427, 470, **471**, 480, 487, 489
Andrena thoracica (Cliff Mining Bee) **122**, **189**, 427, 437
 distribution and habitats 414, 453
 mating behaviour 91, 121, 453
 nests/nesting behaviour 118, 121–122, **122**, 307, **308**, **428**, 432, 453
 parasites/cleptoparasites 307, 307–308, **308**, 318, 319, 323, 384
 pollen collection, flowers visited 189, 229, 293, 437
Andrena tibialis (Grey-gastered Mining Bee) 84, **382**, 497
Andrena tridentata (Pale-tailed Mining Bee) 447
Andrena vaga (Grey-backed Mining Bee) **70**, **118**, **123**, **331**, **382**, 429
 male mating behaviour **70**, 71, 90
 nests/nesting behaviour 118, **118**, 120, 122–123, **123**, 379
 orientation flights 120–121
 parasites 319, 321, **331**
 pollen- and nectar-collecting days 238
 recently established/re-established 449, 451, 475, 478
Andrena varians (Blackthorn Mining Bee) 516
Andrena wilkella (Wilke's Mining Bee) 21, 86, **251**, 251–252, 474
Andrenidae 28, 33, 33, 46–48, 241, 290–297
 characteristics of genera **34**, 46–48, **18**, 57
 genera and species number **33**, 416
 ground nesting 109, 117–125, **118**, **121**, **122**, **123**
 mouthpart structure 241, 242
 oligolectic/polylectic species 269, 270, 272, 290–297
 pollen transport **21**, **255**, 257
 see also *Andrena* (mining bees); Panurgus (shaggy bees)

Anthidium (wool carder bees) 32, **34**, 41, 56, 218
 flower types associated 226, **248**
 polylectic 257, 269, 270, 271
 species number, latitudinal gradient 416
Anthidium manicatum (Wool Carder Bee) 22, 41, **41**, **46**, 81, 82, **97**, **246**, **346**, **523**
 deterrents (hairs) against parasites 345, **346**
 male patrolling/mating behaviour 30, 81, **81**, 82, 96–97, **97**, 100
 nesting (cavity/aerial) 142, 144–145
 plant hairs collection 81–82, **82**, **144**, 144–145
 pollen collection, flower types **246**, **246**
 secretions from *Pelargonium* 345, **346**
 Stelis targeting 399
 territorial behaviour (males) 78, 79, 82, 100
Anthophila 2, **3**, **407**, 408
Anthophora (flower bees) 37–38, 280–281
 buzz of 37, 249
 characteristics/morphology **34**, 37–38, **55**, 281
 cleptoparasites of 38, 367, 391
 emergence timing 31, 430
 flower types/colours associated 37, 202, 226, 280, 281, 282
 hovering without landing 22, 37, 209, 218
 nests/nesting behaviour 17, 37–38, 71, 105, 117, **117**
 pollen collection 37, 281, **281**
 pollen transport 253–254, **254**, **255**, 256, 281
 polylectic 269, 270, 271, 280–282
 roosting 432
 species number, latitudinal gradient 416
Anthophora bimaculata (Green-eyed Flower Bee) 21, **40**, **66**, 73, 117, **117**, **197**, **199**, **469**
 cleptoparasites/parasites of 312, 366
 distribution and habitat 31, 413–414, 480
 males patrolling, territorial behaviour 79, 87
 nectar and pollen collection **197**, **199**
 nesting aggregations 73, 117, **117**
Anthophora fulvitarsis 370
Anthophora furcata (Fork-tailed Flower Bee) 38, 117, **247**
 buzz pollination 249
 habitat and distribution **488**
 hairs 38, 245, **247**, 249, 281
 heat loss/generation 425
 male patrol routes 87
 nests/nesting behaviour 38, 110, 117, 154
 pollen collection, flower types 245–246, **247**, 249, 281, **423**, 423–424

570 · SOLITARY BEES

Anthophora plumipes (Hairy-footed Flower Bee) 10, 11, **14–15**, 19, **20**, 373, 430
 beetles parasitic on 336
 characteristics, and tongue 37–38, **71**, 184, 185, **241**
 cleptoparasites of 367, 368–369, **369**, 371, **372**, 373, 425
 emergence, timing **15**, 28, 428–429, **429**
 female body size, cleptoparasitic pressure 425, 427
 flower constancy 216
 foraging distances 237
 garden flower types visited 185, 228, 229, 281–282
 heat loss/generation 207, 424–425
 hind-leg scopae 253–254, **254**
 life cycle, and brood cells **14–15**, **371**, 373
 lifespan 428–429, **429**
 male territorial behaviour 79
 males patrolling **71**, 78–79, 87, 429
 mating display 71, **71**
 nests, aggregations 37, **37**, 71, 367–368, **368**, 430, **431**, 501, **521**, **523**
 pollen collection/pollination 184, 185, **195**, **196**, 249, **249**, 281, **281**
 predators of 350, 351, **351**, 355
 secretion from Dufour's gland **371**, 372
 size, effect on flight temperature 425
 sounds emitted, signal 37, **249**, 369
Anthophora quadrimaculata (Four-banded Flower Bee) 31, 79, **80**, 87, **240**, 281, 414, 434, 492
Anthophora retusa (Potter Flower Bee) 37, 367, 464, **479**
Anthracite Bee-fly (Anthrax anthrax) 317, 320, **321**
Anthrax anthrax (Anthracite Bee-fly) 317, 320, **321**
Anthriscus sylvestris (Cow Parsley) 78
Apiaceae (parsley family) 196, 199, 200, 232, 474
Apidae 32, 33, **40**, 280–283
 characteristics of genera **34**, 37–40, 55–56, 241
 classification, taxonomic revisions 37
 cuckoo bee genera 32, 362
 genera/species number, latitudinal gradient 32, 416
 ground nesting 109, 117, **117**
 mouthpart structure 241, **242**
 pollen transport **21**, **255**, 256
 polylectic and oligolectic species 269, 270, 271, 280–283
 social behaviour 159–160
Apis see honeybees (Apis)
Apis mellifera (Western Honey Bee) **6**, 37, **40**, 229, 244, 256
apoid wasps 2, **3**, 406, **407**
Apoidea 2, **3**, 406, 408
Argiope bruennichi (Wasp Spider) 351–352

Armed Nomad Bee (Nomada armata) **376**, 378
Armeria maritima (Thrift) 184, **186**, 190
Arthropods (Arthropoda) 13–14
Ash (Fraxinus excelsior) 154
Ashy Furrow Bee (Lasioglossum sexnotatum) 293, 413, 447
Ashy Mining Bee see Andrena cineraria (Ashy Mining Bee)
Asteraceae 196, 198, **198**, 200, 204, 268, 291, 468
Andrena pollen generalists avoiding 290, 291
floral rewards and bee evolution 268–269
flower closure 200, 204, **205**, 206
flower fidelity of solitary bees 220
pollen grain structure **180**
pollen indigestibility 267, 268
pollen rationing 198
pollen toxicity 264, 265, 267, 273, 290
pollination method 198–199, **199**, 206–207, 257, 283, 289
polylectic bees foraging on 264, 267, 283
specialist bees 264, 267, 289, 290, 445, 501
Andrena 220, 264, 290, 291, **291**, 292
Colletes 264, 299
Dasypoda **206**, 206–207, 273
Eucera 282
Heriades truncorum 267, 289
larvae fed on non-host pollen 267
Melitta 276
Osmia 265, **295**, 340, 399, 441
Panurgus 257, 269, 296–297
Sapyga wasps parasitising and 265, 340
Asteroideae 198, 264, 299
Augochlorini 173, 176

Ballota nigra (Black Horehound) 79, 246
Banded Dark Bee (Stelis punctulatissima) **46**, **398**, 399
Bare-saddled Blood Bee (Sphecodes ephippius) 390
Bare-saddled Colletes (Colletes similis) 299, 315, **315**
base-banded furrow bees see Lasioglossum (base-banded furrow bees)
Beaked Hawksbeard (Crepis vesicaria) **205**, 206
Bear-clawed Nomad Bee (Nomada baccata) **376**, 480
bee-flies (Bombyliidae) 10, 185, 305, 317–322
 egg-laying, larvae 317, 318, **318**, 319, 320, 322
 flight 317, 320
 impact on solitary bee populations 321
 species, and host species 319, 320
Bee Orchid (Ophrys apifera) 211

Bee Wolf (Philanthus triangulum) 344–345, 355
beetles (Coleoptera) 333–336, 337–338, 349, 360
Bellflower Blunthorn Bee see Melitta haemorrhoidalis (Bellflower Blunthorn Bee)
Bellis perennis (Daisy) 196
Big-headed Mining Bee (Andrena bucephala) 74, 104, 159, **160**, 489, 495
Bilberry Bumblebee (Bombus monticola) 487
Bilberry Mining Bee (Andrena lapponica) 470–471, 487, **490**
Bilberry Nomad Bee (Nomada glabella) 377, 487
Black-headed Leafcutter Bee (Megachile circumcincta) 93
Black-headed Mining Bee (Andrena nigriceps) **215**, 479, 480, 487, **488**
Black Horehound (Ballota nigra) 79, 246
Black-horned Nomad Bee (Nomada rufipes) 367, 373, 378
Black Mining Bee (Andrena pilipes) 118, **181**
Black Oil Beetle (Meloe proscarabaeus) 334, **334**, 336, 337
Black-thighed Epeolus (Epeolus variegatus) **315**, 363, 364, **365**, 366, 454, 474
Blackthorn Mining Bee (Andrena varians) 516
blister beetle (Stenoria analis) 336, 338
blood bees see Sphecodes (blood bees)
Bloomed Furrow Bee (Lasioglossum albipes) 162, 164, 170, 171, 175, **359**, 415
Blue Mason Bee see Osmia caerulescens (Blue Mason Bee)
Bluebells (Hyacinthoides non-scripta) 472
blunthorn bees see Melitta (blunthorn bees)
Blunthorn Nomad Bee (Nomada flavopicta) 36, 382
Bombus see bumblebees (Bombus)
Bombus hortorum (Garden Bumblebee) 185, 229, 298
Bombus lapidarius (Red-tailed Bumblebee) **21**, **40**
Bombus lapponicus 84
Bombus lucorum (White-tailed Bumblebee) 244, 298
Bombus monticola (Bilberry Bumblebee) 487
Bombus muscorum 474
Bombus pascuorum (Common Carder Bumblebee) 10, **11**, 229
Bombus terrestris (Buff-tailed Bumblebee) **187**, 509
Bombyliidae see bee-flies (Bombyliidae)
Bombylius 305
Bombylius canescens 317, 320

Bombylius discolor (Dotted Bee-fly) 317, 320, **321**, 322
Bombylius major (Dark-edged Bee-fly) 10–11, **11**, 317, 318, 319, **319**, 320, 320, 322
Bombylius minor (Heath Bee-fly) 317, 320, **321**
Boraginaceae 196, 202, 248
bramble(s) (*Rubus*) 155, 189, 301, 354, 474, 518
Bramble (*Rubus fruticosus*) 155, 301
Bramble (*Rubus ulmifolius*) **189**
Brassica napus (Oilseed Rape) 216, 262, 483, 515
Brassicaceae 193, **193**, 354
bristle-headed bees (*Rophites*) 33, 49, 51, 57, 281
Bristly Oxtongue (*Helminthotheca echioides*) 273
Broad-banded Nomad Bee (*Nomada signata*) 495
Broad-margined Mining Bee (*Andrena synadelpha*) 84
Bronze Furrow Bee (*Halictus tumulorum*) 298, **298**, 358
Broom (*Cytisus scoparius*) 195, 467
Bryonia alba 260, 293
Bryonia dioica (White Bryony) 31, 76, 76, 183, 260, 293, **294**, 500
Bryony Mining Bee *see Andrena florea* (Bryony Mining Bee)
Buff-banded Mining Bee (*Andrena simillima*) **48**
Buff-tailed Bumblebee (*Bombus terrestris*) **187**, 509
Buff-tailed Mining Bee (*Andrena humilis*) **6**, 206, **206**, 291, 432, 495, **496**
Buffish Mining Bee *see Andrena nigroaenea* (Buffish Mining Bee)
Buffish Mining Bee, subspecies (*Andrena nigroaenea sarnia*) 453
Bugle (*Ajuga reptans*) 154, 237, **248**, 472
Bulbous Buttercup (*Ranunculus bulbosus*) 208, **209**
Bull-headed Furrow Bee (*Lasioglossum zonulum*) 438, 479
bumblebees (*Bombus*) 3–4, 32, 39, 55, 81, 176
anatomy/structure 3–4, 259
buzz of 249
cuckoo 4, 362, 401
diversity and latitudinal gradient 415
eusocial behaviour 161, 176
flower types associated 200, 226, 298
insecticide impact 512, 513, 514
landscape (spatial scale), habitats 482
nectar for brood cells 252–253
non-resource-based male patrolling 83–84
pollen size limits for 200
as pollinators, of crops/fruit trees 509, 516
predators, spiders 352
short-/long-term memory 218–219

solitary bees *vs* 3–4, 4, 10–11
species in Britain *vs* Europe 420
Bush Vetch (*Vicia sepium*) **227**
buttercup (*Ranunculus*) *see Ranunculus* (buttercup)
Butterwort (*Pinguicula vulgaris*) **209**

Cacoxenus fly 325–328, **327**, 328, **328**, 329
Cacoxenus indagator 325–326, **328**, 431
Calluna vulgaris (heather, Ling) 32, 66, 131, 295, **295**, **300**, 378, 437, 465
Caltha palustris (Marsh Marigold) 228
Calystegia sylvatica (Greater Bindweed) 133
Campanula 87, 261, 269, 280, **298**
pollination mechanism 192, 278, 286–287
Red Helleborine flowers mimicking 212
reproductive cycle, phases 278, 287
scent, 'signature' components 87, 279
specialist bees 31, 191–192, 212, 261, 269
Chelostoma species **286**, 286–287, **287**
Melitta haemorrhoidalis 277, 278, 279, 280, **280**, 288
pollen toxicity overcome 267, 269
Campanula poscharskyana (Trailing Bellflower) 286, 287
Campanula rotundifolia (Harebells) 192, **192**, 278, 280
Campanulaceae 191–193, **192**, 225, 286
carpenter bees *see Xylocopa* (carpenter bees)
Carrot Mining Bee (*Andrena nitidiuscula*) 479
catmint (*Nepeta*) 423, **423**, **433**
Catsear (*Hypochaeris radicata*) 123, **291**, 433, **434**
Catsear Nomad Bee (*Nomada integra*) 487, **488**
Celandines (*Ficaria verna*) 204, 228
Celonites 281
Centaurea (species) 198, **199**
Centaurea nigra (Common Knapweed) 199, 283
Centaurea scabiosa (Greater Knapweed) **199**
Centranthus ruber (Red Valerian) 493
Cephalanthera rubra (Red Helleborine Orchid) 211–212
Ceratina (small carpenter bees) 32, **34**, 38, 56, 160, **255**, 256, 348
flower types associated 226
nests/nesting behaviour 38, 160
pollen transport 38, **255**, 256
polylectic 269, 270, 271
species number, latitudinal gradient 416
Ceratina cyanea (Little Carpenter Bee) 38, **40**, 107, 110, 155, 160, 256, 438
Cerceris 309
Cerceris arenaria (Sand Tailed Digger Wasp) **9**

Cerceris rybyensis (Ornate Bee Fox) 51, 350, 355–357, **356**, **359**
prey (solitary bee) species **356**, **357**, 357–358
stinging of prey 355, 358, **359**
Chaetodactylus osmiae 347
Chalcidoidea 305, 345–346
Chalk Furrow Bee (*Lasioglossum fulvicorne*) 170, **183**, 220, 387
Chalk Yellow-face Bee (*Hylaeus dilatatus*) 153
Chamaenerion angustifolium (Rosebay Willowherb) **183**
Cheilosia albitarsis 230
Chelostoma (scissor bees) 32, **34**, 42, 57, 257
characteristics/morphology 57, 286–287
cleptoparasites of **339**, 339–340
flower types associated 226, 286, 287, 288
foraging behaviour **286**, 286–287, 288, **288**, 340
mandibles **286**, 287, **287**, 288, **288**
oligolectic 269, 270, 271, 286–289
species number, latitudinal gradient 416
Chelostoma campanularum (Small Scissor Bee) 42, **46**, 153, 279, **280**, 286, **286**, **287**, 287–288, 464
flower types (*Campanula*) 31, 192, 212, 279, **280**, 286, **287**
Chelostoma florisomne (Large Scissor Bee) 42, 230, **268**, 288, **288**, 339, **340**, 438
Chelostoma fuliginosum 212
Chelostoma rapunculi 267, 279
Chocolate Mining Bee (*Andrena scotica*) 74, **85**, 104, 159, 311–312, 319, 323, 437, **453**
Chrysididae (jewel wasps) 305, 343–345, **344**, 409
Chrysura hirsuta 344
Chrysura radians 344, **344**
Cichorioideae 198, 204
Cicindella campestris (Green Tiger Beetle) 360
Cirsium (species) (thistles) 232
Cirsium arvense (Creeping Thistle) 301, 474
Cirsium vulgare (Spear Thistle) 79, 283, **506**
Clarke's Mining Bee *see Andrena clarkella* (Clarke's Mining Bee)
Cliff Mason Bee (*Osmia xanthomelana*) 110, 441, 445
Cliff Mining Bee *see Andrena thoracica* (Cliff Mining Bee)
Clover Blunthorn Bee (*Melitta leporina*) **21**, 35, 115, 277, 432, **433**, 464, **496**
clovers (*Trifolium* species) 186, 474
Coast Leafcutter Bee (*Megachile maritima*) **19**, 80, **80**, 93, **94**, 207, 210, 474, **475**, 479
Coelioxys (sharp-tail bees) 32, **34**, 38, 42–43, **57**, 257, 390–394, 401

characteristics (structure/behaviour) 42–43, 57, 392–393, 402–404
host species/recognition 362, 391, 392, 401
life cycle 392–393, 393, 394
males 391–392, 393
nest detection and behaviour 391, 392, 393, 394
number of species 362, 392, 416
scarce species and their hosts 400
thermogenesis 434
Coelioxys conoidea (Large Sharp-tail Bee) 46, 474, 479
Coelioxys inermis (Shiny-vented Sharp-tail Bee) 42, 391, 392, 393
Coelioxys mandibularis (Square-jawed Sharp-tail Bee) 391, 478
Coelioxys quadridentata (Grooved Sharp-tail Bee) 414, 492
Coleoptera (beetles) 333–336, 337–338, 349, 360
Colletes (plasterer bees) 30, 31, 34, 53–54, 58, 401
anatomy/morphology/ characteristics 58, 131, 300, 401
brood cells 53, 128, 129, 129, 130–131, 157, 366
cleptoparasites of *see Epeolus* (variegated cuckoo bees)
emergence timing 30, 31, 65, 66, 156
evolution 128
flower types associated 30, 226, 290
foraging behaviour 299–301, 300
larvae, toxins in pollen and 264
life cycle, and larvae 156–157
losses/gains over recent decades 446
males patrolling emergence sites 65–68, 66, 67
nesting behaviour 129–131, 130, 131, 132, 273
pollen transport 21, 255, 258
polylectic or oligolectic species 264, 269, 270, 272, 290, 299–301
shadow flies parasitic on 306, 312
species number, latitudinal gradient 416
Colletes cunicularius (Early Colletes) 54, 67, 98, 132, 156, 157, 299–300, 421, 425
cleptoparasites (*Nomada*) 381
distribution/recently established in UK 421, 449, 452
female scent/sex pheromones 62, 88, 89
habitats 478, 480, 489, 490
life cycle, and brood cells 156, 157
mating behaviour 65–66, 67, 88, 89, 97–98, 98
nests, aggregations 65–66, 126, 127, 129, 131, 132
overwintering 438
pollen collection 131, 258, 299–300
Colletes cunicularius celticus 421
Colletes daviesanus (Davies' Colletes) 299, 366

Colletes floralis (Northern Colletes) 299–300, 415, 418, 480
Colletes fodiens (Hairy-saddled Colletes) 54, 299, 299, 364, 365, 506
Colletes halophilus (Sea Aster Bee) 31, 68, 127, 130, 132, 199, 300, 430
brood cells, and sex/number 109, 129, 314
cleptoparasites of 366
distribution and habitats 31, 301, 312, 474, 504
emergence, timing 66
life cycle 156–157, 157–158
mating behaviour 66–67, 67, 68, 127
Miltogramma parasitising 312–313, 313–314, 314, 315
nests/nesting behaviour 126, 130, 312–313, 430, 432
pollen collection and transport 31, 31, 132, 199, 199, 300
pollen specialists, flower types 156–157, 199, 299, 300, 300
wasp spiders as predators 352
Colletes hederae (Ivy Bee) 32, 54, 66, 104, 112, 126, 127, 128, 156–157, 299–300, 502
as an autumn bee 66, 104, 450
brood cells and life cycle 157
cleptoparasites of 336, 338, 364, 366, 454
distribution/recently established 32, 364, 449, 450, 451, 453, 454
emergence timing 364, 366
nectar/pollen harvesting, mechanism 243, 243, 244, 245
nesting, nest aggregations 112, 126–127, 127, 128, 129, 454
pollen specialists (oligolectic) 156–157, 232, 243, 300
pollen transport 131
polylectic with strong preference 300
Colletes kincaidii 130, 366
Colletes marginatus (Margined Colletes) 54, 299–300, 364, 364, 414, 480
Colletes similis (Bare-saddled Colletes) 299, 315, 315
Colletes succinctus (Heather Colletes) 53, 156–157, 299–300, 300, 311, 316, 467
brood cells, life cycle 53
cleptoparasite of (*Epeolus cruciger*) 364, 364, 434
distribution, and habitat 415, 467
emergence, and flight period 66, 364, 365, 366
flower types 31–32, 66, 131, 156, 300
nest aggregations, tumuli 126, 128, 311
parasites of 310, 311, 313, 315, 316, 320, 321
pollen collection 131, 156–157, 300
polylectic, or oligolectic 299, 300
predators of (birds) 350
Colletidae 21, 22, 33, 53–54, 156, 299–301, 401

as ancestral group, evolution 408
characteristics of genera 34, 53–54, 54, 58
flight period 156, 300, 467
genera/species number, latitudinal gradient 33, 416
mouthpart structure (and 'brushes') 241, 243, 243
nesting habits 109, 110, 126–131, 129, 148–153
pollen transport 255, 258
polylectic or oligolectic species 269, 270, 272, 299–301
comfrey (*Symphytum*) 202
Common Birds-foot-trefoil (*Lotus corniculatus*) 186, 194, 194–195, 202, 202, 220, 251, 283, 442, 443, 474
Common Carder Bumblebee (*Bombus pascuorum*) 10, 11, 229
Common Catsear (*Hypochaeris radicata*) 123, 291, 433, 434
Common Fleabane (*Pulicaria dysenterica*) 245, 289, 298
Common Furrow Bee (*Lasioglossum calceatum*) 29, 29, 162, 170, 171, 268, 358, 359, 432
Common Knapweed (*Centaurea nigra*) 199, 283
Common Mourning Bee *see Melecta albifrons* (Common Mourning Bee)
Common Poppy (*Papaver rhoeas*) 186, 187, 187, 249
Common Primrose (*Primula vulgaris*) 184, 184, 185, 216, 228
Common Ragwort (*Senecio jacobaea*) 283
Common Reed (*Phragmites australis*) 150
Common Sainfoin (*Onobrychys viciifolia*) 432
Common Tansy (*Tanacetum vulgare*) 315
Common Yellow-face Bee (*Hylaeus communis*) 98, 98, 148–149, 149, 153, 197, 287
conopid flies (fat-headed flies) 305, 322–323, 323, 324, 325
Cornus sanguinea (Dogwood) 232
Cow Parsley (*Anthriscus sylvestris*) 78
Cowslip (*Primula veris*) 216, 249
Crabronidae 406
Crambe maritima (Sea Kale) 181
Crataegus monogyna (Hawthorn) 232, 413
Creeping Buttercup (*Ranunculus repens*) 230, 268
Creeping Cinquefoil (*Potentilla reptans*) 301
Creeping Thistle (*Cirsium arvense*) 301, 474
Crepis capillaris (Smooth Hawksbear) 204
Crepis vesicaria (Beaked Hawksbeard) 205, 206
Cretotrigona (fossil bee) 409

SPECIES INDEX · 573

Cruciferae 230
Ctenocolletes bees 348
Ctenoplectrella viridiceps 406
cuckoo bees see cuckoo bees (solitary bees) (General Index)
Cycads 409
Cymbalaria muralis (Ivy-leaved Toadflax) 493
Cypripedium calceolus (Lady's Slipper Orchid) 212–213, **213**
Cytisus scoparius (Broom) 195, 467

Dactlyorhiza paetermissa (Southern Marsh Orchid) 210
daffodil (Narcissus longispathus) 207
Daisy (Bellis perennis) 196
dandelions (Taraxacum) 189, 196, 228
dark bees (Stelis) 33, **34**, 45, 57, 257
Dark-edged Bee-fly (Bombylius major) 10–11, **11**, 317, 318, 319, **319**, 320, 320, 322
Dasypoda (pantaloon bees) 32, **34**, 35, 55, 273
 characteristics **34**, 55, 273
 flower types, and foraging strategy 226, 273
 nests/burrows 113, 115, **115**, 273, **274**
 oligolectic species 206, 271, 273
 pollen transport **206**, **255**, 273, **274**
 species number, latitudinal gradient 417
Dasypoda hirtipes (Pantaloon Bee) 35, 36, 89, 206, **206**, 207, **274**, 496
 habitats 474, **496**
 hairs on hind legs 115, **255**, **274**
 Miltogramma parasitising 35, 312, **316**, 317
 nests (burrows) 113, 115, **115**, 274
 oligolectic 206, 273
 pollen collection/transport **206**, 207, **255**, **274**
Daucus carota (Wild Carrot) 196, 301, **353**
Davies' Colletes (Colletes daviesanus) 299, 366
dead-nettles see Lamiaceae (dead-nettles)
Devil's-bit Scabious (Succisa pratensis) 200, 203, **203**, 292
Dialictus (sub-genus) 170, 173
Dipsacaceae (teazel family) 196, 199–200, 201
Diptera see flies (Diptera)
Dog Rose (Rosa canina) 518
Dog Violet (Viola canina) **209**
Dogwood (Cornus sanguinea) 232
Dotted Bee-fly (Bombylius discolor) 317, 320, **321**, 322
Downland Furrow Bee (Halictus eurygnathus) 449, 452, 464
Drosophila 326
Drosophila indagator 327
Dufourea (short-faced bees) 33, 49, 57, 416
Dusky-horned Nomad Bee (Nomada bifasciata) 449, 452

Early Colletes see Colletes cunicularius (Early Colletes)
Early Nomad Bee (Nomada leucophthalma) 379, **380**, 381
Early Spider Orchid (Ophrys sphegodes) 211, **212**
Echium 263, 269
Echium vulgare (Viper's Bugloss) 73, 196, **197**, 203–204, 267, **285**, 285–286
Elder (Sambuca nigra) 133
end-banded furrow bees see Halictus (end-banded furrow bees)
Epeoloides coecutiens 366
Epeolus (variegated cuckoo bees) 32, **34**, 38, 53, 56, 363–367
 characteristics (structure/behaviour) 38, 56, 402–404
 egg-laying, and larvae 366, 366–367
 species number, latitude gradient 362, 363, 416
Epeolus compactus 366
Epeolus cruciger (Red-thighed Epeolus) 40, 363, **363**, 434
 distribution 364, 454
 egg-laying anatomy 367
 emergence timing/flight period 364, **365**
 host species **362**, 363, 364, 364, 454
Epeolus fallax 366
Epeolus variegatus (Black-thighed Epeolus) **315**, 363, 364, **365**, 366, 454, 474
Ericaceae 292
Eristalis intricarius (Furry Dronefly) 10
Eristalis tenax 244
Erysimum mediohispanicum 230
Eucera (long-horned bees) 32, **34**, 38, 56 characteristics/morphology 38, 56, 283
 flower types associated 226, 282, 283
 hovering without landing 218
 nectar/pollen foraging together 38, **252**, 283
 oligolectic 269, 270, 271, 282–283
 pollen transport **255**, 256
 species number, latitudinal gradient 416
Eucera longicornis (Long-horned Bee) 40, **227**, **252**, 282, 283, 477, 495, **496**
 flower types 77, **227**, 282–283
 habitat 464, 474, **477**, 495, **496**
 males patrolling 76–77
 nomad bees targeting 373, 400
 oligolectic 282
 pollen-nectar mix 252, 253, **282**
Eucera nigrescens (Tuberculate Long-horned Bee) 282
euglossines (neotropical orchid bees) 63
eumenid wasps 336
European Orchard Bee see Osmia cornuta (European Orchard Bee)
European Paper Wasp (Polistes dominula) 333

Evening Primrose (Oenothera drummondii) 239, 249
Evylaeus (sub-genus) 170, 171, 173

Fabaceae 193, **194**, 194–195, 219, 220, 222, 239, 292, 464, 474
 Andrena species pollinating 219, 292
 bilateral symmetry **194**, 195, 250–251
 Eucera pollinating 282, 283
 Osmia parietina pollinating **194**, 194–195
Fabricius' Nomad Bee (Nomada fabriciana) 373, 376, 377
fat-headed flies (conopid flies) 305, 322–323, **323**, 324
Ficaria verna (Celandine) 204, 228
Field Maple 85
Field Scabious (Knautia arvensis) **51**, 197, 200, 201, **201**, 232, 292, 378, 482
Field Sowthistle (Sonchus arvensis) **205**, 206, 207
figworts (Scrophulariaceae) 222, 245, 280, 281
Five-spined Rophites (Rophites quinquespinosus) 51, **52**
Flame-shouldered Blister Beetle (Sitaris muralis) 336, 355
Flat-ridged Nomad Bee (Nomada obtusifrons) 487
Flavous Nomad Bee (Nomada flava) 373, **376**
flesh fly (Miltogramma punctata) 53, 312–314, 313, **314**, 315, **315**, **316**
flies (Diptera) 181, 228, 305, 306–328
 bee-flies see bee-flies
 bees vs 10
 Cacoxenus (Drosophilidae) 325–328
 conopid see conopid flies (fat-headed flies)
 shadow flies see shadow flies
flower bees see Anthophora (flower bees)
Forget-me-not (Myosotis sylvatica) **203**
Fork-jawed Nomad Bee (Nomada ruficornis) 377, **378**, 488
Fork-tailed Flower Bee (Anthophora furcata) 38, 87, 110, 117, 154, 245–246, **247**, 249, 281, 423, 488
Four-banded Flower Bee (Anthophora quadrimaculata) 31, 79, **80**, 87, **240**, 281, 414, 434, 492
Four-spotted Furrow Bee (Lasioglossum quadrinotatum) 348
Fox and Cubs (Pilosella aurantiaca) 289
Fragaria vesca (Wild Strawberry) 154
Fraxinus excelsior (Ash) 154
Fringe-horned Mason Bee see Osmia pilicornis (Fringe-horned Mason Bee)
Fringed Furrow Bee (Lasioglossum sexstrigatum) 449, 451–452
furrow bees see Halictus (end-banded furrow bees); Lasioglossum (base-banded furrow bees)
Furry-bellied Blood Bee (Sphecodes hyalinatus) 387, 488
Furry Dronefly (Eristalis intricarius) 10

574 · SOLITARY BEES

Garden Bumblebee (*Bombus hortorum*) 185, 229, 298
Gasterosteus aculeatus (Stickleback fish) 60
Gasteruptiidae, wasps 305, 341–342, **342**
Gasteruption assectator (Small Javelin wasp) 341
Gasteruption jaculator (Javelin Wasp) 341, 342, **342**
Geranium, species 78, **188**
Geranium molle (Dove's-foot Cranesbill) **188**
Geranium pratense (Meadow Cranesbill) 186, **188**
Geranium pusillum (Small-flowered Cranesbill) 172, **188**
Geranium pyranaicum (Hedgerow Cranesbill) **188**
Geranium robertianum (Herb Robert) 179, **179**, 183
Geranium x oxonianum 238, 301
German Wasp (*Vespula germanica*) 317
Germander Speedwell (*Veronica chamaedrys*) 191, **191**, 378, **379**
Giant Blood Bee (*Sphecodes albilabris*) 449, **451**, 452
Glechoma hederacea (Ground Ivy) 154, **196**, 228, **281**, 472
Gold-fringed Mason Bee *see Osmia aurulenta* (Gold-fringed Mason Bee)
Gooden's Nomad Bee (*Nomada goodeniana*) 9, 384, **384**
Gorse (*Ulex europaeus*) 6, 195, 467
Greater Bindweed (*Calystegia sylvatica*) 133
Greater Burdock (*Actium lappa*) 283
Greater Knapweed (*Centaurea scabiosa*) 199
Green-eyed Flower Bee *see Anthophora bimaculata* (Green-eyed Flower Bee)
Green-fanged Tube Web Spider (*Segestria florentina*) 351, **351**
Green Furrow Bee (*Lasioglossum morio*) 170, 185, **191**, 287, 293, 298, **298**, 387
Green Tiger Beetle (*Cicindella campestris*) 360
Grey-backed Mining Bee *see Andrena vaga* (Grey-backed Mining Bee)
Grey-gastered Mining Bee (*Andrena tibialis*) 84, **382**, 497
Grey-patched Mining Bee (*Andrena nitida*) **18**, 47, **48**, 78, 84, **85**, 253, 254, 437, 468
Grey-tailed Furrow Bee (*Lasioglossum prasinum*) 413
Grooved Sharp-tail Bee (*Coelioxys quadridentata*) 414, 492
Ground Ivy (*Glechoma hederacea*) 154, **196**, 228, **281**, 472
Gwynne's Mining Bee *see Andrena bicolor* (Gwynne's Mining Bee)

Habropoda pallida 336

Hairy-footed Flower Bee *see Anthophora plumipes* (Hairy-footed Flower Bee)
Hairy-saddled Colletes (*Colletes fodiens*) 54, 299, **299**, 364, **365**, 506
Hairy Yellow-face Bee (*Hylaeus hyalinatus*) 153, **252**, 301
Halictidae 2, 17, 33, 49–53, 175, 297–299
characteristics 29, **34**, 49–53, **57**, 241, **242**
emergence, and flight periods 29, 297, 302
eusocial *see* eusocial behaviour (*General Index*)
foraging behaviour **298**, 298–299
genera and species number 33, 416
life cycle, phases 161–162, 297
nests/nesting behaviour 109, 125–126, **126**, 171–172
oligolectic or polylectic species 269, 272, 298
parasites of 312, 333
phylogeny (evolutionary history) 169–173
pollen transport **21**, **255**, 258
prey species of *Cerceris rybyensis* 356, **357**, 357–358
species number, latitudinal gradient 416
see also individual genera (as per page 33)
Halictinae 173, 176
Halictoxenus 333
Halictus (end-banded furrow bees) **34**, 50, 51, **51**, 57, 161, 297
characteristics and morphology 50, 57
eusocial mode of life 50, 111, 161–162, 163, 176, 297, 385
flower types associated 226
habitats 468
life cycle 297, 435
nesting 50, 161
parasites/cleptoparasites 50, 333, 342, 385
pollen transport 50, **255**
polylectic species 269, 270, 272
predators, spiders 352
species number, latitudinal gradient 416
Halictus albipes **359**
Halictus eurygnathus (Downland Furrow Bee) 449, 452, 464
Halictus farinosa 309
Halictus ligatus 325
Halictus maculatus (Square-headed Furrow Bee) 447
Halictus petrefactus **169**
Halictus rubicundus (Orange-legged Furrow Bee) 163, **164**, **202**, **356**, 438
'facultative' eusocial behaviour 162, 163–164, 176, 438
life cycle options 163–164, 438
N. American *vs* British populations 163–164

nectar/pollen collection 202, **202**
predators and parasites of 324, 358
Halictus tumulorum (Bronze Furrow Bee) 298, **298**, 358
Harebells (*Campanula rotundifolia*) 192, **192**, 278, 280
Hawksbeard Mining Bee (*Andrena fulvago*) 207, 220
hawkweeds (*Hieracium*) 133, 189
Hawthorn (*Crataegus monogyna*) 232, 413
Hawthorn Mining Bee (*Andrena chrysosceles*) 332, **332**, 437
Heath Bee-fly (*Bombylius minor*) 317, 320, **321**
Heath Sand Wasp (*Ammophila pubescens*) 102
heather (*Calluna vulgaris*) 32, 66, 131, 295, **295**, **300**, 378, 437, 465
Heather Colletes *see Colletes succinctus* (Heather Colletes)
Heather Mining Bee (*Andrena fuscipes*) 76, 84, **295**, 295–296, 378, 437, 467, 470, **490**
Hedera helix (Ivy) 66, 131, 232, 243, 244
Hedgerow Cranesbill (*Geranium pyranaicum*) **188**
Hedychrum rutilans 345
Helianthemum nummularium (Rock Rose) 190, 208, **209**, 219, 220, **220**
Helminthotheca echioides (Bristly Oxtongue) 273
Heracleum sphondylium (Hogweed) 196, 232
Herb Robert (*Geranium robertianum*) 179, **179**, 183
Heriades (resin bees) 33, **34**, 43, 57, 257
characteristics 43, **57**
competition between females 147
'cuckooing' attempts 147–148
flower types associated 226, 289
foraging behaviour 289, **289**
oligolecty 257, 269, 270, 272, 289, **289**
resin in nest construction 110–111, **145**, 145–147, **146**
species number, latitudinal gradient 416
Heriades rubicola (Small-headed Resin Bee) 43, 153, 448, 449
Heriades truncorum (Large-headed Resin Bee) 43, **46**, **96**, 267, **289**, 342, **398**, 448
characteristics 43
cleptoparasites of 395, 397, **398**
foraging behaviour 238, 267, 289, **289**
in gardens 238, 502
larvae, and life cycle 158, **158**
mating behaviour 73, 95–96, **96**
nesting 43, 110–111, **145**, 145–147, **146**, 158, 341
oligolecty 289
parasites of **342**, 344
pollen collection, and 'vibrations' 30, 43, 267, 289

SPECIES INDEX · 575

range and number increase 446, 448, 502
Hieracium (hawkweeds) 133, 189
Hippocrepis comosa (Horseshoe Vetch) 220
Hogweed (*Heracleum sphondylium*) 196, 232
Holly (*Ilex aquifolium*) 84, 183
Hollyhock (*Alcea rosea*) 200
honeybees (*Apis*) 3–4, **6**, 32, 55, 81, 352
anatomy/structure/characteristics 3–4, 5, **5**, 23, 55
commercially managed 514
domesticated, out-competing wild bees 462
eusocial behaviour 161
flower types associated 220, 221, 226, 228, 259
insecticide impact 512, 513, 514
lifespan, effect of 'wear and tear' 429–430
nectar for brood cells 228, 252–253
pollen size limits for 200
as pollinators, of crops 509, 515, 516
resource partitioning 221
short- and long-term memory 218–219
solitary bees *vs* 3–4, 4, **5**
species in Britain *vs* Europe 420
threats to (mites, colony collapse) 509
worker 3–4, 4, **5**, **6**, 429
Hoplandrena 159
Hoplitis (lesser mason bees) 33, **34**, 43, 57, 257
flower types associated 226, 285
foraging behaviour 285–286, **286**
limited pollen, size of offspring 108
oligolecty 269, 285–286, **286**
species number, latitudinal gradient 416
Hoplitis adunca (Viper's Bugloss Mason Bee) 73, 196, 267, **285**, 285–286, 344, 448, 449
Hoplitis anthocopoides 263
Hoplitis claviventris (Welted Lesser Mason Bee) **46**, 95, 395, **395**
Hoplitis leucomelana (Kirby's Lesser Mason Bee) **396**, 397
Horseshoe Vetch (*Hippocrepis comosa*) 220
Hot-lips Sage (*Salvia microphylla*) 298, **298**
Hyacinthoides non-scripta (Bluebells) 472
Hylaeus (yellow-face bees) 30, **34**, 53, 54, 58, 301
as ancestral bee (model) 408
cavity nesting 53, 54, 110, 132, 148–150, **149**
characteristics/morphology 21, 30, **34**, 54, 58, 148, 301
flower types associated 188, 226
foraging behaviour 54, 301
pollen collection 54, 202, 253, 301
pollen/nectar mixture 149, 253, 301

pollen transport 21, 54, 107, 149, 253, 258, 301
polylectic or oligolectic species 54, 269, 270, 272, 301
species number, latitudinal gradient 416
Hylaeus communis (Common Yellow-face Bee) 98, **98**, 148–149, **149**, 153, **197**, 287
Hylaeus dilatatus (Chalk Yellow-face Bee) 153
Hylaeus hyalinatus (Hairy Yellow-face Bee) 153, **252**, 301
Hylaeus pectoralis (Reed Yellow-face Bee) 72, 149–153, **151**, **152**, 156, 301, 341, 350, 473, **473**
Hylaeus pictipes (Little Yellow-face Bee) 12, **54**, 447
Hylaeus signatus (Large Yellow-face Bee) 21, **54**, 225, 301, **301**, 488, 520
Hymenoptera 2, **3**, 17–18, 61–62, 181, 406
parasitic on solitary bees 305, 339–346
see also specific groups (see page 305)
Hypericum calycinum (Rose of Sharon) 133
Hypochaeris radicata (Common Catsear) 123, **291**, 433, **434**

Ilex aquifolium (Holly) 84, 183
Impunctate Mini-miner (*Andrena subopaca*) **5**, 78, 191
Ivy (*Hedera helix*) 66, 131, 232, 243, 244
Ivy Bee (*Colletes hederae*) see *Colletes hederae* (Ivy Bee)
Ivy-leaved Toadflax (*Cymbalaria muralis*) 493

Jacobaea vulgaris (Ragwort) 43, 238, 289, **289**
Jasione montana (Sheep's-bit) 225
Javelin Wasp (*Gasteruption jaculator*) 341, 342, **342**
Jersey Mason Bee (*Osmia niveata*) 448, 449

Kirby's Lesser Mason Bee (*Hoplitis leucomelana*) **396**, 397
Kirby's Mining Bee (*Andrena afzeliella*) 47
Knapweeds 199, 232
Knautia arvensis (Field Scabious) **51**, 197, 200, 201, **201**, 232, 292, 378, 482

Lady's Slipper Orchid (*Cypripedium calceolus*) 212–213, **213**
Lamb's Ear (*Stachys byzantina*) 30, **41**, 81, **82**, 144, **144**, 246, **246**, **247**, 249
Lamiaceae (dead-nettles) 193, 195, **195**, **196**, 222, 249, 472
Anthophora pollenating **195**, **196**, 249, 280–281
nectaries 239, 245
Lamium album (White Dead-nettle) 195, **195**, 249
Lamium purpureum (Red Dead-nettle) 78, 195, **195**, **196**, 282, 369

Large Bear-clawed Nomad Bee (*Nomada alboguttata*) 449, 452
Large Gorse Mining Bee (*Andrena bimaculata*) 84, 293
Large-headed Resin Bee see *Heriades truncorum* (Large-headed Resin Bee)
large-horned bees (*Eucera*) 32, 38
Large Meadow Mining Bee (*Andrena labialis*) 474
Large Sallow Mining Bee (*Andrena apicata*) 381, 489
Large Scabious Mining Bee see *Andrena hattorfiana* (Large Scabious Mining Bee)
Large Scissor Bee (*Chelostoma florisomne*) 42, 230, **268**, 288, **288**, 339, **340**, 438
Large Shaggy Bee (*Panurgus banksianus*) **48**, 89, 123, **125**, 206, 207, **296**, 433, **434**
Large Sharp-tail Bee (*Coelioxys conoidea*) **46**, 474, 479
Large Yellow-face Bee (*Hylaeus signatus*) 21, **54**, 225, 301, **301**, 488, 520
Lasioglossum (base-banded furrow bees) 13, **34**, 50–51, 57, 161, 297
characteristics 49, 50–51, 57
cleptoparasites of 51, 168, 385
eusocial 51, 68, 162–163, 167, 173, 176, 297, 325, 385
evolutionary history 170, 171, 173, 297
flower types associated 188, 226
habitats 468
life cycle, and overwintering 51, 168, 297, 435
losses/gains over recent decades 446
mating behaviour 89, 98–99
nest aggregations 68, 125, **125**, 432, **494**
nesting habits 68, 69, **125**, 125–126, **126**, 165, 167, 171
oligolectic or polylectic species 269, 270, 272
parasites 51, 323, 324, 325, 333, 342
pollen collection 51, 202
predators, spiders 352
species number, latitudinal gradient 416
sub-genera 170–171, 173
Lasioglossum albipes (Bloomed Furrow Bee) 162, 164, 170, 171, 175, **359**, 415
Lasioglossum calceatum (Common Furrow Bee) 29, **99**, 162, 170, 171, 268, 358, **359**, 432
Lasioglossum cinctipes 325
Lasioglossum forbesii 325
Lasioglossum fratellum (Smooth-faced Furrow Bee) 438
Lasioglossum fulvicorne (Chalk Furrow Bee) 170, **183**, 220, 387
Lasioglossum leucopus (White-footed Furrow Bee) 287, **298**
Lasioglossum leucozonium (White-zoned Furrow Bee) **49**, 187, **187**, 220, **326**

Lasioglossum malachurum (Sharp-collared Furrow Bee) 69, **69**, 99, **165**, **166**, 170, 385, 474
brood cells 126, 171–172
cuckoo bees/nest parasitism 168, 169, 386
evolutionary history 171
habitat 474
males patrolling 68–69, **69**
nest aggregations 68, 125, **125**, **126**, 432
nest usurpation, 'alien' workers in 166–167
nests/nesting 68, **69**, 125, 125–126, 165, **165**, 386
'obligate' social behaviour 68, 125, 162, 165–169, 171, 174, 385
overwintering 69, **165**
pheromones, and mating 69, 89–90, 98–99
sociogenetic structure of nests 166, 167–168
worker bees 69, 165–166, **166**, 167, 172, 174, 386
Lasioglossum marginatum 170
Lasioglossum minutissimum (Least Furrow Bee) 387
Lasioglossum morio (Green Furrow Bee) 170, 185, **191**, 287, 293, 298, **298**, 387
Lasioglossum parvulum (Smooth-gastered Furrow Bee) 112, 333, 487
Lasioglossum pauperatum (Squat Furrow Bee) 474
Lasioglossum pauxillum (Lobe-spur Furrow Bee) **21**, 162, **168**, 168–169, 170, **170**, **171**, **172**, 187
nest, and 'turret' entrance **170**
obligatory eusocial species 162, 168, 169, 171, 175
resource partitioning 221
Lasioglossum prasinum (Grey-tailed Furrow Bee) 413
Lasioglossum puncticolle (Ridge-cheeked Furrow Bee) 474
Lasioglossum quadrinotatum (Four-spotted Furrow Bee) 348
Lasioglossum rufitarse (Rufous-footed Furrow Bee) 212–213, **213**
Lasioglossum sexnotatum (Ashy Furrow Bee) 293, 413, 447
Lasioglossum sexstrigatum (Fringed Furrow Bee) 449, 451–452
Lasioglossum sextrigatum 293
Lasioglossum villosulum (Shaggy Furrow Bee) 242
Lasioglossum xanthopus (Orange-footed Furrow Bee) 387
Lasioglossum zonulum (Bull-headed Furrow Bee) 438, 479
Lathbury's Nomad Bee (*Nomada lathburiana*) 382, **382**, **383**
Lathyrus pratensis (Meadow Vetchling) 76, **282**, 283, 474
leafcutter bees *see* Megachile (leafcutter bees)

Least Furrow Bee (*Lasioglossum minutissimum*) 387
Leontodon hispidus (Rough Hawkbit) 206, 220
lesser mason bees (*Hoplitis*) *see* Hoplitis (lesser mason bees)
Leucophora 305, 306, 307, **307**, 309, 310–311
bee defensive action against 309–310, 311
hosts and host specificity 309, 310–311
at nesting aggregations 306, 307, 311, 312
tracking bees and egg laying 306–307, **307**, **308**, 308–309
Leucophora cinerea 310–311
Leucophora grisella 53, 309, 310–311, 311, 315
Leucophora obtusa 307, **308**, 308–309, 310–311, 311
Leucophora personata 307, 309, 310–311, 311–312
Leucophora sericea 309, 310–311
Leucophora sponsa 309, 310–311
Linaria purpurea 249
Ling (*Calluna vulgaris*) 32, 66, 131, 295, **295**, **300**, 378, 437, 465
Lipara lucens (fly) 150, **151**, 153, 473
Liptotriches australica 432
Little Carpenter Bee (*Ceratina cyanea*) 38, **40**, 107, 110, 155, 160, 256, 438
Little Dark Bee (*Stelis breviuscula*) 153, 397, **398**, 448, 450
Little Nomad Bee (*Nomada flavoguttata*) 382
Little Sickle-jawed Blood Bee (*Sphecodes longulus*) 387
Little Yellow-face Bee (*Hylaeus pictipes*) **12**, **54**, 447
Lobe-spur Furrow Bee *see Lasioglossum pauxillum* (Lobe-spur Furrow Bee)
Long-fringed Mining Bee (*Andrena congruens*) 489, **491**
Long-horned Bee (*Eucera longicornis*) *see* Eucera longicornis (Long-horned Bee)
long-horned bees *see* Eucera (long-horned bees)
Lotus corniculatus (Common Birds-foot-trefoil) 186, **194**, 194–195, 202, **202**, 220, 251, 283, 442, 443, 474
lungworts (*Pulmonaria* species) 154, 248
Lysimachia 274
Lysimachia vulgaris (Yellow Loosestrife) 36, 77, 116, 225, 275, **276**, 474

Macropis (oil-collecting bees) **34**, 36, 55
characteristics/morphology 36, 55, 275
flower types associated 226, 274, 275
foraging behaviour 274–275
oil collection 105, 116, 274–275, **276**
oligolectic species 77, 271, 274–275, **275**

pollen collection 36, 275, **276**
species number, latitudinal gradient 417
Macropis europaea (Yellow Loosestrife Bee) 36, **36**, 77, 274–275, **276**, **473**
cleptoparasites of 366
habitats **473**, 474, 520
mating behaviour, female response 77, 77–78, 88–89
mutualism with flowers 225, 275
nesting behaviour 77, 116, **116**, 474
oligolectic 77, 225, 474
plant oil collection 105, 116, **116**, 225, 253, 275, 474
pollen and oil collection/mix 116, 253, 274–275, **276**
Maidstone Mining Bee (*Andrena polita*) 449, **450**
Malvaceae 200
Margined Colletes (*Colletes marginatus*) **54**, 299–300, 364, 364, 414, 480
Marsh Marigold (*Caltha palustris*) 228
mason bees *see* Osmia (mason bees)
Meadow Buttercup (*Ranunculus acris*) 220
Meadow Cranesbill (*Geranium pratense*) 186, **188**
Meadow Vetchling (*Lathyrus pratensis*) 77, **282**, 283, 474
Megachile (leafcutter bees) 1, 12, 33, 40, 43, 57
aerial/cavity nests 73, 132–134, 135, 341
buzz of 73, 249, 284
chalcid parasites of 345
characteristics/morphological features 34, **34**, 45, 57, 283–284
cleptoparasites of, Coelioxys 391–394
emergence, timing 30
flower types associated 226, 283, 284
foraging behaviour 250–251, 257, 283–284, **284**
ground nests 73, 134, 346, 393, 394
leaf cutting/holes 132–133, 135, 346
mating behaviour 73, 88, **89**, 93–95, **94**, **95**
parasite-repelling substance 346
pollen transport **21**, **255**, 257
polylectic and oligolectic species 269, 270, 272, 283
species number, latitudinal gradient 416
Megachile centuncularis (Patchwork Leafcutter Bee) **21**, 88, **89**, 94, 95, 134, 135, 187, **205**, 287, 391
Megachile circumcincta (Black-headed Leafcutter Bee) 93
Megachile leachella (Silvery Leafcutter Bee) **46**, **135**, **250**
basking 207
beetle predator 360
buzz, whilst foraging 249, 284
cleptoparasites, Coelioxys 391, 392, 400
foraging, flower types **250**, 250–251, 284, 360

SPECIES INDEX · 577

ground nesting 73, 133, 312, 351, 474
habitat and distribution 414, 474, 479
leaf segment collection 133, **135**, 351, 474
males harassing females 83, 94, 133
Miltogramma species parasitising 312
morphological features 94–95
Megachile ligniseca (Wood-carving Leafcutter Bee) 42, 44, 74, **95**, 135, 197, 256, 284, 393
 anatomical features 94, **95**, **256**, 284
 cavity nesting 110, 134, **135**, 393
 Coelioxys association 392, 393, **393**, 394
 leaf segment cutting 44, 134, **135**, 154
 male patrolling, mating behaviour 73–74, **74**, 94, **95**, 100
 nectar/pollen collection **197**, **256**, 283, **284**, 287
 as wood-borer 154
Megachile maritima (Coast Leafcutter Bee) 19, 80, **80**, 93, **94**, 207, **210**, 414, 474, **475**, 479
Megachile pluto (leafcutter bee) 12
Megachile pyrenaica 397
Megachile rotundata 345
Megachile willughbiella (Willughby's Leafcutter Bee) 44, 46, **86**, 95
 anatomical features 93–94, **95**
 flower types and pollen collection 283, 287
 leaf segment cutting 133, 134
 mating behaviour 80, **86**, 87, 93–94, **95**
 nests 134
Megachilidae 28–29, 32–33, 33, 283–289
 cavity nests (burrows) 107, 110, 132–148, 341
 characteristics of genera 34, 40–45, 46, 56–57
 cleptoparasite genera included 32, 33, 362, 394
 cocoons, spinning 157
 free-standing nests above ground 110
 leaf segment cutting 132–133, 135, 137
 mouthpart structure, long tongue 241, **242**
 oligolectic and polylectic species 269, 270, 272, 283–289
 parasites of, *Gasteruption* 341
 pollen collection 256, 283–289
 pollen transport/scopae 28–29, 245, 253, 256–257, 283–284
 species number, latitudinal gradient 416–417
Melecta (mourning bees) 32, **34**, 38–39, 56, 367–373
 characteristics **34**, 39, 56, 367, 369, **370**, 402–404
 cleptoparasitic 38–39, 367, 373. 370–371
 egg-laying and larval development 370–372, **372**, 373

emergence, timing and conditions 367–368
host species **362**, 367, 368, 400
mating behaviour 369–370, **370**
scarce species and their hosts 400
species **362**, 367
species number, latitudinal gradient 416
Melecta albifrons (Common Mourning Bee) **39**, 367, 368–369, **369**, 425, 521
 behaviour in host nest 370–371
 egg-laying, larval development 370–372, **372**, 373
 thermogenesis 434
Melecta luctuosa (Square-spotted Mourning Bee) 367, 400
Meliponini 412
Melitta (blunthorn bees) 32, 35–36, 55, 253, 270, 276–279, **278**
 as ancestral bee 408
 buzz of 249, 277
 characteristics/morphology **34**, 35, 55, 277, 279
 flower types associated 35–36, 226, 270, 277, 278, 279, 288
 foraging behaviour 277, **278**, 278–279
 ground nests 114–115
 losses/gains over recent decades 446
 mesolectic species 276, 278
 oligolectic species 271, 276, 277, 278, 279, 302
 pollen/nectar collection/mix 35, 107, 253, 277, 278–279
 pollen transport 21, **255**, 277, 279
 roosting strategy 432
 scent of flowers, 'signature' compounds 279
 species number, latitudinal gradient 417
Melitta dimidiata (Sainfoin Blunthorn Bee) 35, **36**, 114, 277, 432, 464
Melitta haemorrhoidalis (Bellflower Blunthorn Bee) **30**, **36**, 278–280, **280**
 anatomy/morphology 35
 flower types visited 31, 35, 192, 277, 278, 280, 464
 habitats 35, 464, 500, 520
 mesolectic 276, 278
 oligolectic on Campanulaceae 31, 277, 278, 500
 pollen collection, nectar mixed 278–279, **279**
Melitta leporina (Clover Blunthorn Bee) 21, **35**, 115, 277, 432, **433**, 464, **496**
Melitta tricincta (Red Bartsia Blunthorn Bee) 32, 35–36, **36**, 76, **76**, 277, **278**, 414
Melittidae 33, 35, 273–280
 as ancestral group, evolution 408
 characteristics of genera **34**, 35–36, 55
 cocoons, spinning 157
 genera and species number 32, 417
 ground nesting 109, 114–116, **115**
 mouthpart structure 35, 241, **242**

oligolectic species 269–270, **271**, 271–272, 273–280
pollen collection 35, **255**, 273–280
pollen mixed with nectar/oils 35, 107, 278–279
pollen specialism as fixed trait 35, 273
pollen transport 35, 255
species number, latitudinal gradient 417
Melittobia acasta 345, 346
Melittosphex burmensis 408–409
Mellitidia tomentifera 432
Meloe 334–335, 337
Meloe franciscanus 336
Meloe proscarabaeus (Black Oil Beetle) 334, **334**, 336, 337
Meloidae see Oil beetles (Meloidae)
Micrandrena 28, 47, 48
Mignonette (*Reseda lutea*) 225, 301
milkworts (*Polygala* species) 234
Miltogramma 305, 306, 313–314
 bee' defensive action against 315, **316**
 solitary bee species associated 313
 tracking bees, larvae deposition 307, 312, 313, **314**, 315
Miltogramma germari 312, 313, **316**, 317
Miltogramma punctata (flesh fly) 53, 312–314, 313, **314**, 315, **315**, 316
mining bees *see Andrena* (mining bees)
Misumena vatia 352, **353**, 355
Mites (Acarina) 346–348, **347**, **348**
Monodontomerus obscurus 345
Monosapyga clavicornis 339, **340**
Mountain Mason Bee *see Osmia inermis* (Mountain Mason Bee)
mourning bees *see Melecta* (mourning bees)
Mouse-ear Hawkweed (*Pilosella officinalis*) 220
Mutilla atra 342, 343, **343**
Mutilla europaea 342
Mutillidae (wasps) 305, 342–343, **343**
Myopa 322, **323**
Myopa buccata 323, **323**
Myopa hirsuta 323, **323**
Myosotis sylvatica (Forget-me-not) **203**

Narcissus longispathus (wild daffodil) 207
Nepeta (catmint) 423, **423**, **433**
Noble False Widow Spider (*Steatoda nobilis*) 351, **352**
Nomad bees (*Nomada*) *see Nomada* (nomad bees)
Nomada (nomad bees) 32, **34**, 39, 48, 56, 373–385
 antennae 381, **382**, 382–383, **383**, 384
 attraction to nests by sight, then scent 383, 384
 body size 377
 Bombylius major (bee-fly) parasitising 322
 characteristics (structure/behaviour) 9, **34**, 39, 56, 377, 402–404

chemical sense/signal/pheromones 381–382, 383, **383**, 384
colours 9, 39, 373, 377, 402
distribution and new arrivals in UK 449, 452
eggs, and larvae 384–385
emergence, timing 39, 376–377, **378**, 381
fanning behaviour, scents and 381, 383, 384, **384**
female **376**, 377, 381, 382, 383, 384–385
flower types, associated with hosts 378–379, **379**
host species 39, 48, **362**, 373, 374–375
Andrena 48, 373, 374, 377–381, **378**, **379**, **380**
losses/gains over recent decades 446
males, Batesian mimicry 377
mating behaviour 382, **383**
mutual adaptation (with host) 378, 379
nectar collection 378
nest excavation, egg-laying 379, **380**, 383, 384, 385
scarce species and their hosts 400
scent mimicry 63, 381, 382, 383
species, and number of 39, 362, 373, 374–375, 416
sub-genera 373
wasps, distinguishing 9, 39
Nomada alboguttata (Large Bear-clawed Nomad Bee) 449, 452
Nomada argentata (Silver-sided Nomad Bee) 378
Nomada armata (Armed Nomad Bee) **376**, 378
Nomada baccata (Bear-clawed Nomad Bee) **376**, 480
Nomada bifasciata (Dusky-horned Nomad Bee) 449, 452
Nomada fabriciana (Fabricius' Nomad Bee) 373, 376, 377
Nomada flava (Flavous Nomad Bee) 373, **376**
Nomada flavoguttata (Little Nomad Bee) 382
Nomada flavopicta (Blunthorn Nomad Bee) 36, 382
Nomada fucata (Painted Nomad Bee) 40, 376, **376**, 382
Nomada fulvicornis (Orange-horned Nomad Bee) **382**
Nomada glabella (Bilberry Nomad Bee) 377, 487
Nomada goodeniana (Gooden's Nomad Bee) **9**, 384, **384**
Nomada guttulata (Short-spined Nomad Bee) 191, 377, 378, **379**
Nomada integra (Catsear Nomad Bee) 487, **488**
Nomada lathburiana (Lathbury's Nomad Bee) 382, **382**, **383**
Nomada leucophthalma (Early Nomad Bee) 379, **380**, 381
Nomada obtusifrons (Flat-ridged Nomad Bee) 487

Nomada panzeri (Panzer's Nomad Bee) 311, 377
Nomada pseudops 383
Nomada ruficornis (Fork-jawed Nomad Bee) 377, **378**, 488
Nomada rufipes (Black-horned Nomad Bee) **367**, 373, 378
Nomada sheppardana (Sheppard's Nomad Bee) 51
Nomada signata (Broad-banded Nomad Bee) 495
Nomada zonata (Variable Nomad Bee) **376**, 449, 452
Northern Colletes (*Colletes floralis*) 299–300, 415, 418, 480
Northern Mining Bee (*Andrena ruficrus*) 415, 418, 470, 487

Oak Mining Bee (*Andrena ferox*) **48**, **64**, 74, 104, 159, 439, 472
Odontites vernus (Red Bartsia) 32, **76**, **76**, 277
Oenothera drummondii (Evening Primrose) 239, 249
oil beetles (Meloidae) 333–336, **334**, **335**, 337–338, 349
defence methods **335**, 335–336
oil-collecting bees *see Macropis* (oil-collecting bees)
Oilseed Rape (*Brassica napus*) 216, 262, 483, 515
Oligochlora eikworti 406
Onobrychys viciifolia (Common Sainfoin) 432
Ononis spinosa (Spiny Restharrow) **250**, 251, 284
Ophrys apifera (Bee Orchid) 211
Ophrys species (bee orchids) 211, 225
Ophrys sphegodes (Early Spider Orchid) 211, **212**
Orange-footed Furrow Bee (*Lasioglossum xanthopus*) 387
Orange-horned Nomad Bee (*Nomada fulvicornis*) **382**
Orange-legged Furrow Bee *see Halictus rubicundus* (Orange-legged Furrow Bee)
Orange-tailed Mining Bee (*Andrena haemorrhoa*) **28**, 84, **85**, 120, 213, **214**, 377, **378**, 478
Orange-vented Mason Bee *see Osmia leaiana* (Orange-vented Mason Bee)
orchids 63, 210, **210**
Ornate Bee Fox *see Cerceris rybyensis* (Ornate Bee Fox)
Osmia (mason bees) 16, 45, 57, 439, 440–441
as carpenters 17
cavity nesting 45, 132, 134–142, 439, 440–441
cocoons, and conversion rate for 106
distribution and habitat 439–445, 440–441
emergence, flight period 28–29, 30, 430, 440–441, 444

female *vs* male, appearances 19, 57
flower types, pollen sources 226, 285, 439, 440–441, 445
foraging behaviour 248, 257, 284–285, **285**
losses/gains over recent decades 446
males patrolling female emergence sites 73
morphological features 19, **34**, 45, 57, 248, 284–285
neonicotinoid effects 513
oligolectic species 257, 262, 265, 270, 284, 285
parasitism by Aculeate parasites 440–441
parasitism by chalcids 345
parasitism by mites 347, **347**
parasitism by *Sapyga* wasp 265, 339, **339**, 440
pollen transport **255**, 257
polylectic species 269, 270, 272, 284–285, 445
species number, latitudinal gradient 417
Osmia aurulenta (Gold-fringed Mason Bee) **143**, **193**, **247**, 414
habitats, and distribution 414, 440, 445, 464, 478
nests 440, 445
parasites 339, 341, 440
pollen source, nectar collection **193**, 440, 445
Osmia bicolor (Red-tailed Mason Bee) 139–142, **141**, **142**, **143**, 440
habitats **140**, 414, 440–441, 445, 464, **481**
nests in snail shells 139–142, **140**, **141**, **142**, **143**, 341, **341**, 440, 445
polylectic 445
Osmia bicornis (Red Mason Bee) 25–26, **26**, 28–29, **29**, 45, **187**, **203**, **285**, 439, 440
ageing effect on nest provisioning 430–431
in bee hotels 28, 521
brood cells 107, 158, 326–328, **327**
cavity nesting 29, 45, 73, 134, 136, **137**, 158, 326–327, **327**, 440
characteristics and colours 45
cleptoparasites of 326–327, **327**, **328**, **339**, **339**, 399
emergence, timing 28, 435, 438, 444–445
female sex pheromones 89
habitat and distribution 439, 440, 445
life cycle, and timing 158, 327, **329**, 435
males patrolling, mating behaviour 73, 91–92, 99
Osmia cornuta competing 448
parasites 347, **347**, 349, 440, 445
pollen collection, flower types **187**, **187**, **203**, 262, 440, 445
pollen transport (under abdomen) 28–29

SPECIES INDEX · 579

polylectic 262, 435, 445
predators of (birds) 350–351
protandry 328, 438
response to spring temperatures 444–445
sensory learning 219
supercooling point and cryoprotectants 443–444, **444**
'two-horned' 25, 26, **26**
Osmia caerulescens (Blue Mason Bee) 46, **89**, 187, **247**, 284–285, 427, 440
basking, and heat loss/generation 421, **422**, 423, 425
males patrolling, mating behaviour 73, 78, **89**
morphology 285
nesting 440, 445
pollen collection 187, 400, 445
polylectic 284
Sapyga cleptoparasitic on 339
Osmia cornifrons 264
Osmia cornuta (European Orchard Bee) 414, 430, 440, 443
distribution/arrival in Britain 440, 443, 448, 449
polylectic 445
response to spring temperatures 444–445
supercooling point and cryoprotectants 443–444, **444**
Osmia inermis (Mountain Mason Bee) 110, **111**, 344, 418, 437, 440, **471**
nests 440, 442–443
Osmia leaiana (Orange-vented Mason Bee) 45, **72**, **92**, 92–93, 399, 440
brood cell provisioning 137, 138, 139
delayed emergence 437–438
Heriades using failed nests 147–148
male mating behaviour **72**, 73, 78, **92**, 92–93
nesting and leaf nibbling 136–138, **138**, 139, **139**, 440, 445
parasitism by *Sapyga* species 340, 399
parasitoids of, *Chrysura radians* 344
pollen collection 265, 440
Osmia niveata (Jersey Mason Bee) 448, 449
Osmia parietina (Wall Mason Bee) **194**, 194–195, 344, 418, 441, 442, **442**, **443**
Osmia pilicornis (Fringe-horned Mason Bee) **155**, **248**, 439, 441, **442**, 472, **472**
habitat and distribution 441, 442, **472**
nests 110, 154, **155**, 441, 442, 445
pollen and nectar collection 285, 441, 442
Osmia spinulosa (Spined Mason Bee) **72**, 73, **245**, 340, 437, 441
habitats 441, 464
males, and mating behaviour 95
nests in snail shells **72**, 73, 399, 441, 445
pollen collection (Asteraceae) **295**, 340, 399, 441

Osmia uncinata (Pinewood Mason Bee) 344, 418, 441, 443
Osmia xanthomelana (Cliff Mason Bee) 110, 441, 445
Oxlip (*Primula elatior*) 184

Painted Nomad Bee (*Nomada fucata*) 40, 376, **376**, 382
Pale-tailed Mining Bee (*Andrena tridentata*) 447
Palemochartis liparae (parasitic wasp) 153
Pantaloon Bee (*Dasypoda hirtipes*) *see Dasypoda hirtipes* (Pantaloon Bee)
pantaloon bees (*Dasypoda*) *see Dasypoda* (pantaloon bees)
Panurgus (shaggy bees) 33, **34**, 48, 57, 296
emergence, timing 439
flower types associated 226, 296
foraging behaviour 296, 296–297
ground nesting 123–125, **124**
oligolectic 269, 270, 272, 296–297
pollen transport 257
species number, latitudinal gradient 416
Panurgus banksianus (Large Shaggy Bee) 48, **89**, 123, **125**, **206**, 207, **296**, 433, **434**
Panurgus calcaratus (Small Shaggy Bee) 74, 89
Panzer's Nomad Bee Bee (*Nomada panzeri*) 311, 377
Papaver rhoeas (Common Poppy) 186, 187, **187**, 249
Patchwork Leafcutter Bee (*Megachile centuncularis*) **21**, 88, **89**, 94, 95, 134, 135, 187, **205**, 287, 391
Peacock Butterfly (*Aglais io*) 101
Pelargonium 345, **346**
Perkin's Mining Bee (*Andrena rosae*) 470, **471**
Philanthus triangulum (Bee Wolf) 344–345, 355
Phragmites australis (Common Reed) 150
Pilosella aurantiaca (Fox and Cubs) 289
Pilosella officinalis (Mouse-ear Hawkweed) 220
Pinewood Mason Bee (*Osmia uncinata*) 344, 418, 441, 443
Pinguicula vulgaris (Butterwort) **209**
Plain Dark Bee (*Stelis phaeoptera*) **398**, 399
plasterer bees *see Colletes* (plasterer bees)
plum (*Prunus* species) 232
Polistes dominula (European Paper Wasp) 333
Polygala species (milkworts) 234
Polyommatus bellargus (Adonis Blue butterfly) 464, **465**
Potentilla erecta (Tormentil) 155, 190, 191, **191**
Potentilla reptans (Creeping Cinquefoil) 301

Potter Flower Bee (*Anthophora retusa*) 37, 367, 464, **479**
primroses (*Primula vulgaris*) 78, 184, **184**, 185, 228
Primula elatior (Oxlip) 184
Primula veris (Cowslip) 216, **249**
Primula vulgaris (Common Primrose) 78, 184, **184**, 185, 216, 228
Primulaceae (primrose family) 184, 185
Prosopis see Hylaeus (yellow-face bees)
Prunus (species) (plum) 232, 413, 516
Pulicaria dysenterica (Common Fleabane) **245**, 289, **298**
Pulmonaria species (lungworts) 154, 248

Ragwort (*Jacobaea vulgaris*) 43, 238, 289, **289**
Ranunculaceae 286
Ranunculus (buttercup) 207, **208**, **209**, 262, 264, 267, 288, 340
toxicity of pollen 264, 267, 288, 340
Ranunculus acris (Meadow Buttercup) 220
Ranunculus bulbosus (Bulbous Buttercup) 208, **209**
Ranunculus repens (Creeping Buttercup) 230, 268
Rape, Oilseed (*Brassica napus*) 216, 262, 483, 515
Red-backed Mining Bee (*Andrena similis*) **491**
Red Bartsia (*Odontites vernus*) 32, 76, **76**, 277
Red Bartsia Blunthorn Bee (*Melitta tricincta*) 32, 35–36, **36**, 76, 277, **278**, 414
Red Campion (*Silene dioica*) 183, **214**, **215**
Red Dead-nettle (*Lamium purpureum*) 78, 195, **195**, **196**, 282, 369
Red-girdled Mining Bee (*Andrena labiata*) 191, 207, **208**, 377, 378, **379**, 474, 487, **496**
Red Helleborine Orchid (*Cephalanthera rubra*) 211–212
Red Mason Bee *see Osmia bicornis* (Red Mason Bee)
Red Poppy (*Papaver rhoeas*) 186, 187, **187**, 249
Red-tailed Bumblebee (*Bombus lapidarius*) **21**, **40**
Red-tailed Mason Bee *see Osmia bicolor* (Red-tailed Mason Bee)
Red-thighed Epeolus *see Epeolus cruciger* (Red-thighed Epeolus)
Red Valerian (*Centranthus ruber*) 493
Reed Yellow-face Bee (*Hylaeus pectoralis*) **72**, 149–153, **151**, **152**, 156, 301, 341, 350, 473, **473**
Reseda lutea (Wild Mignonette) 225, 301
Reseda luteola (Weld) 301, **301**
Resedaceae 225
resin bees *see Heriades* (resin bees)

Reticulate Blood Bee (*Sphecodes reticulatus*) **386**, 387
Rhingia campestris 185
Ridge-cheeked Furrow Bee (*Lasioglossum puncticolle*) 474
Rock Rose (*Helianthemum nummularium*) 190, 208, **209**, 219, 220, **220**
Rophites (bristle-headed bees) 33, 49, 51, 57, 281
Rophites quinquespinosus (Five-spined Rophites) 51, **52**
Rosa canina (Dog Rose) 518
Rosaceae (rose family) 190, 346
Rose family (Rosaceae) 190, 346
Rose of Sharon (*Hypericum calycinum*) 133
Rosebay Willowherb (*Chamaenerion angustifolium*) **183**
Rough Hawkbit (*Leontodon hispidus*) 206, 220
Rowan (*Sorbus aucuparia*) 154
Rubus (brambles) 189, 301, 474, 518
Rubus rolifolius 354
Rubus ulmifolius (Bramble) **189**
Rufous-footed Furrow Bee (*Lasioglossum rufitarse*) 212–213, **213**

Sainfoin Blunthorn Bee (*Melitta dimidiata*) 35, **36**, 114, 277, 432, 464
Salix (willows) 66, 183, 232, 472, 473
Salvia microphylla (Hot-lips Sage) 298, **298**
Sambucus nigra (Elder) 133
Sand Tailed Digger Wasp (*Cerceris arenaria*) **9**
Sandpit Blood Bee (*Sphecodes pellucidus*) **390**
Sandpit Mining Bee (*Andrena barbilabris*) 69, 90, 118, 119, 452, 479
Sapyga 339
Sapyga pumila (*Sapyga* wasp) 265, 339
Sapyga quinquepunctata 265, 339, **339**, 340, 341
Sapygidae 305, **339**, 339–341, **340**, **341**
satellite flies *see* shadow flies
Scabiosa columbaria (Small Scabious) 200, **201**, **201**, 292, **292**, 378
scissor bees *see* *Chelostoma* (scissor bees)
Scrophulariaceae (figworts) 222, 245, 280, 281
Sea Aster (*Tripolium pannonicum*) 66, 198–199, **199**, 216, **300**, 314
Sea Aster Bee *see* *Colletes halophilus* (Sea Aster Bee)
Sea Kale (*Crambe maritima*) **181**
Segestria florentina (Green-fanged Tube Web Spider) 351, **351**
Senecio jacobaea (Common Ragwort) 283
shadow flies 305, 306–317, 349
  bees' defensive action 309–402, 310, 311, 315
  egg laying and behaviour 306, 307, **307**, 308, **308**, 309

impact on bee populations 309, 313–314
predators of 315, 317
see also *Leucophora*; *Miltogramma*
shaggy bees *see* *Panurgus* (shaggy bees)
Shaggy Furrow Bee (*Lasioglossum villosulum*) **242**
Sharp-collared Furrow Bee *see* *Lasioglossum malachurum* (Sharp-collared Furrow Bee)
sharp-tail bees (*Coelioxys*) *see* *Coelioxys* (sharp-tail bees)
Sheep's-bit (*Jasione montana*) **225**
Sheppard's Nomad Bee (*Nomada sheppardana*) 51
Shiny-vented Sharp-tail Bee (*Coelioxys inermis*) **42**, 391, 392, **393**
short-faced bees (*Dufourea*) 33, 49, 57, 416
Short-fringed Mining Bee (*Andrena dorsata*) 84, 437, 452
Short-spined Nomad Bee (*Nomada guttulata*) 191, 377, 378, **379**
Shrubby Seablite (*Suaeda vera*) 474
Silene dioica (Red Campion) 183, **214**, **215**
Silver-sided Nomad Bee (*Nomada argentata*) 378
Silvery Leafcutter Bee *see* *Megachile leachella* (Silvery Leafcutter Bee)
Sitaris muralis (Flame-shouldered Blister Beetle) 336, **338**, 355
small carpenter bees *see* *Ceratina* (small carpenter bees)
Small Flecked Mining Bee (*Andrena coitana*) 415, 427, 470, 487
Small-flowered Cranesbill (*Geranium pusillum*) **172**, **188**
Small Gorse Mining Bee (*Andrena ovatula*) 47
Small-headed Resin Bee (*Heriades rubicola*) 43, 153, 448, 449
Small Javelin Wasp (*Gasteruption assectator*) 341
Small Sallow Mining Bee (*Andrena praecox*) 84, 478
Small Sandpit Mining Bee (*Andrena argentata*) 69, 71, 90–91, **91**, 113, **113**, 118, **118**, 467, 480
Small Scabious (*Scabiosa columbaria*) 200, **201**, **201**, 292, **292**, 378
Small Scabious Mining Bee (*Andrena marginata*) **48**, 120, 200, 201, **201**, 257, **292**, 292–293, 378, 464
Small Scissor Bee *see* *Chelostoma campanularum* (Small Scissor Bee)
Small Shaggy Bee (*Panurgus calcaratus*) 74, 89
Small Tortoiseshell Butterfly (*Aglais urticae*) 101
Smicromyrme rufipes 342, 343
Smooth-faced Furrow Bee (*Lasioglossum fratellum*) 438
Smooth-gastered Furrow Bee (*Lasioglossum parvulum*) **112**, 333, 487

Smooth Hawksbeard (*Crepis capillaris*) 204
Smooth-saddled Dark Bee (*Stelis odontopyga*) 399, 448, **450**
Smyrnium olusatrum (Alexanders) 253
Solanaceae 248
Sonchus arvensis (Field Sowthistle) **205**, 206, 207
Sorbus aucuparia (Rowan) 154
Southern Marsh Orchid (*Dactlyorhiza paetermissa*) **210**
Spear Thistle (*Cirsium vulgare*) 79, 283, **506**
Sphecidae 406
Sphecodes (blood bees) 13, **34**, 50, 51, 53, 57, 176, 385–390
  absence of chemical adaptation 386–387
  attacking/resistance from guard bees 386, 387
  characteristics (structure/behaviour) **34**, 53, 57, 402–404
  colour 385
  egg-laying 169, 386
  emergence, timing 386, 387
  evolutionary tree 390
  generalist species, and host constancy 390
  habitat, specialisation 387
  host induced to be 'akinetic' 387
  host intimidation 386, 390
  host species 50, 51, 343, **362**, 385–386, 387, 388–389
  invasion of *Lasioglossum* nests 168, 386
  killing of host 386, 403
  losses/gains over recent decades 446
  mandibles, and cavity for 386, 387, **390**
  mating **386**
  overwintering 386
  parasites of (Mutillidae) 343
  scarce species and their hosts 400
  size 386, 387
  species number, latitudinal gradient 53, **362**, 410
Sphecodes albilabris (Giant Blood Bee) 449, **451**, 452
Sphecodes crassus (Swollen-thighed Blood Bee) **168**, 168–169
Sphecodes ephippius (Bare-saddled Blood Bee) 390
Sphecodes hyalinatus (Furry-bellied Blood Bee) 387, 488
Sphecodes longulus (Little Sickle-jawed Blood Bee) 387
Sphecodes monilicornis (Square-headed Blood Bee) **52**, 386, 387, 390
Sphecodes pellucidus (Sandpit Blood Bee) **390**
Sphecodes reticulatus (Reticulate Blood Bee) **386**, 387
Sphecodes spinulosus (Spined Blood Bee) 387
Sphecodogastra 173

SPECIES INDEX · 581

Spined Blood Bee (*Sphecodes spinulosus*) 387
Spined Mason Bee *see Osmia spinulosa* (Spined Mason Bee)
Spiny Restharrow (*Ononis spinosa*) 250, 251, 284
Spiraea douglasii (Steeple Bush) 133
Spotted Dark Bee (*Stelis ornatula*) 153, 395, **395**, 396, **396**, 397
Square-headed Blood Bee (*Sphecodes monilicornis*) 52, 386, 387, 390
Square-headed Furrow Bee (*Halictus maculatus*) 447
Square-jawed Sharp-tail Bee (*Coelioxys mandibularis*) 391, 478
Square-spotted Mourning Bee (*Melecta luctuosa*) 367, 400
Squat Furrow Bee (*Lasioglossum pauperatum*) 474
Stachys byzantina (Lamb's Ear) 30, **41**, 81, **82**, 144, **144**, 246, **246**, 247, 249
Steatoda nobilis (Noble False Widow Spider) 351, **352**
Steeple Bush (*Spiraea douglasii*) 133
Stelis (dark bees) 33, **34**, 45, 57, 257, 349, 394–399, 401
  behaviour at host nest 396, 397
  characteristics (structure/behaviour) 45, 57, 402–404
  host species 362, 395
  life cycle 396, 396–397
  losses/gains over recent decades 446
  resemblance to host 396
  scarce species and their hosts 400
  species number, latitudinal gradient 362, 394, 417
Stelis breviuscula (Little Dark Bee) 153, 397, **398**, 448, 450
Stelis nasuta 397
Stelis odontopyga (Smooth-saddled Dark Bee) 399, 448, 450
Stelis ornatula (Spotted Dark Bee) 153, 395, **395**, 396, **396**, 397
Stelis phaeoptera (Plain Dark Bee) **398**, 399
Stelis punctulatissima (Banded Dark Bee) 46, **398**, 399
Stenoria analis (blister beetle) 336, 338
Stickleback fish (*Gasterosteus aculeatus*) 60
Streptsiptera (twisted wing parasites) 305, 329–333
Stylops 305, 329–333, **330**, **331**
  impact on host bee 332
  larvae 331, 332–333
  mating 330, **331**, **331**
Stylops ater **331**
Stylops kirbii 330, 333
Stylops melittae 333
Suaeda vera (Shrubby Seablite) 474
Succisa pratensis (Devil's-bit Scabious) 200, 203, **203**, 292
sweat bees *see* Halictidae
Swollen-thighed Blood Bee (*Sphecodes crassus*) 168, **168**, 168–169

Sycamore (*Acer pseudoplanatus*) 232
Symphytum (comfrey) 202
Syrphidae 10

Tanacetum vulgare (Common Tansy) 315
Taraxacum (dandelions) 189, 196, 228
Tawny Mining Bee (*Andrena fulva*) 10, 69, 118, **118**, 204, 311, 312, **312**, 377, 516
Teucrium scorodonia (Wood Sage) 79, 472
Thecophora 325
Thecophora atra 323, **325**, **326**
Thecophora occidentis 323, 325
Thick-margined Mini-miner (*Andrena falsifica*) 489
thistles (*Cirsium* species) 232
Thomisus onustus 354
Thrift (*Armeria maritima*) 184, **186**, 190
Thrips 407, 409
'thunderflies' 407
Tormentil (*Potentilla erecta*) 155, 190, 191, **191**
Tormentil Mining Bee (*Andrena tarsata*) 191, **191**, 309, 415, 427, 470, **471**, 480, 487, 489
Trailing Bellflower (*Campanula poscharskyana*) 286, 287
Trichrysis cyanea 344
Trifolium repens (White Clover) 35
Trifolium species (clovers) 186, 474
Tripolium pannonicum (Sea Aster) 66, 198–199, **199**, 216, **300**, 312
Tuberculate Long-horned Bee (*Eucera nigrescens*) 282
twisted wing parasites (Strepsiptera) 305, 329–333

Ulex europaeus (Gorse) **6**, 195, 467

Variable Nomad Bee (*Nomada zonata*) 376, 449, 452
variegated cuckoo bees *see Epeolus* (variegated cuckoo bees)
Varroa mite 348
velvet ants (Mutillid wasps) 305, 342–343, **343**
Veronica chamaedrys (Germander Speedwell) 191, **191**, 378, **379**
Vespula germanica (German Wasp) 317
Vespula species (social wasps) 351 *see also* wasps (General Index)
Vespula vulgaris (common wasps) 244
Vicia sepium (Bush Vetch) 227
Viola canina (Dog Violet) **209**
Violet Carpenter Bee (*Xylocopa violacea*) 12, **12**, 40, **40**, 110, 153–154, 448, 450
Violet-winged Mining Bee (*Andrena agilissima*) 312
Viper's Bugloss (*Echium vulgare*) 75, 196, **197**, 203–204, 267, **285**, 285–286
Viper's Bugloss Mason Bee (*Hoplitis adunca*) 73, 196, 267, **285**, 285–286, 344, 448, 449

Wall Mason Bee (*Osmia parietina*) **194**, 194–195, 344, 418, 441, 442, **442**, 443
wasp-bees *see Nomada* (nomad bees)
Wasp Spider (*Argiope bruennichi*) 351–352
wasps *see* wasps (General Index)
Weld (*Reseda luteola*) 301, **301**
Welted Lesser Mason Bee (*Hoplitis claviventris*) 46, 95, 361, **395**
Western Honey Bee (*Apis mellifera*) **6**, 37, **40**, 229, 244, 256
White-bellied Mining Bee (*Andrena gravida*) 452, 495, **496**
White Bryony (*Bryonia dioica*) 31, 76, **76**, 183, 260, 293, **294**, 500
White Clover (*Trifolium repens*) 35
White Dead-nettle (*Lamium album*) 195, **195**, 249
White-footed Furrow Bee (*Lasioglossum leucopus*) 287, **298**
White-tailed Bumblebee (*Bombus lucorum*) 244, 298
White-zoned Furrow Bee (*Lasioglossum leucozonium*) 49, 187, **187**, 220, **326**
Wild Carrot (*Daucus carota*) 196, 301, **353**
Wild Mignonette (*Reseda lutea*) 225, 301
Wild Strawberry (*Fragaria vesca*) 154
Wilke's Mining Bee (*Andrena wilkella*) 21, 86, **251**, 251–252, 474
willows (*Salix*) 66, 183, 232, 472, 473
Willughby's Leafcutter Bee *see Megachile willughbiella* (Willughby's Leafcutter Bee)
Wolbachia 439
Wood-carving Leafcutter Bee *see Megachile ligniseca* (Wood-carving Leafcutter Bee)
Wood Sage (*Teucrium scorodonia*) 79, 472
Wool Carder Bee *see Anthidium manicatum* (Wool Carder Bee)
wool carder bees *see Anthidium* (wool carder bees)

Xylocopa (carpenter bees) 16, 17, 32, 40, 56
  flower types associated 226
  mites on 348
  nesting habit 110
  species number, latitudinal gradient 416
Xylocopa violacea (Violet Carpenter Bee) 12, **12**, 40, **40**, 110, 153–154, 448, 450

yellow-face bees *see Hylaeus* (yellow-face bees)
Yellow-legged Mining Bee *see Andrena flavipes* (Yellow-legged Mining Bee)
Yellow Loosestrife (*Lysimachia vulgaris*) 36, 77, 126, 225, 275, **276**, 474
Yellow Loosestrife Bee *see Macropis europaea* (Yellow Loosestrife Bee)

GENERAL INDEX

abdomen of solitary bees 7, **8**, 245, **245**
  conopid flies inserting eggs 322
  scopae on 29, 245, **256**, 283, 289
  Stylops parasitic on 329–330, **330**
acarinaria (mite pouches) 348
actinomorphic flowers 193
adaptations (bee)
  host manipulation by parasites 332–333
  mutual, of bees and flowers 190, 222
  for nectar collection 22, 240–243, 302
  for oligolecty/specialism 263, 290–291
  for pollen collection **21**, 244, 245–248, 302
adaptations (plant) 200, 201, 202, 222
adaptive host manipulation 333
'aerial' nesters see nest(s), aerial/cavity
ageing, of bees 430–431
aggregations, nests see nest(s), aggregations
agricultural chemicals 512, 513, 514, 515
agriculture
  bee nesting habitats, importance 518
  on chalk downlands, impact on bees 464
  conservation strips, pollinator strategy 515, 516, 517, **517**, 519
  conversion of land from, wild flower mixes 505
  crop pollinators and, importance 508, 509, 513–514, 515, 528–529
  buzz pollination 249–250, 509
  decline in bee numbers 508–509, 510, 511, 512, 514, 516
  by few dominant species 528–529
  floral resources, monitoring/enhancing 510, 514–515
  integration with semi-natural habitat 515, 516, 517, **517**, 519
  intensive/industrialisation of **456**, 464, 483, 509, 511, 512, 514, 525
  pollination need, as barrier to 513–514
  threat of/decline in insects 455, 456, 508, 509, 510, 511, 516, 519
  seed mix to include plants for solitary bees 517, 518
  'sustainable farming' 519
  wild flower strips 514, 516, 517, **517**, 518, 519
allotment sites 497
'altruistic' behaviour 161, 174
anatomy of solitary bees 4, **5**, 7–8, **8**, 367
  male vs female **17**, 17–20, **18**
  modifications to restrain female 18, 93–94, **94**, 95, **95**, 96
  of protobees (in bee evolution) 408
  see also specific structures (e.g. antennae, mandibles)

antennae 7, **8**, 19, 62, 175
  chemical stimuli, sensitive to 381–382, **382**
  Nomada 381, **382**, 382–383, **383**, 384
anthers (flower) 179, **179**, 198, **198**, 203
  heterostylous plants 184, **184**
  'poricidal' 248
antiaphrodisiacs 63, 99, 120, 211, 382–383
antifreeze compounds 443
apomixis 189
arolium (pad between claws) **8**, **43**, **44**, 45
artizans 16
  see also cuckoo bees (solitary bees)
'Asteraceae paradox' 290–291
  see also Asteraceae (Species Index)
auditory cues 62
autumn, emergence in 66, 104, 437, 450

bacteria 105–106, 439, 462
Baldock, David 111, 140, 468, 497–498
Bartinney Nature Reserve, Cornwall 470, **471**
basitarsus 4, **5**, **8**, **9**, 256, 275, 277, 283
basitibial plates 46, **47**, 113
baskets *see* pollen baskets (corbiculae)
basking 207, **208**, 394, 421–424, **422**, **443**
  preferred places, duration 421, **422**, 423
Batesian mimicry 208, 377
bats, pollination syndrome 224
Beavis, Ian 437, 495
bee(s) 2–5
  distinguishing *see* distinguishing solitary bees
  species number *see* number of species of all bees
  studies on 23–25
  'bee friendly' policies 460
bee hotels 23, 28, **42**, 132, 145, 461, **521**, 521–522
  competition by *Heriades* females 147
  constructing and dimensions **521**, 521–522
  males 'staking out' 73, 74, **74**
  Osmia bicornis and *Cacoxenus* 326–328, **327**, **328**
  parasitism concerns and rates 328, 349
  species found in 448, 522, **523**
bee syndrome, Vogel's 222, 223
Bees, Wasps and Ants Recording Society (BWARS) 23, 25, 33, **418**, 511
Belgium, bee species/genera 420, 420
'bilabiate' flowers 245
biodiversity
  conservation *see* conservation
  green space management to enhance 497, 505, 506, **507**, **508**

loss, concerns 458–460, 497, 525, 526, 528
  drivers 502
  see also decline in bee numbers/species
  planning changes to enhance 525–526
  protection, demand for 530
  see also diversity and distribution of bees
Biodiversity Action Plan (UK) 527, 528
Biodiversity Net Gain (BNG) 470, 526–527
'biophilia' 530
birds, as solitary bee predators 317, 350–351
birds pollination syndrome 224
bivoltine species 48, 102, 173, 261, 292, 376, 435, 437
body plan (bees) 7–8
books published on bees 13, 23–24
Breckland, Norfolk and Suffolk 230, **231**, 232–233, 413–414, 480, **481**
Brockholes Nature Reserve, Lancashire 487
brood(s), produced by foundress 161, 162, 164, 165, 173, 176, 286
brood cells 1, 16, 60, 103, 236
  aerial/cavity nests 133–134, **139**, 326–328, **327**
  construction, mastic use 45, 133, 134, 137, 138, 139, **141**, 142, 154
  construction, mud use 134, 136, **137**
  construction, resin use 110–111, **145**, 145–146, 341
  in galls of *Lipara lucens* 150, **151**
  in snail shells 110, 139, **143**
  clustering, in halictids 171
  completion rate 106, 107–108, 120
  contents, assessment of specialism 260
  female, completed before male cells 109, 439
  ground nesters 16, 113, **113**, **114**, 115, 116, 119, 120, 126, **129**
  lining 107, 113, 120
    with cellophane-like substance 54, 129, 130, 148, **149**, 150, 366
    with leaf segments 133–134
    with mastic 45, 133, 134, 137, 138, 139, **141**, 154
    with plant hairs 81, 82, **144**, 144–145
    with resin 110–111, **145**, 145–146, 341
    with secretions *see* Dufour's gland secretions
  water-proofing 16, 113, 115, 116, 120, 126, 129, 134
  liquid food-store (Colletidae) 53
  male 109, 431
    switch to, by older bees 431

opening for inspection 172
permeability 103
protection 103, 104
provisioning for 62, 102, 103,
  104–108, 113, 115, 147, 236, 461
  adjusted for size of bee 108
  ageing (of bee) effect on 430–431
  conversion rate 106
  demands on females 104–105, 236
  diurnal pattern 122–123
  dry weight 106
  flight duration 122, 123
  foraging trip time/number 107,
    115, 120, 123, 131, 135, 147, 236
  limiting factors 105, 106
  load-carrying capacity 106–107
  mass provisioning of cluster
    171–172
  minimum pollen estimation
    119–120
  nectar 16, 105, 109, 113, 115, 236,
    252–253
  number per day 107
  plant oils, pollen with 105, 116
  pollen 16, 105, 109m136, 113, 119,
    120, 236, 253, 281
  pollen-nectar mass 16, 105, 113,
    115, 120, 124, 253
  time constraints 104, 106, 107, 135,
    237
  size/volume 108, 109
  time required to complete 120
  walls 113, 115, 120
  wood-borers/nests in dead wood
    154
  see also nest(s); under specific species
    (Species Index)
brownfield sites 484, 484, 485, 487–489,
  488, 489, 490, 491, 493
  development on, threat to bees 502,
    505
brushes, pollen transfer 22, 255, 258,
  283–284
  see also scopa (scopae)
Buglife 483, 485
building development, on grassland
  469, 470, 485
bumblebees see bumblebees (Bombus)
  (Species Index)
bush-crickets 60
butterflies 223, 464
butterflies pollination syndrome 223
'butterfly meadow' sites 505–506
buzz of bees 7, 22, 62, 73, 248, 249,
  277, 284
buzz pollination 248, 249–250, 281, 509

Cambridge University Botanic Garden
  494, 495
cantharidin 335, 336
Canvey Island, south Essex 484, 486,
  487–488
Canvey Wick 484, 485, 486
capitulum (capitula) 197, 197, 197–198,
  199–200, 203
carinate species, *Lasioglossum* 170

Carson, Rachel 458, 512
caudicle 210, 211
cellophane-like substance, brood
  cell lining 54, 129, 130, 148, 149,
    150, 366
cemeteries, bee species in 493, 494, 495
central place foragers/foraging 179,
  236
chalk downland, habitat 463–465,
  465, 487
chemical communication
  courtship and mating 62–63, 93, 94,
    369–370, 370
  eusocial behaviour (halictids) 174,
    175, 176
  mining bees (*Andrena*) 64–65, 88,
    381–382
  nest detection by cuckoo bees 383,
    384, 391
  see also pheromones; scent(s)
chemical senses 7, 20–21, 62–63, 279,
  280, 381
chitin 7
chorion (eggshell) 371, 372
chromosomes 17–18
cities, bees favouring 414
'citizen science' 458, 528
classification of bees 2–3, 3, 13,
  32–33, 33
cleptoparasites (nest-parasites) 2, 16,
  104, 304
aculeate wasps (Sapygidae) 305,
  339–341
blister beetles (*Stenoria analis*) 336,
  338
Cacoxenus 325–328, 327, 328, 328, 329
Chrysididae (jewel wasps) 305,
  343–345, 344
cuckoo bees see cuckoo bees (solitary
  bees)
defensive behaviour by hosts
  309–310, 311, 368–369, 369
gasteruptiid wasps 341–342, 342
genera 32, 33, 37, 38, 42
host specialism 349
Leucophora (shadow flies) 306, 307
not tolerant to plant toxins 265
oil beetles (Meloidae) 333–336
protection from, nest aggregations
  311
scent mimicry by 63
shadow flies see shadow flies (*Species
  Index*)
Sphecodes see *Sphecodes* (blood bees)
  (*Species Index*)
climate 439–445
  British 411–412, 439–445
  distribution and range of bees 415,
    439–445, 463
  latitude and see latitudinal
    gradient
  Osmia species 439–445, 440–441
  supercooling points and 443–444,
    444
  diversity of bees 439–445, 440–441,
    463

effects on bee activity 415
Mediterranean-type 410, 411
warm temperate, bee abundance/
  species 410
see also weather
climate change 445, 458, 463, 497
  Buffish Mining Bees, orchid flower
    timing 211
  halictid transition to eusociality 173
  solitary bee losses/gains 445–447,
    446–447, 454
clypeus 8, 143, 246, 246, 295
  hairs on, pollen-collecting 245, 246,
    246, 247, 281, 285
co-extinction, of bees and parasites
  349
coastal habitats 414, 474–480, 475, 476,
  477, 478, 479
cocoon(s) 102, 103, 137, 157, 236
Cerceris rybyensis 359
Coelioxys (sharp-tail bees) 393, 393
conversion ratio 106
nutritional requirements for
  spinning 106, 157
Osmia species 106, 158
overwintering in 158, 158, 430
species spinning 106, 158
supercooling points (*Osmia*)
  443–444, 444
coevolution of bees
  with flowering plants 20, 265, 266,
    268–269, 290, 296–297, 302, 409,
    460–461
  with parasites 306, 349
cognitive and learning abilities 21,
  236, 263
  loss, by neurotoxin insecticides 513
  see also learning by bees
Colchester, wildlife sites 498–499
Colne Point Nature Reserve, Essex
  474, 475
colour of flowers see flower(s), colours/
  colourful
colour of solitary bees 4, 9
combs, pollen transfer 255
common names, for bees 26
communal species 16–17, 74–75, 104,
  159, 312
'communalism', species 159
communication, bee 7, 175, 176
  see also chemical communication;
    vibrations
competition between females, nest
  sites 108, 121–122, 168
competition between male solitary
  bees 59, 60, 61
  advantageous morphological
    features 18, 65
  in basking spots 422
  males patrolling female emergence
    sites 67–68, 75, 87
  males patrolling flowering plant
    sites 67–68, 75, 78–79, 80, 81, 87
  'scramble' competition 65, 70
  see also territorial behaviour of
    solitary bees

competition by plants for pollinators 190, 208, 209–210
'complete' metamorphosis 14
composite flowers, solitary bee genera 226
conservation 455–530
  amateur naturalists and 458–459, **459**
  biodiversity 458–460, 529–530
    economics and 460, 527–528, 529
    ecology insights to justify 529
  heather heath losses and concerns 467
  mosaic of habitats needed, 'living landscapes' 482, 483
  national organisations 458–459
  protected spaces, habitat survival 482
  from small to landscape scales 482
  understanding dispersal/range expansion 452
  see also biodiversity; habitats for bees
Convention on Biodiversity 460
conversion ratio 106
copulation (bee) 61–62, **67**, 88, 91, 92
  interval between mounting and 92–93, 97
  mounting without 78, 88, **92**, 92–93
  pheromone compounds stimulating 89, 90
  copulatory phase 88
corbicula(e) see pollen baskets (corbiculae)
corbiculate bees 4, 200, 252–253
  solitary bees vs 4, 4
  see also bumblebees (*Bombus*); honeybees (*Apis*) (Species Index)
corolla 179, **179**, 239
corolla tube 193–194, 250, 251, 252
costly information hypothesis 221
courtship 18, 61, 88
  Megachile willughbiella 93, 94
  Osmia species 78, 91–92, **92**, 92–93
  see also mating/mating systems (bees)
crab spiders 352, **353**, 354
'creative destruction' 530
Cretaceous period
  bee evolution 2, 406, **406**, 408
  flowering plant diversification 410
crop (anatomical), pollen/nectar transport in 4, **21**, 54, 107, 120, **131**, 253, 301
crops (agricultural) see agriculture
cross-pollination (and cross-fertilisation) 178, 182–190, 510
  flower colour change triggered 202
  importance/advantages 183, 185
  orchids 210
  strategies to promote 183–184, 186–187
cross-species usurpation, of nests 147–148
cryoprotectants, *Osmia* species 443–444, **444**
cuckoo bees (solitary bees) 2, 4, 9, 16, 103, 361–406, 399, 462

anatomy, for egg-laying **367**
bad weather effect 434
competition between species 399
diversity and distribution 401
  latitudinal gradient 401, 415, **416**, **417**, 417
  new arrivals in UK 449, 452
eggs and egg-laying **366**, 366–367, 371, **372**, 373, 386
emergence, timing 364, 367–368, 376–377, **378**, 381
evolution 361, 401, 402
females **9**, 100, **376**, 377, 381, 382, 383, 384–385
finding 'correct' hosts 371, 378–379
genera 32, 33, 37, 38, **362**, 362
  characteristics, summary 401–402, 402, 403
host genera 362, 401
larvae feeding on food stores 361, 367, 371–372, **372**, 385
life cycle 168, 367
males seeking nectar, as pollen vectors 191
mating 61, 100
number of species 361, 362, 399, 401
protandry reduced/absent 439
scarce species (UK) 399, 400
scent mimicry 63
variegated see *Epeolus* (Species Index)
cuckoo bumblebees 4, 362, 401
'cuckooing'
cross-species 147–148
intra-specific 147, 166
cuticle 7, 14
  casting off, by larvae 102, 156
  pheromones/volatile substances on 69, 89, 90, 92, 98–99, 381

Danforth, Bryan 13, 74, 109, 110, 111, 153, 163, 172, 173, 175, 265
Darwin, Charles 59–60, 83, 178, 183, 184, 185, 210, 210, 409, 468, 530
deception 190, 208–214
  competition by 209–210
  by complex (zygomorphic) flowers 209
  by orchids 210, **210**, 211–212, 212–213, **213**
  rewards lacking to pollinators 210
  scents, as signals 63
  see also mimicry
decline in bee numbers/species 445–447, 446–447, 454, 456, 463, 508–509, 510
  agricultural landscapes 508–509, 510, 511, 512, 514, 516
  amateur/scientific concerns over 458, 459–460
  oligolectic bees, before/after 1980 509–510
  see also biodiversity
definition/description, of bees 2–5
Defra (Department of Environment, Food and Rural Affairs) 526
dehydration 434

Delpino, Federico 222
diapause (dormant stage) 14, 103, 156, 164, 435, 438, 439, 444–445
dietary breadth 108, 269–270, 302, 461
  Andrenidae 290–297
  Apidae 280–283
  Colletidae 299–301
  Halictidae 297–299
  Megachilidae 283–289
  Melittidae 273–280
  see also oligolecty/oligolectic bees; polylecty/polylectic bees
digital photography 12
dioecious plants 183, 213
disorganised behaviour 430
dispersal behaviour 452–453
  see also diversity and distribution of bees
distinguishing solitary bees 9–11
  from bee-flies 10–11
  from corbiculate bees 3–4, 4, **5**
    bumblebees 3–4, 4, 10–11
    honeybees 3–5, 4, **5**
  from flies and hoverflies 10
  from wasps 3, 9, **9**
diversity and distribution of bees (general) 410–412
  amateur naturalist concerns 458–459
  English Channel as barrier 420
  'hotspots' (global) 410
  latitudinal gradient (global) 410–411, **411**
  measurement difficulties 412, **413**
  scientific concerns 459–460
  species number (global) 182, 410, **411**
  see also biodiversity
diversity and distribution of bees, in Britain 27–58, 410–412, 412–418, 463
  10,000 to 8,000 YBP 419
  as an island 418–421
  limitations/barriers 418–419, 420
  Breckland 230, **231**, 232–233, 413–414
  climate effect see climate
  by county 463
  decline see decline in bee numbers/species
  increasing range/area 448, 450–451, 452, 463
  latitudinal gradient see latitudinal gradient
  losses/gains/extinct, recent decades 445–447, 446–447, 454
  Mesolithic–Neolithic transition 419
  multiple colonisation 453, 454
  new arrivals (species) 443, 448–454, **451**
  northern bias, species with 415, 418, 442
  ratio of number of genera to species 415
  recently established/re-established species 448, 449–450, 450, 480

southern bias, species with 414, **414**, 415, 416–417, 463
species density, by area size sampled 412, **413**
species density, map of UK **418**
species/genera number in UK *vs* Europe 419, 420, 420
in urban areas 414, 492–493
see also biodiversity; habitats for bees
diversity of floral resources, decline 510, 511
dormant stage *see* diapause (dormant stage)
'drumming', pollen release 245
Dufour's gland, anatomy **131**
Dufour's gland secretions
Andrena-Nomada pairing 381–382
composition 131, 175
for lining brood cells/nests 105, 115, 120, 126
Andrena 90, 120
Anthophora 169, **371**, 372
Colletes 54, 129, 130, 131
Hylaeus 148
Lasioglossum 69, 126
Melitta 115
workers, discrimination of kin/non-kin 174
see also pheromones (bees secreting)
Dungeness headland, Kent 475, 478

ecology and ecosystems 455–530
decline in species, bleak outlook 456, 458
sources of hope, to avert crisis 458
see also biodiversity; habitats for bees
economic value of wildlife 527, 528
economics, biodiversity conservation 460, 527–528
'category mistake' and risk 529–530
'ecosystem service' 509, 528, 529
ectoparasite 345
ectoparasitoids 304, 343
Edwards, Mike 13, 25, 27, 276
eggs (insect) 101–102
aculeate wasps (*Sapyga*) 339, 340
bee-flies (*Bombylius*) 317, 318, **318**, 319
Cacoxenus 326
conopid flies 322
Leucophora 307, **307**, **308**, 308–309, 312
oil beetles 334, **335**
eggs (solitary bee) 14, 17, 60
of cuckoo bees 366, 366–367, **367**, 371, **372**, 373, 386
fertilisation 60, 61
fertilised, female solitary bees from 17, 61–62, 174, 176
laying 102, 113, 124, 134, 136, 149, 156, 439
number, and placement 60, 101, 102
in pollen loaf 115, 139

production rate/time to mature 107–108
unfertilised, male solitary bees from 17, 61–62, 174, 176
electrostatic charge, for pollen attachment 3, 239, 244, 411
Else, George 13, 25, 27, 276
emergence 435–437
bivoltine species 48, 102, 173, 261, 292, 376, 435, 437
cuckoo bees, timing 364, 367–368, 376–377, **378**, 381
delayed (parsivoltinism) 437–438
males first, then females 18, 61, 63, 65, 66, 109, 438–439
see also protandry
seasons for 12, 27–32, 104, 435, 437, 461
late summer/autumn 66, 437, 450
spring 28–29, 30, 104, 156, 435, 437, 461, 472
summer 30–31, 156, 437, **451**
times, control/factors affecting 437–438, 439
univoltine bees 102, 363, 435, 437, 510
emergence traps, parasitism rate 311, **312**
enantiomers, bees discriminating between 86
endoparasitoids 304
energy 235
English Nature 485
environmental changes 510
eusocial behaviour, transition to 163–164, 166, 167, 173
see also biodiversity; climate change
'environmental services', concept 527–528
enzymes, allozymes, phylogeny and 170
Essex nature reserve 230, **231**, 232–233, 484
Europe, bee species/genera 420, 420
new arrivals from, in Britain 448–454, 449
eusocial behaviour 104, 161, 174, 302
conditions favouring transition to 173–176
geographical/environmental 163–164, 166, 167, 173
overwintering, fertilised females 68, 69, 161, 165, **165**, 176, 385, 386
evolution of 161, 169–173, 175, 297, 430
facultative 163–164, 175, 176, 438
Halictidae 2, 17, 29, 50, 104, 159–160, 161, 173, 297, 302
ancestrally solitary species 175
communication systems 174, 175, 176
formerly eusocial, currently solitary 175
impact of conopids reduced 323, 325

innate plasticity 162–163
Leucophora cleptoparasitism 312
life cycle 161–162, 176
phylogeny (evolutionary history) 169–173, 175
'queens' *see* foundress females ('queens')
Sphecodes targeting 385–386, 387
workers *see* workers
maintaining order in nest 174, 386
obligate 165–169, 171, 174, 175
see also Lasioglossum malachurum (*Species Index*)
retained despite environmental change 164
reversal to solitary lifestyle 171, 173, 175
solitary-to-eusocial transition 170, 171, 172–173, 176
evolution of bees 2, 262, 405–408, **407**, 410
adaptation to flowers 190, 280
ancestors (bee) 2, 128, 406–408
coevolution *see* coevolution of bees
convergent 402
cuckoo bees 361, 390, 401, 402
of eusocial mode of life, Halictidae 161, 169–173, 175, 297, 430
evolutionary tree 390, **407**, 408
fossils from Cretaceous period 2, 406, **406**, 408, 409
of haplodiploidy 174
of male and female differences 59–60, 174
of nest aggregations 159
origin, from wasps 2, 406, 408–409
of pollen specialism 261, 262, 268–269, 280, 302
protobee (*Melittosphex burmensis*) 408–410
sexual selection, as 'second agency' 59
of *Sphecodes* (blood bees) 390
transitional form of wasp/bee 408–409
evolution of flowers 177, 215, 221, 225, 230, 409–410, 412
evolution of wasps 406–407, **407**
exoskeleton 7, 13, 14
extinct solitary bees 446, 447, 511
extinction rates 511, 512
eyes, bee-flies 318
eyes, solitary bees 5, **5**, 7, **8**, 30
hairy 5, **5**, 394
vs honeybees 4–5, **5**

facial foveae **8**, 46, 47, **47**
Falk, Steven 13, 23, 25, 26, 33, 463, 489, 526
families of solitary bees 32–33, 33–58
characteristics 55–58
farming/farmland *see* agriculture
farnesol/farnesyl hexanoate 90, 120, 211
Farnham Heath, Surrey 468
fecundity 493

586 · SOLITARY BEES

female solitary bees 1–2, 17–20, 60
  anatomy 3, 4, **5**, 7, **8**, **50**, **51**
  appearance, *vs* male bees 17, **18**, 18–20, **19**
  central place foraging 179, 236
  competition for nesting sites 108, 121–122, 168
  corbiculate bees compared to 4
  diploidy 18
  emergence timing, after males *see* emergence
  features by family, *vs* males 55–58
  fertilised, overwintering 51, 68, 69, 161, 162, 164, 165, **165**, 176, 438
  from fertilised eggs 17, 61–62, 174, 176
  foraging and demands on/for 16, 60, 62, 104–108, 236
    see also brood cells; eggs (solitary bee); nest(s)
  foraging disrupted by males 76, 77, 79, 104
    see also male mate-finding strategies
  foundress *see* foundress females ('queens')
  hairs 3, 4, **5**, **8**
    see also scopa (scopae)
  harassment by males *see under* mating
  longevity 427, 428, 429, 438
  mating *see* mating/mating systems (bees)
  nectar intake 107, 236, 238
  nectar- or pollen-collecting days 238–239
  pheromones secreted *see* pheromones
  ratio to male bees 17, 61
  size, *vs* male bees 17, 82
  sperm storage 61–62, 176, 453
  sterile 17
  stings 2, 7, 8, 23
fertilisation of flowers 179–180, 189
  see also cross-pollination (and cross-fertilisation); self-fertilisation
fertilisation of solitary bee eggs 60, 61
fertilisers, artificial 512
fidelity of pollinators *see* flower fidelity
finding solitary bees 11–13, 27
Fingringhoe Rangers (MoD), Essex 474
'fixed preference' 216–217
flagellum (of antenna) 7
flies pollination syndrome 224
flight by bees 22–23
  heat loss/generation during 424–425
  male patrolling *see* male mate-finding strategies
  minimum temperature 12, 421, 425
  orientation *see* orientation flights
  quick/darting, *Anthophora* 22, 37
flight by flies 317, 320
flight muscles 424
flight period (phenology) 12, 27, 103, 104, 292

linked with flowering time 28, 103, 131, 204, 207, 265–266, 461, 467, 516
  mismatch 201, 258
  oligolecty or polylecty 302
  timing/seasonal 12, 27–32, 104, 156, 282
  see also emergence; *individual species*
floccus 257, 258
'floral clock' 207
floral constancy *see* flower constancy
floral resources 103, 178, 244, 510, 514–515
  loss, decline in pollinators 511
  see also flower(s); nectar; pollen; rewards (floral)
florets 197–198
florivores (flower-eaters) 190, 192, 193, 197, 204, 354
flower(s) 178–181, 190–200
  abundance, reason for flower fidelity 217
  actinomorphic (radially symmetrical) 193
  adaptations for pollination style 222
  attractants for pollinators 177, 189, 190, 239
  changes after pollination 190, 202, **203**
  bee adaptation to 190, 222
  bee manipulation by *see* deception; mimicry
  bees sheltering in 433, **434**
  bell-shaped 191–192, **192**, 226, 228
  bowl-shaped 228
  classification, pollination categories 222
  colours/colourful 177, 182, 202, 221, 238
    change, after pollination 190, 202, **203**
    crab spiders (predators) matching 352, **353**, 354
    recognition by bees 238
    red and blue ranges 202, 211–212
  competition by, for bee pollinators 190, 208, 209–210
  diversity, importance 461
  evolution 177, 215, 221, 225, 230, 409–410
  explosive, pollen spread 195
  factors increasing production 204
  fertilisation 179–180, 189
  fidelity *see* flower fidelity
  foraging by bees *see* foraging by solitary bees
  functional groups, pollinator groups (garden) 227–228, 229
  heterostylous ('pin' and 'thrum' forms) 184, **184**, 185, 228
  honeybee/bumblebee preferences 220, 221
  importance for bees 178–179, 181–182, 460–461
  importance of bees for 181–182, 460–461

influencing solitary bee behaviour 190–200, 202, **203**
  mimicry/deception *see* deception; mimicry
  timing of flowering 202–207
  matching by *Nomada* and host bees 378
Mediterranean region, seasons 411
mirror-image 184
morphology 190–200
  pollination *see* pollination by solitary bees
  selection pressures on 207, 266
nectar and pollen source *see* nectar; pollen
nototribic 226, 245, 277, 281
opening/closing 204–207
  Asteraceae 200, 204, **205**, 206
  closing as pollination response 204, **205**, 206, 207
  conditions causing 204
pollinator interactions 227–228
  see also pollen; pollination; pollination syndromes
  recognition by bees 236, 238
  as 'rendezvous' sites 63, 64, 65, 75, 235
  response to wingbeats and buzz 249
  rewards for bees *see* rewards (floral)
  scents *see* scent(s), flower
  shapes, bee species visiting by 228, 229
  sharing with other bee/insect species 263
  size, breeding system relationship 188, **188**
  structure 178–181, **250**
  time spent at, by Ivy Bee 244
  timing of appearance 103, 131, 202–207, 265–266, 461, 467, 516
    extended season, benefits 202–203
    spring flowering 28, 472
  types for specific bee genera 220, 226
  warming up of solitary bees in 207, **208**
  wild *see* wild flowers
  zygomorphic (bilateral symmetry) *see* zygomorphic flowers
flower constancy 216, 217, 217, 220, 221, 259, 263
  explanation/description 216, 259
  learning/memory affecting 218, 219
flower fidelity 215–218, 259, 294
  bumblebees/honeybees 220, 221
  description/definition 216
  field studies, summer pasture 219–220
  'fixed' *vs* 'labile' preference 216–217, 217
  monocultures/crops 216
  reasons for 216–217, 217
  for Rock Rose flowers 219–220, **220**
  transient 216

flowering plants
  bee dependence on 178–179, 181–182, 460–461
  coevolution with bees 20, 266, 268–269, 290, 296–297, 302, 409, 460
  evolution/origin 177, 215, 221, 225, 230, 409–410, 412
  in gardens, exotic/suggested species 520, 524
  species number in Britain 227
food, for bees 3, 178–179, 235, 406, 460–461
  bee survival for days without 433
  collection 20–22
  in gardens 520–521
  latitudinal gradient in availability 415, 418
  limited, shift to male offspring 108
  risk minimisation by flower constancy 221
  transporting see transport of pollen/nectar
  see also nectar; pollen
food security, concerns 510
foot/feet, brushes on 22, 245
foraging by bumblebees 220
foraging by honeybees 220, 221, 237
foraging by solitary bees 237, 303
  after emergence from nest or hibernation 215
  behaviour of *Andrena wilkella* **251**, 251–252
  for brood cells see brood cells, provisioning
  conflict with protection against shadow flies 309, 315
  demands of (distance/specialism) 236, 237
  duration, at each flower (Ivy) 244, 278
  effect on lifespan of bees 429
  efficiency, by specialist bees 263
  fidelity of pollinators see flower fidelity
  first flight of day 237
  flower colour and 202, **203**, 238
  flower scent, scent gradient 237, 238
  genetic preference for specific flowers 215–216
  groups of rewarding flower species 215
  as innate and learned behaviour 236, 238
  landmarks for, visual map and 237, 238
  at low temperatures 104–105
  oligolectic vs polylectic bees 216, 236, 237, 259–269
  see also oligolecty/oligolectic bees; polylecty/polylectic bees
  optimal 217
  Optimal Foraging Theory 216–217
  posture whilst (*Megachile*) 257, 283
  rate, resource availability and 107
  slowing, by switching flower type 217

species-specific differences 260
  on summer pasture 219–220, **220**
  time constraints, brood cell provisioning 106, 107, 135, 237
  trip duration 107, 244, 278, 303
  trip number 107, 115, 120, 123, 131, 135, 236
  trip ranges/distances 237
  see also flower(s); nectar; pollen
forewings
  length, prey of *Cerceris rybyensis* **357**, 357–358
  venation 2, 23, **34**
fossil bees 406, **406**, 408, 409
foundress females ('queens') 50, 160, 161, 162, 163, 165, 166, 173
  broods of workers/sexuals 161, 162, 164, 165, 173, 176, 386
  lifespan 430
  maintaining order in nest 174, 386
  pheromones/secretions 175, 176
  size 174
  Sphecodes monilicornis attacking 386
  subsocial behaviour 160, 161, 162, 163
France, bee species/genera 420, 420
fruit trees, pollinators 509, 516
functional groups, flowers/pollinators 227–228

galea **8**, 239, **241**, **242**, 295
galls, nesting in 149–153, **151**, **152**, 341, 473
gametes 180
ganglion 7
garden(s) 497, 501
  bee nesting places see bee hotels
  flower types, bee species 227–228, 229, 493, 500, 524
  habitat-specialists (bees) 501
  helping bees, advice 500, 519–522, 524
  houses without or little, threat to bees 502
  importance for bees 497, 501, 502, 519, 520
  pollination networks in 232
  solitary bee diversity, measurement 412
  solitary bee species 493, 500, 520, 520
    'lockdown' survey 500–502
    as 'stepping-stones', gene flow 501
  gardeners, role 522
Gelli Tips, Rhondda Cynon Taf, Wales **490**
genera of solitary bees 32–33, 33–58
  characteristics 55–58
  increase/decrease over recent decades 445–447, 446–447
  number, and number of species 32–33, 416–417
generalism and generalist bees 225, 236, 259–269, 287
  less efficient foraging than specialists 263

trend towards 451
  in urban landscapes 504, 505
  see also polylecty/polylectic bees
generalist plants 225, **231**, 232, 287
'genome downsizing' 409–410
geographical distribution see diversity and distribution of bees
germ pores **180**
Germany, bee species/number 269–270, 271–272, 273, 282, 456
glaciers 418, 419
glands
  Dufour's see Dufour's gland
  epithelial oil (*Macropis*) 275
  on modified tarsi, *Megachile willughbiella* 94
global warming see climate change
glossa 240, 241, **241**, **242**, 243
'grab and stab' procedure 424
grass/lawns, maintenance
  in gardens 497, 520–521
  in green spaces **457**, 458, 502, **503**, 506
  shift to reduced management 506, **507**, 508
grass stems, covering nests 140, **141**
grassland/pasture
  ancient, bee community (in Breckland) 230–234, **231**
  bee diversity (UK) 413–414
  field observations 219–220, 230, 232–233
  limestone, bee types on 219–220
  lowland dry acid 468–470, **469**
  Northwick, Canvey Island 484, 485, **486**
  plant species (in Britain) **231**, 232
  Thames Gateway sites 484–485, 492, 508
  Gravel Hole, county Durham 487
grazing, by animals 464, 465, 468, 482
green spaces see urban landscapes and habitats
gregarious nature 16
Grimston Warren, Norfolk 480
gymnosperms 409
gynandromorphs 19, **20**
gynes (fertile females) 166, 167, 176

habitats for bees 413–414, 455, 463–480
  Breckland 230, **231**, 232–233, 413–414, 480
  for cavity nesters 461–462, 500, 521, **521**
  chalk downland 463–465, **463–465**, **465**, 487
  cities 414
  coastal 414, 474–480, **475**, **476**, **477**, **478**, **479**
  farmland see agriculture
  fragmentation, impact 504, 509
  grassland see grassland/pasture
  habitat-specialists, visiting gardens 501
  heather heath (lowland with acidic soil) see heather heath

'ideal' 411
latitudinal gradient in UK see latitudinal gradient
loss of 455, 456, 511
lowland dry acid grassland 468–470, **469**
Machair grassland **483**
mosaic/mix of patches 482, 484
natural 463–480
'partial' (not all bee requirements) 481–482
requirements for/of 461, 462, 481–482
semi-natural 463–480, 481–482, **518**
  characteristics 480–483
  conservation strip in agriculture 515, 516, 517, **517**, 519
  decline/destruction, restoration 480
  disconnected fragments 481, 484
  importance 482–483
  MoD ranges 480, **481**
  pollination network 516–517
  spatial scales (local, landscape, regional) 482
  upland sites (mountain/moorland) 470–471, **471**
  urban see urban landscapes
  urban–rural divide and boundary erosion 492
  warm, dry 410, 411
  wetlands **473**, 473–474
  woodland 471–473, **472**, 480, 526–527
  see also diversity and distribution of bees
Hadleigh Downs Country Park, Essex **487**
hair bands **29**, 43, 50, 51
hairs of bees, and pollen collection 3, 4, **5**, **8**, 18
  on abdomen (under side) 4, 245, **256**
  age effect 430
  around eyes 5, **5**, 394
  on base of tergites 50, **50**, 51, **51**
  branched, on body 4, 22, 244, 245, 256, 293
  brushes of on trochanter (*Megachile*) 283–284
  on clypeus 245, 246, **246**, **247**, 281, 285
  colour 11, 18–19, 453
  electrostatic charge, pollen attachment 3, 239, 244, 411
  on face 22, 245, 246, **246**, 248
  on front legs 245, 248
  growth, bee development 14
  on hind legs 4, **5**, **8**, 254
  see also scopa (scopae)
  insulating 424, 461
  long 3, 4, **5**, 18, 207, 256, 257, 258, 283
  long plumose 200, 244, 257, 258, 273, 277, 283, 293
  male *vs* female bees 11, 18–19
  plumose curved (floccus) 257, 258
  shape/structure **247**, 254, 256, 257
  wasps *vs* bees 3
hairs of plants, bees collecting 81–82, **82**, **144**, 144–145

hand lens use 13
haplodiploidy 18, 61–62, 174, 176
harassment of female bees see under mating
head, of bees 7, **8**, 18, 65
  of larvae 156
heather heath 465–468, **466**
  importance, but losses with time 467
  restoration and tree removal 468, 480
hedgerows 232, 518
Herbert, William 183
herbivores 204, 228, 230, 465
hermaphrodite flowers 183, **183**, 186, **188**
heterostyly 184
hibernaculum 176
Hislop, Louise 487–488, **489**
honeybees see honeybees (*Apis*) (*Species Index*)
Höppner, Hans 397
horns, *Osmia bicornis* 25, 26, **26**
host intimidation, by *Sphecodes* (blood bees) 386, 390
'hotspots', bee abundance/species 410
hoverflies 10, 185, 187, 191, 228, 230, 244, **454**
hovering, without landing 22, 37, 208, 218
hydrocarbons 89
hypopus 347

Ice Ages 364, 418, 454
identification of solitary bees 12–13, 27
  guides for 23
  keys for 13, 25, 27
  see also distinguishing solitary bees
immature stages of bees 14
immune system, evasion 331, 333, 349
imported species (in plants/timber/ accidental) 448, 450
imprinting 262, 390
'incomplete' metamorphosis 14
inquilines see cleptoparasites
Insect Pollinator Initiative 511, 528
insecticides 512, 513, **514**
insemination 61–62
'interference competition' 244
intra-species usurpation, of nests 147, 166
Ireland, bee species/genera 420, 420, 482

Johnston's organ 7, 62

keel, of Fabaceae flowers **194**, 250, 251, 252
Kelling Health, Norfolk 356, **466**, 467
Kenfig Dunes, Bridgend, Glamorganshire 478
Kew Royal Botanic Gardens 468
keystone species 445
killing of bees (for research) 13
Kirby, Reverend William 23–24, **24**, 330, **330**, 397
Kirk, William 409

Knapweeds 199, 232
Knight, Thomas 183
Kölreuter, Joseph 222

labial palpi **8**, 240, 241, **241**, **242**
'labile preference' 216–217
labium **8**, 239, 240, **241**
labrum **8**, 240, **242**, 281, 283
'land use' and 'land management' 525
landscape recovery projects 519
landscape scale, habitats 482
large solitary bees **12**
  selective advantages 108
  temperature regulation 12, 424, 425, **425**, 427
larvae (bee) 14, 102, 155–156, 304
  anatomy 156
  cuckoo bees 361, 367, 371–372, **372**, 385
  fed on non-host pollen 266–267
  growth/development 155–158, 156, 435
  male *vs* female, conversion ratio 106
  nutrition 105–106, 156, 157
  overwintering as 157
  as parasite target 303, 304, 435
  pollen grain structure preventing digestion 267
  predators of (birds) 350–351
  protection from harmful microbes 105
  Stelis species **396**
  survivorship, in urban areas 493
  toxins in pollen and 264, 265, 267, 288
larvae (flies and other insects)
  aculeate wasps (*Sapyga*) 339, 340
  bee-flies 317, 319, 320, 322
  Cacoxenus 326, **327**, 328
  Chrysura wasps 344
  conopid flies 322, 325
  Gasteruption wasps 341
  instar 319
  Myrmosa atra wasps 343
  oil beetles (Meloidae) 333–334, **334**
  shadow flies, *Miltogramma* 307, 313
  Stylops 331, 332–333
  triungulin, of blister/oil beetles 334, **334**, **335**, 336
latitudinal gradient
  bee size and temperature regulation 427
  effects on bee activity 415
  global, diversity of bees/organisms and 410–411, **411**
  solitary bee diversity in UK **414**, 414–415, 418, **418**, 463, 470
  cuckoo bee species 416, **417**, 417
  by family/genus 416–417
  northern bias, species with 415, 418, 442
  Osmia species 440–441, 442
  southern bias, species with 414, **414**, 415, 416–417, 463
lawns see grass/lawns, maintenance

leaf/leaves
  cutting by *Megachile* species 132–133, 135, 346
  hairs, cutting by *Anthidium manicatum* 81–82, **82**, **144**, 144–145
  nibbling by *Osmia leaiana* 137–138, **139**
leaf mastic *see* mastic
learning by bees 218–219, 238
  foraging strategies and 236, 238, 244, 261, 263
legs 7, **8**
  front legs, hairs on, for pollen 245, 248
  hind legs 4, **5**, **8**, 254
    plant oil collection 116, 275, **276**
    scopae on *see* scopa (scopae)
  mid-legs 71, **71**, 253
lethargic behaviour 329, 332
life cycle of bees 13–17, **14–15**, 101–158, 435
  climatic conditions and 411–412
  diapause *see* diapause (dormant stage)
  stages 14, **14–15**, 16, 102
  time constraints 104, 435
  see also eggs (solitary bee); larvae (bee); pupa
lifespan of solitary bees 12, 18, 427–431, 438
limestone grassland, bee types on 219–220
linalool 66, 89
'living landscapes' 483
load-carrying capacity 106
long-term memory (LTM) 218
longevity of bees (lifespan) 12, 18, 427–431, 438
lowland dry acid grassland 468–470, **469**

Machair grassland, Inner Hebrides **483**
macrocyclic lactones 126, 131
male gamete 180
male mate-finding strategies 63–87, 87, 99–100
  changes during day/season 65
  communally nesting species 74–75
  competition *see* competition between male solitary bees
  digging for female mates 6, 66, 67, **67**
  harassment of females *see* mating/mating systems
  male emergence before females 18, 61, 63, 65, 66, 438–439
  non-resource-based strategies/patrolling 64–65, 83–87, 87, 100
  patrolling female emergence sites 18, 63, 65–75, **66**, **67**, 87, 99, 236, 462
    Andrena species 69–71, **70**, 427
    Anthophora species 71, **71**, 79, 87, 429
    cavity nests, 'staking out' 73–74, **74**, 100

Colletes species 65–68, **66**, **67**
  competition between males 67–68, 75, 87
  Heriades truncorum 73, 95
  Hoplitis adunca 73
  Lasioglossum species 68–69, **69**
  Megachile species 73–74, **74**, 100
  at nest aggregations 18, 63, 65–75, 100
  Osmia species **72**, 73, **89**, 91–92, 99, 421
pheromones released *see* pheromones
pouncing on females 61, 65, 66, 68, 71, 73, 75, 88, 97, 421
at rendezvous sites 63, 64, 65, 75–83, 235
resource-based (patrolling flower sites) 64, 65, 75–83, 87, 100, 235, 236, 421, 462
  Anthidium manicatum 30, **81**, 81–82, 100
  competition between males 67–68, 75, 78–79, 80, 81, 87
  diurnal patterns, cooler periods 76
  females seeking mud, resin or leaf cuttings 83
  leafcutters (*Megachile*) 73–74
  male 'queues' 79
  non-resource-based, distinction 87
  oligolectic species 64, 75, 76–78
  plant hair collection and 81, 82, 144
  polylectic species 75, 78–79
  scent-marking with pheromones 64–65, 83, 86, 422
search strategies 63–65
male solitary bees 1, 3, 17–20
  anatomy 3, 4, 8
  for coercion/restraining female 18, 93–94, **94**, 95, **95**, 96
  for competition between males 18, 65
  appearance, *vs* female bees 17, **18**, 18–20, **19**
  basking 421, **422**
  competition *see* competition between male solitary bees
  conversion ratio, male larvae 106
  corbiculate bees compared to 4
  as dispersive sex 69
  emergence, timing 18, 61, 63, 65, 66, 438–439
  features by family, *vs* female 55–58
  haploidy 18
  longevity 427, 428, 429
  mating *see* mating/mating systems
  nectar collection 18, **31**, 236
  patrolling *see* male mate-finding strategies
  pheromones produced 63, 64–65, 83, 86
  ratio to female bees 17, 61
  resource limitations leading to shift to 108

sheltering in flowers 433, **434**
size *vs* female size 17, 82
switch to male brood cells, by older bees 431
from unfertilised eggs 17, 61–62, 174, 176
Mallorca, pollination network 233
mandibles **8**, 17, 19, 22, 35, **242**, 287, **287**
  burrow digging 113
  cuckoo bees 386, 387, **390**, 397, 401
  holding grass stems **141**, **143**
  large 40, 43, **288**
  leaf cutting 133, 137, 138
  mud carrying 134, 136
marginal cells **34**
'mass provisioners' 62, 102, 171–172
mastic, leaf 346
  brood cell construction 45, 133, 134, 137, 138, 139, 142, 154, 155
  cap for aerial/cavity nest 137–138, 140, **141**, **143**
  collection, for snail shell nests 140, **142** 142, **143**
mating balls 66, **67**, 67–68, 69
mating behaviour (of parasites) 330–331, **331**, 336
mating/mating systems (bees) 18, 63–87, 88–99, 100
  coercion/restraining female 18, 93–94, **94**, 95, **95**, 96
  cuckoo bees 100
  duration 97, 98
  female secretions inhibiting 88, 89–90, 99, 100
  first male to mate, monandrous female bees 18, 61, 63
  last male to mate, polyandrous female bees 61
  male harassment of females 60, 77, 79, 83, 88, 90–91, 95–96, 133
  inhibition, and repulsing males 63, **77**, 88–89, 90–91, **91**, 99
  mate-finding *see* male mate-finding strategies
  mating with more than one female 64, 65
  mounting a female 61, **67**, **70**, 77, 79, 81, 88, 97, 100
  but not copulating 78, 88, **92**, 92–93
  gap between mating and 92–93, 97
  phases 61–62, 63
  pheromones secreted *see* pheromones
  'post-copulatory embrace'/phase 88, 99
  pouncing on females by males 61, 65, 66, 68, 71, 73, 75, 88, 97, 421
  preference for unrelated females 69
  rapprochement 61, 63–64
  unsuccessful, *Osmia leaiana* 78, **92**, 92–93
  vibratory taps during 62, 97, **98**
  see also copulation; courtship; sexual behaviour; *specific species* (*Species Index*)

maxilla 156, 239, **242**
maxillary palpus (palpi) 240, **241**, **242**
Maze Park, County Durham 487
Mediterranean region 410, 411, 434
memory, in insects 218, 237, 238
mentum 240
mesolecty 236, 260
Mesolithic–Neolithic transition 419
mesosoma 7
metamorphosis 14, 435
metasoma 7, **8**
metatarsus (basitarsus) 4, **5**, **8**, **9**, 256, 275, 277, 283
protobee (evolution of bees) 408, 409
Middlewick MoD Range, Essex **469**, 470
mimicry 208–214
　Batesian 208
　floral, pollination and 208–214, 209, **209**
　　by bee orchids (*Ophrys*) 211
　　by Early Spider Orchid 211, **212**
　　by poorly rewarding plants 208, 209, 215
　　pseudocopulation and 63, 211, **212**
　　within-species, Red Campion 213–214
　Müllerian 208
　see also deception
mirror-image flowers 184
Mitcham Common, Surrey 468, 470
Mites (Acarina) 346–348, **347**, **349**
'mitigation hierarchy' 526, 527
MoD ranges, bee habitats 463, **469**, 470, 474, **476**, 480
monandrous solitary bees 18, 60–61, 63
monoecious plants 183
monogynous nests 166
monolecty 236, 259, 260
　obligately or facultative 260
monoleges, definition 259
moths pollination syndrome 223
motor learning 218, 219, 263
mountains, habitat 470–471, **471**
mouthparts 7, 22, 239–240, 240–241, 302
　anatomy **8**, 239–240, **241**
　of larvae (bee) 156
　solitary bee families 241, **242**
　see also mandibles; tongue
Mucky Mountains, Newton-le-Willow **488**
mud, for brood cells in aerial nest 134, 136, **137**
mud-brick nesting site 37, 521, **523**
Müller, Hermann 178
Müllerian mimicry 208
mutualisms/mutual adaptations 60
　cuckoo bee and hosts 379
　mite larvae and bees 348
　solitary bees and flowers 36, 222, 225, 265, 460–461

naming of bees 25–26
National Pollinator Strategy 514–515, 526
native plant species (British) 227
'natural capital' 528, 529
natural history, amateur 458–459, **459**
natural selection 409–410
　cleptoparasites, female bee body size 425, 427
　plants, by pollinators 221–222
　pollen collection by protobee 409
naturalists, amateur 458–459, **459**
nectar 4, 18, 22, 178, 302, 511
　advertising location, nectar guides **188**, 191
　Brassicaceae 193, **193**
　for brood cells 16, 105, 113, 115, 236, 252–253
　for bumblebees/honeybees 219, 220
　collection (by solitary bees) 4, 22, 119, 238–244, **243**, 252
　　adaptations for 22, 240–243, 302
　　by cuckoo bees 369, 378
　　male solitary bees 18, **31**, 236
　　with pollen 35, 38, 238, 243, 244, 252
　separate days from pollen days 122, 238–239
　temperature effect 423, **423**
　tongue for *see* tongue
　composition and sugars 105, 181, 239, 243
　dilute 105, 214, 218
　female bees imbibing 107, 236, 238
　pollen mixed with 16, 107, 113, 115, 252, 253, 283, 301
　in brood cells 16, 113, 115, 120, 124, 253
　corbiculate bees 4, 252–253
　in crop, for transport/storage 4, 54, 107, 120, 253, 301
　Melitta blunthorn bees 35, 107, 253, 277, 278–279
　scent paired with, learning 219
　secretion as disadvantageous to orchids 211
　secretion by plants, and function 179, 180–181
　storage 253
　toxins in 181, 235
　transport *see* transport of pollen/nectar
'nectar days' 122, 238–239
nectar guides **188**, 191
nectaries 179, **179**, **188**, 238, 243, 245, 279
　hidden 239
Neff, John L. 106, 107
neonicotinoids 512, 513
nest(s) 1, 16–17, 109–155, 134
　aerial/cavity 17, 132–153, 461
　　architecture/design 133–134, 136, **137**, **139**, 145, 148, **340**
　　in bee hotels *see* bee hotels
　　body size variability and 108, 109
　　in built-up urban areas 492

　　in chalk downlands 465
　　competition for sites 108
　　construction method 133–134, 136, 148, **149**, **152**
　　in galls 149–153, **151**, **152**, 341, 473
　　in gardens/garden walls 461, 500, 521, **521**
　　genera and species 109–110
　　leaf hair lining 81, 82, **144**, 144–145
　　leaf segment cap 134, **137**, 150
　　leaf segments to line/divide up 133, 135
　　mastic cap 137–138, 140, **141**, **143**
　　mud lining of cells 134, 136, 521
　　portals, construction 148
　　resin cap 110–111, **145**, 145–146, **146**
　　size, variable nutritional input 108–109
　　in snail shells *see* snail shells
　　in stems 109–110, 132, 155, **339**, 341, 345, 351, 439, 445, 462
　　vestibule (gap) 134, 136, **139**, 150, **340**, 341
　　weather effect/protection 432
　　in wood, or pith extraction 17, 110, 117, 153–155
　ageing (bee) effect on 430
　aggregations 16, 50, 53, 104, **112**, 159, 306, 431
　cleptoparasitic flies at 306, 307, 311, 312, 313–314, **314**, 315
　density, genera 111
　males patrolling *see* male mate-finding strategies
　towards morning sun **431**, 431–432
　brood cells in *see* brood cells
　burrows *see* nest(s), ground/underground/burrows
　buzz of bees at 249
　cavity *see* nest(s), aerial/cavity
　classification of types 109–110, 154
　communally nesting species 104
　competition for sites 108, 121–122, 168
　construction by females 60, 104, 113–114, 423
　digging, lining 107, 113–114, 115, 117, 118, 120, 123, 125, **127**, 128
　resting period after 121–122, **122**
　'reverse digging' after 117
　speed 104, 117, 120, 130, 134, 136, 154, 423
　time spent, foraging trip number 134, 135
　construction in glass tubes, *Colletes kincaidii* 130–131
　in dead wood 17, 40, 54, 110, 117, 154, **155**, 442, 472
　entrances 16, 47, 113, **113**, **114**
　cavity nests, navigation, visual cues 134
　closing, protection from flies 315, **316**, 317, 319–320

GENERAL INDEX · 591

closing, protection from *Nomada* 379
closing in bad weather 432
concealed 114, 116, 121, **121**, 134
covering of 120, **121**, 237
'false'/'accessory'/'decoy' 121, **121**, 125, 237, 379, **380**, 381
grass stems covering 140, **141**
ground/underground/burrows 113, **113**, 115
identification, orientation flights 117
second, *Panurgus* 123, **124**
sharing (communal) 16–17, 74–75, 104, 159–160, **160**, 312
tumulus at 113, 123, **124**, 125, 128
'turret' **170**
waiting at, before foraging trip 120
foraging/provisioning for *see* brood cells; foraging
free-standing above ground, resin nests 17, 110–111, 145–146, 341, 412
in galls 149–153, 150, **151**, **152**, 341, 473
ground/underground/burrows 6, 16, 47, 111–131, **112**, 113, 121–122, **122**, 154, 461
architecture 16, 113, **114**, 119, 123, **124**, 126, 128, 171
on banks/dunes 65, 66, 126, **494**, **496**
body size and cell size 108
Colletidae 65, 66, 126–131, **127**, **128**, **129**
depths 111, 129
in a ditch **525**
in gardens 522
genera and species 109
Halictidae 125–126, **126**
on heather heath **466**
lining, composition 107, 113, 120
on lowland dry acid grassland 468, 470
Melittidae 114–116, **115**
protection from flooding/weather 103, 129, 412, 432
'pyramids' (tumuli) next to 113, 123, **124**, 125, 128
'reverse digging' after completion 117
in root plates 111, **112**, 121
side-burrows 16, 128
vertical, or horizontal ground 111, 118, 126
walls 16, 113, 115
waterproof lining 16, 113, 115, 116, 120, 126, 129, 134
of honeybees, in hedgerow 6
isolated, shadow fly visits 315
materials used 16, 83, 462
mixed-lineage, eusocial bees 166–167
monogynous 166
orientation flights and *see* orientation flights

protection from inclement weather 103, 129, 412, 432
resin 17, 110–111, 145–146, 341, 412
resin cap 110–111, **145**, 145–146, **146**
in root plates 111, **112**, 121
sealing, by nest foundresses 165, 168
secretions for 16, 22, 115, 120
sites, range of 431–434
sites and distance from flower patch 237, 261
in snail shells *see* snail shells, nesting in
solitary 104, 315
in stems 109–110, 132, 155, **339**, 341, 345, 351, 439, 445, 462
stones in 146, **146**
urban *vs* rural locations 492
usurpation *see* usurpation of nests
wood-borers and pith excavators 17, 110, 117, 153–155
nest box, observation 132, 133, 136
nest foundress *see* foundress females ('queens')
nest-parasites *see* cleptoparasites
Netherlands, bee species/genera 420, 420
neurotoxin insecticides 512, 513
nomenclature of bees 25–26
non-resource-based patrolling *see* male mate-finding strategies
non-resource-based polygyny 64–65
Norway, bee species/genera 420, 420
'nototribic' deposition, of pollen 226, 245, 277, 281
number of species of all bees 410
global 182, 410
UK 33, 183
UK *vs* Europe 420
number of species of solitary bees 2, 11–12, 27, 32–33, 33, 182
in Britain *vs* Europe 420, 420
cuckoo bees, distribution in UK 415
nutrition, of bees
of larvae 105–106
see also food, for bees; nectar; pollen
nutritional quality of pollen *see* pollen
nymph life stage 347

observing solitary bees 11–13
β-ocimene 354
oils, floral, collection *see* plant oils
'old age' changes 430
Olds, Liam 488–489
oligolecty/oligolectic bees 198, 201, 216, 217, 236, 245, 262, 263, 302, 461
adaptations for 263, 290–291
advantages/benefits 262–263, 264, 265, 267–268, 302
coincidence of flight period and flowering 263, 265, 461
decline in (before/after 1980) 509–510
definition/criteria 259, 260
disadvantages/risks/handicaps 259, 260, 261, 263, 265–266, 302
emergence, timing 435, 437

evolutionary history 261, 262, 268–269, 280, 302
families, genera and species 200, 269–270, 271–272, 273–301
species and pollen sources 436
species number 261, 437
in gardens 500, 520
'imprinting' hypothesis 262
innate attraction to specific plants 261–262
larval immunity to plant toxins 264, 265, 273
male mating strategy, resource-based 64, 75, 76–78
more efficient foraging 263
'narrow' and 'broad' 236, 260
non-genetic mechanisms predisposing 262
nutritional reward for 264
responses to plant visual/scent cues 238, 261, 262
seasonal (summer), *Andrena bicolor* 261
shift to other plants, and costs of 262, 263, 282, 290
shift to polylecty 105, 261, 262, 266, 290, 302
threats to, sparse distribution of plants 504
univoltine bees 437
see also specialism and specialist bees
oligoleges, definition 259, 260
O'Neill, John 526, 527
optimal foraging 217, 221
Optimal Foraging Theory (OFT) 216–217
orchids 63, 210, **210**
organochlorines and organophosphates 512
orientation flights 117, 120–121, 133, 134, 237, 306, 356
when leaving nests 120–121, 134, 142, 237, 306
when returning to nests 117, 133
Orkney and Shetland 191, 415, 416–417, 427
outbreeding *see* cross-pollination (and cross-fertilisation)
Outer Hebrides 191, 415, 520
ovary 179, **179**
overheating, of bees 425, 429
overwintering 430
adult fertilised females 51, 68, 69, 161, 162, 164, 165, **165**, 166, 176, 438
as adults 155, 156, 430, 435, 437, 438
Colletes cunicularius (Early Colletes) 156, 438
eusocial halictids 68, 69, 161, **165**, 176, 297, 385, 386
in larval stage (*Colletes hederae*) 157
Nomada, as adults 385
as prepupae 158, **158**, 430, 438
Sphecodes species 386, 387
ovipositor 23
ovules 179, **179**, 180

papillae, pollen grain 180
parasites and parasitism (general) 303–304, 306
predators vs 304
parasites of solitary bees 304, 306–349, 462
adaptive host manipulation by 332–333
of adult bees 322, 323, 325, 330, **330**, 331, **331**
of bee larvae/eggs 304, 306–307, 309, 313, 319, 331, 345, 435
bee toxins produced against 335, 336, 345, 346
conservation of 349
cuckoo bees *see* cuckoo bees
egg-laying 102
emergence time influenced by 439
emergence traps, parasitism rate 311, **312**
evasion/prevention by bees 136, 306, 309, 315, 322, 323, 333, 335–336
host-specific, and niche specialists 306, 349
increase, with age of bee 431
newly established populations and 454
orders, groups, and genera 305
see also specific groups (page 305) *(Species Index)*
spiders catching 355
Wolbachia (bacteria) 439
parasitic wasps 136, 153, 265, 339, 340
parasitoids 237, 304
bee-fly larvae 317
Chrysura (wasp) 344
egg laying 102
generalist, in bee hotels 349
Stylops 331, 331–332
parental investment 60
Parker, Rob 140
parsivoltinism (delayed emergence) 437–438
patrolling flights *see* male mate-finding strategies
Paxton's framework 63–64, 65, 74, 78
pedicel 7, **8**
Pembry Country Park, Carmarthenshire 478
Perkins, R.C.L. 24
pesticides 504, 511, 512, 513, **514**
absence on MoD sites, bee recovery 480
petals, flower 179, **179**, 182, 433
phenology *see* flight period (phenology)
pheromones (bees secreting) 62, 63
antiaphrodisiac 63, 99, 120, 211, 382–383
'calling', secreted by females 88, 89
courtship phase 88
on cuticle 69, 89, 90, 92, 98–99, 381
Early Spider Orchids releasing 211
female 63, 66, 69, 88, 89, 92, 99
changes after mating 89–90, 99
inhibiting mating 62, 88, 89–90, 99, 120

foundress females ('queens') 175, 176
linalool 66, 89
males releasing 54, 63, 83, 88, 382–383
males scent-marking a trail 64–65, 83, 86, 422
release by *Nomada* 63, 382–383, **383**
see also chemical communication; scent(s)
phylogeny (evolutionary history), eusocial behaviour 169–173
Pilkington, G. 88, 132, 134, 136, 137, 148
pith extractors (bees) 17, 110, 155
planidium 319, 320, 331
planning system, and bees 525–527
recent changes to enhance biodiversity 525–526
plant(s)
generalist species 225, **231**, 232, 287
hairs, collection by bees 81–82, **82**, 144, 144–145
specialisation 225, 227, 234
see also flower(s); flowering plants
plant-bee networks *see* pollination networks
plant breeding systems 182–190
apomixis (seeds, without fertilisation) 189
cross-pollination *see* cross-pollination (and cross-fertilisation)
self-pollination *see* self-pollination
plant oils 36, 102, 253, 302
collection method 275
pollen mixed with/agglutination 105, 116, 275
see also Macropis (Species Index)
plant-pollinator relationships 221
see also pollination syndromes
pollen 4, 178
anthers producing 179, **179**
attachment, electrostatic charge and 3, 239, 244, 411
in/for brood cells and nests 16, 105, 236, 253, 281
brushing into scopae 22, 120, 245, 253–254, **254**, 273, 283–284, **295**
setae 253, 256, 273, 275, 284, 301
collection 4, **6**, **21**, 22, 119, 244–252
adaptations for **21**, 244, 245–248, 302
from Asteraceae, presentation in florets 198, **198**
from Boraginaceae 248, **248**
from Fabaceae 250–251, 292
hairs for *see* hairs of bees
incidental, positive charge and 244
from Lamiaceae 245–246, **246**, 247, 280–281
large loads, bee species 131, 273, 277
mechanism 244–252
mesolecty 236, 260
monolecty *see* monolecty
with nectar 35, 238, 243, 244, **252**

oligolecty *see* oligolecty/oligolectic bees
polylecty *see* polylecty/polylectic bees
properties limiting range of bees 200
by protobee, in evolution 409
rate, and load-carrying capacity 106–107
separate days from nectar days 122, 238–239
sonication (buzz pollination) 248, 249–250, 281, 509
speed, ambient temperature and 423, **423**
warm, dry habitats 411
see also specific bee families, genera, species
concealed, buzz pollination 249–250
cost of production for plants 193, 244, 266, 302
reducing, but attracting bees 192, 193, 200, 266
difficult to determine pollen diet of bees 260
fate of/destiny 200–202, 201, 360
forms, in Thrift flowers **186**
generalist bees *see* generalism and generalist bees
grooming 200, 244, 245, 253, 254, **295**
interference between (plant) species 180, 195
loads, basis of pollination network 233
magnitude of bee demand for 244
morphology, in heterostylous plants 184
narrow pollen diet *see* oligolecty/oligolectic bees
nectar mixed with *see under* nectar
non-host, effect on larvae 266–267
'nototribic' deposition 226, 245, 277, 281
nutritional composition/quality 105, 108, 245, 263–264, 302, 406, 409
nutritional needs of female bees 235, 236, 252
oil mixed with 105, 116, 275
properties limiting bee species 200
protection from hazards 192
proteins 105–106, 263, 264, 267, 280
rationing 190, 192, 193, 198, 200, 244
release (by flowers)
buzzing of bees 248, 249–250, 281, 509
short-tongued bees 187, 245
staggered **203**, 203–204, 266
synchrony with flight period 265–266
scent, 'imprinting' oligolecty 262
secondary presentation (SPP) 192, **192**
specialist bees *see* specialism and specialist bees
'superabundant' sources 265

symbiotic bacteria transferred via 105
toxins in 200, 201, 264, 265, 267, 288, 290, 340
transfer from face 246, **246**, 248, 253
transfer into brood cells 136
transfer to scopae *see* pollen, brushing into scopae
transport *see* transport of pollen/nectar
viability 182, 192–193
wastage reduction 194, 200, 215
see also pollen grains
pollen baskets (corbiculae) 3–4, **5**, 10, **21**, 182, 253, 256
'baskets' in *Andrena* females 257
bees in tropics (Meliponini) 412
corbiculate bees 3–4, **5**, 10, **21**, 182, 253, 256
femoral 257, 258
propodeal 257, 258, 293, 296
pollen grains
Asteraceae, structure **180**
coated in pollenkitt **180**, 267
digestion 267
exine and intine (walls) 267
size, and structure 180, **180**, 192, 200, 248, 267
very large, species with 200
pollen thieves 190, 196, 200, 299
pollen tube 180, **180**
pollenkitt **180**, 200, 267
pollination 20, 22, 177, 179
by animals 181, 222, 508
by birds 181
cross- *see* cross-pollination
floral attractants for 177, 189, 190, 193, 197
by insects 178, 189, 221, 222, 302, 508, 529
crops *see* agriculture
decline in insect number 508–509, 510
importance 410, 508, 529
by wasps, of Ivy 244
see also pollination by solitary bees; pollinators
as key ecosystem service 528, 529
natural selection of plant traits 221–222
networks *see* pollination networks
reduction, spiders causing 354
self- *see* self-pollination
styles *see* pollination syndromes
in urban areas 492
by wind, of crops, neonicotinoids and 513
see also pollinators
pollination by solitary bees 20–22, **21**, 181–182, 190–200, 302
of Apiaceae 199, 200
of Asteraceae *see* Asteraceae (*Species Index*)
attractants for 177, 189, 190, 193, 197, 211, 239
attracting bees, reducing pollen costs 192, 193, 200, 266

of Boraginaceae 196, **197**
buzz 248, 249–250, 281, 509
of Campanulaceae 191–192, **192**, 278, 286–287
coevolution of bees and plants 20, 266, 268–269, 296
of Common Poppy (*Papaver rhoeas*) 186, 187, 249
competition between plants 190, 208
of Dipsacaceae 199–200, 201, 203
effectiveness, *vs* non-bees 182
evasion of plant adaptations 200, 201, 202
of Fabaceae 193, **194**, 194–195, 239, 250, 282, 283, 292
fidelity *see* flower fidelity
field studies, in summer pasture (UK) 219–221
flower closing response after 204, **205**, 206, 207
flower colour change after 190, 202, **203**
flower constancy *see* flower constancy
flower morphology and 190–200, 207, 229, 266
bellflowers (Campanulaceae) 191–192, **192**, 226
bowl-shaped flowers 226
clustered flowers 196–200, 226
complex flowers 245
deep tubular flowers 193–196, **194**, **195**, **196**, 226, 228
small symmetrical flowers 191, **191**
types/shapes and bee genera 226
flower opening/closing 204, **205**, 206, 207
flowering timing 202–207
functional groups of flowers/bees 227–228
in gardens 500
of heterostylous plants 184, **184**, 185
importance of solitary bees 181–182, 460–461
of Lamiaceae 195, **195**, **196**, 239, 245, 246, 249
less-/more-rewarding plants and 190, 208
limiting duration of visits 189, 190, 237, 244, 278, 303
long-tongued bees 184, 185, 193, **194**, 194–195, 222, 228
mimicry (floral) *see* deception; mimicry
of orchids 210, **210**, 211–213, **213**, 226
plant types and specific bee genera 226
of poorly rewarding plants 190, 208, 209, 215
portion size, increasing 192
precise placement of pollen 193–194, 195
rationing pollen 190, 192, 193, 198, 200, 266
reduced, by mixture of nectar/

pollen 253
reward from plants *see* rewards (floral)
scabious bees, beneficial or harmful? 201
selection pressure on flower form 207, 266
short-tongued bees *see under* tongue
staggered release of pollen **203**, 203–204, 266
temperature affecting 204, 207, **208**, 228
time of day 203, 204, 206–207
'pollination crisis' 234, 460, 508, 509
pollination networks 230–234, **233**, 510, 516
Essex nature reserve **230**, **231**, 232
habitat degradation/species loss 234
inter-connectedness 234
nested 232
nodes and edges **231**, 232
resilience in, and 'redundant' species 510
studies, challenge of 232
trait-based 232–233
pollination syndromes 221–230
bee genera and flower types 223, 226, 227
close specialisations (harmony) 225
functional groups, British garden 227–228, 229
list of syndromes and features 223–224
pollination webs *see* pollination networks
pollinators
conserving/enhancing, proposals 514–515
of crops *see* agriculture
decline in, concerns 508–509, 510, 511, 512
diversity and importance of 515–516
as economically important service 528
insecticide impact 512
international survey 515
monitoring of abundance 509
research, funding 528
pollinium 210, **210**
pollution, in urban landscapes 504–505
polyandrous solitary bees 61, 65
polygyny 64–65
polylecty/polylectic bees 216, 217, 220, 236, 260, 302, 461, 504, 505
advantages/disadvantages 262–263
bivoltine species 437
broad 260–261
definition/criteria 260–261
eusocial halictids 173
families, genera and species 257, 258, 269–270, 271–272, 273–301, 435
'imprinting' hypothesis and 262
labile preference 216–217
seasonal (spring) (*Andrena bicolor*) 261

shift to from oligolecty 105, 261, 262, 266, 290, 302
in urban landscapes, advantage 504, 505
'with strong preference' group 300
see also generalism and generalist bees
polyleges, definition 259–260
post-copulatory embrace 88, 99
predators of bees 315, 317, 350–355
  beetles 360
  birds 317, 350–351
  parasites vs 304
  small mammals 360
  spiders 351, 351–355, 352, 353, 355
  wasps 317, 355–357, 356
prementum 240, 242
prepupa 14, 102, 150, 152, 156, 411, 435
  in cocoons 158, 158, 430
  diapause phase 435
  overwintering as 158, 158, 430, 438
propodeum 4, 5, 7, 8
  pollen baskets 5, 257, 258, 293, 296
protandry 63, 65, 109, 151, 183, 184, 201, 328, 377, 438–439
protection of bees, importance 527–530
proteins, in pollen 105–106, 263, 264, 267, 280
protoanemonin 267
protobee (bee evolution) 408–410
protogyny 184
pseudocopulation 63, 211, 212
ptilinum 328
punctures 7
pupa 14, 102–103, 157, 158, 158
  bee-fly 319, 319, 322
  metamorphosis to 435
puparium 330
pygidium 8, 49, 113, 128

'quasisocial' mode 160
queen honeybees, lifespan 429–430
'queens' see foundress females ('queens')

rain/raining 432, 433, 434, 523
rapprochement (phase in mating system) 61, 63–64
reed stems, use for nests in galls 150, 151, 152, 153
'rendezvous' sites (flowers as) 63, 64, 65, 75, 235
reproductive performance, determination 106
reproductive strategy/behaviour 60
  see also mating/mating systems (bees); sexual behaviour
resin
  abdominal secretions softening/hardening 110
  nest construction 110–111, 145–146, 146, 412
  brood cell partitions 110–111, 145, 145–146, 341
resource(s) see food, for bees; nectar; pollen

resource-based male strategies see male mate-finding strategies
resource-based polygyny 64
resource partitioning 221, 262–263, 264, 302, 462
rewards (floral) 235, 238–252, 302, 514–515
  competition by deception and 209
  complex flowers 245
  factors influencing 218, 239
  flower response to wingbeats/vibrations 239
  lacking from orchids 210
  plants improving floral signals 209–210
  pollination without, mimicry and 208, 209, 215
  quantity/quality, as flower fidelity reason 217, 218
  short-term memory and 218
  see also nectar; pollen
'rima' 49, 50
roosting, by bees 432, 433
rose leaves 132, 137, 346

Salisbury Plain, bee species 463, 464, 480, 481
Saunders, Edward 24
scape 7
scent(s)
  flower 87, 237, 238, 239, 279, 280, 286
  change after pollination 190, 211
  use for foraging by bees 237, 238
  learning by honeybees 219
  marking vegetation by males 64–65, 83, 86, 422
  mimicry by cleptoparasites 63
  Nomada (nomad bees) 381
  obligate/facultative eusocial species 175, 176
  of pollen, imprinting oligolecty 262
  produced by bees, mating and 62–63
  non-resource-based male patrolling 83, 84, 85, 86, 87
  resource-based mating strategy 79
  species-specific (enantiomer) discrimination 86
  see also pheromones (bees secreting)
scientific names 25–26
Scilly Isles 416–417, 453
scopa (scopae) 2, 3, 4, 5, 19, 22, 107
  on abdomen 256, 283, 289
  absent in cuckoo bees 401
  absent in Hylaeus 21, 54, 107, 258, 301
  absent on male solitary bees 18, 19
  on Andrena with parasitic Stylops 330, 330, 331
  brushing pollen into see under pollen
  contents, assessment of specialism 260
  on hind legs 3, 4, 5, 8, 22, 253–254, 254, 256, 257, 273, 279, 296
Scotland 319, 415, 442, 450, 463, 470, 480

'scrabbling', pollen release 245, 278, 283, 286
'scramble' competition 65, 70
'scrambling', pollen collection 251, 252, 289
sea walls, habitat 474–475, 476
Secondary pollen presentation (SPP) 192, 192
secretions see Dufour's gland; pheromones (bees secreting)
seed(s) 180
seed dressings 513
seed mixes, to support solitary bees 517–518, 519
selection pressures
  flower morphology 207, 266
  for/against pollen specialism 261–266, 267–268
  see also natural selection
self-fertilisation 186, 188, 188, 189
  see also self-pollination
self-incompatible plants 186
self-pollination 186, 188, 188, 189
  avoidance strategies 183–184, 186, 198
  disadvantages 183
  as fall-back 186, 188, 198, 198, 204
  flower closure not observed 206
  response to weather 204, 204
  specific flowers/sizes 188, 211, 293
self-sterile plants 186, 189
semi-natural habitats see habitats for bees
'semisocial' mode 160
sensilla 175, 341
sensory learning 218, 219, 236
sepals 179, 179
setae (bristles) 136, 241, 245, 253, 256, 273, 275, 284, 301
  on galeae 241, 248
sex determination 17
sexual attractants 63
  see also pheromones (bees secreting)
sexual behaviour 59–100
  cleptoparasites 61
  male vs female (bee) differences 59, 60
  mate-finding see male mate-finding strategies
  mating see mating/mating systems
  Stylops (parasite) effect on bees 332, 332–333
sexual dimorphism 20
sexual selection 59
sheltering, by bees 433, 434, 434, 461
'shivering' 104–105, 424, 432
short-term memory (STM) 218
Shuckard, W.E. 24
Site of Special Scientific Interest (SSSI) 470
size of solitary bees 4, 12, 108–109
  ground- vs cavity-nesting species 108
  impact of habitat fragmentation 504
  large see large solitary bees

GENERAL INDEX · 595

linked to tongue length 232
male vs female 17, 82
pollen limitation and offspring size 108
small see small solitary bees
specialists vs generalists 108
wing length, as proxy for body mass **426**, 426–427
small mammals, as bee predators 360
small solitary bees **13**, 14
temperature regulation 425, **425**, 427
Smith, Frederick 24
snail shells, nesting in 45, **72**, 73, 110, 139–142, **141**, **142**, 440, 445, 478
covering with grass stems/budscales 140, **141**, **142**
nest architecture 139, **141**, **143**
social behaviour 159–160, 161
see also eusocial behaviour
'socially polymorphic' species 163, 164
sonication, pollen collection 248, 249
sounds 62, 369
buzz 7, 22, 62, 73, 248, 249, 277, 284
wingbeats 22, 239, 249
South Downs nature reserve **456**, 464, **465**
specialism and specialist bees 225, 227, 234, 236, 259–269, 273–301, 302
of Asteraceae see Asteraceae (Species Index)
classification/categories 236, 260, 290
difficulties in assessing 260
evolution 261, 266, 302
on island of Mallorca 233
larvae fed on non-host pollen 267
nutritional quality of pollen 263, 264
proportion, by habitat type 261
resource partitioning as benefit 264
'superabundant' pollen sources/flowers 265
tolerance to plant toxins 264, 265, 273, 290–291
see also oligolecty/oligolectic bees
specialist plants 225, 227, 234
species of all bees **531–537**
see also number of species of all bees
sperm 17–18, 60, 61, 88, 99
storage by female solitary bees 61–62, 176, 453
spermatheca 62, 172
Spetchells, Prudhoe 487, **489**
spiders, as solitary bee predators **351**, 351–355, **352**, **353**, 355
spiroacetals 86, 279
spoil heaps 487–489, **488**, **489**, **490**, **491**
spray drift (of pesticides) 511
Sprengel, Christian Konrad 177, 178, 192, 210
spring
solitary bee appearance 28–29, 30, 104, 156, 435, 437, 461, 472
temperature, Osmia cornuta response 444–445

stamens 179, **179**
stems (plant)
covering nests 140, **141**
nests in 109–110, 132, 155, **339**, 341, 345, 351, 439, 445, 462
reed, use for nests in galls 150, **151**, **152**, 153
Step, Edward 16, 25
sterile females 17
sternites **8**, 258
stigma 179, **179**, 198
in heterostylous plants 184, **184**
mimicry by Red Campion 214, **214**
sting(s) 2, 23, 406
Cerceris ryhyensis 355, 358, **359**
Sphecodes (blood bees) 387
stipes 239
studies on bees 23–25
style (flower) **179**, 184, 198, **198**
sub-genus, definition 373
submarginal cells **5**, **34**, 37, 38, 47
'subsocial' mode 160
summer, bee appearance 30–31, 156, 437, **451**
sunlight, flower closing 204
supercooling point 443–444, **444**
sustainable development 459–460, 519
'sustainable farming' 519
symbiotic bacteria 105–106

'tapping', pollen release 245
tarsi **8**, 93, 94, **94**, 95
taxonomy of bees and wasps 2–3, **3**, 13, 32–33, 33, 406
temperature (ambient) 12, 421–427
bee activity 12, 228, 423
bee lifespan, effect on 429
Bombylius major flight and 320
flower opening/closing 204
flower visiting/foraging increased with **423**, 423–424
gradual adjustment to changes by bees 445
high, damaging 424
'ideal' bee habitat 410, 411
low 12, 228, 418
foraging at, 'shivering' and 104–105, 424
minimum for flight 12, 421, 425
nest sites and **431**, 431–432
Osmia cornuta vs O. bicornis response 444–445
in urban spaces, impact 504
temperature (bee body) 421
body size effect on regulating 12, 425, 425
for flight 421, 425
hotter, faster movements 422, 423, 424
measurement in flight ('grab and stab') 424
strategies to control 3, 421–424
roosting/clustering together 432
see also basking; thermogenesis
supercooling points, of Osmia 443–444, **444**

warming up 105, 424
bees using corollas 207, **208**
large vs small bees 12, 424
tergites 8, **8**
hairs on base 50, **50**, 51, **51**
territorial behaviour of solitary bees 18, 65, 78–79, 100
Anthidium manicatum 81, 82
Anthophora species 71, 79, 87
against bumblebees/honeybees 80, 81
males patrolling flower sites 67–68, 75, 78–79, 80, 81, 82, 87, 100
Thames Gateway sites/landscape 484–485, 492, 508
thermogenesis 104–105, 418, 424–427, 461
cuckoo bee vs host 425, 434
large vs small bees 424, 425
'shivering' 104–105, 424, 432
thermophilous species 445
thorax 7, **8**
threatened species 529
threats to bees
in countryside 508
houses with little/no gardens 502
by intensive agriculture 455, 456, 509, 510, 511, 516, 519
by tree-planting 497, 502, **503**, 507
in urban landscapes see urban landscapes and habitats
tibia 4, **5**, **8**
tongue 3, 19, 22, 238–244
anatomy 239–240, **241**
bilobed (Colletidae) 53
folding away 22, 240, **242**
function 22
hairy 29, 50
length correlation/technical meaning 241, 242–243
length linked to bee body size 232
long 37, 185, 240, 241
flower species visited 229
pollination 184, 185, 193, **194**, 194–195, 222, 228
solitary bee families with 37, 229, 241, **242**
for nectar collection 3, 10, 22, 238–244
mechanism 241, 242–243
short 35, 53, 240, 241, **242**
flower types visited 193, 228, 229
pollen release, methods 187, 245
solitary bee families with 35, 53, 229, 241, **242**
torpid bees 424, 434, **434**, 435
Town and Country Planning Act (1947) 525
towns/cities, bee habitat see urban landscapes and habitats
toxins 264
in nectar 181, 235
in pollen 201, 264, 267, 288–289, 290
specialist bee larvae adaptations 264, 265, 289

transport of pollen/nectar 3, 4, 20–22, 28, 182, 252–258
  in crop 4, 21, 54, **54**, 107, 120, 253, 301
  of nectar 22, 182, 252–253
  of pollen 253–254, **254**
  brushes, baskets and combs 21, 254, **255**, 255–258
  dry pollen 107, 182
  grooming before transport 200, 245, 253
  ingestion by *Hylaeus* bees 54, 149, 253, 256, 258, 301
  scopae *see* scopa (scopae)
  by solitary bee family 21, **255**, 255–258
  pollen baskets *see* pollen baskets (corbiculae)
tree-planting, threat to bees 497, 502, **503**, 507
tree removal, heather heath restoration and 468, 480
tropics, social stingless bees in 412
tumulus, at nest entrance 113, 123, **124**, 125, 128
Tunbridge Wells, Kent 495, **496**

UK Biodiversity Action Plan (BAP) 468, 485, 489, 527, 528
ultraviolet light 238
UN Convention on Biodiversity 510, 528
univoltine bees 102, 363, 435, 437, 510
upland sites (mountain/moorland) 470–471
urban landscapes and habitats 457, **457**, 483–489
  bee abundance linked to floral resources 504, 505
  bee species/populations in UK 492–493, **494**, 495, **496**
  bee species/populations in USA 493
  boundary erosion 492
  brownfield sites *see* brownfield sites
  densely 'built-up' landscapes 492
  diversity comparison with rural sites 492
  gardens *see* garden(s)
  generalist *vs* specialist bees 504
  green spaces, bee species in 492, 493, **494**, 495, **496**, 497
  bee hotels in 521, **521**
  biodiversity-friendly, transition to 497, 505, 506, **507**, 508
  reduced grass management 506, **507**, 508
  threats to bees 502, **503**, 506
  habitat breadth and species survival 504

heat island effect 504
insect diversity loss, drivers 502
Thames Gateway 484–485, 492, 508
threats to bees 502–508, **503**
  brownfield site development 502, **505**
  impervious surfaces 502, 504
  mowing grass and tree planting 502, **503**
  pollution 504–505
  towns/cities with long history 493, 495
  'urban sprawl' 492
usurpation of nests
  cross-species 147–148
  intra-specific 147, 166
  Lasioglossum malachurum 166, 167, 168
  prevention, by nest closure 168
  see also cuckoo bees (solitary bees)

venation, wing 2, 23, **34**
vibrations 7, 62, 92, 432
  buzz pollination 248, 249
  sugar concentration of nectar and 239
  taps during mating 97, **98**
visual maps, of landmarks 237, 238
voltinism 435–437
von Linné, Carl 207

waist 7
warming up (bees) *see* basking; temperature (bee body); thermogenesis
Warne, Frederick 25
Waser, N.M. 216, 217, 218
wasps 2, 3, 102, 244
  aculeate 3, 102, 305, 339–341, 355, 406, **407**
  apoid 2, **3**, 406, **407**
  bees distinguished from 3, 9, **9**
  eggs and provisioning for larvae 102
  evolution, and fossils 2, 406, **407**
  gasteruptiid 305, 341–342, **342**
  Leucophora targeting 311
  mutillid 305, 342–343, **343**
  parasitic 136, 153, 265, 339, 340
  solitary 102
  taxonomy 2, **3**, 406
water, imbibing 434
weather 461
  effect on cuckoo bees 434
  effect on lifespan of bees 429
  hot dry, bee activity reduced 434
  inclement, nest protection from 103, 129, 412, 432

parsivoltinism (delayed emergence) 437–438
poor, survival without feeding 433
rain/raining 432, 433, **434**, **523**
self-pollination response to 204, **204**
sheltering by bees 433, 434, **434**, 461
see also climate; temperature (ambient)
weight of solitary bees 106
wetlands, habitat **473**, 473–474
wild flowers 28, 228, 505
  in gardens/on lawns 521, 524
  grown for commercial purposes 517–518
  margins of fields 514, 516, 517, **517**, 518, 519
  pollination, importance, 'ecosystem service' and 529
pollination concerns 509, 529
wildlife refuges, urban 484
wingbeats 22, 239, 249
wings 2, 7, 22–23, 37
  ageing effect 430
  length **357**
  Cerceris rybyensis prey **357**, 357–358
  as proxy for body mass **426**, 426–427
  range in north *vs* south UK 427
  veins, pattern 2, 23, **34**
wood, dead, nests in 17, 40, 54, 110, 117, 154, **155**, 442, 472
wood-borers (bees) 17, 110, 117, 153–155
woodland habitat 471–473, **472**, 480, 526–527
worker honeybees 3–4, 4, **5**, **6**, 429
  see also honeybees (*Apis*) (Species Index)
workers (solitary bees - halictids) 17, 50, 161, 162, 164, 165, **166**, 172, 385
  'alien' 166, 167, 174
  attacks from *Sphecodes* 386
  chemical analysis of secretions 175
  division of foraging labour **172**, 174
  'drift', to adjacent nests 166, 167
  foundress producing broods 161, 162, 164, 165, 173, 176, 386
  'obligately' eusocial bees 165, 166, 167, 174
  reproduction, sexual brood 172, 174
  size, *vs* queen size 174
working memory, bee 218, 219

zygomorphic flowers 193–194, 195, 196, 209, 222, 226, 228, 229, 250, 277
solitary bee genera pollinating 226, 229
zygote 180

The New Naturalist Library

1. Butterflies — E. B. Ford
2. British Game — B. Vesey-Fitzgerald
3. London's Natural History — R. S. R. Fitter
4. Britain's Structure and Scenery — L. Dudley Stamp
5. Wild Flowers — J. Gilmour & M. Walters
6. The Highlands & Islands — F. Fraser Darling & J. M. Boyd
7. Mushrooms & Toadstools — J. Ramsbottom
8. Insect Natural History — A. D. Imms
9. A Country Parish — A. W. Boyd
10. British Plant Life — W. B. Turrill
11. Mountains & Moorlands — W. H. Pearsall
12. The Sea Shore — C. M. Yonge
13. Snowdonia — F. J. North, B. Campbell & R. Scott
14. The Art of Botanical Illustration — W. Blunt
15. Life in Lakes & Rivers — T. T. Macan & E. B. Worthington
16. Wild Flowers of Chalk & Limestone — J. E. Lousley
17. Birds & Men — E. M. Nicholson
18. A Natural History of Man in Britain — H. J. Fleure & M. Davies
19. Wild Orchids of Britain — V. S. Summerhayes
20. The British Amphibians & Reptiles — M. Smith
21. British Mammals — L. Harrison Matthews
22. Climate and the British Scene — G. Manley
23. An Angler's Entomology — J. R. Harris
24. Flowers of the Coast — I. Hepburn
25. The Sea Coast — J. A. Steers
26. The Weald — S. W. Wooldridge & F. Goldring
27. Dartmoor — L. A. Harvey & D. St Leger Gordon
28. Sea Birds — J. Fisher & R. M. Lockley
29. The World of the Honeybee — C. G. Butler
30. Moths — E. B. Ford
31. Man and the Land — L. Dudley Stamp
32. Trees, Woods and Man — H. L. Edlin
33. Mountain Flowers — J. Raven & M. Walters
34. The Open Sea: I. The World of Plankton — A. Hardy
35. The World of the Soil — E. J. Russell
36. Insect Migration — C. B. Williams
37. The Open Sea: II. Fish & Fisheries — A. Hardy
38. The World of Spiders — W. S. Bristowe
39. The Folklore of Birds — E. A. Armstrong
40. Bumblebees — J. B. Free & C. G. Butler
41. Dragonflies — P. S. Corbet, C. Longfield & N. W. Moore
42. Fossils — H. H. Swinnerton
43. Weeds & Aliens — E. Salisbury
44. The Peak District — K. C. Edwards
45. The Common Lands of England & Wales — L. Dudley Stamp & W. G. Hoskins
46. The Broads — E. A. Ellis
47. The Snowdonia National Park — W. M. Condry
48. Grass and Grasslands — I. Moore
49. Nature Conservation in Britain — L. Dudley Stamp
50. Pesticides and Pollution — K. Mellanby
51. Man & Birds — R. K. Murton
52. Woodland Birds — E. Simms
53. The Lake District — W. H. Pearsall & W. Pennington
54. The Pollination of Flowers — M. Proctor & P. Yeo
55. Finches — I. Newton
56. Pedigree: Words from Nature — S. Potter & L. Sargent
57. British Seals — H. R. Hewer
58. Hedges — E. Pollard, M. D. Hooper & N. W. Moore
59. Ants — M. V. Brian
60. British Birds of Prey — L. Brown
61. Inheritance and Natural History — R. J. Berry
62. British Tits — C. Perrins
63. British Thrushes — E. Simms
64. The Natural History of Shetland — R. J. Berry & J. L. Johnston
65. Waders — W. G. Hale
66. The Natural History of Wales — W. M. Condry
67. Farming and Wildlife — K. Mellanby
68. Mammals in the British Isles — L. Harrison Matthews
69. Reptiles and Amphibians in Britain — D. Frazer
70. The Natural History of Orkney — R. J. Berry

71. *British Warblers* — E. Simms
72. *Heathlands* — N. R. Webb
73. *The New Forest* — C. R. Tubbs
74. *Ferns* — C. N. Page
75. *Freshwater Fish* — P. S. Maitland & R. N. Campbell
76. *The Hebrides* — J. M. Boyd & I. L. Boyd
77. *The Soil* — B. Davis, N. Walker, D. Ball & A. Fitter
78. *British Larks, Pipits & Wagtails* — E. Simms
79. *Caves & Cave Life* — P. Chapman
80. *Wild & Garden Plants* — M. Walters
81. *Ladybirds* — M. E. N. Majerus
82. *The New Naturalists* — P. Marren
83. *The Natural History of Pollination* — M. Proctor, P. Yeo & A. Lack
84. *Ireland: A Natural History* — D. Cabot
85. *Plant Disease* — D. Ingram & N. Robertson
86. *Lichens* — Oliver Gilbert
87. *Amphibians and Reptiles* — T. Beebee & R. Griffiths
88. *Loch Lomondside* — J. Mitchell
89. *The Broads* — B. Moss
90. *Moths* — M. Majerus
91. *Nature Conservation* — P. Marren
92. *Lakeland* — D. Ratcliffe
93. *British Bats* — John Altringham
94. *Seashore* — Peter Hayward
95. *Northumberland* — Angus Lunn
96. *Fungi* — Brian Spooner & Peter Roberts
97. *Mosses & Liverworts* — Nick Hodgetts & Ron Porley
98. *Bumblebees* — Ted Benton
99. *Gower* — Jonathan Mullard
100. *Woodlands* — Oliver Rackham
101. *Galloway and the Borders* — Derek Ratcliffe
102. *Garden Natural History* — Stefan Buczacki
103. *The Isles of Scilly* — Rosemary Parslow
104. *A History of Ornithology* — Peter Bircham
105. *Wye Valley* — George Peterken
106. *Dragonflies* — Philip Corbet & Stephen Brooks
107. *Grouse* — Adam Watson & Robert Moss
108. *Southern England* — Peter Friend
109. *Islands* — R. J. Berry
110. *Wildfowl* — David Cabot
111. *Dartmoor* — Ian Mercer
112. *Books and Naturalists* — David E. Allen
113. *Bird Migration* — Ian Newton
114. *Badger* — Timothy J. Roper
115. *Climate and Weather* — John Kington
116. *Plant Pests* — David V. Alford
117. *Plant Galls* — Margaret Redfern
118. *Marches* — Andrew Allott
119. *Scotland* — Peter Friend
120. *Grasshoppers & Crickets* — Ted Benton
121. *Partridges* — G. R. (Dick) Potts
122. *Vegetation of Britain & Ireland* — Michael Proctor
123. *Terns* — David Cabot & Ian Nisbet
124. *Bird Populations* — Ian Newton
125. *Owls* — Mike Toms
126. *Brecon Beacons* — Jonathan Mullard
127. *Nature in Towns and Cities* — David Goode
128. *Lakes, Loughs and Lochs* — Brian Moss
129. *Alien Plants* — Clive A. Stace and Michael J. Crawley
130. *Yorkshire Dales* — John Lee
131. *Shallow Seas* — Peter J. Hayward
132. *Falcons* — Richard Sale
133. *Slugs and Snails* — Robert Cameron
134. *Early Humans* — Nicholas Ashton
135. *Farming and Birds* — Ian Newton
136. *Beetles* — Richard Jones
137. *Hedgehog* — Pat Morris
138. *The Burren* — David Cabot & Roger Goodwillie
139. *Gulls* — John C. Coulson
140. *Garden Birds* — Mike Toms
141. *Pembrokeshire* — Jonathan Mullard
142. *Uplands and Birds* — Ian Newton
143. *Ecology and Natural History* — David M. Wilkinson
144. *Peak District* — Penny Anderson
145. *Trees* — Peter A. Thomas